KNIGHT'S CROSS

By the same author

NON-FICTION

Alanbrooke (1982)
And We Shall Shock Them (1983)
The Christian Watt Papers (1983)
In Good Company (1990)

FICTION

August 1988 (1983)

Treason in Arms:
The Killing Times (1986)
The Dragon's Teeth (1987)
A Kiss for the Enemy (1985)
The Seizure (1988)
A Candle for Judas (1989)

The Hardrow Chronicles:
Adam Hardrow (1990)
Codename Mercury (1991)
Adam in the Breach (1993)
The Pain of Winning (1994)

KNIGHT'S CROSS

A LIFE OF
Field Marshal Erwin Rommel

David Fraser

HarperPerennial
A Division of HarperCollins*Publishers*

This book was originally published in Great Britain in 1993 by HarperCollins Publishers.
A hardcover edition was published in 1994 by HarperCollins Publishers.

First HarperPerennial edition published 1995.

The Library of Congress has catalogued the hardcover edition as follows:

Fraser, David, 1920–
Knight's cross : a life of Field Marshal Erwin Rommel / David Fraser. — 1st ed.
p. cm.
Includes bibliographical references and index.
ISBN 0-06-018222-9 (cloth)
1. Rommel, Erwin, 1891–1944. 2. Marshals—Germany—Biography. 3. World War, 1939–1945—Campaigns—Africa, North. 4. Africa, North—History, Military. 5. World War, 1939–1945—Germany. I. Title.
DD247.R57F73 1994
940.54'23'092—dc20
[B] 93-43832

ISBN 0-06-092597-3 (pbk.)
95 96 97 98 99 **HC** 10 9 8 7 6 5 4 3 2 1

Contents

PART 5 1943–1944

Illustrations

Maps

(Maps drawn by Steven Maison)

Acknowledgements

The considerable number of published works about Erwin Rommel (in America, Britain, France and Germany) have provided for me an admirable starting point in writing *Knight's Cross*; and the research represented by them and quoted in them has helped me enormously. I am grateful to all those authors, living and dead, who have preceded me in the study of this remarkable soldier; and the selective bibliography records them, together with many others from whose labours I have derived benefit. In the latter category, and not listed in the bibliography, I have drawn on many interesting articles on various aspects of Rommel and his campaigns in a number of specialist publications, notably the *Journal of the Royal United Services Institution*, *Armor*, *Kommando*, *Oase* (the journal of the Afrika Korps), *Deutsche Soldatenzeitung*, *Frontsoldat erzahlt*, *An Cosantoir* and the *Vierteljahrshafte fur Zeitgeschichte*.

Documents relating to Rommel exist in profusion. He himself wrote an account of his experiences in the First World War, published in 1937 as *Infanterie greift an*, and the original amended typescripts are extant and available. He also wrote a thoughtful analysis and account of his campaigns as a General, in France and North Africa, drawing on his diaries. Those relating to North Africa were published in 1955 as *Krieg ohne Hass*, and were largely incorporated, in translation, in *The Rommel Papers*, edited by Sir Basil Liddell Hart, together with edited excerpts from his correspondence and from his reflections on other campaigns. Where Rommel's own words have been quoted without other attribution I have used the text of *Krieg ohne Hass*.

An enormous mass of material both about Rommel and the circumstances in Germany in his time is kept in the National Archives in Washington ('NAW' in my notes). I am grateful to Mr Robin Cookson for sending me what I requested and to Colonel James Hamilton-Russell, at that time of the British Embassy in Washington, for his good offices in the matter. An equally enormous archive (including most of the relevant war diaries) exists at the Imperial War Museum in London ('IWM' in my notes). I extend my sincere thanks to the museum's Director, Dr Alan Borg, to Mr Robert Suddaby and Mr Philip Read (Keeper and Deputy Keeper of Documents), and to Dr Glyn Bayliss (Keeper of the Department of Printed Books) for their unfailing assistance not only in making available to me the hundreds of documents requested, but for giving me most helpful advice and space in which to study. I am also most grateful to Mrs Hilary Roberts of the museum's Photographic Department for her help in guiding me through the photographic archive, as well as to Lady (Rosamund) Thornton for the loan of photographs in her possession.

Interesting records and correspondence exist in the Bundesarchiv-Militararchiv in Freiburg ('BAMA' in my notes), and I am particularly appreciative of the help received from Oberst A.D. Dr Kehrig (Leitender Archiv Direktor) and Herr Meyer, who rapidly and efficiently dealt with my requests. I am also most grateful to Signor Crispo, of the *Amminstrazione Centrale Archivistica* in Rome, for dealing so expeditiously with my enquiries. Dr Hermann Weiss, Director of the *Institut für Zeitgeshichte* in Munich ('IZM' in my notes), responded most helpfully to my questions and in Munich his staff, on my visit, were models of efficiency and courtesy. I owe them thanks for access to documents.

I acknowledge with gratitude the permission of both the Trustees of the Imperial War Museum in London and the authorities of the Bundesarchiv in Freiburg to reproduce photographs in their possession.

A considerable and important archive exists on microfilm at EP Microfilm Ltd, Wakefield, Yorkshire ('EPM' in my notes). This includes a record of numerous interviews conducted by Mr David Irving in the course of researching his own book on Rommel, and I must pay tribute to his huge and meticulous industry in placing so much on record.

As ever I am much in debt to the patient, courteous and cheerful staff of the London Library, an institution without which I cannot imagine writing any book at any time.

To many individuals I owe thanks for their help, advice, information, personal reminiscence and encouragement. I must especially record my appreciation of the kindness of Herr Manfred Rommel, who gave generously of his time and whose personal recollections and perceptions were invaluable, as was the loan of photographs. I must also express my very particular gratitude to Mr Robin Edmonds, to Major General John Strawson, to the late Freiherr Axel von dem Bussche-Streithorst, to Dr Georg Meyer and to Oberrat Dr Stumpf who read the book in whole or part in typescript and whose wise comments were immensely helpful. Finally, I owe this book's production entirely to my wife, who not only mastered the word processor, typed and organized, but throughout commented with her own very particular shrewdness and good judgement.

David Fraser
Isington
1993

PART 1

1891–1918

CHAPTER ONE

'Keep the Right Wing Strong'

ERWIN ROMMEL's name stands as one of the great masters of manoeuvre in war, one of that select company whose personalities transcend time and whose energy still communicates. The victories of these men once depended on their ability to signal their intention, impose their will, and act with alarming speed amidst the confused conditions and across the variable distances of the battlefield. In a comparable way their personalities cut like sabres through the curtains of history, penetrating to successive generations with an immediacy which quickens the blood. Living legends, they project, each in his way, the classic image of the warrior: brave, vigorous, sharp of eye and mind, rapid in decision, alert in danger, faster and bolder in the fight than his enemies. Of this extraordinary brotherhood is Rommel – the brotherhood of Hector, of Rupert of the Rhine, of those who can only be described as heroes; and it is curious that so determinedly practical a modernist as Rommel – the least fanciful of men – should have joined a company so bonded by myth.

Attitudes to warfare have been greatly affected by time, and the reputation of warrior heroes rises and falls with fashion. In the present day people are more repelled than fascinated by the energy of a character such as Charles XII of Sweden, another master of the battlefield, a man who pursued his enemies, making war without interruption or remorse, for eighteen years. When a prisoner of the Turks in 1714, Charles invited death by organizing a bloody resistance by his fellow captives against all odds, killing and killing amongst his jailers until overpowered. Earlier, in mid-campaign, the King had met remonstrance from his civil councillors by telling them that their duties lay simply in raising money for his wars, and that his wars were fought for the glory of the Royal House of Vasa, some small reflection of which they should be proud to share. In our day such a figure appals, is incomprehensible in his arrogance and amorality. Yet when we read of how Charles conquered with phenomenal speed in the Northern War, subduing Poles, Russians, Germans, establishing new states, overturning old ones; when we learn that he began his brief and stupendous career at the age of fourteen, a boy-king, and had ended it at thirty-six; we acknowledge, however uneasily, something remarkable in that pale face and tall, slender figure which burst upon Europe, sword in hand as (in Winston Churchill's phrase) 'the most furious warrior of modern history . . . dauntless and implacable, with cold calculation and, for a long spell, a charmed life'.[1] Charles, a hero, reaches us, though he may repel us, and we respond.

It is the sheer energy of these men which leaps across the years, and which made them in their own day centres of myth. Men such as Jeb Stuart, the great Confederate cavalryman of the American Civil War, inspiring his troops whenever he appeared; and appearing at so many different and widely separated points on the battlefield (each of them crucial) as to give a near-supernatural impression, 'constantly in the saddle . . . everywhere present at all hours of the day and night . . . gaiety and humour unfailing',[2] galloping along the ridges of Chancellorsville before Lee's great victory over the Army of the Potomac: or the incomparable Bedford Forrest in the same war, a man as brilliant and elusive as mercury. These men, whether sovereigns or subordinates, were beyond doubt heroes, captains of the host, masters of the field. Their broader judgement of politics and events, inevitably affected by the circumstances and sentiments around them, may have been limited or distorted; and in studying any of them it is necessary to be wary of verdicts deriving from wholly different eras, places, cultures. These men, fallible attitudes and all, were children of their times, and their times must be seen through their eyes as well as objectively; but their impact survives. We may think of our admiration as immature crudity, but it will not go away.

In an age of airpower, of large-scale mechanization, of the destructive capacity of firepower increased to a previously unimagined degree; in an age where war between modern, industrial societies has seemed inevitably to imply a conflict of depersonalized masses operating over huge areas; in such an age, masters of the battlefield in the heroic manner appeared not long ago to have had their day, their achievements splendid but primitive and irrelevant. Here is Churchill, again, ruminating in the aftermath of the First World War on the military character of Marlborough:

> In the times of which we tell the great commander proved in the day of battle that he possessed a combination of mental, moral and physical qualities adapted to action which were so lifted above the common run as to seem almost godlike. His appearance, his serenity, his piercing eye, his gestures, the tones of his voice – nay, the beat of his heart – diffused a harmony in all around him. Every word he spoke was decisive . . . that age has vanished for ever.[3]

Nine years after publishing that sentence, its author would find himself striding up and down his temporary quarters in Cairo calling out again and again the name of a certain German general and crying: 'What else matters but beating him?' For a figure seemed to have leapt from the pages of history – in one sense a wholly contemporary figure, technically expert and inventive, progressive in his ideas, singularly adept at twentieth-century campaigning – yet one whose personal mastery of the battlefield recalled the achievements of an earlier, heroic age; a mastery which had, indeed, led Churchill himself to refer in the House of Commons to 'a very daring and skilful opponent . . . and, may I say across the havoc of war, a great general.' Erwin Johannes Eugen Rommel.

The heroic type, to which undoubtedly Erwin Rommel belonged, fails utterly if charisma is unsupported by professional competence. These men, however their lists are composed or argued, knew their trade. Whether kings or soldiers by career, whether students of war from family tradition or impelled into it by accident of history or of revolution, they mastered their business, they used their minds, they were diligent as well as naturally attuned to the demands of battle. They sprang, and will always spring, from widely differing backgrounds; but whatever the upbringing, the familial influences, the heredity, there has always been a combination of three qualities which together compose the battlefield commander's skill. The first could be called temperament.

The master of the battlefield has always relished the challenges of combat. He himself (and his biographers *de rigueur*) has often deprecated this relish, at least in modern times. War can be a gruesome business. Its incidentals are mutilation, death and destruction, its atmosphere is one of violence and pain, its consequences are suffering and bereavement, and it generates – although not necessarily among the fighting troops – casual brutality at best, vicious cruelty at worst. The severest condemnations, in our own day, are reserved for those who plot or glorify warfare, and since the hero's natural element is battle, our attitude towards him has to be ambivalent. The ambivalence is generally resolved today by the concept of duty – of the image of the soldier stoically, even brilliantly (in the technical sense) obeying orders, discharging a task odious by definition but doing so with purity because the quarrel, the cause of the war, was outside his power to affect. In our age the sovereign or dictator-general has certainly existed, but has seldom been found on the battlefield or in the front line. There the field commander can be and has been (at least by historians) absolved from guilt of producing the circumstances which gave him his employment and his chance.

We should not, however, imagine from this innocence that the hero-figure does not enjoy his work – were it so, his temperament would unfit him for it. It is impossible for a man to do something with skill and artistry without deriving satisfaction from it, and the victorious captains of history, whatever the justice or injustice of their causes, have fought with a zest which only failure or the overriding obtuseness of a superior has diminished. Zest does not imply heartlessness. The loss of friends, the suffering of subordinates (even of enemies) have often* acutely touched the sympathies of a great commander; they certainly touched those of Erwin Rommel. But it would be humbug to pretend that relish was absent, any more than it was absent in Alexander of Macedon or The Great O'Neill. The temperament of the successful warrior, however careful he may be to express it in a manner deferential to the conventions of his day, is a temperament eagerly attuned to the peculiar demands of battle. Battle is his element. And although in the cool aftermath of war he stigmatized it as a stupid business, battle was Erwin Rommel's element.

If temperament is indispensable to skill in battle so, too, is understanding, knowledge of war. Every victorious captain has possessed it – that sense of what will and will not work in the often extraordinary conditions produced by large numbers of men trying to kill each

* It would be agreeable but, I am afraid, mistaken to write 'always'.

other. Such understanding must, to an extent, be acquired by study and reflection (although 'study' implies an academic process, often inapplicable). It must, certainly, derive much from experience, from what has happened before – and experience can be, often has to be, the experience of others, history itself. It must, without question, be rooted in practical, technical, professional factors, mastered and appreciated; the capabilities and shortcomings of weapons, equipment, vehicles; above all, the capabilities and shortcomings of men. But understanding and knowledge of war, at the sort of pitch where it distinguishes the great battlefield commander, in some way surpasses the cerebral. It becomes a sixth sense, an instinct, a gut reaction beyond such phenomena (equally essential but more easily describable) as the ability to judge a situation and an opportunity shrewdly and instantly. It transcends, although it is close to, the *coup d'oeuil* which enabled Wellington at Salamanca suddenly to leap up as he watched Marmont's columns with a cry of 'By God! That'll do!' and to gallop off for perhaps the most brilliant victory of his career. It lies in that which men called Erwin Rommel's '*Fingerspitzengefuhl*', his almost animal response to the dangers, the chances, the currents of battle. Such is knowledge, understanding, refined into instinct and applied with instancy. It can be recognized in all the great masters of manoeuvre in varying guises and under many names.

'Applied with instancy'; the third, the culminating, quality of the three which make the great battlefield commander,is the ability to think and act clearly, resolutely and above all fast. Temperament may bring a necessary enthusiasm, knowledge may produce a sure, even an uncanny, insight into what should be done, but the victorious master of manoeuvre is he whose actions are so rapid, sure and energetic that they set the pace and direct the course of battle. It is Rupert (despite the myth of impetuosity, a highly experienced and professional commander) charging at Powick Bridge in the earliest days of the English Civil War, charging before any, and particularly the enemy, were ready; charging and routing superior forces before they could deploy, before they knew what hit them; charging so that the action 'rendered the name of Prince Rupert very terrible indeed'.[4] So it remained. And so, three centuries later, it was with Rommel.

Erwin Rommel was born a Swabian. The Duchy of Swabia had long been absorbed by the Kingdom of Wurttemberg, and *Schwaben* was a territorial rather than a political expression. The Swabian, nevertheless, had and has recognized characteristics. He was stable, stolid rather than emotional, in sometimes self-consciously dour contrast to his ebullient neighbour in Bavaria. He was careful with money, thorough, prudent. He was a loyal Wurttemberger, but felt himself superior to anybody else in the kingdom. Above all he had common sense, a level head, shrewdness. Rommel, despite his panache, his later flair for showmanship, the romantic aura his achievements created, was a Swabian through and through.

His father, also Erwin Rommel, was a schoolmaster in Heidenheim in Wurttemberg, some fifty miles east of Stuttgart and twenty north of Ulm; his mother, Helene von Luz (who lived until 1940), was daughter of the local *Regierungs-Prasident*. As a child young Erwin, born on 15 November 1891, was small, pale-faced, fair-haired and blue-eyed, sometimes dreamy but always even-tempered, an 'easy' child. Academically as well as athletically he was, when young, unremarkable, although during adolescence he discovered a taste for mathematics (both his father and his grandfather were distinguished mathematicians) which remained with him until his death. The remarkable physical toughness and energy which characterized Rommel the soldier, in every rank, were not in evidence in the early days. His family however (he had one elder sister and two younger brothers*) remembered one thing about him; he seemed afraid of nobody. In adolescence he began to develop unexpectedly fast, mathematical ability being accompanied by enthusiastic application to skiing, bicycling and exploration. In reading he always preferred practical books to works of the imagination.

Erwin senior had served in the army as an artillery officer before adopting a schoolmaster's career, but there was no particular military tradition in the family. The young Erwin, having passed his basic examinations, began to study aircraft, and had some idea of applying to the Zeppelin works at Friederichshafen. His father, however, was against this and advised him to look towards the army.

The German Empire had been declared in 1871 in the victorious aftermath of the Franco–Prussian War. The independent sovereigns, the Kings, the Grand Dukes, Dukes, Princes, of the various states of Germany combined in allegiance to the King of Prussia as German Emperor. Thereafter, in most departments of political life, it was a question of reconciling the conflicting and legitimate needs of

* One, a pilot in the First World War, returned from it seriously wounded; the youngest became an opera singer.

diversity and unity – of balancing that diversity which the peoples as well as their rulers valued, feeling, as Bavarians or Wurttembergers, utterly remote from Prussians or Silesians, against that unity which the imperatives of economic progress and military security dictated.

The problem – recurrent and by no means inapplicable to our own times – was recognized by the 1871 constitution which established relationships within the Empire. By its provisions, on the military side, the troops of the Kingdom of Wurttemberg formed the XIII Army Corps of the Imperial German Army. The Imperial army – regulated by an Imperial General Staff whose ancestors were the great Prussians of the War of Independence, Scharnhorst and Gneisenau, and whose most distinguished chief, the elder Moltke, had directed the campaign against France which had brought all this about – was massive in size, would unite on mobilization, was organized on an Imperial basis and recognized the Kaiser as Supreme Commander. It enjoyed all (or, in the view of its detractors, some) of the advantages of unity. But it also enjoyed (again, in the view of its critics, to a fault) the benefits of diversity. Wurttembergers, at least in theory, joined the Wurttemberg Army, and the King of Wurttemberg retained certain patronage, at least over the lower ranks of officers.

The resolution of the tug between unity and diversity appears – at least to a foreigner – to have worked admirably. It was no doubt helped by the tradition of the 'nation in arms', under which a man served, in one capacity or another, as embodied conscript, first-line reservist, *Landwehr* or whatever, until the age of forty-five; a tradition applied impartially throughout the Empire. The German soldier therefore, whatever his homeland, shared the privilege and the burden of service with all. Local patriotism was powerful, sometimes exclusive; but there now existed – assiduously cultivated but growing in congenial soil – the overarching concept of the Fatherland, subject of pride, myth and song.

Erwin Rommel, throughout life, showed a certain wary scepticism, even sensitivity, over matters of social class. There is a frequently held misconception that officers of the Imperial German Army were overwhelmingly members – even if junior and impoverished members – of the nobility, but this was not so. The old Prussian concept of a throne served exclusively by a hereditary caste, a chivalric order, had been enormously modified by the pressures of events, social evolution, and the sheer needs of an expanding army. In Prussia itself, from 1845 onwards, there had been a considerable intake of members of the middle classes into the ranks of the Officer Corps – a movement actually formalized by an Imperial decree of 1890 – while in the southern German states, including Wurttemberg, the bourgeoisie had held commissions as a matter of course for a considerable time. In the Imperial army as a whole in 1912, the year Rommel joined his regiment from Cadet School, slightly more than a quarter of the junior officers were 'of the nobility' – a classification which not only included all descendants of noble (by definition) land-owning families but also the large numbers of the 'ennobled' (by granting of the aristocratic prefix 'von') for distinguished service, legal, administrative or whatever. In Wurttemberg the proportion was significantly smaller and the nobility, as such, exercised little influence in the army. Rommel, therefore, whose family on his father's side was solidly bourgeois (his mother was 'of the nobility'), was entirely in his element, and little suspicion of social resentment attaches to his early life. The influence of the nobility within the elite General Staff – then and later – was a different matter, for when Rommel was commissioned the General Staff was at least half made up of noblemen;[5] but Rommel's vigorous disputes with the General Staff lay far in the future.

Affecting him more immediately was a considerable evolution which had taken place in attitudes to the training as well as the selection of potential officers – an evolution mirrored in other nations and other ages. The old Prussian tradition had been based not only on heredity but on 'character'; education, as such, had been largely disregarded or despised. The revolution in military science and attitudes brought by the Napoleonic Wars had changed all that; in their aftermath military reform was introduced in Prussia, and the foundations laid of an enduring and impressive edifice – the General Staff and its system. The arch-reformer had been Scharnhorst,* and he was impatient with the idea that war could be successfully conducted without trained intellect. He believed in the selection and development of the educated man as officer, and his reforms gave effect to his belief.

There was a reaction. Unsurprisingly – and certainly not uniquely – the nineteenth century saw a conflict between those in authority in Prussia (and, subsequently, in the German Empire) who reckoned

* Scharnhorst, Chief of the Prussian General Staff, led the 'Military Reorganization Commission' established by the King of Prussia in 1807 to reform the army in the aftermath of disaster at Napoleon's hands.

that the only requirement for an officer was that he should come from a 'good home', and show 'character' – the conservative wing of opinion – and those who demanded that he should show, and be encouraged to develop, intellectual and academic qualities as well – the progressive wing. Argument swung to and fro, with compromises and half-measures and reversals of policy. As late as 1897 there had been a circular issued which placed sufficiently exclusive emphasis on the need in candidates for commission of 'upbringing', 'will' and 'common sense' (without reference to academic qualification) to discourage the 'progressives'. But the tide turned, to some extent, in the first years of the twentieth century, and by the time Rommel joined the profession in which he was to die, less than 4 per cent of entrants were excused from the necessity to show educational certificates before acceptance – the device by which uneducated young men of 'character' and good birth had previously penetrated the system.

This would have made little difference to Rommel. It is impossible to believe that he would have failed, whether entry had been determined by the conservative enthusiasts for character or the progressive demanders of intellect. Entry for officers followed various systems. The principal officer academy in young Erwin Rommel's day was at Danzig on the Baltic, an ancient West Prussian port almost as far from Swabia as it was possible to travel without leaving Germany altogether. This, the *Kriegsschule* towards which Rommel's eyes were now directed, was an Imperial institution; but to be recommended to attend it it was first necessary to join the Army of Wurttemberg. Since he possessed the necessary educational qualifications Rommel was entitled to enlist as an '*Avantageur*' – a candidate for a commission; and, had his career foundered, he would have been entitled to serve only one year (instead of two) as '*Einjahrig Freiwillig*' – a one-year volunteer – on the basis of buying his own uniform and equipment, which father Rommel undertook to do.

Erwin Rommel, however, had no intention of being deflected from aiming at a commission or of giving up the army. His first attempt was to join the artillery – from having been socially undistinguished a short time previously, the artillery now ranked next to the cavalry in prestige – but from the local artillery commander there came a discouraging response. Herr Rektor Rommel was informed that there would be no vacancy in the foreseeable future, '*In absehbare Zeit keiner Stelle frei wird*', and Rommel's second choice – the Engineers – were equally discouraging. Young Erwin was thus directed towards the 124th Wurttemberg Infantry Regiment in the 26th (Wurttemberg) Infantry Division; after a four-month setback for medical reasons (he had a hernia and underwent an operation), he joined the regiment in July 1910 as a cadet.

Until comparatively recent times it was the invariable custom in the German army that, before a cadet could be trained and commissioned as an officer, he had to serve in the ranks of his chosen regiment, prove himself as a junior leader,* and – very importantly – be in effect elected by those who would be his brother officers. Such ruthless interior scrutiny, albeit not necessarily formalized, exists in many armies. It can clearly lead to abuse, to the dominance of cliques, and to the self-perpetuation of flawed values; but it recognizes the sovereign importance of mutual confidence between members of a fraternity who may face battle together, dependent on each other's affection and self-sacrifice, totally bonded.

Rommel's service with his regiment as a cadet lasted from July 1910 until March 1911. He satisfied the regimental authorities, and was appointed corporal in October and sergeant at the end of the year – a cadet, nominally receiving some of the perquisites of an officer but generally facing more hardship than either a young officer or a private soldier, not unlike a midshipman in the Royal Navy, was expected to qualify and discharge the duties of non-commissioned rank if he was to reach Cadet School. In March Rommel was sent to the *Königliche Kriegsschule* in Danzig.

Rommel worked hard. He had never found academic study easy or congenial, but in all practical exercises he was increasingly competent and he took duty seriously, determined to succeed. The course lasted eight months, until the end of November 1911. His final report – singularly uninformative – showed him as competent in all subjects tested, including leadership ('*Führung – Gut!*'), and in January 1912 the monocled and dapper young Lieutenant Erwin Rommel rejoined the 124th Regiment in their barracks at Weingarten.

The Imperial army was an enormous and magnificent machine. Its components were drilled and trained with great regularity and thoroughness, and were splendid to the eye. Rommel was photographed in the high-collared tunic and spiked helmet of the Wurttemberg infantryman, neat, compact, elegant; and he would probably

* In wartime, certainly in the Second World War, including a spell at the front.

have echoed the reminiscence of an exact contemporary, a Saxon, writing in modern times of the same age:

> When after two years I was 'officer of the day' for the first time and walked self-consciously through echoing corridors with the brass gorget on silver chains around my neck I felt there could not possibly be a dizzier claim to military distinction. If this be militarism then I plead guilty! One was proud of one's uniform and the words 'esprit de corps' still had a meaning.[6]

All over the German Empire young officers were still exulting in the sense of belonging to the greatest army in the world, the victors of 1870; exulting, too, in the privileged position they held in society. In many a German city civilians were expected to give way on the pavements to an officer in uniform, and in the highest councils of the Empire the influence of the military leaders was supposed to have proper power with the Kaiser, a ruler who, in the Prussian tradition, was regarded as Supreme Commander as much as Constitutional Sovereign. The Kaiser saw the Chief of the General Staff in audience more frequently than he did any Minister – a Chief of the General Staff whose predecessor had once unabashedly asserted the right of the General Staff to have the prime say in formulating German foreign policy.[7] For in this curiously atavistic community all the qualities of a modern and dynamic industrial nation, highly organized, regulated in an enlightened manner, technically progressive and at the head of Europe in many commercial, social, artistic and educational attainments – all these formidable characteristics overlaid like a solid but nevertheless permeable veneer the historic temper of Prussia; that Prussia of which it had been said that whereas in most societies the state possesses an army, in Prussia the army possessed a state. And it was Prussia – austere, dutiful, principled, God-fearing Prussia – which gave its tone to the Imperial army (only Bavarian troops being to some extent exempted), and, to a degree which many German states found odious, to the German Empire itself. Cloaked though it might be by the civil conventions of the twentieth century, Rommel grew up in a warrior society. In Prussia the nobility and the officer corps had been designated the 'first estate of the realm'; and now, throughout the Empire, those rising in the social scale, sometimes achieving commissions as Reserve officers, tended to ape the manners and exaggerate the customs and prejudices which a warrior society exalts.

Few doubted the inherent necessity for this. The Imperial army might be huge, or at any rate become huge on mobilization, but it was smaller than the army of either France or Russia, Germany's neighbours to west and east; and France had formed an alliance with Russia which assumed as potential enemy (offensive or defensive depended on the point of view) the Triple Alliance – the central powers of the German and Austro-Hungarian Empires, the former Dual Alliance, now combined with a somewhat unreliable Italy. In 1912, giving detailed effect to an understanding reached many years earlier, the Russian General Staff undertook, in case of war, to concentrate 800,000 men against Germany and to begin an offensive on the fifteenth day after mobilization was ordered; while the details of a planned French offensive in Alsace in the event of war were no secret from the German General Staff.

To the Germans the army was the guarantor of their security in a Europe in which there were, as there often had been, enemies on both flanks; a Europe, furthermore, in which Germany had on countless occasions provided battlegrounds for external and superior powers, marching and fighting and devastating across German lands. Forty years ago Prussian rather than Austrian hegemony had been established over the German peoples; France had been decisively beaten, thus securing the western frontier and embracing the German or German-occupied provinces of Alsace and Lorraine; and the Empire had been established. Its principal guardian was the Imperial army.

In the two years following Rommel's first commission it appeared increasingly likely that that guardian's strength might at any time be tested. Germans would have indignantly – then or later – refuted accusations of aggressive intent, pointing, not without justice, to a record of peace since 1871 compared to the records of, say, Britain (four wars) or Russia (two). The danger, they would have claimed and believed, lay with others. Across the eastern frontier Russia was in a state of fundamental political instability which might always seek outlet in external adventure; and Russia had, in the Balkans, some troublesome client states, in process of ridding themselves of memories of Turkish suzerainty and flexing their muscles. Of these, from Germany's viewpoint at one remove, the most menacing was Serbia.

Serbia was a small country, fully independent since 1878, and the Serbs were keen to expand and strengthen their position. North and west of them were provinces of the Austro-Hungarian Empire, some of them inhabited by Slavs whom the Serbs regarded as more akin to themselves than to the Teutons and Magyars who ruled from Vienna, and some of them, notably Bosnia and Herzegovina, formally annexed by Austria (which had, at Turkish invitation,* long been administering them) only in 1908. To Austria-Hungary Serbia was a nuisance, a troublemaker, a promoter of internal subversion among a dissident minority in some departments of the Empire; an international petty gangster. But behind the gangster, often calling the tune and certainly orchestrating it, stood Russia. It was this which (to some misgivings in Berlin) made Austria-Hungary's concerns those of Germany also, and on more than one occasion the German Emperor had spoken of his support for his ally (the Dual Alliance had been formally contracted since 1879) in terms whose extravagance had sent anxious shivers through Europe. The younger Moltke, Chief of the German General Staff, had in 1909 assured Vienna that if Russia mobilized, so would Germany. It was a far cry from the days when Bismarck had declared: 'For us Balkan questions can in no case constitute a motive for war.'

The Russian army was very large, although believed to be less efficient than it might be, and certainly requiring time to mobilize if only because of the great distances men and regiments needed to travel. But to the west was the French army, regarded (certainly until 1870) as the premier land force in Europe, still inspired by the spirit of the great Napoleon, and believed by every German, not without reason, to be animated by a desire for revenge: revenge for 1870, revenge for Sedan, recovery of the 'lost provinces' (albeit in the case of Alsace inhabited by people of German race, language and traditions) of Alsace and Lorraine. To the Germans – in whatever quarter of the Empire a somewhat insecure people, volatile to a surprising degree beneath a sometimes phlegmatic exterior – there were dangers on every side. And Rommel would have heard disturbing accounts of the unreliability of some of the forces of Germany's principal ally, Austria-Hungary, of their disaffection from Vienna and from each other. Germany's security might be fragile.

Nor was this all. Some years before, but continuing into and colouring the first decade of the twentieth century, there had taken place the so-called 'scramble for Africa', wherein the principal European powers had secured colonies, dependencies and concessions in the undeveloped territories of the African continent; and the eyes of each of the European contenders were on the others, noting,

* Procured at the Congress of Berlin in 1878.

occasionally supporting, frequently forestalling or frustrating, the moves of rivals. Germany came somewhat late to this game, but played it with a will, and this brought her into competition and near-collision with both France and Britain. To Germans the reactions of others were spiteful and hypocritical – spiteful because deriving from ineradicable anti-German prejudice, as with France, and hypocritical (in all cases) because very black pots were calling kettles the same. Because these territories were distant and their value (if any) lay in their commercial potential, colonial rivalry brought into sharp relief the maritime question.

Throughout the nineteenth century Britain, basking in the warm afterglow of Trafalgar, ruled the waves; but as the century ended this hegemony was increasingly questioned, particularly by Germany. Kaiser Wilhelm II was a devoted student of strategy and a convinced believer in seapower. He was certain that if the German Empire was to have the place in Europe and the world which its position demanded – frontiers secure, overseas possessions defensible, resources and raw materials assured, peace and prosperity beyond the reach of any combination of assailants – then Germany must be a first-class naval power. Supported in this, very naturally, by the northern and Baltic states of the Empire, the German Government had embarked upon a considerable ship-building programme, aimed at producing a high seas fleet which, at the least, could not be faced down by the British and their friends; a fleet which would deny to Britain (if Britain were, in fact, hostile) the opportunity to strangle a Continental power as Pitt had once sought to strangle Napoleon. At the same time, and for parallel if confused reasons, there grew in Germany a very widespread distrust and dislike of England.

The Kaiser's naval ambitions had, again very naturally, aroused considerable misgivings in Britain. The German General Staff were aware of friendly conversations between military staffs in London and Paris, beginning in 1906, while an '*Entente Cordiale*' had been concluded between Britain and France in 1904 – developments which, although not irrevocably committing Britain in case of Continental war, undoubtedly went a long way towards it. Britain, therefore, – Prussia's ally in countless wars of earlier centuries – might conceivably now be numbered among Germany's enemies. Faced with all this, Germans felt hemmed in, misunderstood and very seriously threatened; threatened by the unpredictable impulses of a Russia inclined to exploit the savage and primitive societies of the

Balkans, threatened by the implacable revanchism of France and threatened, at one remove but not impossibly, by the uncomprehending jealousy of Britain.

These matters looked very different from west of the Rhine or east of the Vistula – or from across the English Channel. From those quarters German external policy was generally suspect as brutally adventurist. Germany's size and strength, the notorious volatility of her Emperor and what was excoriated as her acceptance of an unashamedly militaristic philosophy were all perceived as prime menaces to European peace. But, regarded from whatever perspective, the general situation of Europe seemed to contain all the ingredients of conflict. While Rommel was a cadet at Danzig, France occupied Morocco. Germany had, for the previous few years, watched French expansionary moves in North Africa with envious disapproval, and as a signal of this disapproval in 1911 a German warship was ostentatiously sent to Agadir. This move – of a sort undertaken on many a previous occasion by a confident Britain – aroused indignation, and London reacted diplomatically but vigorously (albeit rather inconsistently, since Britain had, some years earlier, invited Germany to take an initiative in North Africa in order to forestall France; but the wind had changed). In 1912, Rommel's first year back with his regiment, France assured Russia of military support *in all circumstances* if it ever came to war. In the Balkans Serbia rumbled. It seemed unlikely that there would, for ever, be nothing for warriors to do.

Moltke summed up the view from Berlin in a memorandum in December 1912. The Triple Alliance, he said, was a defensive union. Russia was ambitious to appear the protagonist of Slavdom in Europe, and to secure an outlet to the Adriatic after the defeat of Austria. France wanted revenge and the lost provinces. England wanted to be rid of 'the nightmare of German seapower'. Germany only wished to defend herself; but the strategic-defensive demanded a defensive-offensive, a requirement imposed by the twin logics of geography and relative strength.[8]

In this tense international atmosphere the German General Staff worked ceaselessly on their plans and movement tables. In case of war against France and Russia, a war on two fronts, strategy ordained a main and decisive effort against one before turning to deal with the other. The main effort was to be made against the French, whose mobilization must be more rapid than that of the Russians; and the plans for a great western offensive with thirty-five army corps –

conceived by an earlier Chief of General Staff, Count von Schlieffen, as a giant wheel with the outer rim turning through Belgium and brushing the Channel[9] – were those which involved the XIII (Wurttemberg) Corps of the Imperial army. Schlieffen himself had given up office in 1906. His enormous concept dominated his later years, and when he died in 1912 his final, muttered, words had been: 'Keep the right wing strong.'

CHAPTER TWO

The Swoop of the Falcon

ROMMEL'S TIME at the Cadet School at Danzig had had one surprising consequence. He had fallen in love – itself most natural: but he had fallen in love with the girl who remained the love of his life until his last day. Lucy Mollin was dark and beautiful, daughter of a west Prussian landowning family, with both Polish and Italian blood, and was studying languages in Danzig. She was a Catholic, and marriage to a Protestant would be condemned by her Church – Rommel was *Evangelisch*. Nevertheless she accepted him.

Rommel's temperament, in private as well as professional matters, was essentially faithful. He valued loyalty almost beyond anything, and disliked the feckless, the changeable, the inconstant. Lucy was not only his wife (they married, a wartime wedding, in November 1916, but were engaged with less or more formality from Danzig days), but for almost thirty years his utterly trusted companion and confidante, devoted, staunch and with a considerable sense of humour. He wrote to her daily – or as near daily as the exigencies of campaigning allowed – whenever they were separated, and his hurried notes from the battlefield to '*Meine liebste Lu*' from '*Dein Erwin*' gave insight into what was uppermost in his mind at that moment. Rommel had only limited interests outside his profession. He enjoyed sport, skiing, and other physical pursuits, and he always had an active and enquiring mind; but his inner absorption was in his soldiering and in his family. The latter provided a sure solace and background of loyalty and love, and it never failed him. His character was wholly dedicated and wholly faithful, wholly true.

In the summer of 1914 Rommel was serving on attachment with an artillery regiment at Ulm, broadening his military experience by commanding a platoon in one of the batteries of the 49th Field Artillery Regiment – the regiment, as it happened, whose commander had written to Rommel's father the first terse rejection of his son's application to become a gunner. His time between first joining his parent regiment, the 124th Infantry (6th Wurttemberger) at Weingarten, and this summer had been spent ceaselessly training successive intakes of recruits who would spend their scheduled time in the ranks, be demobilized, then return to the colours if the clock struck. Rommel's attachment to the Artillery had been an agreeable and instructive change, but in an emergency it would abruptly cease.

To the people of Germany the Austrian alliance made sense. Every German understood that their country had a vital interest in preventing the dismemberment or defeat of Austria-Hungary. Every German appreciated that as far as external dangers were concerned Austria-Hungary was primarily threatened, now that the Ottoman Empire had been emasculated and driven from Europe, by the power of Russia. And every German knew that Russian designs were most likely to be advanced, at least initially, by a Balkan proxy, Serbia.

Serbia was wholly unreconciled to the Austrian annexation of Bosnia and Herzegovina. When the heir to the Imperial throne, the Archduke Franz-Ferdinand, was murdered with his wife during a visit on 28 June 1914 to Sarajevo, capital city of Bosnia, the assassination (carried out by 'nationalists' who aimed to create a Yugo-Slav state independent of Vienna and joined to Serbia) was attributed by the Austrians without hesitation to the design of Belgrade. Austria-Hungary accordingly sent an ultimatum to Serbia, listing a number of demands whose acceptance could barely be reconciled with Serbia's continued existence as an independent state. To Vienna the time, keenly anticipated, had come to settle with Serbia once and for all; the crime provided the spark to fire a long-laid fuse, and war with Serbia would unite – no small factor – Teutons and Magyars within the Empire in support.

The Austrian ultimatum was presented on 23 July, demanding a response within forty-eight hours. On the following day Germany notified Russia, France and Britain that the Austrian demands appeared to Berlin as entirely proper – in effect a notification that Germany would stand by Austria-Hungary.

Serbia accepted all but two of the Austrian demands, but by then matters had acquired a momentum of their own. Dissection of the cause of the coming conflict has little to do with the story of

Lieutenant Erwin Rommel: what affected him and his entire generation of young Germans – and young Russians, Frenchmen and Britons – was that the onrush of events appeared unstoppable. At various instants all the principal participants – the aged Emperor Franz Josef, the Tsar, the Kaiser, even the German General Staff itself – recorded what appeared warnings or remonstrances, but the terrible logic of mobilization and movement plans dictated urgency. To be a day, two days, behind an opponent might mean losing essential time in reaching a critical point, might mean the difference between victory and defeat.

Accordingly – and for reasons which at each stage and in all cases seemed wholly justified to the indignant and ultimately enthusiastic public opinions of Germany, Austria-Hungary, Russia, France, Britain and Serbia – the terrible machine rolled forward. Serbia's reply to the Austrian ultimatum was received in the evening of 25 July, and deemed inadequate. Austria-Hungary declared war on the morning of 28 July. Before then Russian mobilization had been decided upon, but – largely through the intervention of the Tsar – total mobilization throughout the entire Russian army and front was not ordered until the afternoon of 30 July.

At that moment war on a huge scale was inevitable. To Germany – where it was widely held that the Austro-Serbian conflict did not directly threaten Russia – total Russian mobilization had to be met, immediately, by German mobilization. This, indeed, had been made plain to Vienna, and mobilization was ordered at 1.45 p.m. on the following day, 31 July. Early that morning Rommel had ridden with his Artillery battery through Ulm – ridden out to exercise and back to barracks, regimental band playing at the head. That afternoon the mobilization order arrived, and that evening Rommel returned to the 124th Infantry Regiment at Weingarten.

On 1 August Germany, in the face of Russian mobilization, declared war on Russia. And also on 1 August, true to the understandings of the Franco–Russian Treaty and aware of German mobilization orders, France mobilized. On 3 August Germany declared war on France.

One piece still remained to be moved. The Schlieffen Plan had always required free passage of the German right wing through Belgium (and in its earlier form through Holland), and on the evening of 2 August, the day before the declaration of war on France, Germany presented her demands to Belgium – whose neutrality had, since the Treaty of 1839, been guaranteed by the European powers, including

Britain. Belgium rejected the German demands, and the German Imperial Army began its forward march on 4 August in accordance with the routes and timetables worked out so meticulously during the preceding decade. Belgian resistance was bludgeoned down as it was met.

The British had been initially dubious about involvement; the Cabinet had been divided. An early newspaper billboard had simply advised 'To Hell with Serbia'. But complex though the issues were, the invasion of Belgium was the last straw. Britain had made an attempt to secure from Germany an undertaking not to violate Belgian neutrality, and the attempt had been rejected. The Schlieffen Plan was a military necessity and a political disaster. Britain, somewhat irresolute until that moment, declared war on Germany at eleven o'clock on the same 4 August; midnight by German time.

Next evening Rommel began his journey by train towards the western front. Leading echelons of the 124th Regiment had preceded him by several days, newly kitted out in the field-grey uniform which would become so familiar to the peoples of Europe during the next thirty years.

Schlieffen had proposed to avoid the international recrimination which invasion of the Low Countries would invite (and, in the event, the intervention of Britain) by inducing a French move into Belgium. He hoped that the arrival of massed German forces opposite the Ardennes would lead the French Command to enter southern Belgium, in order to occupy a favourable defensive position forward of French territory. This would provide a pretext, and would actually assist his plan. For Schlieffen always envisaged a grand encirclement, a decisive battle, an *Entscheidungsschlacht*, an envelopment from the right which would be made the more effective the further the French army was lured forward in the centre, while in the east Germany would stand on the defensive.

The French Plan XVII, which envisaged a concentrated French offensive towards the Saar and into Lorraine, was concerned with the area of the Franco–German border, well south of the main German *Schwerpunkt*. The French, well aware of German intentions, had indeed meditated a move forward into Belgium once it was known that German troops were massing opposite Belgium and the Maastricht Appendix,* but had discarded the idea, not least through

* In a move very similar to that actually executed, with depressing results, twenty-five years later.

well-founded anxiety about the international reaction and the views of Britain. Schlieffen's hopes were unrealized.

In the event the French offensive excluded Belgium; and the German offensive, while following the original concept, included a fatal dilution of the strength of the right wing,[1] a dilution which, combined with an equally fatal (but perhaps inevitable) premature inward move of the right-hand German army, von Kluck's First Army, led to the Battle of the Marne, the successful Franco-British countermove from the outer flank against the German right.

Rommel threw himself into his own small part in these opening scenes of the drama with the greatest possible ardour. His regiment, in the XIII (Wurttemberg) Corps of General von Fabeck, was a part of the German Fifth Army under the command of the Imperial Crown Prince. The Fifth Army formed the pivot of the giant wheel, responsible for the area north of Metz and south of Luxembourg: the southern Ardennes. Here the function of the Germans was to engage and hold the French troops while to the south, east of a line between Metz and Nancy, the German Sixth Army under Crown Prince Rupprecht of Bavaria was intended (under Schlieffen's concept) to meet a French attack and then be prepared to withdraw, to entice the enemy into a 'sack' as a preliminary to total encirclement when the powerful right wing completed its enormous march.

As anticipated, and as ordained under the French Plan XVII, the French First and Second Armies advanced into Lorraine. Their offensive, however, was broken by the German Sixth Army defences at Sarrebourg and Morhange, east of Nancy and Metz. This successful defensive – inflicting, as it did, huge casualties on the French – led Prince Rupprecht to plead for a change of plan, to deviate from the policy of withdrawal and enticement, and to pass instead to the offensive, to advance and cross the Moselle. Moltke acquiesced, and by this additional dilution of Schlieffen's concept (for the right wing, already weakened, could now expect no early reinforcement from the centre-left) the imminent stalemate was the more assured. Meanwhile, on and in the days immediately following 20 August, the German Fifth and Sixth Armies attacked. The Schlieffen Plan had been modified beyond repair and the whole German front was now attempting an offensive movement, sufficiently strong at no point.

No doubt the bitter consequences of neglecting certain essential principles of war struck hard into Rommel's consciousness when he became older and more senior, and had time to reflect on the lessons

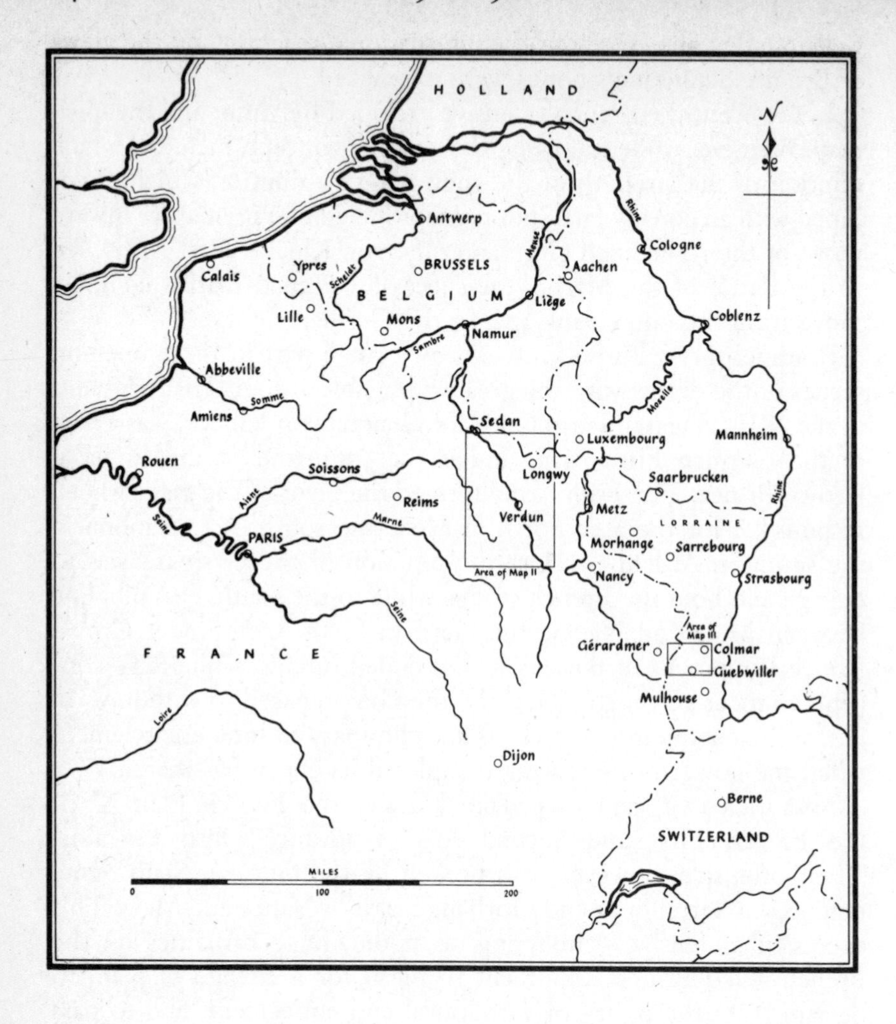

I The Western Front 1914–1918

of that immense conflict with the benefit of hindsight and study. There were plenty of historic precepts at issue here: 'Strike with a closed fist not an open palm'; 'He who seeks to defend everything defends nothing'; 'Not overall numbers but strength at the decisive point is what counts'; and so forth. Rommel – never a member of the General Staff but a first-class, self-taught military theorist and writer – would have learned the larger lessons of 1914 just as surely as he absorbed the smaller, the tactical, the direct. Meanwhile he was a young officer, knowing little beyond his immediate surroundings. He first encountered the enemy near Longwy in the area of the border between Belgium and France.

This was open warfare, each side seeking to advance, with fire positions adopted hastily, regiments sweeping forward in long lines, guns being galloped forward or back and engaging at comparatively close quarters. When Erwin Rommel first came under enemy fire he was already totally exhausted after nearly twenty-four hours on foot or in the saddle on missions of reconnaissance and liaison; throughout his life, although he made enormous demands on the physical and moral stamina of his men, he drove himself harder than any – yet never ignored the human factor, the limits of the soldier's endurance and the effect on his performance when pushed near or beyond those limits. Rommel was and remained a hard man and a hard commander; but he was the reverse of a mechanical or mechanistic commander of the kind that can draw arrows on a map without intuitive sense of what it feels like to be the flesh and blood at the point of those arrows.

It was 22 August 1914. Rommel's battalion – he commanded a platoon in the 2nd Battalion of the 124th Regiment – had left their peacetime station in Weingarten by train on 2 August and had moved to the area of Ruxweiler. After some hard training to make a team of a unit now brought up to strength with reservists, the regiment – consisting of three battalions each of four companies – had started a northward march on 18 August, crossed the Luxembourg frontier on the same day, and then swung south-westward and marched into Belgium and thence across the border into France, marching towards the Meuse as part of the Crown Prince's Fifth Army on the inner flank of the 'Schlieffen wheel'. 21 August had been planned as a battalion day of rest after the hard marching of the previous three days, but for Rommel it had been the reverse. He had been ordered to take a five-man reconnaissance patrol forward to discover whether and where enemy were deployed to the immediate front. There were, in fact, no enemy deployed, but Rommel felt – and recorded – the

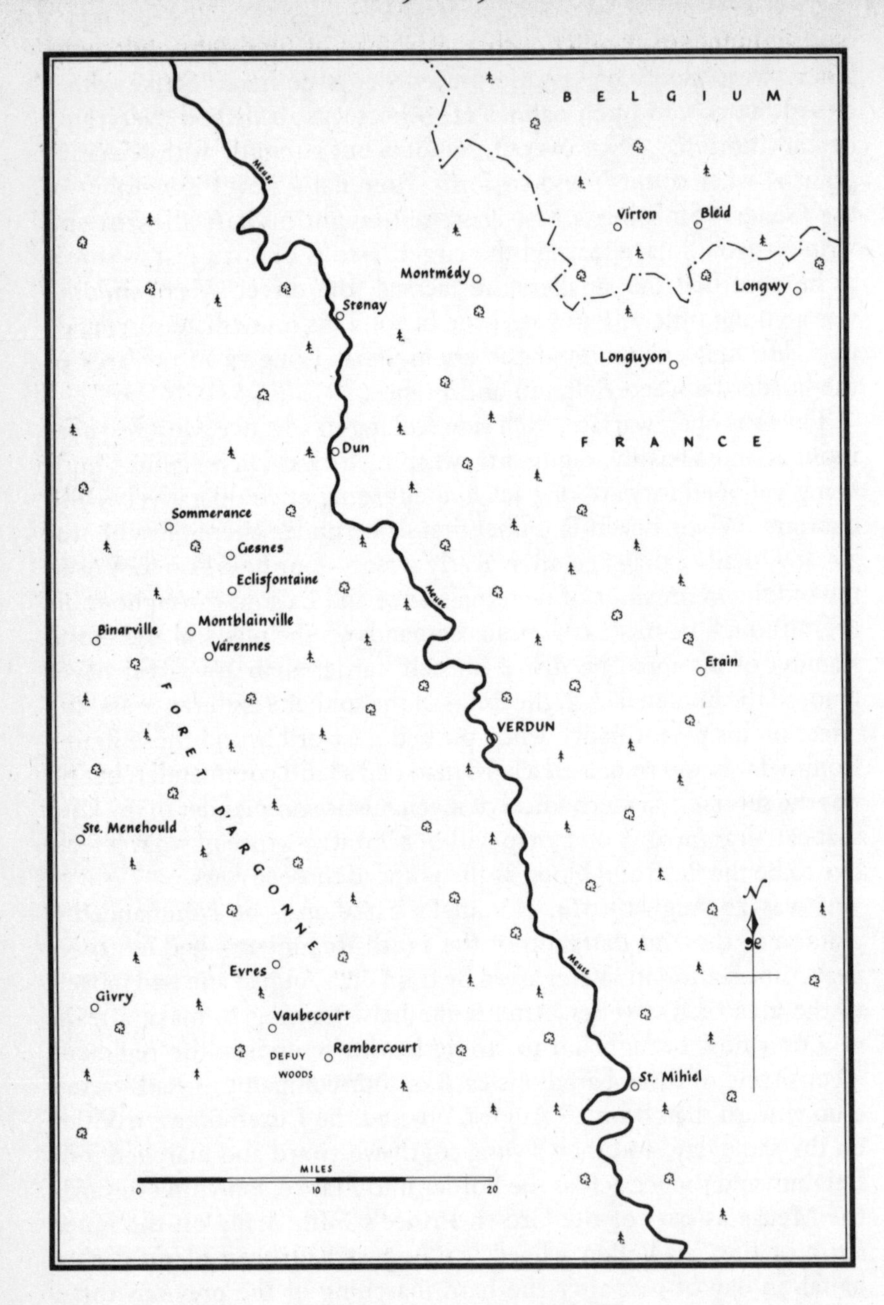

II Area of operations, 124th Regiment, Fifth Army, August 1914–September 1915

responsibility of possible danger to his men, the need to move with care and often laboriously and exhaustingly, the tingle of nerves when, for the first time, war has replaced peacetime manoeuvres. He also felt for the first time the extraordinary fatigue of warfare, the lack of sleep, the need to keep the body going and the senses alert. Furthermore – and this was to plague him often – his stomach was in disorder.

Immediately he had made his (negative) report, Rommel had been sent on another mission by his Regimental Commander, using this energetic and resourceful young officer to what seems like excess. Rommel was ordered to contact the neighbouring battalion (the 1st) to his own 2nd with withdrawal orders, and to act as guide thereafter. The 1st Battalion Commander, however, explained that he was not free to comply. He had been placed under command of an improvised brigade (brigades in the German army were constituted, on occasion, for especial tasks, with battalions drawn from regiments as ordered, rather than permanently established). When visited by the twenty-three-year-old Lieutenant Rommel the General commanding the brigade in question told him bluntly that he couldn't spare the 1st Battalion, 124th Regiment. They were not to obey their Regimental Commander's order.

Rommel made his way back, totally exhausted, to report this confusion among his seniors; to be at once ordered to find the superior General to both disputants and request a decision. He found the superior General – von Moser by name – provoked his anger by what he recounted, and obtained the decision the 124th Regiment wanted; that its 1st Battalion was to revert to Regimental command by daybreak 22 August. This had meant, for Rommel, a twelve-hour round trip, most of it in the dark, and represented his day of rest on 21 August – twenty-four hours spent continuously in the saddle or on dismounted reconnaissance. And on 22 August Rommel was again sent ahead of the battalion with his platoon. The regiment – with three others in line – was advancing towards a place called Bleid. As they moved forward there were periodic volleys from in front, and Rommel and his men threw themselves down among the concealment given by some growing potatoes. No casualties. No enemy visible. The platoon picked itself up and repeated the manoeuvre once or twice more until, through the fog, the first roofs of Bleid loomed.

On the outskirts of Bleid Rommel saw some prominent farm buildings. Leaving the bulk of his platoon in position he went cautiously forward with three men. He moved behind the cover of the nearest building (which held no sign of enemy occupation) and put his head round the corner of the wall. Standing on the roadway beyond were fifteen or twenty French soldiers, drinking coffee, careless of danger.

Rommel wrote afterwards, without complacency or bravado, of his reactions. Should he send back for his platoon? No point, he instantly decided. He and his companions could deal with this. Sheltered by the buildings, he told them what to do – safety catches off, move out from cover suddenly, open fire from the standing position. It was not Rommel's first contact with the enemy, but as he and his patrol jumped from behind the farm buildings and opened up on the astonished Frenchmen, it was the first time he had killed.

Rommel's trio dropped about half of their opponents. Soon, however, fire was returned from other buildings and he retreated with his little party to where his platoon were in position. He had shown a typical audacity – a readiness to act rather than wait, to attack at once and in person rather than make a prudent plan and form up appropriate forces. In that tiny instant at a farm on the outskirts of Bleid it is possible to discern the man who would one day, against or without orders, launch an offensive against the British in Cyrenaica while most of his troops were still in the process of arriving in Africa from Europe.

Thereafter Rommel, with his whole platoon now, began clearing the east end of Bleid, building by building. He did it methodically – German battle drills for assaults of this kind were a matter of peacetime training, and Rommel was a profound believer in battle drill and training, a belief fortified by every experience from this time forward. He moved one section into covered positions from which they could rake an enemy-held house with fire, smothering doors and windows with bullets. At the same time he led another section, the assault section, to a flank and then, at a signal, stormed into the objective, hand grenades and small arms at the ready. And then, leap-frogging, he repeated the movement. More troops from the rest of the regiment began shooting their way into Bleid and soon it became a blazing inferno of, in Rommel's words, 'thick, suffocating smoke, glowing beams and collapsing houses'.

Part of Bleid was still in French hands. Rommel, who was by now separated from his company and his battalion, withdrew his men three hundred yards north-east of the village and took stock from the crest of the nearest shallow ridge. Eight hundred yards ahead he saw the red breeches of French infantry, on the edge of a wheatfield and having, apparently, just dug trenches. He had little idea of the whereabouts of the rest of 2nd Battalion.

Rommel's account of the moment is characteristic: 'Since I didn't want to remain inactive with my platoon I decided to attack the enemy deployed opposite us, in the 2nd Battalion sector, to the south.'[2] Rommel knew the boundaries of his own battalion's sector of responsibility. He was still within them, and the enemy were within view and range. He manoeuvred his men into position and opened fire. After fifteen minutes he was delighted to see another part of his own battalion coming up on his platoon's right. The French had themselves been returning fire but now, said Rommel, 'the platoon could attack without second thoughts'. Rushing forward, groups supporting groups with well-rehearsed fire and movement, Rommel reached dead ground and ordered the fixing of bayonets. He then charged, but when he reached the French position it was deserted. The enemy had gone.

Rommel decided that by now, again well ahead of the battalion, he had best wait until the main body came up, and himself went ahead, once more, to explore. His enterprise was rewarded. To the north he saw columns of French infantry, driven from their positions by German artillery, moving across the German front from right to left. Once again he wondered – for a moment. 'Was I quickly to bring up the rest of the platoon? No, best for us that they stay where they are.' Rommel, two men with him, opened fire on the heads of the French columns and was rewarded by seeing them disperse and change direction westward. Other French soldiers appeared, running from the clump of bushes behind which Rommel had been standing, and were picked off by him and his companions. Soon troops of another of the division's regiments, the 123rd, appeared. And at that moment Rommel, sick in the stomach and totally exhausted, lost consciousness. When he regained it he found himself in the middle of a confused firefight. 'I took over part of the defence line,' he said, 'and occupied the slope, on the Gevimont–Bleid road.' Gradually the position hardened and the regiment's companies assembled. They had lost 15 per cent of their men and 25 per cent of their officers. This was the battle of Longwy.

The small incidents of this battle, memorable as must be a young man's first experiences of warfare, his first experiences of making decisions under fire, already showed the quintessential Erwin Rommel. Twenty-six years later, again in France, he wrote that in encounter actions the day goes to the first to plaster opponents with fire. This was Rommel's instinct, the instinct of the natural warrior; but it was rooted in a whole range of experiences, and the first was near Longwy in August 1914. Furthermore – and this, too, was quintessentially Rommel – he was suffering at the time from food poisoning and feeling thoroughly unwell. Ill health often plagued him on active service but it seldom if ever suppressed his energy or dulled the immediacy of his fighting reactions.

From the fight at Bleid the 124th Regiment followed up to the Meuse and deployed in woods east of Dun; and here they came up against the remarkable superiority and effectiveness of French artillery fire. The Meuse valley was the scene of an intensive artillery duel. On the last day of August the regiment crossed the river by pontoon bridge to the ridges beyond, where they again came under heavy artillery bombardment.

Soon the advance was resumed, and once again Rommel found himself commanding the point platoon of the battalion advance guard. After riding for five minutes he heard heavy firing and shouting from his right and turned towards it – another characteristic reaction. He was moving along a forest trail and was soon confronting 'a number of dark objects a hundred metres ahead of us' which swiftly showed themselves to be packets of French infantry. Bullets sang round his head, and his company deployed astride the trail and crawled through undergrowth to what was reckoned to be assaulting distance. 'I charged with the platoon,' Rommel wrote, 'and from the other side of the clearing we came under rapid fire.' A few minutes later the company was under fire from the rear as well as the front; part of Rommel's own battalion had started firing early from further back, and Rommel found himself between the fires of Germans and French. There was considerable confusion in the woods and around them, and before nightfall Rommel had repeatedly thrown himself down and lain, pressing himself into the ground, to avoid the attention of German bullets. He was – once again – cut off from the rest of his company. With only twelve men left alive with him, he had set out to attack a French gun battery discovered ahead of them, but was forced to go to ground and then to withdraw by fire from all directions. He was reported as killed that day.

As Rommel withdrew his men through the woods he was shaken, more than hitherto, by the huge numbers of wounded men, of both sides, in piteous condition. 'We helped them regardless of whether friend or foe,' Rommel said, and this was assuredly true. Throughout his life as a soldier Rommel showed chivalry. His instinct was to treat an enemy as a fellow human being. He wrote, with amused incredulity, that many Frenchmen had been unwilling to surrender, having been taught that they would immediately be beheaded by their German captors.

After a short rest in the first days of September the regiment marched south, then west, then south-east again to the area of Gesnes, north-west of Verdun. And after a few more skirmishes at the head of his platoon Rommel found himself appointed Battalion Adjutant. Although, as such, executive assistant to his commanding officer, he was as often as not employed as reconnoitring scout, liaison officer or emergency detachment commander. There is no swagger or bravado in Rommel's own account of these early soldiering adventures, but the reader has the impression of an officer who was used and over-used whenever a doubt or a problem arose.

Rommel's 2nd Battalion was still engaged in open warfare; the siege conditions of the ensuing years on the Western Front had not yet set in, and although Rommel had quickly learned – and insisted upon – the necessity for digging, for quick entrenchment, these were still protective trenches, dug quickly against the ubiquitous and accurate French artillery fire rather than the huge defensive system into which warfare would soon degenerate. His account of action on 7 September gives a taste of the times. The regiment had marched some way south and was now south-west of Verdun.

> We now got the order from Regimental Headquarters – 'Second Battalion to advance no further. Stay where you are.' I first passed on the order and then galloped back to the regimental command post, about 260 metres to the rear, to find out how long this was going on. Colonel Haas wanted to delay the attack until Grenadiere III* had come up with us. It was impossible to say when that would be.
>
> Meanwhile the French artillery was getting busy. Their fire was especially harassing our reserve companies which were still in close order, and the forward companies which had no sort of cover.
>
> My horse was fast and I rode forward again to get the leading companies entrenched in some potato and root fields. A French

* Sister regiment in the division.

battery ranged on me with shrapnel as I rode back, zigzagging and escaping their attentions without difficulty.

Two kilometres north of Vaubecourt the battalion and regimental command posts had set up next to each other in a cutting, and before long the place was under very heavy French artillery fire from several batteries. This was unsurprising! The coming and going of messengers, the clusters of people, the streams of traffic and the various observation posts gave the location away. This fire was maintained for some hours and in these circumstances an attack was unthinkable. I lay down in the cutting, dead tired, and tried to catch up on the night's sleep I'd missed. Shellfire in my immediate vicinity now bothered me not at all.

And so on. Whenever the battalion advanced or attacked in those days, the place of its commander or his adjutant was mounted and moving with the assaulting companies. On the following day:

I rode a few metres behind the assaulting line on the left wing, behind 7 Company. It was already getting dark and we were approaching the edge of a wood about 150 metres away. Had the enemy abandoned it? Or would he plaster us with fire as we covered this next stretch? Flashes along the edge of the wood soon answered the question, as rapid fire cut into us and a firefight was soon under way. The reserves were soon deployed and crowded up to join the forward line where men were lying bunched together, while the second-line echelons of the regiment, in rear, were trying to get some sort of cover in the earth from the enemy's furious fire.

Then part of the machine-gun company offloaded weapons from vehicles and opened up on the French on the edge of the wood, firing over our own front line several hundred metres ahead. There were soon yells from the leading riflemen that our own machine guns were shooting into them!

I was mounted on the left wing of the battalion and I galloped over to the machine guns and stopped their fire. Then I dismounted, handed my horse to the nearest man and led one of the machine-gun platoons back to the left wing of 7 Company where my brave fellows were engaged in a hot firefight and where the machine gunners soon joined in.

Rommel, as so often, then saw his chance. With a yell of '*Zum Sturm, auf Marsch, Marsch*' he led forward the troops with him, smashing into the woods only to find that the enemy had fled. He decided that the French could be cut off. Taking two 'squads' and a heavy-machine-gun platoon, he climbed to higher ground to the left of the woods which had held the French enemy, and was rewarded by finding that from the ridge he could see several hundred yards beyond the woods. The retreating French must surely emerge this way.

Rommel's initiative – surely inappropriate to a Battalion Adjutant except in grave emergency – was risky. His conscience was uneasy because nobody had given him leave to take the heavy-machine-gun platoon on this independent venture – he had simply seen what he thought was an opportunity and grabbed it. Time passed, and it began to grow dark. No Frenchmen appeared. Rommel eventually sent the machine-gun platoon back; his enterprise seemed to have misfired.

The machine gunners had just departed when, 130 yards away, moving over the bare crest of a hill in front of him, there appeared a column of French. It was as Rommel had hoped, backing his hope with action. The French were withdrawing across his field of fire. Withdrawing, furthermore, in close order. His machine gunners had gone but Rommel ordered rapid fire from his sixteen rifles.

The French column did not oblige by rapid dispersal as often hitherto. They turned, formed, and charged Rommel's impertinent little detachment. There were two companies of them. Rommel kept up rapid fire, having decided to run for it when the French were poised for a bayonet charge but not before. The French kept coming on, but German fire had dropped a good many of them, and forty yards out their advance stopped, with Rommel still holding his ground. The French then withdrew, and Rommel sent out a patrol from his detachment, which counted thirty enemy dead and brought in a dozen prisoners. The little action was over. Rommel, making his way back to recover contact with his battalion, met the Regimental Commander, who was wholly unimpressed by the Rommel manoeuvre and accused him of firing on a German regiment by mistake. 'Even the prisoners,' Rommel said, 'didn't convince him.'

Now the German Fifth Army was passing to the defensive, although only temporarily, and digging began in earnest. The violence of the enemy artillery fire impressed them all with the value of the spade and Rommel wrote that even the battalion staff, Battalion Commander, Adjutant and four messengers dug themselves a twenty-foot trench. The ground was very hard, the work back-breaking and progress slow. That night, 7 September, the battalion was in position near the woods of Defuy and received mail for the first time since the campaign opened. All that night, and all the next day, they dug. The French artillery opened up on them at six in the morning and kept up a harassing fire throughout the day, raising it to an intensive bombardment towards evening. Whenever shells were not actually falling the men dug. By 9 September the battalion felt itself well entrenched, with trenches at least six foot deep. So far there had been no French infantry attack, but the French guns had inflicted heavy casualties on the flanking battalions and Rommel's 2nd found themselves isolated. A massed French assault could only be a matter of time.

Rommel, from personal reconnaissance and observation, discovered a position where the French reserves were massing, six hundred yards away. He decided that if machine guns could be deployed on a knoll to the battalion's left, the French infantry could be reached by flanking fire, with likely disruptive effect. He suggested this to the machine-gun Platoon Commander, who had doubts, and declined. Rommel, as so often in life, took matters into his own hands. He took command of the battalion machine guns himself (no doubt invoking, as Adjutant, his own Battalion Commander's unsought authority), and soon had them deployed and firing into the formed-up French reserves, pulling out smartly before the inevitable French artillery retaliation.

That night Rommel had a severe shock. He had woken, in pouring rain, at midnight to hear a rattle of musketry amidst the incessant gunfire, and assumed the possibility of a French night attack. His Battalion Commander was away at the regimental command post and Rommel moved out to discover what he could. He saw a column approaching in the darkness, and borrowing a platoon of men from the nearest company he deployed ready to fire. When the column was fifty yards out Rommel challenged. He found that he was challenging one of the battalion's own companies which, led by a young lieutenant, was withdrawing – for reasons which seemed to Rommel wholly inadequate. 'I read him,' Rommel wrote, 'an unfriendly lecture about his incorrect appreciation!' But he had been close to firing on his own men, and the recollection afterwards made him shudder.

Now it was time to go forward again. In the small hours of the morning the battalion was ordered to attack in the darkness and clear the enemy from its front. Attacking in massed formation, with all four companies in line, the 2nd Battalion charged forward towards a place called Rembercourt; but when light came the French guns began once again to take their toll. It was 10 September. That day Rommel reported severe losses – four officers and forty men killed, four officers and 160 men wounded, and eight missing. The great French fortress of Verdun was now almost encircled by the march of the German Fifth Army; but it had been a costly march.

Throughout those September days of mobile operations Rommel had been, as so often, suffering agonies from the state of his stomach. He was also, no doubt like all of the German troops, suffering from enormous fatigue. On the afternoon of 12 September, two days after the fight in the darkness and dawn at Rembercourt, he collapsed into a sleep from which nobody could wake him, and was reprimanded for this lapse the following day. That day, 13 September, the battalion withdrew from the front encircling Verdun, retracing its steps and marching north-west towards Varennes; and thence on northward to Eclisfontaine and Sommerance on 18 September, where a few days' rest were promised. As often happened alarm, order and counter-order disturbed the rest, but Rommel had a short while in a dry bed. They were now on the edge of the forest of the Argonne, and on 22 September the 2nd Battalion received orders to attack.

To Rommel, who had collected the orders from the regimental command post, the attack plan seemed unimaginative and likely to lead to unnecessary casualties unless conducted with greater ingenuity. He suggested to Major Salzmann, his Battalion Commander, a withdrawal and a covered approach to an intermediate position, from which the whole battalion could reach a concealed forming-up place and assault from an angled direction. Rommel believed that by this manoeuvre the battalion could achieve surprise, and the French position would be outflanked.

And so it happened. 'Our powerful thrust in flank and rear surprised them. Panic swept the defenders and the reserves behind them. Fifty French, several machine guns and ten ammunition limbers fell into our hands; and some field kitchens with the French evening meal warm, ready and cooking on open fires.' German losses were four killed and eleven wounded; and the entire French brigade abandoned its positions.

The 124th Regiment was now manoeuvring west of Varennes, meeting detachments of enemy, engaging in sporadic firefights in the forested country. Rommel was everywhere. The Battalion Staff, he wrote on 24 September, followed 7th Company – but shortly afterwards, in order to get 7th Company moving forward again, Major Salzmann and Rommel got into the front line, where Rommel took a rifle and ammunition from a wounded man and assumed command of 'a couple of squads'. The function of authority was not simply to command but to lead, where leadership was needed.

A few moments later Rommel saw five Frenchmen immediately to his front. He shot two of them, missed another and found his magazine empty. He had been a keen and proficient bayonet fighter in peacetime exercises, and the bayonet was his best hope now. Fixing it, Rommel charged. A French bullet crashed into his left thigh, making a hole the size of a clenched fist. It was his first wound. At the base hospital at Stenay a few days later he was decorated with the Iron Cross, Second Class.*

The German armies were now bogged down. The French continued to hold Verdun as constituting the eastern hem of a sack into which, from north and north-east, the great advance had been pushing; while at the western end of the enormous front, the French and British had managed not only to stem the German advance but to counter-attack in the valley of the Marne. The crisis had come on 8 September, more than two weeks before the fighting round Varennes in which Rommel had been wounded. After that crisis the German right wing had withdrawn to the River Aisne.

In the following month each side attempted to outflank its opponent's seaward flank, leading to what, on the Allied side, became known as the first battle of Ypres, in Flanders. These attempts failed, and as winter set in the German Imperial forces and the Franco–British armies opposing them settled down to the depressing compulsions of positional warfare. When Rommel, his wound still unhealed but his impatient temperament driving him to return before the medical authorities approved, rejoined his regiment in January 1915, he found a very different scene.

Wherever Rommel appears in those early encounters of his fighting life there is the sense of a man endowed with extraordinary powers

* It has been unkindly remarked that the Iron Cross II could only be avoided by suicide. Such witticisms are seldom profound, and in Rommel's case awards for valour – including, ultimately, the highest award of all – followed thick and comparatively fast.

of rapid decision, and remarkable self-confidence and courage, both moral and physical, in following decision with action. Young Rommel, to a degree which must often have infuriated slower or more senior heads, made up his own mind, saw and decided in an instant, and then grabbed the nearest troops (persuading or ignoring their own local commander) and moved like lightning to inflict maximum damage on the enemy. His nose for an opponent's vulnerability was like the scenting of a thoroughbred hound, while his lethal rapacity in striking at that vulnerability became legendary, and so remained. His was indeed, as Sun-tzu described the quality of decision, like the swoop of the falcon.

CHAPTER THREE

Gebirgsbataillon

JANUARY 1915. The stalemate in the west had now set in, and all along the vast front from Switzerland to the sea entrenchments had been dug, barbed-wire entanglements were spread and preparations were set reluctantly in hand for what, to the consternation of governments and armies alike on both sides, might conceivably be a very long haul. Warfare had become indistinguishable from siege, and in the west, at least, manoeuvre was a thing of the past.

It could not be otherwise. The length of the front was great, but the size of the armies committed to it by both sides was also great, and overall pretty evenly matched. The development of small arms, and particularly the weight of small-arms fire which could be brought to bear by machine guns, meant that the bullet dominated the immediate, tactical battlefield, drastically constricting movement by men and horses; while the continuing development of artillery and shell-fuses, as well as means of communication, gave a new and immensely powerful influence to the artillery arm. For the first time, on such a scale, great industrial nations were in the field against each other – nations backed by scientific inventiveness and enormous engineering and productive capacity, as well as by populations greatly increased by over a century of industrialization and prosperity. The stalemate in France and Belgium, once the great initial manoeuvres had failed in their objects, was an inevitable consequence of the size of the armies committed and the effectiveness of their weapons. To break it would require either new weaponry or extraordinary and audacious ingenuity; or the exhaustion through attrition of one side or the other. In such warfare – siege warfare – a breach can only be effected after long, methodical and expensive preparation. Surprise, if not impossible, is remarkably difficult to achieve. Thus it was in January 1915. Meanwhile the soldiers dug, sat, shivered and suffered, inflicting such damage as they could on the other side from static positions, made as secure as possible.

Rommel returned to the 2nd Battalion as a twenty-three-year-old Company Commander, and in his view the company's position was very far from being as secure as possible. He found the trenches were in some places too shallow (the difficulties of drainage were an awful constraint on entrenchment below a certain level). He found the dugouts – excavated out of the trench walls to give shelter to eight or ten men – were inadequately roofed; roofs were too high, and thus could be targeted by the enemy, and were too weak, so that the men were sadly vulnerable to shell fire. Rommel set about correcting all this with his usual energy and forthrightness. His 9th Company (the three battalions of the regiment now had five companies each, numbered serially) consisted of two hundred men, and the company sector was a quarter-mile of south-facing front-line trench in the western part of the forest of the Argonne, near Binarville. Rommel soon made it more secure, by sheer hard work. 'The commander,' he wrote of that time, 'can win the trust of his subordinates very quickly if his orders are clear and prudent, if he takes constant care of them and if he is hard with himself, living under the same austere conditions as his men.' He shared a damp and freezing four-foot-high dugout, which needed bailing out every four hours, with one of his Platoon Commanders. The reserve trenches – companies alternated spells in the front line trench with a few days in reserve, some hundreds of yards in rear – were, if anything, worse, and the companies in reserve were used for fatigue parties rather than being able to work on their own line. They returned to the front line with relief.

Rommel wrote that once a commander had won his soldiers' trust they 'will follow him through thick and thin', and so it was with his own men. The French, in positions three hundred yards away, would harass working parties with small-arms fire and with sporadic shell fire. Life was, in physical terms, extraordinarily disagreeable in that first winter of the war, on both sides and in all sectors. Yet morale – probably near universally – was remarkably high. Nobody had yet lost the sense that rapid victory could come with the spring. Every soldier was sure that he was defending his own country, and defending its rights and liberties, against a ferocious and menacing enemy. And Rommel's presence, as ever, acted as a tonic on his troops. Anybody who once came under the spell of his personality, a brother officer wrote, turned into a 'real soldier'. However tough the strain Rommel seemed inexhaustible, seemed to know exactly how the enemy would probably react. The same officer wrote that Rommel had an exceptional imagination, seemed to know no fear whatsoever, and that his men 'idolized him'.[1]

In the last week of January it snowed and rained on alternate days. A small German diversionary attack was planned for the twenty-ninth of the month – an attack intended to attract and hold French strength to the area. Rommel's division – the 27th – with all its regiments was to take part, and in Rommel's 124th Regiment sector there was to be a raid, a limited attack on what was known as the 'Central' position. Rommel's own 2nd Battalion were to 'hold', and fire on enemy escaping from the German assault which was to be made by the 3rd Battalion, on the 2nd Battalion's right.

The 3rd Battalion attack went well, and after a short while Rommel saw that battalion's Adjutant approaching his company: 'The Battalion Commander of 3rd Battalion wishes to know whether 9th Company wants to be left out of the action or to join in.'

Rommel said that to join in would be a pleasure. In fifteen minutes he had crawled with his company up a trench which ran forward from the right of their position and had assembled his men in an area a hundred yards nearer the French. Now, though, Rommel's men had been spotted, and small-arms fire began singing over their heads as they flattened their bodies on the frozen ground. Ahead of them was a hollow which could give slightly more cover and in which they might form up for an assault. To reach it would mean covering sixty yards of bare slope in full view of the enemy, but Rommel decided that they must try it – the French fire, although indiscriminate, was heavy, and soon casualties would mount. As he decided, however, he heard a bugle far to the right. It was the 3rd Battalion signal to attack. Rommel turned to his own company bugler and ordered him to sound the charge.

9th Company leapt to their feet and charged forward, cheering. They raced down the slope, crossed the hollow whose shelter had beckoned them seconds before, and found themselves up against the French wire. They were rewarded by the glimpse of red trousers and blue coat-tails as the French took to their heels, and they pressed on, through a first, second and third French defensive line, each abandoned before 9th Company reached it, with no loss whatsoever.

The woods were thinner now, and in their enthusiastic pursuit

Rommel's men could soon see large masses of French infantry withdrawing, and were able to pause and shoot, hurry forward, pause and shoot again. After a while, and after at least one change of direction, 9th Company came under heavy fire from their left. Soon they found themselves confronted by the strongest wire obstacle they had ever encountered – a tangle of wire fences over a hundred yards deep. Rommel started to crawl through. He turned, waved and shouted to the rest to follow.

Nobody followed. More shouting and more waving. Still nobody followed. Rommel, alone now, explored inside the wire obstacle and found a passage through. He crawled back and found his leading Platoon Commander: 'Obey my orders instantly or I shoot you.'

The Platoon Commander decided to obey. Rommel had observed of an earlier occasion that soldiers sometimes lose their nerve, and that then the commander must take vigorous action, using his personal weapons if necessary. Following their commander the whole of 9th Company crawled through the obstacle.

The wire, they discovered, was covered by observation and fire from blockhouses sixty yards apart. The blockhouses were joined by a breastwork, behind which riflemen could fire from a fire step. Between this breastwork and the wire was a water-filled (and now frozen) ditch, five yards wide. The entire formidable system extended through the Argonne. This bit of it appeared unoccupied.

Rommel pushed his company out into a semi-circle, ordered digging and sent a written message back to 2nd Battalion headquarters:

> 9 Company has penetrated to a strong French earthwork position 1½ km south of our start line. Request urgent support, machine-gun ammunition and hand grenades.

Rommel did not record in very informative detail the response to this. 9th Company, after all, was now deep into the French lines. It had reached this position through the extraordinary energy shown by its commander; and he had started the day and set out on the enterprise in response to the neighbouring Battalion Commander's invitation to 'join in the action'! The reactions of his own commanding officer can only be guessed.

Very soon Rommel's company came under heavy fire. They fought their way to another part of the French position which Rommel reckoned would make a defensible strongpoint, and, as he later recorded with a certain ingenuousness, 'began getting anxious about our reinforcements and supplies'. He was holding a semi-circle, including four French blockhouses, with the wire behind him, and he placed one platoon as a reserve between the wire and his main position. Digging was proving near impossible in the frozen ground, and Rommel knew how vulnerable he was. Suddenly, only fifty yards to Rommel's right, French infantry began withdrawing through the wire, in full view. 9th Company opened fire.

At first the result was encouraging, but the situation soon became alarming. The French, in battalion strength, first moved sharply westward, away from Rommel's position. Having assembled (using yet another path through the obstacle) on Rommel's side of the wire, they wheeled and advanced towards him on a broad front. Rommel knew that the French were holding, undisturbed, on his other flank; they could thus fire from both flanks on the passage from his own position back through the wire. The advancing enemy were now close, it had been impossible to dig in and ammunition was low. The extreme right blockhouse fell again to the French infantry. At that moment a messenger yelled across from the German side of the wire entanglement. Support was not possible. 9th Company were to withdraw – 2nd Battalion had moved forward and taken up positions half a mile to the north, half a mile back! 9th Company were, in effect, surrounded.

Rommel reckoned that he had three alternatives. The first, he later recorded (but clearly never contemplated), was to keep shooting until his ammunition was exhausted and then surrender. The second was to obey the battalion order exactly and move back through the wire. Since the passage of the wire was now covered by French fire from both flanks Rommel assessed the cost of this as a likely loss of half his company. The third alternative was to attack – to hit the enemy, somewhere, so hard that they would be disorganized and stopped in their tracks. And taking advantage of that disorganization 9th Company would withdraw.

Rommel was in no doubt whatever. He directed his reserve platoon at the blockhouse just taken by the French and drove the enemy from it in double time. He then hurried his whole company eastward and doubled them back through the wire in single file. The Frenchmen ejected from the blockhouse had the anticipated discouraging effect on their comrades on that flank; the French advance wavered and stopped. On the other flank the French opened fire, scoring a few hits on 9th Company, but the range was three hundred yards and Rommel's men were running fast. He cleared the wire obstacle and ultimately found the 2nd Battalion in their new position. He had had five men wounded, all of whom he had managed to bring with him.

Later in the evening the battalion beat off a massive French infantry attack; and throughout the night they suffered from French artillery harassing fire. Twelve men were lost – more, Rommel noted sharply, than had been lost during the entire attack. He commented that it was unfortunate that neither the battalion nor the regiment had been able to exploit 9th Company's success, and he certainly recorded no sort of concern at having acted with such blithe independence in attacking at all. In this, certainly unaffected, attitude lies a key to Rommel's military character. He was a fanatical believer in the importance of bold initiative by individuals. His own triumphs – and some setbacks – generally reflected this belief. Rommel believed that in battle success goes to the commander who seizes opportunity and exploits it; and that only he, rather than his superior, can perceive opportunity in time. His military philosophy, therefore, was one of encouraging, to the maximum, independence of judgement and action within an overall plan; and it was an independence which he exhibited from the first days of combat until the very end. For it he felt, then or later, not the smallest impulse to apologize.

This independence of tactical judgement was a principle of German training, so that although Rommel excelled, he did not outrage the system he served. His exploits on 29 January were rewarded not by rebuke but by the Iron Cross, Class I. He was the first lieutenant in the regiment to be thus honoured.

Unsurprisingly, Rommel by this time was a famous regimental personality, his name a byword. He was slightly built, young-looking and young-seeming, but invariably zealous. He fired everyone with his infectious enthusiasm and his bravery marked him as one completely out of the ordinary. He seemed tireless. He became, and remained, physically very tough, shaming less robust characters by his example. Beyond these human attributes, however, Rommel impressed every German soldier who knew him with his extraordinary instinct for battle. He had a 'feel' for battle, for the enemy's likely plans and reactions, for what might or might not work, which all recognized as being – even in that highly trained and self-consciously military environment – something peculiar, something apart. It seldom deserted him, and, not surprisingly, it infused utter faith in the men he led. 'Where Rommel is,' they said, 'there is the front.'

Rommel, however, tempered these heroic qualities with judgement. Gallant, and leading always from the front, he distinguished between that practice and foolhardiness. He was not born without fear – he later told his son that it had to be overcome, and the method was to face it and defy it on the first possible occasion. But courage was not the antithesis of prudence. 'To live as a hero,' he remarked long afterwards, smiling, 'one must first survive!'[2] Rommel, brilliant infantryman, had all the qualities of a brilliant cavalryman – the swiftness of eye and judgement, the instancy in action and reaction; but he was a Swabian, with a Swabian's coolness, stolidity, liking for calculation, touch of dourness. It was a superb combination. And Rommel was wholly devoted to his men. Casualties upset him deeply. He cared for every soldier as an individual.

In May – it was an incident which can happen in any army, but it is invariably unpleasant – Rommel was told that since he was still very junior he must hand over command of his beloved 9th Company to another. A senior Lieutenant who had just joined the regiment and who, as Rommel permitted himself to observe, 'had not yet had field duty', took over company command. The Regimental Commander knew that this would be painful for Rommel and suggested transferring him, but Rommel resisted. He wanted to stay with the men of 9th Company. 9th Company was his home. Rommel thus again found himself a Platoon Commander – a highly decorated and regimentally famous Platoon Commander.

Trench warfare, more sophisticated, more lethal, more frustrating, continued; with the German and French positions in the Argonne within grenade reach of each other in many places, and with the troops manning small sandbagged posts from which to fight, forward of the main trenches. Mining and tunnelling added danger and excitement to that extraordinary phase of that extraordinary campaign.

At the end of June Rommel's battalion again took part in an attack over the same ground as on 29 January. Rommel, during the fighting, found himself commanding not only his own platoon but elements of the 3rd Battalion as well. At one point he was largely out of touch with other parts of the battalion or regiment. 'I felt it inadvisable to continue . . . to the south,' he wrote afterwards. 'Additionally the day of the last attack [29 January] in which I was so far beyond our own front that I was given up for lost was still fresh in the memory.' It was a rare acknowledgement. Next day, 1 July, he was appointed temporary commander of 10th Company for a spell in the trenches.

And after a richly deserved period of leave – his first since August 1914 – Rommel was given command of 4th Company, in the 1st Battalion. Again it was only to be for a short time. Seniority asserted itself and Rommel returned to platoon command and to another offensive battle – a well-rehearsed offensive battle after which he noted the great benefits of thorough preparation and practice when attacking strongly fortified positions. Rommel, although adept beyond most men at battle in a fluid situation, was the reverse of slapdash. Where time allowed and the situation demanded method he was utterly methodical; but he never let method constrict his mind or slow his actions. Given the choice between speed with its attendant risks and method with its inevitable delays his instinct was for the former – an instinct which only occasionally betrayed him. In September, he was promoted *Oberleutnant*; a major change in his life was imminent.

In September 1915 Rommel left the 124th Regiment which had been his home since he was commissioned, and left it with something of a heavy heart, for he was saying goodbye, as he wrote, to gallant comrades and to the 'bloodsoaked yet by now almost homelike earth of the Argonne'. Rommel was joining a new organization, the *Königliche Wurttemberg Gebirgsbataillon*, a Mountain Battalion forming at Munsingen in Germany and due to move to the Arlberg in Austria for ski-training in December. The battalion was composed of six rifle companies and six machine-gun platoons, and included officers and men assembled from many different regiments and branches of the service. As such *ad hoc* organizations tend to be, it was a highly motivated unit, enjoying the prospect of a different sort of experience. Rommel took command of 2nd Company – permanent command now. Training was arduous and he was out on the slopes, generally carrying a heavy backpack, from dawn to dusk, with music and singing in the evenings, and the development of company spirit. Everybody hoped that the battalion's Alpine training meant that their next move would be to the Italian front.

They were disappointed. On the last day of 1915 the *Gebirgsbataillon*, having left their agreeable quarters in the Arlberg on 29 December and taken train westward, assumed responsibility for a southern sector of the Western Front – a sector very different from that in the Argonne. The battalion was responsible for six miles of front on the steeps of the Hilsen ridge in the high Vosges in Alsace. At this point of the front German and French trenches were a considerable distance from each other, and defensive positions were based on widely separated strongpoints, organized and stocked for all-round defence. Offensive action was limited to raids. The battalion remained in this area for ten months.

Rommel was sceptical about raids. His experience hitherto had taught him that they could be difficult and expensive, and bring casualties incommensurate with their benefits. In October 1916, however, his company was ordered to prepare a raid. The object, as so often, was to take prisoners and, no doubt, obtain identifications. Rommel opened preparations by carrying out a personal reconnaissance. He crawled up a ditch towards the forward French sentry position – both front lines ran through woods and over hills some 3,500 feet high – and cut his way through the dense mass of French protective wire. On the way back he snapped a twig and was at once rewarded by heavy French small-arms fire from all along the enemy position. Rommel reckoned that in this sector, at least, the enemy were alert and any raiding force would suffer heavily.

He next turned his attention to another sector, where he carried out personal reconnaissances over several nights. He decided that in this sector he could get a small party of twenty men through the French wire at a point halfway between two French posts, and then strike outwards at the defenders, taking them in flank or rear. When this had started other wire-cutting detachments, lying up near the known French posts, would take advantage of the disturbance to cut different escape routes for the raiding force in another part of the wire.

Rommel rehearsed all this carefully and took time over it. He wanted bad, preferably stormy, weather, which would or should reduce French alertness; and on 4 October he got it. An icy northwest wind turned into a storm with heavy, driving rain. The French sentries should be using much of their energy in seeking protection from the elements, and the incidental noises of the raiders' movement would be altogether drowned. With three detachments Rommel crawled forward from the company trenches at 9 p.m., sent off the wire-cutting parties to either flank and himself, with twenty soldiers, crept towards the enemy with his own wire cutters. They had practised again and again the technique which would prevent cut wire, its tension released, from flying back and by its twang betraying what was up.

The French wire was high and closely meshed, and the task of cutting through it took several hours. Eventually, using the depth

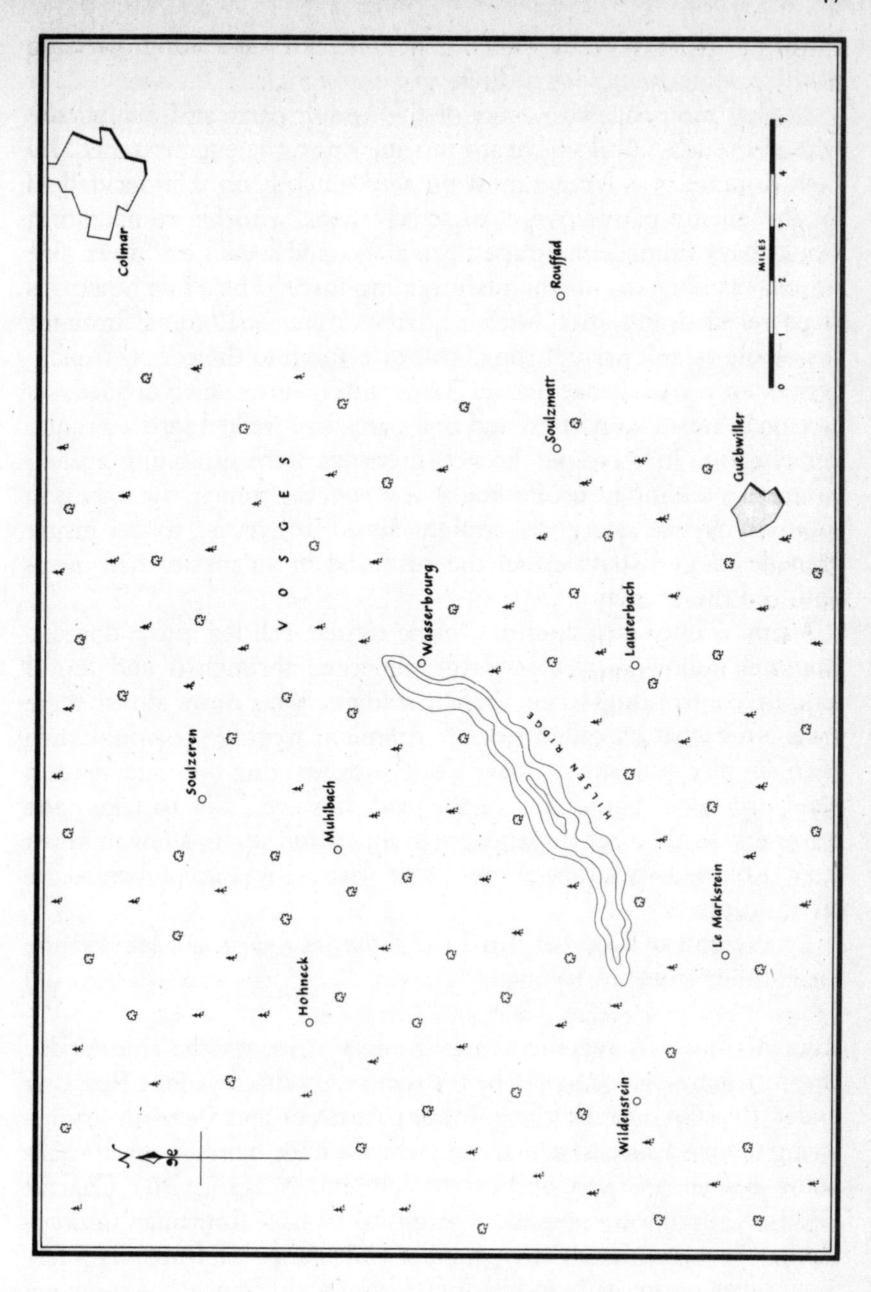

III Hilsen Ridge, high Vosges. Area of operations, *Gebirgsbataillon*, January–October 1916

fortuitously provided by a shell crater, he and the leading two men of the raiding force inched their way through.

At that moment, with most of the assault party still behind the wire, a French patrol was heard moving down the enemy trench. By now Rommel was lying almost on the trench's rim. He decided to let the enemy patrol pass – to attack them, with his companions, would have immediately raised the alarm and would call down fire which was likely to hit the main raiding force. The French patrol's steps receded and after further anxious minutes Rommel brought the whole assault party through the wire and into the enemy trench.

Suddenly hell broke loose. The raiding force had divided, as arranged, into two parties, and one party had walked into a French detachment. In a second French grenades were exploding among Rommel's raiding force. Packed as it was in the trench, the only way to anything like safety was straight ahead, to charge, to get inside grenade range. Rommel led the rush and in an instant had overpowered the French.

A tiny twenty-inch aperture in the trench wall led into a dugout. Rommel, followed by a sergeant, squeezed through it and found himself confronting seven French soldiers, who threw down their arms after what he called a brief argument, writing: 'It would have been simpler and safer to have dealt with this dug-out party with a hand grenade.' The object of the raid, however, was to take back prisoners, so the Frenchmen were grabbed and when, a few minutes later, Rommel withdrew through the wire he was accompanied by eleven captives.

By the end of October 1916 the *Gebirgsbataillon* had left France for another front: in Rumania.

Rumania had entered the war in August 1916, on the side of the Western Allies. Encouraged by the temporary success of the Russians under Brusilov against the southern Austrian and German armies facing them – a success which had yielded a huge number of prisoners – the Rumanians had decided to join forces against the Central Powers. There were possible pickings to be had. Rumanian minorities in Transylvania were living under Habsburg rule; Rumanians felt themselves Latin, with an historic affinity with France; the moment seemed propitious. Unfortunately for Rumania, Brusilov's offensive was halted at the Carpathians and the Austro–German armies, at first severely outnumbered but quickly reinforced, had been able to check a Rumanian advance in Transylvania and themselves pass to the offensive. When Rommel's *Alpendivision* reached the Rumanian theatre of operations at the end of October the Austro–German forces were mounting a great concentric offensive operation into Rumania – south-east from Hungary and north from Bulgaria across the lower Danube. The Rumanians had been driven back, and in the part of the front whither Rommel's division travelled they were holding in the Transylvanian Alps.

The *Gebirgsbataillon* moved by truck through the usual appalling traffic conditions inseparable from war, and ultimately dismounted at a place Rommel called 'Hobicauricany'.* They were attached to a Cavalry Corps and immediately ordered to start marching and climbing, to adopt a position in some extremely high and inhospitable country to the front, and then to reconnoitre forwards.

This was mountain fighting, mountain movement appropriate to the training of the *Gebirgsbataillon*, and Rommel's men were hardy. They had, however, no pack animals, and no winter equipment appropriate to living and fighting in the high hills – hills of six thousand feet. Officers and men marched and climbed carrying heavy loads on their backs. Darkness soon fell. It was a pitch-black night and the rain came down without a break. There was no possibility of rest, warmth or shelter. In wretched circumstances the battalion trudged up and on, and soon after dawn they reached the snow line. A blizzard began blowing. There was no cover of any kind and the men were soaked to the skin. A fellow Company Commander of Rommel's, an experienced mountaineer, represented the situation to the Battalion Commander and suggested withdrawal – sickness and frostbite would soon punish men shivering in damp clothes in these conditions. The response was a threat of court martial. Next day the battalion medical officer directed the evacuation to hospital of forty men.

The weather improved. Positions were adopted which gave a certain amount of shelter to men and horses. Tents were erected. Telephone cable was laid. Living became near tolerable. Rommel never forgot, however, the debilitating effect on even the toughest and best-disciplined troops of really frightful physical conditions.

After three days he reported the company back in good shape; and the beauty of their surroundings inspired him. 'Mist lay over the plain far beneath us,' he wrote, 'and like an ocean breaking on the shore beat against the sunlit peaks of the Transylvanian Alps. A superb sight!'

* Probably Uricani, ten miles west of Petrasani.

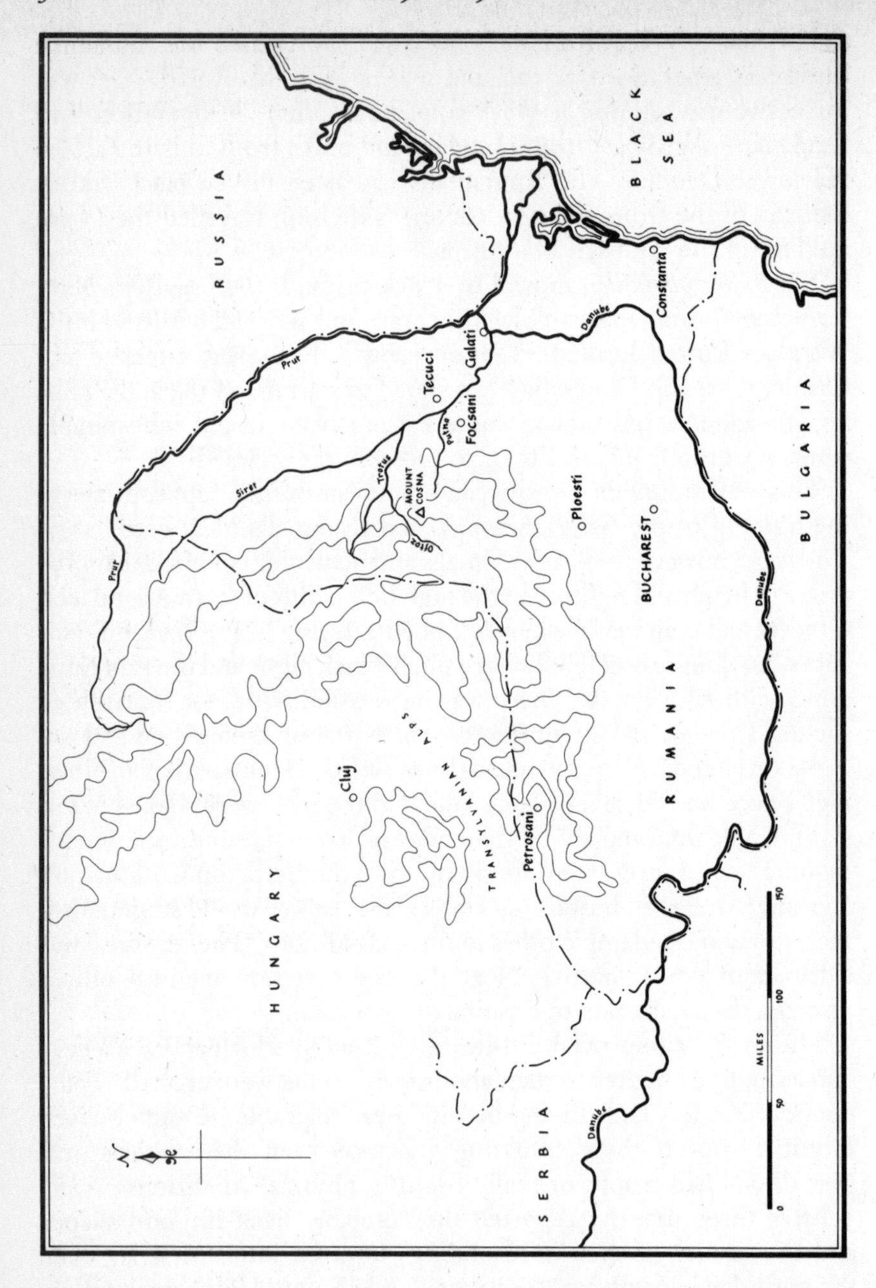

IV Rumania. Area of operations, *Gebirgsbataillon*, October 1916–August 1917

In November, moving to beyond the mountains now, the battalion took part in an attack. Rommel, again exulting in action, in the offensive, reported that 'Our assault troops broke from the cover of the bushes and raced like a torrent down the slope. There was no opposition.' The company continued to advance, moving from village to village, reconnoitring, taking precautions but maintaining momentum, periodically lining up for a coordinated attack. This was again mobile warfare, reminiscent of 1914, warfare in which the enemy might be found anywhere at any time, warfare which would always reward the quickest of eye, mind and decision. Rommel's hectic accounts give the flavour. He went with an advance guard

> of one group, the point of the advance, the company about 150 metres behind. Mist swirled around us and visibility was sometimes a hundred, sometimes only thirty, metres. At the south end of the place our head bumped into a column of Rumanians, marching in close order.
>
> Within seconds we were engaged in a violent firefight at not more than fifty metres. The first volleys were delivered from the standing position. The Rumanians outnumbered us. Rapid fire stopped them, but from both left and right of the road new enemies were emerging among the hedges and bushes, creeping, shooting, getting ever nearer. The situation of my advance group seemed desperate . . .

This sort of warfare was meat and drink to Rommel. But so, perhaps, was almost any sort of warfare. He had been married, during a brief leave, on 27 November.

As the Rumanian army, reinforced by a number of Russian divisions, withdrew towards the plain east of the mountains, resistance stiffened. Bucharest fell on 6 December. The Rumanians were now fighting with the Black Sea at their back and hanging on to a shallow position at the mountains' eastern rim. Rommel's company spent December and early January skirmishing in the snow and mist, laying ambushes, suddenly coming upon Rumanian detachments and engaging, struggling to keep alive and warm in the bitter conditions, taking large numbers of Rumanian prisoners. It was a campaign which rewarded only physical hardihood, ingenuity and boldness. Rommel, as ever, took infinite pains to disguise a line of approach, to come at the enemy from an unexpected direction, to delay fire against a known enemy position until he had prepared a surprise; then he always struck with extraordinary speed and ferocity.

When Rommel stormed the village of Gagesti* on 7 January 1917, he first advanced through the mist to a piece of high ground overlooking the place, shot a thin Rumanian outpost line out of the way and made good the ground as an attack base. Then he waited; scouted forward; and changed his plan, taking his company on a long march through the night, north-west, then north, and then east, marching round identified Rumanian positions and establishing the company on another commanding hilltop. He then lay up for a while, and edged various assault groups to positions opposite selected targets in the village. When he ultimately charged in, supported by his heavy machine guns – his company was by now four miles ahead of the German line – he took 330 prisoners, suffering not a single loss in the process. He noted a facet of this attack which was to remain a characteristic of his technique – the psychological impact of fire, even where conditions prevent it being physically effective. At Gagesti his machine guns were for a while masked by the movement of his assaulting riflemen, but Rommel directed their rapid fire high and to the right, to spray the roofs of part of the village; the impression made on the Rumanian defenders was significant, even though they suffered few casualties from it.

During these days Rommel was commanding not only his own company but a wing of the battalion – two rifle companies and a machine-gun company. Whenever he planned an attack he put into effect what might be called – or what became – classical principles applied in a modern manner. He concentrated fire at the selected point of attack – machine-gun fire; he broke into the enemy defences on a narrow front; he immediately established positions within the enemy defence fabric from which enfilade fire could be brought to bear outwards; and when he advanced he taught his subordinates to maintain momentum and disregard threats to flank or rear. In all this the early Rommel was father to the later Rommel, the General, the Panzer leader; but to a student of war – and throughout his life Rommel was a thoughtful, practical, articulate student of war – it was Napoleonic; and Rommel admired Napoleon above all men, above even the great Frederick himself. As a young officer he had bought a print of the exiled Emperor on St Helena and hung it in his room.

* In the Putna valley, west of Focsani.

The organization of the *Gebirgsbataillon*, based on a large number of strong rifle and machine-gun companies, enabled detachments of varying size to be formed, dependent on task and circumstance. It encouraged flexibility, so that a group appropriate to one battle could be, as a natural process, reformed, reinforced, or recast for another. In a later war their enemies were often to admire the facility with which the German army appeared to set up *ad hoc* groups, remarkably well-suited to particular encounters yet conforming to no specific establishment, uniting or dissolving on the battlefield with what seemed little loss of effectiveness. To a large extent this was to derive from excellent communications: to a very large extent, too, it derived from that unity of tactical doctrine on which the German army has always set store and which reduced time spent on the laborious business of 'familiarization', whereby in other armies units initially strange to each other can struggle towards mutual comprehension and tactical harmony, possibly in the middle of combat. Rommel, within his own command, was later to profit greatly from this flexibility and treat it as essential, natural, a virtue to be expected. And it is likely that his experiences in the *Gebirgsbataillon* impressed on his mind the importance of avoiding organizational straitjackets, of throwing together forces as required for particular actions, of the merits of improvisation.

In the early part of 1917 the battalion was once more transferred to France – back to the area of the Hilsen ridge whence they had moved to Rumania the previous October. This was a return to the trenches, the artillery duels, the never-ending work on fortification, drainage, revetment and repair. But in August 1917 yet another train ride – this time in intense heat rather than near-insufferable cold – again took the battalion eastward to the Rumanian front.

They arrived at Bereczk and then drove to Sosmazo, near the Hungarian–Rumanian border, on 7 August. For the next two weeks Rommel was engaged in the battle for Mount Cosna, one of the hardest of his own personal war. Mount Cosna was in Rumanian hands. It had to be taken – it affected the whole situation on the Eastern Front. Although the campaign in Rumania had taken the Germans to Bucharest and driven the defenders back towards the Black Sea, leaving most of the country under the occupation of the Central Powers, there had been no final collapse. In 1916 the Russians had been engaging more German and Austrian divisions than were deployed on the Western Front, and it was obviously a dominant interest of Germany's enemies to keep Russia in the war. The internal stability of Russia was, of course, a prime factor in this; the Russian Revolution of March 1917 had alarmed the Entente, although there were hopes that a more secure regime might ensue. Thus in the summer an Eastern Front still held, and it was a main German object to eliminate that front so that forces could be transferred westward for a final decision in France and Flanders. In the event this was to be largely achieved by the Bolshevik Revolution in November; but meanwhile there was a strong motive to clear up the situation in Rumania.

The Rumanian line was still established on the eastern edge of the mountains, near where the *Gebirgsbataillon* had conducted operations at the start of the year. The front was familiar; Mount Cosna was its highest – and most easterly – point. Beyond it the mountains fell away to the valley of the Trotus and the Oftoz rivers. East of their confluence the plains ran to the Black Sea, backstop of the Rumanian front.

For the first stages of the battle Rommel was given command of virtually the whole battalion – six rifle companies and three machine-gun companies. His task was to climb into the mountains from the Oftoz valley and establish the attacking force on a plateau from which there could be an assault on the enemy line. Information about the Rumanians was sketchy and, true to form, Rommel sent a party forward to discover what he could. Very soon, and surprisingly, he learned that the enemy appeared to have abandoned a forward position on the plateau; he instantly sent two companies forward to grab it. He then sent out further reconnaissance parties. The brigade plan was that the *Gebirgsbataillon*, together with a Bavarian Reserve Infantry regiment on its left, was to attack in mid-afternoon on 9 August, Rommel leading.

Before this could happen Rommel had an astonishing stroke of luck. One of his reconnaissance patrols, commanded by a sergeant, ran into a party of Rumanians who had stacked their weapons and were resting, with no sentries posted. The patrol returned with seventy-five Rumanians and five machine guns, without having fired a shot. Rommel – untypically, but correctly, requesting prior permission, because a deliberate attack with artillery support was pending – decided to take a detachment of two companies forward to where the Rumanians had been surprised, and thence to lead a convergent attack at the same time as the main body's assault. This, regrettably, involved losing height and then climbing again. When, as he reckoned was inevitable, his party ran into fire during the climb, Rommel gave orders that as far as possible his men were to break off the engagement, make a detour as necessary and keep climbing.

Rommel thus managed to get behind and lever out of their positions an enemy platoon, and then another, and then another. The heat was intense, and men were collapsing as they drove the enemy from five successive positions, never giving them time to collect themselves. Ultimately Rommel found himself with only twelve men, still struggling uphill, pursuing the Rumanians; he finally reached the edge of some woods and what seemed the top of the steepest part of the hill, where his two companies panted up to join him. The laid-down objective was the crest line and Rommel, who had in effect already broken into the heart of the enemy position by his climbing and skirmishing, climbing and skirmishing, now found that crest line only eight hundred yards away. Between his detachment and that objective, however, was a line of Rumanian riflemen and machine gunners who now opened up. Even the few moments he needed to collect his command had given the enemy a chance to get troops in hand and form a new front. The speed of his operations, however, had left the planned mid-afternoon attack far behind.

Rommel had no machine guns with him – the wild weaving and climbing among brushwood on the steep slopes had made that impossible. With his riflemen, he now began working forward, using the dips and breaks in the ground, darting in small bodies, closing the distance between himself and the enemy. Before dark he had manoeuvred his men into a position only ninety yards from the Rumanian defences – a position which he quickly organized for all-round defence against the inevitable Rumanian counter-attack. Rommel by now was eleven hundred yards behind what had been the Rumanian front, and had no means of contact with his own battalion until six o'clock next morning, when a telephone line was run through.

Rommel, by his audacity and perception, had established a perfect jumping-off place for an attack on the first line of the main Mount Cosna position. In effect his detachment had driven in a steel probe which could now be reinforced, and in the early hours the other companies of the *Gebirgsbataillon* tracked forward and joined him. By then, through vigorous patrolling, he had established the whereabouts and limits of the enemy and had decided exactly how to attack them. He was given three rifle companies and two machine-gun companies and ordered to do so. He personally deployed ten machine guns, taking each of them a long way round in order to remain concealed, and giving each exact targets and timings. Then he moved his three assault companies into a fold in the ground and divided them into one force which would make a feint, and another which would deliver the main attack. At midday he gave the signal. The machine guns opened up, and the assaulting companies went in. Thereafter there was a wild, close-quarter firefight, with Rumanians everywhere, local counter-attacks mounted with considerable spirit, and a general mêlée. Ultimately, the battalion was established on its objective, Rommel (who had been wounded in the arm and lost a good deal of blood) holding the forward edge of the new position with two companies.

Rommel learned a good deal from this small but arduous battle. He had attacked – the circumstances had necessitated it – without artillery or mortar support, but he had used heavy machine guns in a support role with devastating effect. He had taken time to reconnoitre and build up a thorough and accurate picture of the enemy. He had correctly estimated how and where to achieve surprise, taking advantage of a certain indolence in the Rumanian preparations, a certain dullness in their awareness. He had, at the immediate tactical level, shown ingenuity, feinting with part of his force in order to draw fire and provoke the premature movement of Rumanian reserves; and then he had delivered his punch with overwhelming strength. The battalion was now established on the near ridge of Mount Cosna, but there was still plenty of fighting ahead.

On the following day Rommel was given, in effect, six rifle companies and two machine-gun companies for the next phase of the offensive battle. He had slept little. His wound was painful, the day had been exhausting and nerve-jangling.

And next day, too, the heat and the climb were to be demanding in the extreme. The Rumanian positions were in considerable depth, line after line with mountain ravines between them, and so far only one line had been taken, deep though it had been. The crown of the mountain was still in enemy hands.

Rommel was given, as objective, the summit of Mount Cosna. The direct approach to it ran along a bare ridge, and the Rumanian positions were, in the main, astride this ridge. Rommel deployed some of his machine guns and two companies – the machine gunners were each carrying loads of 110 pounds and the sun was merciless

– to fire intensively and straight to their front, to distract the enemy from any movement round either flank. He proposed to lead his other four rifle companies and one machine-gun company to the enemy's right flank, to the north, where the wooded slopes of another hill should afford some cover. When he assaulted from the Rumanians' right and, as he hoped, achieved surprise, his supporting-fire group would also advance, straight down the ridge road.

It was a hard march, steep and uneven, the hillside laced with ravines. As Rommel finally assembled his companies for the attack, however, he made a disagreeable discovery – his telephone cable had been cut or damaged. Rommel invariably set high store by his communications, particularly in order to obtain artillery support, and in this instance also to maintain contact with that part of his command which had been ordered to hold the enemy by fire from the front. His signals detachment, therefore, was vital to him: and now contact was lost. There was nothing to be done about this, or nothing quick, and Rommel made his assault plan. Machine gunners and one rifle platoon were to crawl towards the enemy position – the right, as Rommel assumed, of the Rumanian Mount Cosna summit position. When the enemy opened fire these troops would open up at close range, and immediately thereafter the remaining companies would storm in, drop off rifle squads to secure the shoulders of the break-in, and advance, fighting their way right through the objective, exploiting south-eastward thereafter.

The first part of this plan went admirably. Rommel charged with his companies as the supporting machine guns ceased fire, broke in and reached the limit of the first objective with hardly a man lost. When it came to exploitation forward, however – and the summit still lay well ahead – Rumanian machine guns and a strong line of rifles stopped the Rommel detachment in its tracks. Fire was intense, and losses mounted, while Rumanian infantry, supported now by artillery fire, began to move towards Rommel's companies in a counter-attack.

Rommel, as ever flexible, realized that he could not continue the movement with his left, flanking detachment. He began moving with some of his troops to his right, towards the centre line running along the ridge from his original position towards the enemy. He hoped now to attack the left flank of the Rumanians confronting his own flanking movement. At that moment, true to original orders, his supporting fire group under a Lieutenant Jung appeared, attacking straight along the ridge. It decided the day. The Rumanians had committed everything they had to deal with Rommel's threat to their right, and at the appearance of fresh troops from yet another direction they fled. Rommel's force was now established on a lesser hill, on the way to the summit, a hill which had formed the central strongpoint of the forward Rumanian position. Ahead, across a ravine and overlooking it, was Mount Cosna itself. They named this lesser hill 'Headquarters Knoll'.

Rommel and his men were exceedingly tired. He decided, however, that he would be ready to attack in one hour and his Battalion Commander, who had now joined him, agreed. Rommel intended to repeat the morning's manoeuvre, leading four companies on a left-flanking movement into and out of the ravine ahead of them, covered by fire from the knoll they now occupied. It worked; and after several hours of arduous climbing and skirmishing with Rumanian outposts Rommel and his detachment stood on the summit of Mount Cosna, whence the enemy had mostly fled.

Although they had withdrawn from the summit, however, the Rumanians still held strong positions east and north-east of it; and it would not be possible for the Germans to advance into the plain – the strategic object – until these were cracked. Furthermore the Rumanians could, until the Mount Cosna position was deepened by exploitation eastward, both harass it with artillery fire and form up to counter-attack it from uncomfortably near. An attack eastward to deal with these positions would involve movement on the eastward slopes of the mountain in full view.

Rommel nevertheless decided it must be done, afterwards writing, characteristically, that it was better to be a hammer than an anvil. The first part of the attack – a move off the summit into a hollow covered from Rumanian view – had been completed and Rommel was determined to press forward when he received a shock. His telephone was now working and the order from his Battalion Commander came clear and unequivocal. The Rommel detachment was to withdraw immediately to the ridge half a mile west of Mount Cosna.

From which, in fact, they had attacked on the previous day. The Russians had transformed the situation – they had broken through on the northern flank of the Rumanian defenders and were now threatening the German deep left flank.

At a conference with his company commanders at midnight that night, 12 August, the *Gebirgsbataillon* commander, Major Sprosser (who had commanded them from their first formation), asked for views on how to deal with the threat. They were, or were soon likely to be, nearly surrounded, and the expectation was that the first Rumanian – or Russian – attacks would not be delayed long beyond dawn. Rommel made proposals which were adopted; and for the next few dramatic days he can be seen as the organizer and inspirer of defence in a particularly desperate situation.

Rommel said that the summit of Mount Cosna should be abandoned. It should be covered by fire – machine-gun and artillery fire should be able to sweep it when the Rumanians, as anticipated, tried to occupy it – but the Germans should not occupy it themselves. They should occupy reconnaissance positions, to be given up if necessary but not prematurely, south of the mountain, and to the north of it they should take up two company positions on the steps of the slope, organized for all-round defence; these positions should be simulated as being stronger and wider than they were – fires should be kept alight at numerous points to deceive the enemy in this respect.

The fulcrum of the battalion position would be the lesser hill – Headquarters Knoll – Rommel had taken before storming Mount Cosna itself. This should be held by machine gunners – and one rifle platoon – able to cover Mount Cosna. The main force of four companies would be held in reserve near Headquarters Knoll, dug in, ready to counter-attack or reinforce wherever required. Rommel gave himself the maximum ability to affect the battle with reserves, straitened though his defensive resources looked like being.

The Rumanian attack began pretty well as predicted, and it was sustained, from all directions, with considerable fury and in very great strength throughout 13 August. One of Rommel's reconnaissance forces south of the mountain withdrew – as he thought unnecessarily – and he led it into a counter-attack. His outpost platoons were driven from their positions. His reserve companies, in his temporary absence, were committed to the defence of Headquarters Knoll, which was assaulted by huge numbers of Rumanian infantry. All the time Rumanians – taking heavy punishment from the German machine guns and artillery – were streaming westward down from the reoccupied summit of Mount Cosna, as Rommel had expected and for which he had prepared. The enemy were attacking the German positions from east, north and south; and ammunition was running alarmingly low. Again and again engagements at grenade range were going on throughout the battalion area (Rommel's responsibility was the immediate Mount Cosna and Headquarters Knoll sector, while Major Sprosser had a wider remit). When night fell, Rommel reckoned that the *Gebirgsbataillon* could not afford losses of the kind they had suffered in another such day's fighting – and another such day undoubtedly threatened. He was determined, whatever the difficulties, to fortify and reorganize; and an Engineer company was brought up before dark to start work on the heart of the position, Headquarters Knoll.

Rommel had not had his boots off for five days. His feet were swollen and the bandage on his wounded arm was unchanged. His clothes were saturated with blood, and by contrast to the scorching heat of the day the night was cold. All through that night the Rommel detachment worked on its positions, and before dawn, to his huge relief, several reinforcement companies from other regiments arrived. In the following two days the strengthened positions on and forward of Headquarters Knoll held without the same degree of difficulty. A thunderstorm broke on 16 August, relieving the heat but soaking the soldiers. They dried out their wet clothing on their bodies, by the heat of a fire tended by Rumanian prisoners. Rommel wrote that all were in fine spirits!

The situation to their north had been stabilized. On 19 August Rommel, on the offensive once more, again led an attack with three rifle and two machine-gun companies into the Rumanian lines and, yet again, drove the enemy from Mount Cosna. Again he practised the same technique – careful preparation, an arduous but as far as possible concealed approach, devastating supporting fire with machine guns and artillery at the point of attack, and an assault with sufficient troops to fight through the objective, to keep going, to maintain momentum, never to give the enemy time for recovery. Another Rumanian counter-attack failed to drive the *Gebirgsbataillon* from Mount Cosna. They had suffered five hundred casualties in two weeks of summer fighting and thirty of their number had been killed. The battalion was withdrawn into reserve on 25 August.

CHAPTER FOUR

Pour le Mérite

After the battle on Mount Cosna Rommel was given a few weeks' leave and was able to go to the Baltic with his wife for a complete change and rest. He undoubtedly needed it. His notes record a sense of total exhaustion, and at the end of the battle he had a high fever and wrote afterwards that he began to babble 'the most idiotic stuff'. He realized he was unfit to remain in command. He was replaced, a battle casualty without question, and five days later the struggles for Mount Cosna were over. It was October 1917 before Rommel returned to duty, and another chapter in the story of the *Gebirgsbataillon* was starting.

The position of the Central Powers in the autumn of 1917 had resembled a neck within a slowly tightening noose; Germany and Austria-Hungary were surrounded on every flank by enemies, and although they operated on interior lines over a good railway network, those enemies had been growing stronger.

The pressures around the Central Powers differed from one front to another. On one arc of the noose – the western arc – the great and expensive battles known to the Allies as Third Ypres had started in July 1917 and continued until 4 November. Little ground had changed hands, and that little was mainly composed of mud and morass. But in the south-east part of the noose – in Rumania and the Balkans – the German offensive had, in effect, demolished the southern anchor of the Russian front.

Then, in November, would come the Bolshevik Revolution in Russia, followed by a near-immediate demand for peace with

Germany.* Thus the eastern and south-eastern arcs of the noose were soon to be slackened and vanish, and the noose itself dissolve. Ludendorff's vision of being able to concentrate the entire strength of the German Empire for an ultimate blow in the west looked to be within reach. It was very necessary; the strain on Austro–Hungarian and German manpower over the last three years had been terrible, and Allied seapower, by blockade, was having a profound attritional effect on the health and morale of the home front. And the United States – fresh to the war, a huge, untried industrial nation – was now in arms against Germany. It was essential to assemble sufficient strength to reach a decision in the west before American forces – still small and inexperienced – could build up to a point where they must constitute the strongest piece left on the board.

This lay in the immediate future. Meanwhile there remained the southern front, in Italy. And when Rommel returned to the *Gebirgsbataillon* in October 1917 his comrades had moved from Rumania, taking train through Macedonia, and were now deployed in Austrian Carinthia, ready to take part in the next battles of the Italian campaign.

The Italian campaign had begun in May 1915, when Italy had declared war on Austria. Relations between Rome and Vienna were, of course, coloured by the long ages of Austrian Imperial hegemony in north Italy, and there were ethnically mixed populations in Trieste, the Trentino, and the southern parts of Tirol. Austrian annexation of the Adriatic provinces of Bosnia and Herzegovina had been regarded in Italy as offensive, and had influenced Italy in her decision to desert the Central Powers. By the spring of 1915 it was clear that there could be no swift German–Austrian victory. Italy had cast her lot.

The defence against Italy had been conducted in the earlier stages by Austria. The Italian offensive – for the Italians conceived themselves as being on the strategic offensive throughout – was planned as an eastward movement in the southern part of the Isonzo valley, where that river runs towards the Gulf of Trieste. The northern and north-eastern mountain flank were to be held defensively. Trieste itself was a main objective.

* The actual peace of Brest–Litovsk was not imposed by Germany until 3 March 1918, but all threat from the east had disappeared long before that, in the bloody turbulence of the 'October' (Russian calendar) Revolution.

The first Italian advance had soon bogged down into trench warfare, and thereafter a series of Isonzo battles followed a similar pattern to those on the Western Front. Huge numbers of guns were assembled; huge casualties were suffered by both sides; and very small gains in territory resulted. In what was known as the eleventh battle of the Isonzo in August 1917, however, the Italians had managed to push the Austrians a significant distance, had taken Gorizia and were at last within sight of Trieste. They had also taken the high plateau of Bainsizza between Gorizia and Tolmin.* The Austrian front was seriously threatened, the Austrians requested German assistance, and Germany, although hard-pressed, agreed to transfer an army to the Isonzo front – the Fourteenth Army, of seven divisions. As soon as possible it was intended to begin a counter-offensive and drive the Italians back into Italy. The sector selected was Tolmin and the object was to cross the Isonzo river – the high ground on both sides of which was in Italian hands – to force the Italians out of the mountains west of the river, and to advance into the plain of the Veneto and the Tagliamento valley. It was an ambitious programme, and the Italian positions were known to be in considerable depth, with the Isonzo itself forming part of their forward line. Behind it was an intermediate line, and a third line on the high mountain ridges of Mounts Matajur and Kuk. The task of the German Alpine Corps was to capture this final and highest objective; and the task of the Wurttemberg *Gebirgsbataillon* was to move on the right of another regiment, the Bavarian Infantry Life Guards; to eliminate certain identified Italian battery positions; and then to move, behind the Life Guards, to the crest of Mount Matajur itself.

Rommel was given three mountain companies and a machine-gun company – as always, named 'The Rommel Detachment' – and ordered to follow the leading troops for the first phase of the action, and then, after the crossing of the river and the passing of the Italian first line, to act as Battalion advance guard. Rommel was an *Oberleutnant*, soon to be a captain, nominally a company commander; but on most occasions in this campaign, as in Rumania – as often, indeed, in France – he was exercising command of battalion or near-battalion size. Nor was this unusual. In the German army, then and later, rank was given for length of service, qualification, degree; it was not assigned, whether on a permanent or temporary basis, because of

* Now, like much of this theatre of operations, in Yugoslavia (Slovenia), but in 1917, of course, within Austria-Hungary.

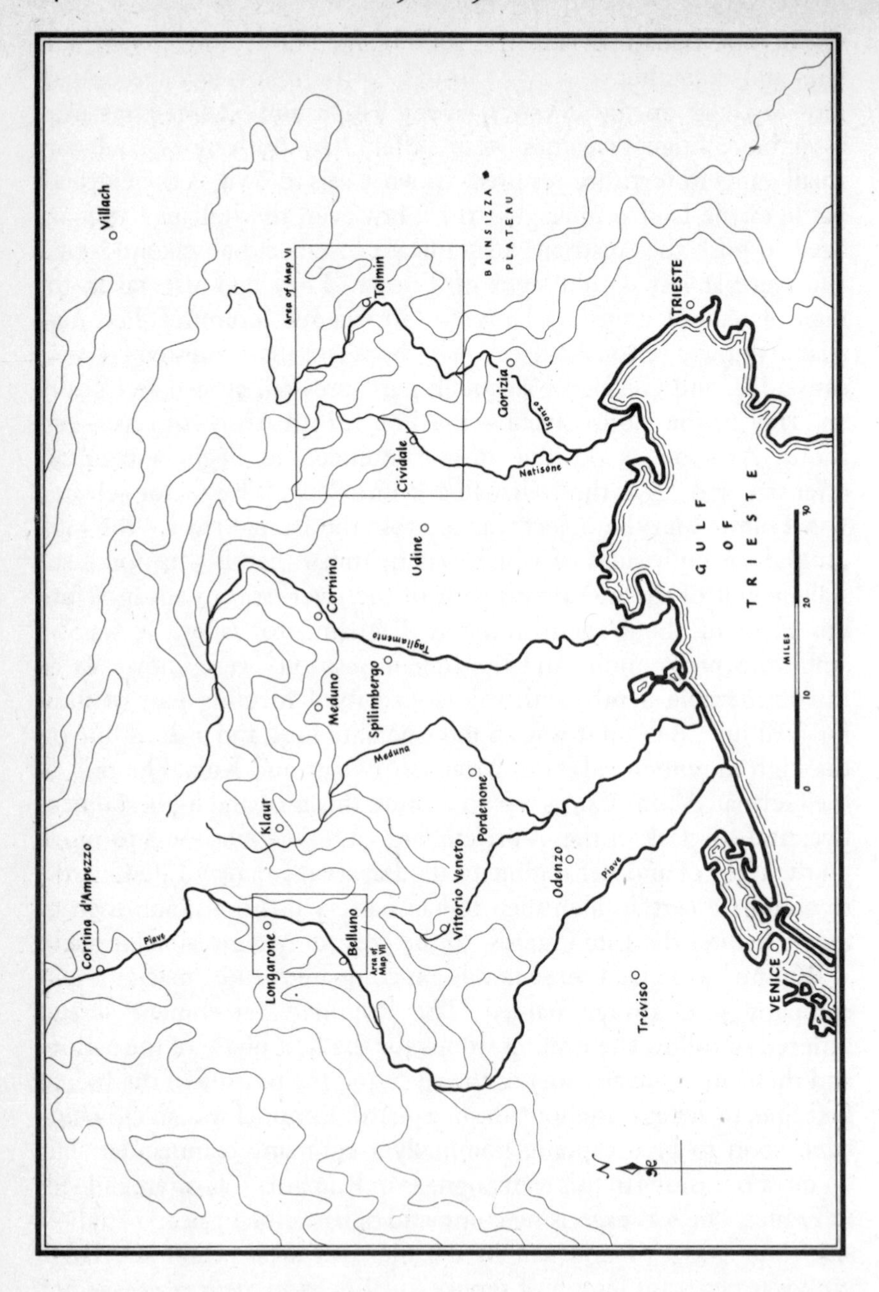

V North Italy. Area of operations, *Gebirgsbataillon*, October–November 1917

the spread of responsibilities a man might for a while assume. If a lieutenant found himself commanding a battalion, so be it; he would do it according to his abilities. In consequence both commissioned and non-commissioned officers in the German army tended to exercise responsibilities considerably greater than might happen in the Services of other nations, without any sense of disparity. Because rank was accorded more sparingly it was more prestigious, and – again in consequence – fewer commanders exercised authority over relatively larger numbers of troops than in other armies. The pyramid of command was broader at base, and responsibility was assumed naturally by comparatively junior ranks. Rommel, a lieutenant or captain running a battle with several companies, was often giving orders to a large detachment commanded by a sergeant: while in the Isonzo battle Major Sprosser – still *Gebirgsbataillon* Commander – had eleven companies, to be combined flexibly as occasion demanded. Each company, furthermore, was established at some two hundred rifles. Whatever the German army's shortcomings, it was certainly not topheavy.

The German attack started on 24 October, preceded by a great artillery preparation. A thousand guns began the programme at 2 a.m. and the sound echoed round the mountains with dramatic effect. It was pouring with rain and the Italian searchlights, seeking to identify attacking troops – since such a bombardment could only herald an attack – were largely ineffective. Just after dawn the battalion advanced under heavy supporting artillery fire, climbed down the steep slopes and crossed the Isonzo without much difficulty. Soon, under cover of the devastating artillery barrage and the appalling weather, they were through the Italian first position and advancing up a steep track towards the second. The slopes were wooded, the trees and bushes still in leaf. Rommel's Detachment, now advance guard, was climbing steeply. They soon ran into an enemy position which opened fire. This, Rommel knew, must be the Italian second position, and his point section reported wire and entrenchments.

Rommel reckoned that by leaving the axis-path and climbing up a steep gully to his left he might find a less obvious approach to the Italian position. He set off, the detachment following, and was soon rewarded by finding a 'continuous, strong and well-wired enemy position . . . but silent as the grave'. It was reminiscent of the early stages of the assault on Mount Cosna. Rommel saw a few Italians near the position, obviously unaware of the proximity of his force.

He found a hidden path, presumably used by the Italians, which

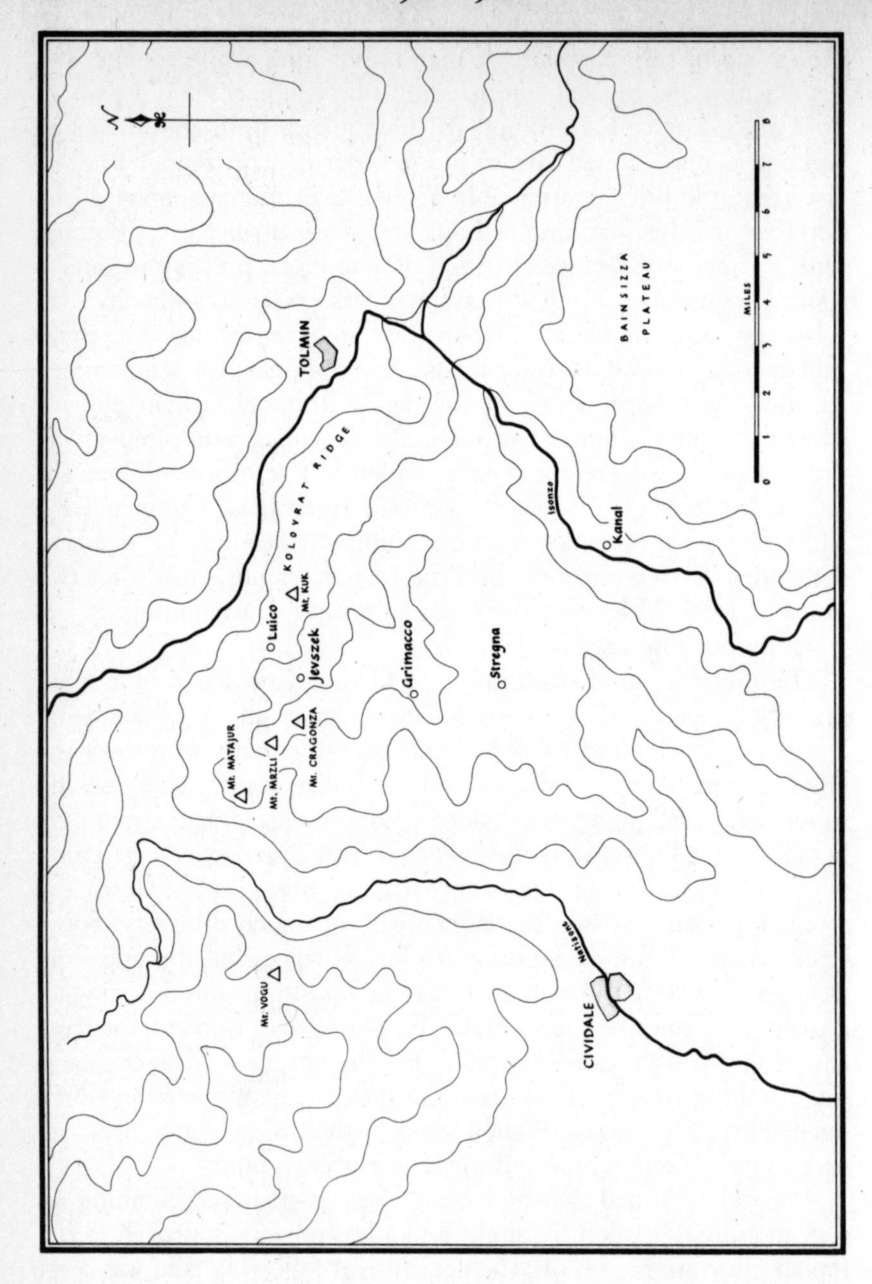

VI *Gebirgsbataillon* area of attack, 24 October 1917

looked as if it ran into their position, and sent an eight-man patrol along it, with orders to move quietly, firing only if necessary. The rest of the Rommel Detachment lay up, ready to open fire if the patrol got into trouble. Soon a member of the patrol hurried back, and reported to Rommel that they had taken seventeen Italians in a dugout and were inside the position. There had been no reaction.

Rommel led the whole detachment along the same path, pushing out flank protection parties when he judged he was into and through the Italian second line. He found a number of Italian prisoners sheltering in dugouts. It was still pouring with rain. And still no battle.

Moving in single file, the Rommel Detachment then covered a thousand yards, the machine gunners carrying loads of ninety pounds. Rommel was now well ahead of any sort of schedule, and felt periodic anxiety that he would walk into German artillery fire. He moved forward, largely disregarding this hazard, and stumbled on a number of Italian patrols and outposts, which were overcome without difficulty. By mid-afternoon he was moving in contact with the Bavarian Life Guards on his left, as originally ordered. Ahead lay a high ridge, the Kolovrat ridge. The main Italian position ran along the highest part of the mountain from this ridge to the – several hundred feet higher – ridge of Matajur.

At seven o'clock that evening, with the Germans now confronting the Italians on their third position, Rommel was summoned to the command post of the Life Guards 3rd Battalion Commander, on his immediate left. 'He demanded,' Rommel said, 'that we come under his command. I refused. I said that my orders came from Major Sprosser, who, to my knowledge, was senior to the Commander of the Life Guards Battalion.' It was clearly an unfriendly interview, and ended with Rommel being ordered to follow the Life Guards and occupy the height. Rommel said that he would inform his own Commander. He was 'none too happy'.

That is easy to believe. Rommel was unlikely to relish the role of following up a Bavarian or any other regiment and occupying ground taken by it. He got a few hours' rest, disgruntled. And as he lay, dozing, he made a plan.

At five o'clock next morning his own Battalion Commander, Sprosser, arrived, leading the rest of the *Gebirgsbataillon*, and Rommel told him his plan. This was to move west, well clear of the Bavarian Regiment, and carry out an independent attack on a part of the front eleven hundred yards to a flank, on a different section of the ridge. Major Sprosser agreed, and with two rifle companies and a machine-gun company, Rommel set out just before dawn on 25 October on the venture which would lead to the capture of the Kolovrat ridge and of Mount Matajur, an exploit to which he was to look back with perhaps more satisfaction than to any other in his crowded fighting career.

Rommel's first move, as so often in similar approach marches, was to overpower a number of Italian outposts and sentries who were watching the Italian front towards the Isonzo far below. Working along the slope, traversing, Rommel came on them from an unexpected direction, using the convexity of the hill (later, he was to become a superb stalker). He had, when starting, drawn a certain amount of rather desultory machine-gun fire, but he was aiming to take his detachment to a bushy part of the slope some two hundred yards from the top of the ridge. He gained it, the whole detachment having apparently avoided observation from the Italians on the ridge above. Rommel had been moving parallel to the Italian position. Behind him he heard the sound of a considerable firefight from the area where he had spent the night, where the Bavarians were still facing a particular high point on the Kolovrat.

Eventually Rommel reached the point he had marked for break-in. It was a matter of turning the Rommel Detachment to its left, climbing the immediate slope and assaulting. He paused while a reconnaissance patrol of five men under a lieutenant climbed the slope to inspect the wire and discover what they could. After a very few minutes they reported back that they had managed to get into the enemy position and take some prisoners, without firing a shot. This was all Rommel needed. Surprise had not been lost. As fast as possible he took the whole detachment up to the saddle, fifty yards wide, which crested the Kolovrat ridge; there they found every Italian in a dugout and very ready to surrender. Soon they had taken several hundred prisoners.

This was too good to last. Rommel decided next to exploit westward, towards Matajur, but he soon found that the ridge, along which ran a track, was covered by an area of high ground west of the saddle. From this point an Italian line of fire trenches ran at right angles to the Kolovrat ridge, blocking the Rommel Detachment's westward movement, while Rommel could see fresh Italian infantry moving up from his left, up the southern slopes of the ridge. And from the enemy in front, covering the ridge track, there now came heavy rifle and machine-gun fire. Rommel, with his companies, was up on the first part of the final objective, the ridge. But the ridge was very bare.

Rommel saw that bull-headed advance westward along the ridge would be futile. He ordered his lead company to hold its position, and doubled the rest back to the saddle as fast as possible. The lead company, engaged now in a savage close-quarter fire and grenade fight, was threatened not only from its front – from the west – but also from the flanks. It was the sort of situation which Rommel was to face again and again. First, his own energy and ingenuity had achieved surprise and astonishing preliminary success; then he had found that the very scope of his success had exposed his command, or part of it, to overwhelming danger; next he had to decide how to cope with that danger – whether to cut his losses, withdraw, sacrifice part of his command to save the rest and change his plan, or to extricate his men by offensive action, somehow, somewhere. He seldom decided on any but the last alternative. On this occasion, he wrote, 'it was instantly clear to me that 2nd Company could only be got out of trouble by a surprise attack, delivered by the rest of the detachment against the enemy's flank and rear'. A hundred yards ahead of him, 'the enemy was massing for attack'.

Rommel got one heavy machine gun down into position to his right, facing west along the ridge, and moved a rifle company to a piece of low ground on his left, from which he reckoned he could take the enemy in flank. His two companies – he had originally asked for four – were committed now, and the lead company was already weakened by having dropped off a detachment to protect the rear, facing eastward along the ridge. Short of men, therefore, he could only rely on the shock of offensive action from an unexpected quarter.

It worked. The Italians attacking his beleaguered point company were taken in flank; and that point company then themselves moved to the offensive, taking the Italians from two directions. Within minutes an entire Italian battalion of twelve officers and five hundred men had surrendered to the Rommel Detachment. This increased the count of prisoners on the Kolovrat position to 1500. But now Rommel saw another group of Italian troops forming up and moving towards him from both front and rear.

Rommel now embarked on one of the boldest strokes of his career. The ridge track running along and behind the Italian positions towards Mount Matajur had been blasted out of the hillside by Italian engineers, and was difficult to cover from the hill positions themselves. Rommel reckoned that he could either 'clean up' the troops defending the high ground which had held him up – the 'Kuk' peak – or press on, ignoring the high ground, and attack enemy troops round a village, Luico, by rushing his detachment as far and as fast as possible down the ridge track; Luico lay below the ridge, to Rommel's right or north side. If he advanced deep along the track he had the chance of annihilating a considerable concentration of enemy troops and thus making capture of the entire objective all the easier.

He, and behind him the troops of the Rommel Detachment, started running. They were, in effect, running along and behind an enemy front line, tearing into astonished parties of Italian soldiers, wagons, guns. As they ran unsuspecting groups of Italians met them, and here and there were taken prisoner – 'This was fun!' Rommel wrote, 'and no shooting! We had over a hundred prisoners and fifty vehicles. Business was flourishing!' Soon he was dropping down from the ridge and approaching a valley up which ran another road, clearly used for Italian resupply. If he could occupy the valley and block that road, he reasoned, a large part of the forward troops, the centre of the Italian final position, would be effectively surrounded – caught between his detachment and the rest of the battalion. On some of the Italian vehicles were supplies of bread, fruit, eggs and wine to refresh the hot and exhausted soldiers of the Rommel Detachment. Rommel reached the point at which he was aiming, and his men faced in all directions and dug in. He and his leading soldiers were two miles behind what the enemy thought of as their front.

But this hectic advance had left much of the Rommel Detachment behind, and when an Italian column, totally unaware of danger, came marching up the road from Luico Rommel had only 150 men under his hand. He deployed them astride the road and then attempted, with the use of a white handkerchief and by sending an officer with a white armband forward, to persuade the marching column to surrender. Unlike some of Rommel's *coups* this one did not immediately work. The Italians grabbed the white-armbanded intermediary, deployed and opened fire. After ten minutes of firing, however, they signalled surrender.

Fifty officers and two thousand men of a Bersaglieri brigade now marched into Rommel's custody. The 'bag' was becoming enormous. One of Rommel's missing companies now caught up and joined him, and shortly afterwards he moved into Luico itself and met the rest of the *Gebirgsbataillon*, who had approached from the other direction.

Rommel was given three companies and three machine-gun companies for the next stage.

Night had fallen, but the moon was up and by its light the Rommel Detachment set out on the very steep climb back up the ridge. Rommel was aiming at the next high point, Mount Cragonza. On the way was the village of Jevszek: and also on the way, quite certainly, were Italian positions. Jevszek would be overlooked, in daylight, from Mount Cragonza.

Rommel occupied Jevszek, avoiding the Italian garrison without difficulty in the darkness. He moved on just before dawn on 26 October but as daylight broke he found that his detachment was under fire from what were obviously strong positions on the near slopes of Mount Cragonza. Behind him part of his detachment had been detailed to deal with the now-surrounded defenders of Jevszek, who found the Germans behind them – a subsidiary battle which went well and yielded another thousand prisoners. But ahead the enemy on Mount Cragonza were well and effectively deployed. Unusually – but having no alternative – Rommel decided on frontal attack. The slopes between him and the Italians on Mount Cragonza were bare and he was attacking uphill, without cover, in full view of the enemy. There was nothing to do but go forward and go hard. Rommel moved with the centre company; in spite of losses the mountain troops reached the enemy trenches, killed their opponents where they failed to surrender, and by quarter past seven in the morning were masters of Mount Cragonza. Now only one summit, the Mrzli, rose between the attacking Germans and their ultimate objective, Mount Matajur.

Rommel and his men were exhausted by the night march, the *coup* in Jevszek, and the alarming frontal assault, but he gave himself and them no rest. After some skirmishing with Italian detachments scattered over the high ground he was able to collect two rifle and one machine-gun companies and to send up light signals, as requests for artillery fire. Within minutes he got it, on the enemy positions on the south-east slopes of the Mrzli. The crown of the Mrzli held another saddle, and in this saddle, Rommel estimated, some three battalions of Italian troops were assembled.

Rommel decided to walk slowly towards them. Nobody fired. He walked further, calling now and waving a handkerchief. Still nobody fired.

He reached a point 150 yards from the nearest Italians.

Suddenly the Italians started tearing away in different directions. Hundreds rushed up to Rommel and hoisted him on their shoulders with cries of '*Evviva Germania!*' Hundreds of others threw down their arms and surrendered. These were fifteen hundred men of the Salerno Brigade. Another regiment of that brigade – with a high reputation – was known to be in position on Mount Matajur, the final objective. And now Mount Matajur lay ahead.

At this moment a distraction and a difficulty arose. Rommel received an order from his commanding officer to withdraw. Major Sprosser, seeing the immense number of prisoners (over three thousand) now taken by the Rommel Detachment, had assumed that the whole objective was in German hands and the battle over. Furthermore, Rommel's easternmost companies had received the order and started to act on it; he found himself with only one hundred men and six machine guns.

With these, he decided to disobey orders. Major Sprosser, he reasoned, had not known the true situation – that Mount Matajur was still in Italian hands. An order given by a commander without full possession of the facts may be overridden by a subordinate who knows those facts! Provided, of course, that that subordinate has the courage of his convictions.

Rommel deployed his machine guns and opened fire. A few minutes later, covered by this fire, he spread out his hundred riflemen and began moving forward – handkerchief at the ready, for he had found this article a useful tool of combat. As his men came within sight of Mount Matajur itself – the only firing, so far, having been from his own machine guns, now silent – he saw a remarkable sight three hundred yards ahead. The renowned 2nd Battalion of the Salerno Brigade was assembling in the open, and twelve hundred men, defying their officers, were laying down their arms.

The summit, however, still lay ahead, and the Rommel Detachment kept going. A few bursts of German machine-gun fire, a four-hundred-yard advance over rocky ground, and a further 120 Italians surrendered. At twenty minutes before midday on 26 October Rommel fired the signal flares to show that the final point of the objective was now in the hands of the *Gebirgsbataillon*. He had been moving and fighting for fifty-two hours and had taken, in all, over nine thousand prisoners. He had lost six men dead and thirty wounded. The achievements of the Rommel Detachment were specifically named in the Alpine Corps Order of the Day. Rommel later learned with irritation that a brother-officer, Schorner, had been awarded the *Pour le Mérite*, the highest award for gallantry in the German army, for allegedly being the first to seize the summit of Mount Matajur. Schorner, later to be a General with a controversial reputation, had gained the palm Rommel reckoned was his; and Rommel was highly competitive and had a considerable ambition for personal glory.

Rommel was not one to exult or sneer at a fallen enemy, and he paid exact tribute on occasion to Italians who had fought well, just as he tried, in retrospect, to analyse the reasons why those who had failed had done so.

The German army had now broken through the mountain barrier and was in full pursuit of the Italians towards Cividale and beyond. They crossed the Tagliamento at Cornino, north-west of Udine and, brushing aside Italian rearguards, marched as fast as possible through Meduna and Klaut towards the upper Piave north of Belluno. This was mountain country again, and on 7 November Rommel, with three companies and a machine-gun company, was ordered to eject the enemy from a pass, five thousand feet high, from which Italian fire was covering the German axis of advance. The mountain walls below the well-prepared enemy positions were sheer. It was a matter of climbing by a circuitous route, making height and attacking the Italians, under cover of heavy machine-gun fire, from a surprise direction.

The attack failed. This was, Rommel said, the first attack since the beginning of the war in which he had failed, and he analysed it honestly in retrospect. The failure was in synchronization – the assaulting companies needed to attack supported by the fire of the machine guns on a timed basis, and they missed their opportunity and attacked too late, when the machine-gun fire was already over. The reason, or one reason, was that they had waited for Rommel in person to join them, and he had waited too long with the machine-gun fire support platoons. It was a personal error – for once he was not in the right place at the right time, and he recorded the fact exactly.

The *Gebirgsbataillon* marched on, taking hill position after hill position and driving the Italians headlong before them. To the Western Allies this was the disaster known as Caporetto, scene of the German *Schwerpunkt*. At last, driving men and horses on as fast as physically possible, the Germans approached the Piave, at a place called Longarone. It was, Rommel wrote,

> an overwhelmingly beautiful sight. In the brilliance of midday sunshine lay the valley of the Piave. A hundred and fifty metres beneath us the clear, green mountain stream rushed over its stony bed in many different courses. Beyond lay Longarone, with 2000-metre peaks towering above it.
>
> Over the Piave bridge was driving the motor vehicle of the Italian demolition squad. On the west bank of the river an endless column of the enemy, of all arms, was marching along the main road, moving south from the Dolomites in the north, through Longarone.

Rommel was at the point of the battalion advance guard, racing to forestall the Italian retreat and if possible capture a bridge unblown. When he first looked over the superb Piave valley he had only ten riflemen with him, riding captured bicycles. Soon he was able to bring long-range rifle fire to bear on a point where the Italians' south-running escape road was hemmed between river and rockface, stopping the Italian column thereby. The Rommel Detachment, breathless, began arriving. Rommel hurried them to Dogna, south of Longarone, determined somehow to cross the Piave and get a position on both banks. Everywhere there were huge columns of Italian prisoners moving from other parts of the front, and by their presence confusing the situation and complicating requests for artillery fire.

Eventually Rommel managed to get across the several branches of the Piave – most of them comparatively shallow but some swift-running – using a weir, and profiting from the freely-given advice of Italian prisoners. Further north some of his men had swum the river. Soon, on the west bank, Rommel had the south-running road out of Longarone blocked at Fae, and prisoners began to run into the trap in large numbers.

Rommel could have remained content with having crossed the river and having trapped huge numbers of Italians by the block he had established; he could have then awaited reinforcements. He decided instead that the greatest results would only accrue if he stormed Longarone itself, packed with Italian troops as it was. He had, he knew, the initiative, but the initiative might pass. Italian artillery was still active and the situation could change. The Italians had a huge numerical superiority. They might turn this episode in a headlong German pursuit into a stiff and costly battle. Rommel decided – it was late on that November afternoon – to attack in the dusk, marching northward towards Longa-

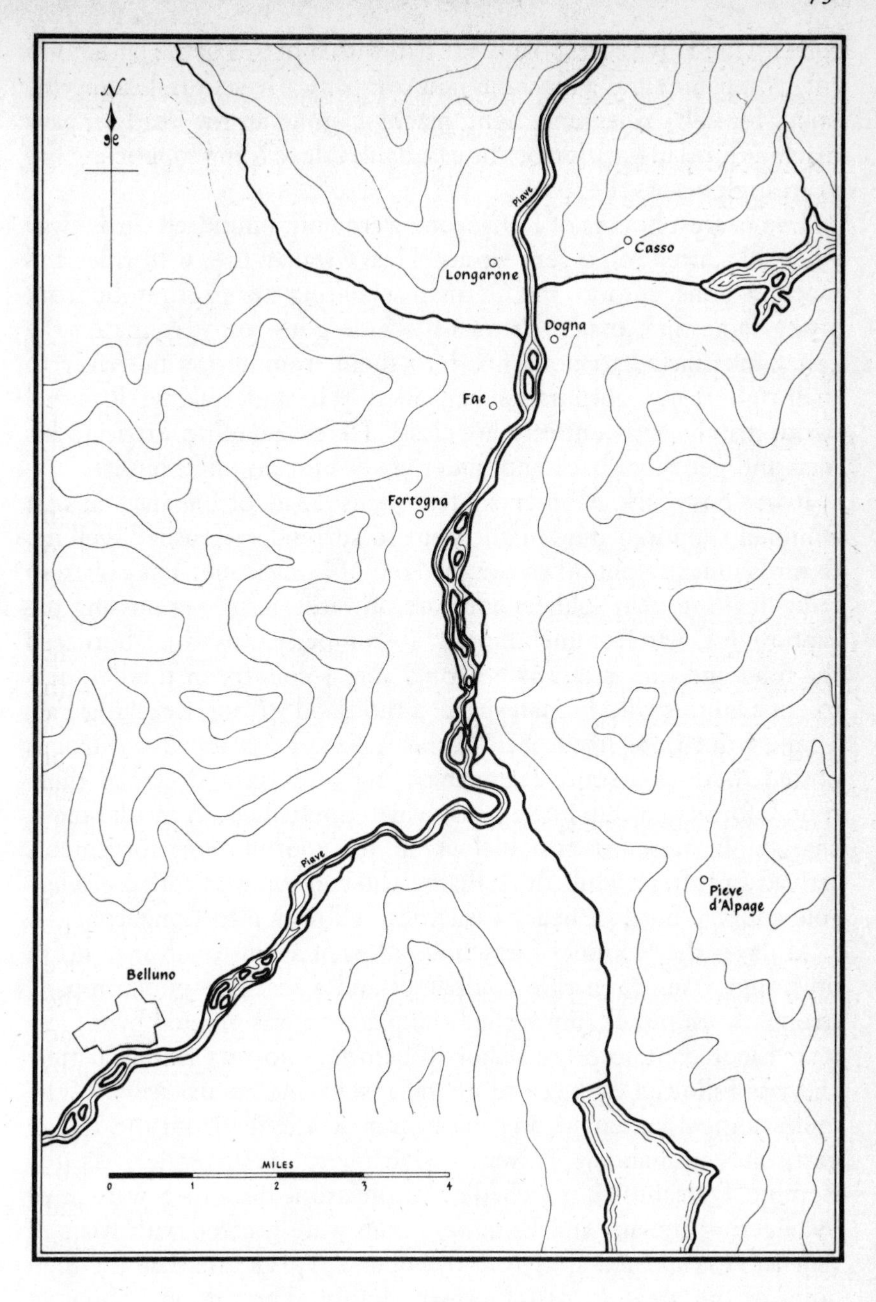

VII Area of *Gebirgsbataillon* crossing of the Piave, November 1917

rone up the Piave's west bank. He had with him two of the battalion's rifle companies and a machine-gun company. He set off, leading the point himself, rifles and light machine guns at the ready, heavy machine guns in position on the east bank able to bring supporting fire at an angle across the river.

The nearest houses of Longarone were only a hundred yards away when all started to go very wrong. Heavy Italian fire, with rifles and machine guns, cut into the German assaulting troops from the front at the same time that their own machine guns from Dogna on the east bank started to pour fire into them from across the river to their right rear. Coordination had failed. 'The undertaking,' Rommel wrote grimly, 'was entirely hopeless!' He managed to extricate his men and get them back and under cover of walls and houses.

It was now dark. Moments later a dense mass of Italians – at first Rommel supposed they might want to surrender – rushed, yelling, towards him straight down the road out of Longarone. They overran individual German soldiers and tore onwards, narrowly missing the capture of Erwin Rommel himself. He jumped a wall which bordered the road and ran as fast as he could across country in the darkness to outstrip this wild human surge, a thousand strong. Reaching Fae, a mile from Longarone, before the Italians he grabbed the troops posted there as a south-facing blocking position and turned them round, forming a thin line against which the Italians, in the darkness, charged in vain. At two o'clock in the morning reinforcements arrived and after a while the Italians, who had come on with considerable courage but lost heavy casualties, fell back into Longarone.

At dawn the Rommel Detachment began again to advance north on Longarone, three rifle companies and a machine-gun company strong. As Rommel approached the place he was greeted by a German Lieutenant, captured shortly before, who was riding a mule and was followed by a crowd of Italians waving handkerchiefs. The mule-riding Lieutenant was the bearer of a letter from the Italian garrison commander. It was a document of surrender. As the Rommel Detachment marched into Longarone the streets were lined by cheering Italians and Rommel's men were greeted with roars of '*Evviva Germania!*' It was 10 November 1917, and marked the high point of the great Austro–German victory. The tide was about to turn. French and British reinforcements, including a British force of five divisions, had been rushed to Italy and by the end of the year had taken the offensive.

Rommel was sent on leave at the end of the first week in January

1918. To his great sadness he was then posted away from the *Gebirgsbataillon*, which was shortly to be transferred again to the Western Front. Before leaving the battalion, he – and his Battalion Commander, Major Sprosser – was at last awarded the *Pour le Mérite*, the medal he had deemed his by right after Matajur.* He was now posted as a junior staff officer to the headquarters of LXIV (Wurttemberg) Army Corps on the Western Front, work he found wholly uncongenial. He was attached to several other formation staffs for short periods, and was not to return to his 124th Wurttemberg Regiment, his home, until Christmas 1918. He would not command soldiers in battle again for over twenty years.

Reading of Rommel in the years from 1914 to 1918, reading his own accounts and the impression made by him on others, the picture drawn is indisputably and almost exaggeratedly heroic. He saw, decided, acted, with an energy which would one day be acclaimed as genius. His was indeed 'the swoop of the falcon'. And Rommel in the years from 1939 – unlike some contemporaries in all armies – profited from the fact that at least some of his campaigns in that earlier war had included fluid and mobile operations. This was congenial to his temperament. It also helped shape his destiny.

Rommel had a pronounced taste for personal glory and the renown which surrounds it, but to suppose that, at any age, he was simply a gifted adventurer, a daredevil, a *sabreur*, would be mistaken. He was also reflective – and articulate. Furthermore, although by no means self-effacing, he was invariably natural, simple, unpretentious in his dealings with all men. He leapt at every experience of battle, but he turned experience into military wisdom by his shrewdness and his objectivity. There thus emerged not only a superb man of action but a military philosopher. It was this gift of Rommel's for distilling experience, for translating individual achievement, recollected, into the language of enduring – and universal – operational and tactical lessons which made him remarkable and lay at the heart of his success. He could, throughout life, decide fast, act boldly, remember clearly, narrate vividly, ponder and deduce wisely. Basil Liddell Hart described Rommel's genius as combining the conceptive and the executive talents, and also referred to him as 'one of the few eminent commanders who have gained distinction as military thinkers and

* Awarded on 18 December 1917. Rommel afterwards said that he had actually complained at the award of the medal to Schorner after Matajur.

writers'.[1] Liddell Hart was, in the main, writing of the Rommel of another war, but the verdict certainly stands when we read Rommel's account of 1914 to 1918, when we consider the Rommel of the Argonne, the Hilsen ridge, the Carpathians, Mount Matajur and the Piave. Other successful commanders from Caesar to Wavell have both fought and written with distinction. Rommel showed early signs of standing in their company.

Already perceptible, too, was the rashness of which critics were later to accuse Rommel, with some justice – the deliberate risks taken with logistic provision, the determination that his own boldness should set the pace of events beyond the dictates of prudence. The perception is fair – Rommel the calculator and organizer of conflict was never to be as obtrusive as Rommel the leader and the inspiration. But to Rommel – at any age – the essence of victory was to win the tactical battle. He believed – by instinct but also with his reason – that method and careful provision may be essentials at certain stages of campaign, but that they are without decisive effect unless the actual tactical battle 'at the point of the spear' is fought better than by the enemy.

This was Rommel's creed. To win the tactical battle may not settle a campaign, but to fail in it ensures the accomplishment of nothing. The Erwin Rommel of those early days, the twenty-three-year-old Lieutenant on the Western Front, the twenty-six-year-old who stormed into Longarone, is instantly recognizable as the man who half-hypnotized an enemy army twenty-five years later. There was, invariably, the same instinctive emphasis on leading from the front, on outflanking, on penetrating to the enemy's rear, on keeping battle fluid. 'Where Rommel is, there is the front' – Rommel's character as a commander was essentially heroic, a character thrusting, Rupert-like in its vigour. Such men are not always also reflective. Rommel was.

PART 2

1919–1939

CHAPTER FIVE

Soldier Without Politics

THE COLLAPSE of Imperial Germany in 1918 has been, for the outsider, somehow overshadowed by the more shattering eclipse of the Third German Reich in 1945. The latter marked the utter destruction of a regime, itself revolutionary and radical, which had lasted a mere twelve years: but it was attended by the invasion of Germany by foreign armies, by scenes of horrific plunder, rapine and destruction, by the dissolution of *Grossdeutschland* and by the virtual cessation of the unified German state as such. New institutions could only arise – with the consent of the victors – from broken ruins amongst a people numbed by suffering, terror and loss.

The events of 1918 appear in retrospect less cataclysmic, but to an entire generation of Germans they provided an awful object lesson, dramatic yet confused. The German army, although powerful, cohesive and occupying a central position, had been outnumbered by the sum of its enemies. It had waged an uneven but (with its Austro-Hungarian Allies) on the whole successful series of campaigns on the Eastern Front and in the Balkans; and had been, it seemed, ultimately rewarded with the convulsion of the Russian Revolution followed by a peace concluded with the Bolsheviks at Brest–Litovsk in March 1918.

But in Italy, where Germany had assisted Austria and scored a major triumph in that Alpine fighting and break-through to the Piave which had established young Erwin Rommel's claim to glory, victory had been followed by disaster when in October 1918 the Austrians – by now deprived of support from hard-pressed Germany – were divided and shattered by an Italo–British army and sued for peace at the beginning of November.

Meanwhile on the Western Front – reinforced by a mass of divisions transferred from the now quiescent east – the German army had undertaken one final and initially brilliant sequence of offensives, beginning in March 1918 and designed to break the stalemate and smash the Western Allies before the full weight of American military power could be developed and deployed in Europe. The attempt came near succeeding, but German casualties had been too great in the terrible battles of attrition which the Germans themselves had undertaken at Verdun and which had been forced on them at Ypres, on the Somme and in Champagne. These battles had also cost the Allies appalling losses, and had led to a good deal of Allied self-questioning, wavering of Allied will, and fragility of Allied morale; but in the end the Allies had, albeit indirectly and with delay, succeeded in their prime object – they had weakened the German army beyond the point where a German offensive could be sufficiently sustained if checked with determination. In the summer of 1918 the German offensive had indeed been checked – and turned, after remarkable early successes. In August the great Allied counter-offensive swung into action. On 11 August the Kaiser told his own General Headquarters, 'The war must be ended.' And on 11 November an armistice was signed whose terms left no doubt that peace was being imposed on a defeated Germany. The strains and losses of the war had left little room for generosity. The terms dictated to post-Revolutionary Russia by Germany (admittedly a Germany still at war) had been harsh as to territorial cessions and conditions. Now it was Germany's turn.

The German people were already suffering great hardship from the effect of economic blockade, but the failure of the Imperial army in the field came as a shattering surprise. The majority had not suspected that the military situation was so grave. While in the inner councils of Government hope of decisive victory had in large measure disappeared as early as 1916, the iron grip on policy of the High Command in the shape of the Chief of General Staff, Hindenburg, and his 'First Quartermaster General', Ludendorff, had sustained the confidence of much of the nation. The High Command had attained a ruthless hold on virtually every aspect of Imperial business. In 1917 the helmeted pair had procured the resignation of the Chancellor, Bethmann-Hollweg, when they fancied him 'soft' on the prosecution of the war. They had – even in July 1918, when the Allied offensives were in full swing – been formulating extreme and absurd 'post-hostilities demands' on Belgium. Now they had lost the war.

The Armistice, and the Treaty of Versailles which followed in 1919, provided for the occupation of the western regions of Germany; for enormous (and unpayable) reparations; for what was perceived and intended as the emasculation of Germany as a military power; for the surrender of a large part of the German High Seas Fleet (which scuttled itself at Scapa Flow in an act of defiance that aroused the unwilling admiration of its enemies); for the acceptance by Germany of certain self-accusatory clauses admitting guilt for the war; and for the loss of territory – notably Alsace and Lorraine to France and the eastern provinces, including part of East Prussia, to a resurrected state of Poland. Austria-Hungary was mutilated in an even more drastic way, with a new state of Czecho-Slovakia created from the Habsburg domains of Bohemia, Slovakia and Moravia; while the ambitions of the southern and eastern allies of the West – Italy, Rumania, Serbia – were appeased by the cession of part of the Tirol and Austrian Adriatic provinces to the former, and by the transfer of parts of Hungary, Slovenia and Croatia to Rumania and to a newly created 'Kingdom of the South Slavs', named Yugo-Slavia, dominated to a large extent by the Serbs.* The Hohenzollern and Habsburg empires ceased to be, and the concert of Europe, destroyed by the outbreak and savagery of the war, was succeeded by a condition of political and economic instability, with, in many quarters, dissatisfaction over the new frontiers: with internal rumblings, as the Russian Bolshevik experiment was attempted in other countries; and, above all, with the bitter resentment of the defeated at the imposed new order.[1]

Nor did humiliations stop at that point: in an Allied note of February 1920 the surrender of various 'war criminals' was demanded. The list (which was described as 'not final') included the German Crown Prince and two of his brothers, Duke Albrecht of Wurttemberg, Crown Prince Rupprecht of Bavaria, Field Marshals von Hindenburg and von Mackensen, Grand Admiral Tirpitz, General Ludendorff, former chancellors Bethmann-Hollweg and Michaelis, and nine hundred other officers and non-commissioned officers. In this international sludge of resentment and instability the German Weimar Republic was born.

The reactions and emotions of a patriotic young German officer

* The cessions of territory inflicted wounds which have, in many cases, not yet healed (1993). But it must be remembered that Serbia, for example, had had no less than half her total manpower killed in the war.

of Erwin Rommel's generation can be imagined by a foreigner if, for instance, a comparable military defeat of Britain be supposed – defeat after a long and painful struggle, with the losses of Loos, the Somme, Third Ypres and so forth echoing in every British mind and home. The maritime counties, dominating the Channel and providing the Fleet bases most contiguous to Europe – Kent, Sussex, Hampshire, Dorset, Devon – would have been garrisoned by the victorious enemy. The British African colonies would have been assigned to German protection; and to a resurrected 'Norseland', in alliance with Germany, there would have been transferred the northern and western parts of Scotland, beyond the Tay and the Clyde. The Royal Navy, confined to bases on the North and Irish Seas, would have been magnanimously allowed a continued existence, but forbidden to build any new ships over ten thousand tons, and absolutely forbidden the construction of submarines or the development of naval aviation. Superimposed on all this would have been the deposition of the British monarch; demands for surrender to the Germans of Field Marshals, Admirals and Generals – Haig, Plumer, Allenby, Jellicoe and others including many juniors; the imposition of a reformed electoral system; and certain measures aimed at decisively weakening the influence of London as a financial centre. Accompanying it would have been economic disaster and a considerable degree of social unrest.

Something like these imagined and extrapolated terms were those which coloured the mind of Captain Erwin Rommel when peace came to Germany in 1918. The Kaiser, Imperial warlord and commander-in-chief, had abdicated. So had the other sovereign Kings and Princes of the component states of the German Empire, including Rommel's own Wurttemberg. No figureheads, no foci of loyalty remained. The unity of Germany was itself threatened by suggested separation, a break-up of the Reich. The army, although it marched back from the Western Front into Germany in tolerable order, was made in many cases and places the butt and scapegoat of popular outrage at what appeared a disgraceful turn of fortune. Violent revolutionaries used the fact of defeat to insult and attack the institutions of the Empire; Rommel himself was jostled and threatened when bringing his wife home from Danzig, and officers were assaulted, their uniforms defaced, as twin symbols of unsuccess and reaction, resented by a people feeling itself shamed and deceived. Everywhere there was hunger – the Allied blockade, increasingly effective in the latter years of the war and then maintained after the Armistice as an additional pressure on Germany, had produced famine conditions in many parts, conditions which had themselves greatly contributed to the collapse.

In some regions, notably in Saxony, Berlin and parts of Bavaria, there was Red revolution, mostly short-lived and suppressed with ferocity by whatever forces of order the demoralized authorities could summon, as well as by the various *Freikorps*: these were formed under private enterprise from demobilized and disgruntled ex-officers and ex-soldiers, in order to assert some sort of brutal order within the vacuum of authority which defeated Germany represented. At their height there were 400,000 Germans serving in the *Freikorps*, and by January 1919 the revolution, such as it was, had been extinguished – although trouble smouldered here and there for some time.

For Erwin Rommel, as confused and demoralized by events as his entire generation, the first task was to find employment. After his adventures in Italy he had served, without enjoyment, on a Corps staff but the first post-war winter saw him back with his own Wurttemberg Regiment. His only training was as a soldier, and he already had the reputation of an outstanding officer. His prime creed, like all his fellows, had been patriotism, and patriotism was now discounted, the Fatherland humbled and bleeding, the Empire finished, the Kaiser gone; but there would still, surely, be some avocations in which patriotism would be recognized as a proper motive. The Imperial army had been dissolved, but under the terms of the Versailles Treaty a tiny army of the new republic had been authorized – to be limited to 100,000 men, to be denied modern equipment such as armoured vehicles, to be without military aviation: an army grudgingly conceived by the victors as a sort of gendarmerie, appropriate to the internal security of the state but incapable of producing a threat beyond its borders, and, in German eyes, incapable of defending the frontiers against the enemies on all sides. Erwin Rommel applied to join it.

The ranks of survivors were thin; of 46,000 regular officers of the Imperial army nearly 11,500 had been killed in the war. Nevertheless the number of applicants hugely exceeded the permitted number of places. An officer of Rommel's reputation, however, was difficult to reject, and he thus joined the new German army of the Weimar Republic. The Reichswehr.

Rommel found himself a Company Commander in a newly designated 'Reichswehr Infantry Regiment 13' at Stuttgart, but the situation in Germany meant service in many different districts, as disaffected elements tried their hands at rebellion. The Reichswehr constituted the principal armed force at the Government's disposal; and it was needed. Rommel was at one time or another in action in Westphalia, on Lake Constance, and quelling a riot with fire hoses at Gmund near his own Stuttgart. Wherever he went his presence of mind commanded respect. He had astonishing capacity for calming the emotions of a mob, for avoiding the use of force. Over whatever troops he found himself placed – however unpromising – he instantly exerted not only authority but authority soon accorded with goodwill. Throughout his life he had the gift of imposing discipline in a way readily accepted even by those initially sceptical. His medals – symbols now of a vanished regime and a resented war – were criticized by some mutinous sailors from Kiel he once found himself commanding. His medals, he told them, served to remind him of the times he had spent in combat, times when he had prayed to Almighty God to save the German Fleet – 'And my prayers were heard because here you are!' They were soon an exemplary company.[2]

The times, however, were disturbed and distressing. Various attempts to challenge the regime by *putsch* were made and the risk of German troops being ordered to fire on other German troops appeared imminent on several occasions. In the early years the officers of the Reichswehr were thus brought face to face with the ugliness of a divided and disturbed nation. This disagreeable reality, furthermore, compounded their own resentment at the general situation and the terms imposed on Germany. The latter were judged harsh and humiliating; but, above all, the Treaty clauses were held to deviate materially from President Wilson's 'Fourteen Points' under which the armistice had been requested by Germany. Germans – and in this the officers of the Reichswehr were certainly representative and probably united – considered that the army had sought an honourable armistice, admittedly on most unfavourable terms, but that the nation had then had imposed upon it, without discussion, terms much more savage. It was a short distance from this to the belief that the army had not, in fact, been defeated in the field; that the negotiators of the armistice, themselves pressed by weak-kneed civilians at home and insufficiently dominated by a demoralized Kaiser, had betrayed an unbeaten army and were responsible for the shame and misery of Germany. The legend of the *Dolchstoss*,* the 'stab in the back', was born.

The legend was what psychologists might describe as a defence mechanism, a belief necessary for the believer's mental security but objectively false. It was difficult for German soldiers to accept that the Imperial army had failed, had lost. Its sacrifices, like the sacrifices of its enemies, had been vast. Its discipline, training, competence and battlefield performance had compelled the admiration of the world. It had faced a large part of that world in arms, winning victory after victory in Italy, in Galicia, in the Balkans, in France. And now it had lost, sued for peace, marched back across the Rhine followed by the marching battalions of the victorious Allies. There had to be a reason, there had to have been a betrayal. The label of traitor could grimly and satisfyingly be attached to the 'men of 1918' who had arranged the surrender; and, by extrapolation, be attached to the political establishment of Weimar. The republic was tarred with the brush of humiliation. It was born of defeat and it carried, unjustly, some of the blame men needed to identify for that defeat. Extremist societies and organizations, many of them recruited from former officers and soldiers, organized campaigns of violence against politicians held to have betrayed the army and the nation. Matthaias Erzberger, a Wurttemberger, and one of the negotiators of the Armistice, was murdered in the Black Forest in 1921; Walther Rathenau, a former Minister, in 1922.

Rommel, a hard-headed, practical young man, absorbed in the mechanics and the theory of his profession, not given to political or historical speculation, is unlikely to have accepted the myth of the *Dolchstoss* without reservation, but he was a dedicated patriot, he loved his country, he was unflinchingly proud of the German army, and like all his comrades he needed something of the myth, he needed to believe that there had been at least partial betrayal of the soldiers by those behind with fainter hearts. Yet myth it was. Although the German army could probably have continued the struggle on the Western Front for some little time without physical disintegration, it would not have been long. In 1918 Germany's manpower was near exhausted, Germany's people were starving under the Allied blockade, and Germany's will to fight was diminishing week by week. Both army and nation had had enough; and

* Literally 'daggerthrust'. The phrase, culled by Ludendorff in conversation with a British General, succinctly expressed the myth.

Austria had had more than enough. By whatever name, capitulation could not have been long postponed; and the full power of the United States, which had only entered the war in 1917, was yet to be deployed. Tactically, on the battlefield, German troops were still in most cases fighting well although morale was uneven. Strategically – and strategy involves the whole nation and its resources – Germany was beaten. It was difficult for men who had fought and suffered skilfully and heroically for four years to come to terms with this fact.

For fourteen years Erwin Rommel served in the Reichswehr, the army of the Weimar Republic. For half that period the dedicated and inscrutable genius who guided the Reichswehr was General Hans von Seeckt, Chief of the *Heeresleitung** – the General Staff had (formally) been abolished, and the army command-in-chief (again, formally) belonged to the President of the Republic. In fact, von Seeckt ruled the roost. His influence on the development of the German army between the two World Wars was enormous: an influence which reached every officer and man.

Faced with the facts of German defeat, German economic and political weakness, German disunity opposite the implacability (as it appeared) of the victorious Allies, Seeckt recognized that the German armed forces must resign themselves to an attenuated and potentially frustrating existence. They would be small – absurdly small for a nation with Germany's area and frontiers, in the middle of the Continent. They would appear archaic – the embargo on modern weapons came at a time when military technology in both camps had developed fast, as it always does in war, and Germany was to be frozen out of modernity, fixed in a time warp. And the armed forces would be bound in subordination to a republican and democratic regime for which few had regard or felt natural loyalty and in whose competence to discharge the essential functions of the state few had confidence. Soldiers like authority. They like efficiently (albeit legally) exercised authority. German soldiers in particular – at least in that age – had in many cases grown up under the shadow of a philosophic and Hegelian reverence for the state as a moral, indeed a divinely ordained, institution. The Kaiser had, amidst much criticism and disillusion, abdicated, but the Kaiser had embodied the ideals of a nation which instinctively yearned for authority to be coloured with a certain romantic emotion, with an appeal to the

* Literally 'army management'.

heart as well as to the mind and the stomach. Weimar appealed to no heart, to sceptical minds and, at least at first, to empty stomachs; the morale of a small professional army, deprived of the stimulus of modernity and uncertain of its rationale, all in defence of an unloved regime, would be fragile indeed.

Seeckt had a policy for the times. Himself a Guardsman and to outward appearance almost a caricature of a traditional Prussian officer, with his thin neck, monocle, moustache, cold unwavering stare and upright carriage, he was in fact a man of artistic sensitivity and high intelligence, indicated perhaps by the surprising beauty of his slender hands and fingers. Seeckt recognized that the secret of the Reichswehr's morale must lie in its quality and its potential. So small a force, so circumscribed as to equipment and resources, could only believe in itself if it knew it was an elite and also if, in time of real need, real danger, it held within itself the ability to modernize and to expand, and to do so with minimum delay.

To this end Seeckt stipulated that the quality of the Reichswehr's officers and men must be outstanding. The recruits, he laid down, should be drawn mainly from the agricultural districts – an anachronistic prejudice, perhaps, but one common in most European nations at the time: it was thought, not without reason, that the background of a farm produced more hardy, more robust, and more adaptable men than the town or city, produced better material for the cadre of potential senior NCOs Seeckt had in mind.* From the first he stated that he needed, essentially, a *Führerheer*, an army of leaders. And for his officers Seeckt was determined to accept only the best of each generation.

Seeckt could be highly selective: the entire officer corps was limited by the Versailles Treaty to four thousand. There were, as was natural, a high proportion of General Staff-trained officers among those chosen. The numbers of the nobility among successful applicants were relatively high – in this Seeckt was unabashedly swimming against the democratic tide – and the ultimate proportion of aristocrats equated to that in the old Imperial army. But Seeckt would tolerate no amateurishness. He insisted on a very high standard of intelligence and education, and he was determined to broaden the general knowledge of his officers. He promoted 'current affairs'

* And, of course, the short-lived 'workers' revolution' had, consistent with Marxist precept at the time, derived chiefly from the industrial proletariat, who were thus regarded as politically less reliable.

lectures and seminars, a wholly new departure, and encouraged secondment to universities and to the *Technische Hochschulen*. Officers were to serve for twenty-five years and soldiers twelve – this, essentially, was to be a long-service and professional army. In terms of religion there was a mix of Catholic and Protestant; and while there were few Jews, there were some. In the old Prussian army practising Jews had not been allowed to be commissioned (although they were in Catholic Bavaria). The prejudice was religious, not racial – regulations described the Prussian army as necessarily rooted in Christian faith. Matters had become easier as war approached, and in 1914 twenty-six Jews had been given the prestigious distinction of Reserve Officer; while in the war itself Jews had received every award for gallantry and service, including the *Pour le Mérite* itself.

Although the General Staff as such and by name had been abolished by the Treaty, Seeckt, in the great tradition of Prussia, was determined that an elite of highly trained and qualified officers should continue to guide the new army. It should penetrate, as before, from the central leadership to the *Truppenstab* of field formations and work to one thoroughly discussed and clearly articulated tactical doctrine and set of strategic principles. Because the General Staff had always been, and now remained, so close and centrally directed a body, with instinctive trust and understanding between its members, it could afford, at the level of Field Formation Staffs, to be comparatively small. Explanations, laborious paperwork and the spelling out of detail could to a large extent be dispensed with by thoroughly indoctrinated disciples of the same school; and under Seeckt the General Staff course was actually extended to four years. For the next twenty years – by no means accidentally – German staffs were, by contrast with some others, exceptionally small and exceptionally efficient.

But while operational staffs could be kept tiny, there had to be plenty of planning capacity at the top and in the higher direction. Limited in size though the Reichswehr was, it numbered a substantial proportion of generals (fifty-five), and a significant percentage of its officers (three hundred) served at the Berlin headquarters, where Seeckt reigned as Chief. He also placed importance, however unobtrusively, on continuity. Each of the Reichswehr's small regiments and companies, whether of infantry or cavalry, embodied, by order, a 'tradition' of one of the previous Imperial regiments.* Each

* Rommel's Regiment 13 was listed, in brackets, as '*Alt Wurttemberg*'.

also prepared plans so that, if manpower were ever again authorized, it could expand itself several-fold, based on the cadre of officers and potential NCOs it possessed. Each officer or man, Seeckt ruled, should be capable instantly of filling a post appropriate to a rank at least two higher than that he held in the Reichswehr. Because of its human quality and its underlying philosophy the Reichswehr could expand, train raw intakes and expand again and yet again, while still retaining essential structure and maintaining the continuity of Germany's military and regimental traditions. The heart of the army was to be kept sound.

So much for the constraints of size, constraints against which the German representatives at the peacemaking process had fought bitterly and with total lack of success – protests having been met with demands for even faster reductions of the German military establishment. Modernity posed different problems, and here Seeckt took advantage of the peculiar political condition of Europe in a way which provides one of the more remarkable paradoxes of our time.

The Russian Empire had been convulsed by revolution and civil war, but when, amidst frightful savagery, a sort of peace emerged and an unsteady, vulnerable and bankrupt regime began to re-establish authority throughout the vast expanses of what was now the Soviet Union, the Bolsheviks needed to build a new Russian army: the Red Army. For this they needed professional help, technical, tactical and organizational.

Seeckt proposed to give it to them. Himself a conservative monarchist to whom Bolshevism was utterly repugnant, he was first and last a German patriot. He appreciated that in the society of nations Russia and Germany were alike pariahs; the one because of Bolshevik beliefs and excesses, the other because of a long war ending in military defeat and the hatreds which that war had engendered. The two pariahs should find mutual interest. There was, furthermore, one interest besides common unpopularity. Germany's eastern frontier and the loss to a new Poland of the Prussian and Silesian provinces was insupportable; and the new Poland, after defeating a Russian attempt at invasion in 1920, also incorporated a great deal of land which had once been the Tsar's. Poland was to some extent a French protégé, and Seeckt, thinking now as a politician (or, as he would have no doubt protested, as a strategist and a patriot), openly claimed that a prime and historic German interest was the establishment of a common frontier with Russia; the existence of Poland was as intolerable to the Bolsheviks as to the Germans. 'Poland,' Seeckt said, 'must disappear, and with it one of the strongest pillars of the peace of Versailles will have fallen – the hegemony of France.' The sentiment was widely shared. In Russia Lenin himself had said, 'Germany wants revenge. We want revolution. For the moment our aims are the same.'[3]

To general international astonishment the Rapallo Treaty between the Soviets and Germans was signed on Easter Day 1922, a treaty which waived compensation claims between the two governments and provided for the opening of diplomatic relations and for a certain economic collaboration. Believing as he did, it was natural that Seeckt, with considerable mystery surrounding the actual measures, should use the atmosphere created by the Treaty to authorize a number of secret agreements between the Red Army and the Reichswehr – agreements, however, of which he kept the German Government discreetly informed. Under these there would be exchanges of officers on a significant scale so that Germany, in effect, provided a considerable number of instructors for the Red Army. In return the Reichswehr developed 'forbidden' weapons – armour, heavy artillery, military aircraft – in Russia and sent successive generations of officers to gain experience in their handling.

Furthermore the Russians agreed that factories should be built wherein materiel forbidden to the Germans should be constructed in Russia. Under these agreements the Reichswehr developed, *inter alia*, schools for the future Luftwaffe near Lipetsk and for the future *Panzertruppen* at Kazan; while future Red Army commanders studied alongside young entrants to the German General Staff (under a different name) in Germany.

Throughout his time at the head of the *Heeresleitung* Seeckt thus pursued wholeheartedly an 'Eastern' policy. Only in firm cooperation with a Great Russia, he wrote, would Germany have the chance of regaining her position as a world power.[4] For him the enemy lay to the west. In different circumstances Bismarck had proclaimed that Germany had no enemies to the east, by which he meant at that time the Russian Empire; and as Bismarck, albeit for very different reasons and with very different reservations, so Seeckt.

Accompanying these remarkable measures, all designed to evade the provisions of Versailles, Seeckt vigorously developed the operational doctrine of the army. Himself a believer in the necessity for an offensive mentality in war and for the cooperation of all arms in manoeuvre, he initiated 'mobile' troop exercises as early as 1921, using the 'troop carriers' permitted by Treaty as surrogate tanks;

while a number of far-sighted enthusiasts for the potential of armoured operations started to climb the promotion ladders of the Reichswehr. Seeckt was that formidable phenomenon, a traditionalist with an understanding of modernity, a progressive who appreciated the value of tradition.

Seeckt's internal policy for the army, therefore, was one of rigorous professional standards and rigorously selected human material: a policy of cadres, incorporating historic tradition as appropriate but constituting a framework within which expansion could be smoothly and rationally undertaken; and a policy of equipment development, in secrecy, with the connivance of Moscow. It was a policy which laid a foundation, and the foundation was strong.

Seeckt also appreciated that the economic and industrial requirements of a one-day-to-be-expanded army needed planning. He established, in 1924, a military economic planning office to study the entire equipment and support requirements of an army of sixty-three divisions (the Reichswehr was composed of ten): and he developed links with industry – notably with Krupps – which enabled armaments research to be undertaken without public notice; where necessary in overseas subsidiaries, as in Sweden, uninhibited by the provisions of Versailles.*

There remained, however, the question of allegiance. Here Seeckt's policy – and orders – were uncompromising. Autocratic and intelligent (he later became a Deputy in the Reichstag), he ostentatiously aspired to 'keep the army out of politics'; a laudable ambition, received with relief by German politicians whose skill in management of a democracy was shallow-rooted and who had no desire to see Germany threatened, like some South American republic, by the recurrent military *putsch*. The army, Seeckt enunciated, must serve the state and the state alone; and in this his policy was faithfully followed by his successors. The army must not be associated with any political faction or party. Soldiers, of whatever rank, must play no part in politics and must have no vote at Parliamentary (or any other) elections. They must be well-educated and of broad interests. But they must be loyal to the state and to the army itself, rather than to party or to any particular grouping of politicians. They must be loyal to their own commander and through him to the President

* The later development of military economic planning would have distressed Seeckt. In the Second World War German economic direction and resource allocation was chaotic – a chaos which made its own contribution to defeat.

of the Republic as embodying the state. When an extremely nervous German Government in Berlin, hearing of the abortive *putsch* in Munich in 1923, summoned Seeckt, they asked 'whether the Reichswehr stood behind the Government?' Seeckt, unforgettably to all present, had suddenly appeared in full uniform, monocled, erect, sword at his side. He stood at the door looking at his nominal political superiors and his answer was unequivocal. '*Die Reichswehr, Herr Reichspräsident, steht hinter mir!*' Behind *me*![5]

Seeckt, while he lasted (he was retired, with a good deal of regret, by the President in 1926), *was* the Reichswehr. He personified German determination to rise from the ashes of defeat. Not since Scharnhorst's evasion of the disarmament clauses imposed on Prussia by the Treaty of Tilsit in 1807 had a conquered nation so indefatigably and ingeniously frustrated a victor's will, or so vigorously restored the belief in itself of a defeated army.

But there were defects – inherent, albeit delayed in their consequences.

Non-politicization was admirable in theory. The ban on soldiers' collaboration with political parties was admirable in theory – and was to become essential as the German political parties themselves increasingly turned to extra-Parliamentary activity, to demonstrations, to rowdyism, ultimately to street-fighting by their own uniformed groups. But there was a reverse side. By the rigorous isolation of the army from the political process Seeckt encouraged in his officers and soldiers a certain political naivety. By law they were not to vote; but he also, as it were, ordered them to avert their eyes. He preached as the highest morality absolute loyalty to the state – the Fatherland – unadulterated by any partisan consideration, but he inadequately reflected that this dedication could make such loyalty vulnerable if the state itself ever conceded power to men themselves inherently immoral. He preached the absolute importance of obedience – the war had proved it: loyalty and obedience must not discriminate, must not follow private political inclination, even conviction. The state and the Oath must be binding and must be all. Yet loyalty and obedience can be betrayed.

This political castration was imposed on some of the most able and energetic men in Germany. The perils inherent in such virtuous abnegation were only perceived by a few. A handful of officers – a tiny, largely aristocratic and certainly unrepresentative handful – came increasingly to believe that officers should be more vocal, should with propriety involve themselves in the political process in order, primarily, to oppose political extremism; and they regarded this line of argument as a moral rather than a political – still less a disciplinary – matter. At no time was Rommel of their number. Rommel regarded the insulation of the Reichswehr from political activity as thoroughly patriotic and thoroughly desirable.

The whole matter of the army in politics was inevitably contentious. In Germany, in the 1920s and 1930s, society was under sporadic threat of revolution from left or right. There had been the bloody Communist and Spartacist uprisings in the aftermath of war, bloodily put down. There had been the attempted *coup* of 1923 by the National Socialists, which had fizzled out in the face of military steadiness – a steadiness achieved to a high degree by Seeckt's own cold resolution, for a number of units and staffs of schools openly supported the Nazi attempt, which led to a specific ban on Nazi recruitment in the Reichswehr. And it was said of Seeckt by critics that not only was his firm 'non-politicization' inadequate but that he himself had a certain discreet sympathy with the radical right – at least when compared to their opponents; that his own personal loyalty to individual Ministers and Chancellors was often tempered by his mistrust of policies which failed to accord with his strongly felt and often narrow patriotic convictions.

For in Seeckt's strict non-politicization of the army there was always a strand of ambivalence. Certainly and admirably he was determined to keep 'party' out of the army, but he was equally determined that the voice and interests of the army should carry all the weight he could procure – a very different matter to the political activity of individuals; and in promoting the interests of the Reichswehr (which, of course, he equated with the interests of a Reichswehr as he conceived it, one which could expand, one which could modernize, one with potential) neither Seeckt nor his successors could avoid intrigue, and thus a certain involvement in the political process. Unused to the ways of democracy, they thought, on occasions, that particular politicians or parties were enemies of the Reichswehr (and thus, by their perception, of Germany), and since the Reichswehr was indispensable to public order, to government, they used what they believed was their pivotal position to further some political developments and frustrate others. To a certain extent this dangerous tendency was checked by the election in 1925 of the venerable Field Marshal von Hindenburg to the Presidency – one of their own. But loyalty to the idea of the republic was a fragile thing, barely existent at times.

Nevertheless Seeckt attempted, not without transient success, to fill the void left by the collapse of Empire and its symbols, and to fill it with a simple and comprehensible patriotism, superior to faction. Under Seeckt this to some extent worked, but only on the surface of life. At a deeper level the officers of the Reichswehr felt, in the main, alienated from the republic and its systems. Exhortations to honour only the idea of the Reich 'above party' led, in those confused and disorientated times, to consequential contempt for party; and contempt for party led too easily to tolerance of a political philosophy which itself abused all parties but one.

Thus did Hans von Seeckt do his duty as he saw it. To the victorious Allies, when (as sometimes happened) determined evasion of the Versailles terms was detected and publicized, it was, of course, taken as evidence of German perfidy and incorrigibility – shameless, menacing and malign. To the officers of the Reichswehr like Erwin Rommel it was entirely right; indeed generally to be criticized only where it did not go far enough or fast enough or openly enough. To the Reichswehr, Versailles was imposed, unjust, and dishonouring. To the Reichswehr, Germany could not breathe freely again until there was sufficient armed strength to defy threats or blackmail by enemies on all fronts. Seeckt's nurturing of an army which would one day be restored in size and modernity was regarded by most as the only path a patriot could tread.

The first years of the Weimar Republic and the Reichswehr were, therefore, uneasy years. They were years of poverty, humiliation and resentment, resentment aimed by a natural process both at the victors of Versailles and at the allegedly self-serving and pusillanimous democratic politicians who had conceded so much and were incapable of directing matters towards German salvation. The Allied reparation demands, pitched at what seemed impossible levels (and deliberately so, in German minds), contributed to a catastrophic inflation, wiping out savings and bringing financial chaos. The inflation rate in the summer of 1923 produced a mark likely to increase (to maintain buying value unchanged) to one and a half million marks by the same autumn. The ensuing sufferings, negotiations, reforms and bitternesses marked a generation of Germans and nourished, even when the financial crisis as such had passed, an underlying emotion of something like despair.

Why has this all happened to us? was the dominant thought. And what hope for the future have we? And, of course – Whose fault is it? It was in an atmosphere coloured by such sentiments that Rommel spent the first ten years of so-called peace following the signing of the Treaty of Versailles, and however robust his common sense he, like all his contemporaries, was conditioned by the times.

In April 1925 the aged Paul von Hindenburg was elected *Reichsprasident*, succeeding Friedrich Ebert, a trade union leader of singular courage. Hindenburg, whose conscience was long uneasy at the part he had played in the Kaiser's abdication, was a symbol – physically enormous, solid, paternal, he seemed to convey reassurance and continuity; and the leaders of the Reichswehr venerated the rank as well as the personality of the Field Marshal. But Hindenburg had to deal with political problems, situations and their aftermath well beyond his competence – the infant German Constitution, in an attempt at some sort of balance of powers between legislature and executive, placed a good deal of authority in the President's hands, not least the appointment of Chancellor, a Chancellor who must thereafter secure sufficient support in the Reichstag. Before Hindenburg's Presidency the French, to enforce the reparations clauses of Versailles, had occupied the Ruhr, and stories had spread of atrocities, especially by French Colonial troops. Then a new reparations plan, the Dawes Plan, had been negotiated which seemed to many to offer some hope of reconciliation, and Hindenburg's election coincided with a move by the German Government towards cooperation with the Allies in return for progressive evacuation of the occupied provinces – a cooperation fiercely denounced by the nationalist right but embodied in the Treaty of Locarno in October 1925. The political situation seemed slightly calmer and the economy improved for a few years.

Rommel's career pattern followed the routine of a 'peacetime' army – taking command of a machine-gun company in 1924 (and none knew better how to do that!), attending courses in driving, gas and so forth: qualifying as a ski instructor, and periodically attached to some other command to give concentrated ski instruction. He was not recruited to the General Staff, nor did he attend the War Academy – a matter he somewhat resented at the time. The General Staff, containing a large number of artillerymen and a significant proportion of noblemen, was always looked at somewhat askance by Rommel. There may have been a touch of bitterness, a sense of undeserved exclusion; but he also came to regard too many of its members as remote and over-intellectual compared to the front-line soldier schooled by his own experience. For their part, some in the higher reaches of the General Staff probably knew of Rommel as one with more than a touch of brashness and reputed egotism. Each had a point.

In September 1929 Rommel was sent as instructor to the Infantry School at Dresden. This had been moved to Dresden from Munich by an angry von Seeckt after the abortive *putsch* of 1923 led by Adolf Hitler, leader of the National Socialist Party, when the staff and cadets of the school had been entirely on Hitler's side; a visiting lecturer on political affairs in 1930 found the young men there still strongly of the same opinion. Although there is no record of his having particularly shared their sentiments, Rommel does not seem to have influenced them in a contrary direction. He himself was essentially apolitical, but he probably sympathized with the general frustration which led the young, in particular, to look for extreme and decisive solutions. To Rommel and his generation National Socialism at that time often seemed the *Jugendbewegung*, the movement of youth and the future.

Whatever the reputation for self-promotion he was given by some, Rommel had been described by his previous commanding officer as tough and stubborn, but also as quiet, tactful, modest and with outstanding military talents, including a superb 'eye for country'. At Dresden he was an extremely popular instructor. He spoke with vitality, he talked convincingly from experience, he lectured in such a way that nobody who heard him was ever bored, confused or inattentive. Every report on Rommel referred to his flair for teaching as well as leadership. His stories from his own exploits on the battlefield, and particularly the taking of Mount Matajur, never failed to carry the cadet audience with him, while his enthusiasm and 'heroic' quality communicated themselves and inspired. Remembered for his warm and sympathetic smile and inspiring character,[6] Rommel was described by the senior instructor as 'a towering personality'. He spent four years at Dresden, and during that period he started to assemble the notes and anecdotal illustrations, drawing on his own contemporary accounts, which he later produced in a book, *Infanterie greift an* – 'The Infantry Attacks' – to be published in 1937 and to achieve instant and enormous success.* He became a Major in April 1932, at the age of forty, after twenty years' commissioned service.

Rommel's son, Manfred, had been born in December 1928, on

* The work was 'polished' by Rommel during a later stint as Instructor at Potsdam: see Chapter 6.

Christmas Eve. As a family man Erwin Rommel was devoted, loyal and domesticated, with simple tastes. He always found his principal happiness with his wife and son. He was expert at making and mending things, practical and ingenious.

He kept up his mathematical interest, actually memorizing the logarithm tables. Whenever he and his wife, Lucy, got the chance they rode, skied, canoed, explored – he remained a countryman at heart. Once, in 1927, he took his wife to Italy, to revisit scenes of past glory at Longarone. And in spite of his uncanny aptitude for war Rommel, whenever he referred to the struggle of 1914–18, described it as a stupid activity, a business nobody should wish repeated.[7] He thoroughly enjoyed his soldiering; but he had watched the old Germany's painful death.

CHAPTER SIX

Darkness and Dawn

IN OCTOBER 1933 Rommel was promoted to the rank of Lieutenant-Colonel and given battalion command – 3rd (Jäger) Battalion, 17th Infantry Regiment, at Goslar in the Harz mountains. His time at Dresden had been professionally satisfying – he liked teaching, he enjoyed the sensation of inspiring the young, and he increasingly developed his skill at communication and exposition. Rommel was extremely articulate. His personality expressed more than simple dash. A mathematician, technologically gifted and inventive, he was a thrusting infantry commander who had once wanted to be an engineer, a supremely practical man also interested in theory. He was adept – and became increasingly so – at converting practical experience into military maxims of near-universal applicability and at explaining, convincingly, the link between the two. He did not 'convert theory into practice' but exemplified instinctive practice and later derived theory from it. Few soldiers have so impressively undertaken and recorded action, while simultaneously drawing lessons therefrom.

But, although a rewarding phase of Rommel's life in the narrow context of his profession, the period between 1929 and 1933 had been grim years for Germany, and despite the Reichswehr's theoretic rejection of politics, no German patriot could live ignorant or apathetic about what was happening in the country. Rommel's only participation in the electoral process was through his wife, Lucy. He had no vote (nor, probably, much inclination to use one), but the embargo did not include wives, and Rommel discussed with Lucy which party she should support in the bewildering succession of elections of that time. He directed her vote towards the liberal centre of politics – the conservative nationalists, he said, were too much dominated by the nobility, whose prejudices, whether within the army or outside it, he continued to view with scepticism.[1] Too much should not be made of this. Some of Rommel's closest military associates and most respected friends were noblemen – men like the (later) Generals von Falkenhausen (his commandant at Dresden) and von Stülpnagel (a colleague there). Rommel did not suffer from inverted snobbery, still less from an inferiority complex; but he was wary of deference paid solely to social rank.

Economic life throughout the Western world was, certainly from 1929, dominated by a near-universal depression. In Germany there was election after election. German democratic institutions seemed to exist in a ferment of violence and instability. Parties of the extreme left and right had long formed private armies of uniformed toughs and these dominated the streets, stridently advocating their own remedies for the (very real) misfortunes of the republic. The burden of reparations seemed again to have frustrated attempts at revival. In May 1928, amidst already widespread economic distress, the political left had made considerable gains in elections. Soon there was increased wage inflation to deal with an increased rise in prices; and, swiftly and inevitably, greatly increased unemployment. The Dawes Plan was succeeded by another – the Young Plan – which was attacked by the nationalists with equal bitterness as mortgaging Germany's future for many years. In 1930, during Rommel's first year at Dresden, the economic crisis had been grave and the German Government's economic problems so acute that unemployment benefit could not actually be paid. Everywhere men sought work and food.

And everywhere, of course, Germans blamed their leaders – despite the fact that in successive Chancellors and senior Ministers Germany had been served by men like Stresemann, Groener and Bruning who were as able, as patriotic and as high-principled as any public servants in German history. When there is suffering, however, governments must be blamed, and Germans blamed not only government but the system. Men, in such circumstances, seek simple causes and simplistic solutions. The republic seemed bereft of firm, capable leadership. The processes of democracy appeared actually to frustrate decision. It was clear, many said, that some commanding intelligence and some decisive authority should be sought. The electoral process seemed unlikely to produce either – in 1931 and 1932 there were five elections within a twelve-month, each bringing a reshuffle and a search for political pacts between parties but leading to no strong government or policy. The ancient President Hindenburg, re-elected in 1932, seemed at the mercy of intriguers and favourites, while everywhere, including within the Reichswehr, there was talk of the coming 'national liberation', the necessary 'national uprising of the people'.

In the provincial elections of April 1932 there were considerable gains by the National Socialist Party. These – the 'Nazis' – had, like the Social Democrats, a strong and rough uniformed army of supporters, the *Sturm Abteilungen*, the SA, and the SA had fought many a bloody battle in the streets against the Communists, the *Reichsbanner* (a comparable organization of the Left) and others. The SA were set upon actually duplicating the Reichswehr, and dreamed of a 'people's army' of National Socialism, 400,000 strong, while the small and designedly Party elite *Schutz-Staffel*, the SS, were attracting young leaders from among the junior officers of the army itself.

The conduct of politics in the republic thus seemed to be going from bad to worse. The general election of July 1932 was particularly violent, with a good many deaths. In yet another general election, in November 1932, the Nazis polled a lower vote than previously but emerged in a strong position, actually cooperating with the Communists in supporting certain strikes while fighting against them in the streets; and a Presidential election, also in 1932, and with the Nazi leader, Adolf Hitler, a candidate, saw the SA mobilized and on duty.

Hindenburg, re-elected, formed a Cabinet of the right which soon proved incapable of achieving sufficient consensus to govern. The country was close to civil war, peace and property were everywhere threatened and the Chancellor, von Papen (who had succeeded the admirable Bruning in May), was respected by few – and certainly by few in the Reichswehr, where despite its tradition, officers were being wooed by both left and right. Nazism had made some headway among young officers; they had witnessed the Depression, they were impressed by the strident nationalism of the Nazis and they felt an instinctive desire for radical change and firm government. In 1930 several young artillery officers had started recruiting for the Party and had been prosecuted.

After the November 1932 general election the reality of the streets was that the Nazis should either be confronted or enlisted; while in the Reichstag they were much the largest party, and it was increasingly hard to govern constitutionally without their support. Von Papen resigned and after a brief interlude with General von

Schleicher (a staff officer of many talents, particularly for political intrigue) as Chancellor, von Papen himself persuaded the eighty-five-year-old President Hindenburg in January 1933 to accept what Papen held to be the logic of events, and to summon to the Chancellorship the leader of the National Socialist Party. When Rommel, in October that year, assumed command of his battalion at Goslar, Adolf Hitler, appointed with total constitutional propriety, had for eight months been Chancellor of the German Reich. And in March 1933 he had been accorded by the Reichstag powers to govern by decree for four years.

The battalion of which Rommel assumed command had, like all Reichswehr units, assumed the traditions of an earlier regiment with a longer history; and in the case of the Goslar Jäger Battalion it was a history with a certain unperceived irony for Rommel in view of his own future life. For this was part of a Hanoverian regiment, whose battle honours included the Seven Years' War, Gibraltar, Spain and – very notably – Waterloo; Waterloo, furthermore, not as part of Blücher's Prussian army but in the service of the King of England. The Goslar Jäger Battalion descended from the King's German Legion of Napoleonic times – a legion with a magnificent fighting record, formed from refugees to England from Hanover when Hanover was overrun by the French. The Hanoverian army itself had been forcibly disbanded but the refugees, coming together in the sanctuary of England, resurrected earlier regimental traditions. Rommel's battalion had been raised in 1803 in Portsmouth. Thereafter its record had been second to none, losing half its strength at Talavera in 1809, achieving spectacular glory at Vittoria in 1813, holding the right-centre and defending the farm of La Haye Sainte at Waterloo. Thereafter the Goslar Jägers had been absorbed into the newly-constituted forces of liberated Hanover, had fought in the campaigns against Austria and France as part of the Prussian army, and had lost three thousand men between 1914 and 1918. Commissioned into the battalion in 1908 – and a Company Commander in it in 1920 – had been one, Heinz Guderian, whose father had also commanded it.*

Rommel was an energetic and inspiring Battalion Commander,

* The men of the Goslar Jäger Battalion, after a heroic defence, were mostly killed or captured by the Russians on the Baltic front in 1945. Few returned from captivity to Germany.

always setting a physical example, chasing his officers up steep, snow-covered slopes in the Harz mountains surrounding Goslar, leading them on the ski-run down and leading them up again for another descent before they could collect themselves or relax. Once again his superiors reported his genius for teaching and training as well as for leadership. 'His Jäger Battalion,' wrote Colonel von der Chevallerie, his Regimental Commander, making that point, 'is the "Rommel Battalion" ';[2] and he assessed Rommel as head and shoulders above the average Battalion Commander in every respect. Rommel loved the Harz and became a skilful stalker of game, a sport whose tactical dimension he much appreciated. He had the quality of instantly attracting loyalty, not only with his obvious and outstanding professional competence but with his friendly and unpretentious personality. He could laugh – at himself, on occasion. He was now in his element, the element of command amid tough, challenging conditions. For the army was experiencing, like all Germany, a new, tonic atmosphere. Already the whole country appeared transformed.

Because of the near-incredible scale of the crimes committed under the regime which he inaugurated, succeeding generations find it hard to come to terms with the character of Adolf Hitler. The German dictator has been the subject of a great deal of literature, but in much of it there is a sense of an uneasy circling round the personality of the man himself, as if the monstrous nature of the acts he inspired necessitates, in turn, a monster – a monster impossible to examine under the ordinary canons of human attributes and behaviour. Faced with the undeniable fact that huge numbers of intelligent, decent, ordinary – or, in the case of Rommel, extraordinary – men and women felt total devotion to the German Chancellor and Führer, historians (with some distinguished exceptions) tend to shrug and look the other way. It is hard to explain adoration of a monster; and that Hitler was remarkable, somehow outside ordinary human perception, is generally conceded. Robert Birley,* a dedicated and experienced student of Germany, described him as the most extraordinary phenomenon in all European history.

In fact, however, Hitler's crimes were undoubtedly diabolical, but his qualities and talents were very human; his attraction and success derived from entirely human causes. Firstly, astonishing and banal though the term may be to those only familiar with the evil of his

* Headmaster of Eton 1949–63, and a leading influence in the reconstitution of German education after the Second World War.

record, Hitler had peculiar charm. Of one who, by all accounts, was incapable of listening and was absorbed in his own strange thoughts the description may seem perverse, but although some were instantly repelled by him others, especially those who shared any of his tastes or interests, those with whom he felt an affinity, have testified to a personality which was not only compelling but powerfully attractive.[3] This is unsurprising – a spell-binder of Hitler's capacity generally possesses extraordinary magnetism. Rommel certainly felt it when he came to know the Führer well. The quality only appears bizarre when we know that he was also responsible for the death and degradation of millions; and in 1933 such knowledge lay well in the future, as did the crimes themselves.

Austrians often possess a particular sort of charm, half-Southern, half Teutonic, softness alloyed with strength. Hitler did not lack it. His demagogic power was legendary and he could flatter, enchant, arouse sympathy and protectiveness in an individual as in a vast audience. He could convey – and Rommel later bore generous witness to it – an impression of modesty, candour, even humility. The able, honest and sensible Swabian General Wilhelm Groener – who had had the courage to assume authority over the defeated Imperial army in 1918, who had always struggled to hold the Reichswehr loyal to the republic, whose record as War Minister was one of integrity and moderation – spoke of Hitler as 'modest', 'orderly', one who was worried by his own extremists, wanted only the best for Germany, was essentially decent, in contrast to some of those around him. So hard-boiled a political operator as David Lloyd-George referred to Hitler as 'the George Washington of Germany'.[4] Hitler was a bewitcher.

Second, Hitler was a master of political psychology. He knew that Germans everywhere, whatever their political antecedents and prejudices, had needs beyond those material demands which he set about satisfying with energy, ruthlessness, inventiveness and a considerable measure of success. These immaterial needs represented wounds to be healed, and the first wound was that to pride and self-respect. Hitler knew – and as an ex-soldier of the front line knew with fierce, personal intensity – that Germans felt injured and felt their dead dishonoured by the apparent Allied view that the distress of Germany was self-inflicted and richly deserved. It was almost unimportant that, in fact, many practical German grievances had been or were in the process of being dispelled; the reparations issue, for instance, had been solved, and the discriminatory measures of disarmament had either been annulled or were up for international negotiation. Hitler knew that the wounds had scarred, nevertheless; and he was determined that the scars should itch, that resentment was justified. It was focused and typified by the allegation – spoken or tacit – of war guilt.

Hitler told Germans that they had no conceivable reason for guilt and thus for unhappiness. He told them that they could hold up their heads, that the future was theirs, that they had every right to be happy. Nobody in authority had told Germans this for fifteen years and they listened and, consequently, *were* happy. Hitler was the great consoler, the great optimist. He induced – tempered by well-aimed shafts of alarm to generate wariness of enemies, not least enemies within – a great national euphoria. At last Germany could be built afresh, in pride, unity and contentment. The sun had come out.

To challenge such a mood was intolerable, it was to seek to darken the sun and drive Germans back into the shadows. Hitler had a message for all generations: he comforted the fears and consoled the resentments of middle age, as surely as he appealed, with considerable success, to the idealism of youth, an idealism to which German youth, in particular, was honourably if sometimes dangerously prone.

Third, of course, there was Hitler's seeming success at dealing with the material problems which a series of weak governments had appeared inadequately to address. Mass unemployment, malnutrition and public disorder had marked many of the years of Weimar, as well as a certain well-publicized 'degeneracy' in the worlds of entertainment and the arts. Hitler attacked all this. A new pseudo-Puritanism was vigorously promoted and anything boldly experimental or seemingly daring in art or literature was prosecuted or persecuted (leading, inevitably, to the emigration of some of the most creative talents in Germany). To deal with unemployment huge programmes of public works were inaugurated. Hitler sponsored draconian legislation against disorder and anything which could be described as sabotage. He moved against the Communists who had, he claimed, for too long been setting German against German in the sterility of a class war which could only profit Germany's enemies and detractors. It is a facile error to suppose that the early adherents of National Socialism were all crude bully-boys; some were, but some were among the most decent, the most patriotic, the most courageous men in Germany, anxious to preserve their country from what they saw as the brutal ravages of Bolshevism and drawn to a party which did not shrink from suiting actions to words. He praised industry, destined to make Germany strong and rich; and he preached that by work Germans could simultaneously – and honourably – enrich themselves and promote the greatness of the Fatherland.

Hitler also preached that all Germans were equal, in that all should share in a glorious National Socialist future. He had, with considerable adroitness, balanced the radical, revolutionary wing of his party against the conservative seekers after order; and to many in all classes, including Erwin Rommel, the ostentatious egalitarianism of the National Socialist movement was by no means automatically disagreeable. Nor was concern for the underdog spurious. New housing for the workers was tackled with enthusiasm – and with a good deal of taste. A 'people's car', a *Volkswagen*, intended to bring motoring into lower-income households than hitherto, was designed. New roads, resembling railways in their conception, were devised and named in consequence '*Autobahnen*'. And, at least for a while, all this appeared to be managed without depreciation of the currency, without that unforgettable inflation of the early 1920s which had ravaged Germany and wiped out the savings of millions. Dr Schacht, President of the Reichsbank, seemed to preside over an economy which pundits might whisper was inherently fragile but which to the mass of Germans seemed once again to be built upon stability, labour and national discipline. A good many people saw through much or all of this, but to most men, and blessedly, there was work, there was food, there was order, there was self-respect.

Fourth, and intoxicatingly, Hitler told Germans that Germany would, once again, be held in regard, even in awe, by other nations. The shadow of defeat had been long. Now it should vanish; and the great indicator of its disappearance should be the restoration of strength to Germany's armed forces, confined as they had been by the edicts of Versailles to a level which was humiliating in principle and impotent in practice. Germany, the phrase ran, had been '*Heerlos, wehrlos, ehrlos*' – without army, arms or honour. No longer.

The attitude of the Reichswehr towards the National Socialist Party had been ambivalent. In the early days Seeckt had firmly defended '*Uberparteilichkeit*', the principle of an army above Party, his ostensible guiding principle in the face of Nazi (and all other) attempts to recruit. Later there had been the notorious trial of three officers convicted of promoting the Nazi Party in Ulm in 1930. Then, however, in the frantic political comings and goings which had immediately preceded Hitler's summons to the Chancellorship in January 1933, it had been impossible for the officers of the Reichswehr to avert their eyes as if uninterested. And, of course, the formal attitude of the leaders of the Reichswehr mattered profoundly to the Government, since the Reichswehr represented the ultimate force by which the authority of the state could be maintained. Successive heads of the army had not only realized their pivotal position, whether welcome or not, but had, willingly or unwillingly, played politics. To them, inevitably, there could be a desirable, a patriotic, outcome to the tortuous problems of government-forming. Or the reverse.

Increasingly, and to a large extent influenced by General von Schleicher, *éminence grise* of successive War Ministers, of Chancellors, of the President – and, ultimately and briefly, Chancellor himself – the leaders of the Reichswehr came to believe that no government which excluded the Nazis could attract sufficient popular support; and that without popular support any government must preside over a slither into civil war, so distraught and divided was the nation. One ironic consequence of this was that the banning of the Nazi *Sturm Abteilung*, the SA, by a decree of April 1932, was unpopular with many in the Reichswehr.* Schleicher, and others, were able (with sincerity and not without plausibility) to convince both the President and the Reichswehr that the Nazis would be 'tamed' by office, by association in positions of responsibility with members of other parties. It was widely believed that the excesses of the Nazis in the streets (which were by no means unique) were the work of rough elements which Hitler needed support in disciplining rather than condemnation for condoning in the first place. Hitler – in Groener's words 'orderly' and 'modest' – seemed to offer hope to the Reichswehr of a peaceful, united and (one day) strong Germany once again, and the immoderation of some of his declarations and dreams would soon fade into proportion.

But if this, a sceptical optimism (with many variations of nuance), was the sentiment among many of the leaders of the Reichswehr it was by no means universally reflected among the young. Attitudes differed strongly by generation. To many young officers, children of the post-war inflation and depression, angry at what seemed the stultifying and ignominious restrictions on national dignity under

* And not only among the young or those serving. The German Crown Prince also attacked it bitterly.

which they had been reared, National Socialism was indeed a *Jugendbewegung*, and its combined appeal to the revolutionary and the romantic nationalist fervour of youth was intoxicating. The Reichswehr had served the republic with an aloofness which was often close to hostility. The Conservative Deutschenationale Volkspartei, criticized by Rommel as essentially class-dominated, was regarded as intolerably old-fashioned. The parties of the left were either anti-patriotic or repellently destructive. Now a regime promised all things new – a revolution, but a patriotic revolution. There was, as yet, no inkling of the excesses, the moral dilemmas, the outrages which lay not far off. Many of the best of the young were powerfully attracted. When the Nazis, celebrating Hitler's advent to the Chancellery, paraded through Bamberg on the evening of 30 January 1933 a young cavalry officer, boldly defying the Reichswehr's prohibition on political activity, marched in uniform at their head. His name was Count Klaus Schenk von Stauffenberg.[5]

Hitler had made German military recovery a central plank of his policy from the earliest days, the days when Ludendorff himself had marched at Hitler's side through Munich in the abortive *putsch* of 1923. Now he was Chancellor of Germany, and he could not only preach but act.

In fact, in the matter of rearmament, as in most things, Hitler sought to claim credit and disparage the immediate past beyond what was justified. The republic had itself inaugurated a (slow-moving) rearmament programme. Now, however, rearmament and expansion could be given priority as a fundamental and revolutionary measure. The Reichswehr certainly expected it. General Ludwig Beck, later to become Chief of the General Staff of the Army, spoke of the Nazi assumption of power in January 1933 as the first major ray of light since 1918. Later the leaders of the army were criticized by some, including some leaders of international opinion, for not 'stopping Hitler' early in his career. Leaving aside the astonishing theory, to which such critics must adhere, that an army can, legally or morally, take it upon itself to remove a legally appointed government, the mood within the Reichswehr would have never countenanced such a *putsch* – one totally alien to German tradition and certainly discordant with the feelings of many, probably most, in all ranks.

Nevertheless, there was initially a good deal of scepticism. When Hitler addressed the senior generals of the Reichswehr in his first week as Chancellor, hinting at the remarkable future he envisaged for German arms, he was heard with some misgivings. Where would all this end?

Misgivings or not, it was clear that this man really intended to restore the armed forces to a respectable strength and to promote modernization; if by his boldness and adroitness he could get away with a revisionist policy on the European stage, well and good. The Luftwaffe, in particular, would come out of the shadows and flourish, under the patronage of Hermann Goering, Hitler's right-hand man within the Party and a distinguished air ace from the war. Conscription would in time be introduced. In December 1933 a decision was promulgated increasing the army to a strength of 300,000. In October 1934 the mobilizable strength was approved as twenty-four divisions, and in March 1935 this became thirty-six, a figure which, in turn, was raised and raised again so that by 1939 Germany possessed fifty-two active and fifty-one reserve divisions and could mobilize nearly four million men. All this, in six years, had come about from Seeckt's *Führerheer* of 100,000. Mechanization of part of the cavalry had begun, drawing on the imaginative work of Guderian and others in the 'Motor Transport Inspectorate'.

There was a good deal of muttering within the General Staff at how shallow-rooted much of this growth was, at the lack of experience within a hugely expanded officer and non-commissioned officer corps, at how inadequate was the base on which expansion had taken place. To the world, however – and to Germany – the achievement was phenomenal. The Reichswehr's policies of cadres fit for promotion and extraordinary responsibility, of education, of discipline, seemed to have been triumphantly justified. Expansion inevitably brought periods of vulnerability – in 1935 and 1936 most of the original '100,000 *Mann Heer*' were dispersed in instructional groups, and the order of battle was thin except on paper. Provided no wild adventures or sudden attacks from abroad threatened, however, building could go ahead on a firm foundation.[6]

The Reichswehr was thus, from Hitler's first day, among the favoured of the nation and to a considerable degree shared the national euphoria. One facet of Seeckt's original policy had, of course, vanished completely, as completely as the 'non-politicization' itself. In no way could a hugely expanded army insist on the sort of human quality which Seeckt and his immediate successors had demanded. In the early years of Hitler's Chancellorship 25,000 new officers were commissioned. Quality was generally admitted to be uneven, but undoubtedly Party spirit was strong. Before all these developments had got far, however, there was one deep-seated cause of quarrel between the Reichswehr and the Government. This lay in the increasing pretensions of the SA.

Ernst Röhm, an army Captain of the Reserve, had become head of the SA in 1931, two years before Hitler was called to power. Röhm represented the essentially social-revolutionary aspects of National Socialism. In particular he regarded the Reichswehr as a force which retained far too much of the caste-conscious and conservative spirit of Imperial Germany and monarchical Prussia (as, indeed, Seeckt had designed). A 'true-believer' in the National Socialist revolution, Röhm saw the Reichswehr as a major obstacle to its consummation. The revolution, he loudly proclaimed, was incomplete. The army of the new National Socialist state should not be this conservative hangover from a discredited past but a new and truly radical army of the people. The Reichswehr should in effect and by a process of assimilation be replaced by the SA, child and standard-bearer of National Socialism. The SA, the brown-shirted, seasoned Party street-fighters, boisterous heroes of the turbulent decade now ended, should be the armed force of the New Germany. There was, Röhm taught, not only a change of party in government, there must be a wholly new-born concept of the state, a National Socialist state radically different from all that had gone before. Adolf Hitler, Party leader, was at its head.

When Hitler had spoken to the leaders of the Reichswehr on assuming power, he had given them an assurance on which they placed great significance. The position of the Reichswehr as guardians of the nation's security would never be usurped by others; and by 'others' he could only mean the SA. Hitler's conduct immediately thereafter was faithful to this undertaking. He went out of his way to show ostentatious respect to Hindenburg. In March 1933, in the Garrison Church of Potsdam, burial place of Frederick the Great and his father, in the presence of all the heroes of the German army, past and present, Hitler did homage, as it were, to the Field-Marshal President. The leader of the 'patriotic revolution' bowed before the very incarnation of German military pride and Prussian glory. 'National honour has been restored,' Hitler intoned, 'the marriage has been consummated between the symbols of the old greatness and the new strength.' He showed all outward honour to the great names of the army's hierarchy. He had given them a glimpse of what he, at least, believed was a glowing future, and he appeared entirely ready to accept professional recommendations on military policy, on promotions, on appointments. He demonstrated to the Reichswehr trust and confidence. The leaders of the Reichswehr reciprocated.

By 1934, the year following Hitler's assumption of power, the SA numbered one and a half million men, officially capable of being enrolled in emergency as special auxiliary police. As the head of this considerable armed force Röhm was claiming that officers of the SA should receive army commissions. He demanded that the SA should have its own aviation branch, intelligence service and so forth. An SA 'Office of Defence Politics', a sort of Nazi shadow Defence Ministry was created. In February 1934 the Reichswehr learned of a letter from Röhm to the Ministry stating that the 'conduct of war, and therefore of mobilization, in future is the task of the SA'. The Reichswehr, the letter continued, should be regarded as a training organization.

Within the army, even among those who had formerly been sympathetic to the SA, regarding it as an essentially patriotic and idealistic organization which attracted (as was true) many admirable young men, these pretensions aroused a mixture of alarm and disgust; and from the way he always thereafter spoke of those days it is clear that Erwin Rommel shared this reaction. Relations between Reichswehr and SA were reaching crisis point, and the crisis could only be resolved by Hitler. At the end of February he tried to play the part of peacemaker, and spoke to the leaders of both organizations at a conference in Berlin. At this conference he reaffirmed the army's unique role. The army, Hitler declared, were to be the only bearers of arms in the state, the nation's defenders. The SA should concentrate on what he called 'political work', on pre-military training, and (for a while) frontier protection. In a speech in the following month he said the same.

Röhm's reaction was angry and derisive (and instantly, of course, reported to Hitler). Röhm spoke of the necessity to 'send the Corporal on leave'; the Corporal was ex-*Obergefreiter* Adolf Hitler. The SA, loyal to its own leadership, well-organized from the street-fighting days, was or felt itself ready for a *Putsch*. Furthermore, Röhm knew Hitler very well. He knew that Hitler had always been reluctant to make difficult decisions or face disagreeable facts, and that he would not welcome a showdown with his *alte Kämpfer*, his old trusties who had won the Nazis' early battles, had protected his popular base, had done most to bring him to power. In the earlier days Hitler had promised the SA much. It seemed he now had new friends.

But Hitler knew that his own vision of the future depended, and depended to a degree few of them could then suppose, upon the army. Hitler, furthermore, aspired to succeed Hindenburg as President; and Hindenburg, eighty-six years old, was failing rapidly and would in fact die at the beginning of August. Hitler had needed some confidence that his own assumption of a post-Hindenburg Presidency would not meet the opposition of the Reichswehr. He received sufficient assurance and he, in turn, gave sufficient assurances.

In June Röhm sent the whole of the SA leadership on leave. In a rather curious gesture he had just sent a conciliatory message to the Commander-in-Chief of the army, General von Fritsch. A selected number of leading SA personalities were to meet him at the end of the month at a pleasant resort on the Tegernsee in Bavaria where Röhm, a notorious homosexual, would be relaxing with some of the chosen at the Gasthof Hanselbauer at Bad Wiessee, on the shores of the lake.

What came to be called 'the night of the long knives' in June 1934 did as much as anything to impress the world with the brutal character of Hitler's regime. The signal of what was to come could perhaps be read in an article published on 29 June by General von Blomberg, the War Minister, in the Nazi paper, the *Volkischer Beobachter*. In this article Blomberg – defining the twin pillars of the state as Party and army – clearly indicated that the army stood behind Hitler and that in dealing with a disgruntled SA Hitler would have the army on his side. Also unequivocally on his side – and of growing significance – was the other Party paramilitary organization, Heinrich Himmler's *Schutz Staffeln*, the black-uniformed SS; while at local level discussions between the army and the SS had been initiated by the latter about action in 'emergency' – the emergency, it was made clear, being a possible revolution by the SA. The scene was set, framed by mutual and engineered mistrust. Hitler had set it.

On 30 June, accompanied by Hitler in person, a detachment of the SS *Leibstandarte*, Hitler's bodyguard, descended on Bad Wiessee and intercepted members of the SA driving from Munich or arriving at the Munich railway station en route to Röhm's conference. There was then, in effect, a massacre. Large numbers of the leading personalities of the SA were shot out of hand in Munich's Stadelheim prison. Röhm himself, scornful of Hitler to the last, was shot in his cell on the following day. A firing squad of the SS did the work. The rest of the SA, terrified, were dismissed by Hitler himself, told to go home, for the time being to put aside their uniforms, and to obey the orders of new commanders who would be appointed.

Meanwhile in Berlin the SS began settling scores with others regarded by the Party leadership, notably Himmler and Goering, as suspect. The Vice Chancellor, von Papen, was arrested but left alive while two of his aides were shot. Von Schleicher, who had been trying, as ever, to play a mediating part of intrigue in order to moderate the extremes of Nazi policy, was shot with his wife, as were General von Bredow and many others. The SA leadership was largely extirpated, and the opportunity was taken to deal with many thought to be potentially hostile to the way the regime was going.

Throughout those appalling hours no move was made by the army. At first it is likely that the Reichswehr leaders were not particularly *au fait* with what was happening, minute by minute: the killing was done by the SS. Soon, however, it became clear that the SS were indulging in an orgy of licensed shooting, with no gestures made whatever to the processes of law. Taxed with the army's inactivity the Commander-in-Chief, von Fritsch, said later that the Minister, von Blomberg, was opposed to any military intervention, and that he, Fritsch, could not move without Ministerial approval.

Such an attitude was not unreasonable. To bring the army, by order, on to the streets of the capital in a police role and in opposition to small parties of SS would have been a highly political decision and an extremely tricky operation. Considering that the SS were operating (in Berlin) under the direction of senior members of the Government it would have amounted to something like a military coup, however virtuous. Fritsch, nevertheless, and the whole hierarchy of the German army at that time, have been blamed for forty-eight hours of inactivity while great crimes were undoubtedly committed. There were many, then and later, who believed that inactivity to be entirely deliberate; believed that the Reichswehr were delighted the SA were being eliminated, and if a few unfortunates like Schleicher got in the way it was of little significance.[7]

Rommel was perfectly clear in his views on this matter.[8] He was commanding his battalion at Goslar and had no direct involvement of any kind, but like every officer in the Reichswehr he ultimately learned what had happened. A man of essential decency and high moral principle, Rommel remained throughout his life grateful to Hitler for his boldness and decision in emasculating the SA. When he learned the details he expressed misgivings about the actual methods used, but of the achievement he had no doubt. He took what was probably the orthodox and majority view within the middle ranks of the Reichswehr – that the SA's pretensions were intolerable and had been threatening Germany with civil war. Hitler had prevented it. Hitler – dependent as he had been on the SA in his early years and owing them much affection – had acted with moral courage and saved Germany. Rommel already felt profound and sincere gratitude to Hitler for what he saw as a swift and extraordinary restoration of national morale. At least on the surface economic problems were on the way to solution, people's heads were high and there was a spring in their step. A sickness seemed to have passed, and the physician – persuasive, optimistic and successful – was Adolf Hitler. Death, indeed, had been followed by resurrection; darkness by dawn.

Not all German officers of his generation felt the same. Reactions were mixed, especially, perhaps, among those who were acquainted with opinion abroad, which was generally aghast. One whose path was to cross Rommel's in very different circumstances ten years later, Freiherr Geyr von Schweppenburg, sought a British friend in private and burst into tears. 'You must believe,' he said, 'that this sort of thing will quickly pass, be extinguished. This is not the real Germany.'[9]

Real Germany or not, Hitler received from the War Minister the congratulations of the Cabinet, and from Hindenburg a telegram of approval.* Rommel was one who believed that both were deserved. Less than five weeks later, on 2 August 1934, Hindenburg died, and within an hour Hitler, uniting the offices of head of state and head of government in his own person, became President, Chancellor, and self-styled Führer. On the afternoon and evening of the same day, throughout Germany, every member of the armed forces, of whatever rank, took an oath:

> I swear by God this holy oath that I will render to Adolf Hitler, Führer of the German Reich and People, supreme commander of the armed forces, unconditional obedience, and that I am ready, as a brave soldier, to risk my life at any time for this oath.

On 15 August Hindenburg's will, dated 11 May 1934, was published. It included the sentence: 'My Chancellor, Adolf Hitler, and his movement have together led the German nation above all professional and class distinctions to internal unity,' and concluded with the words: 'In

* By whom this was drafted cannot be certain. The President was ancient and unwell, in fact virtually comatose.

this firm belief in the future of the Fatherland I close my eyes in peace.'[10]

Rommel first met Hitler in September 1934, a Hitler who since the previous month had been both Chancellor and President, head of the Government and head of state, supreme commander of the armed forces of Germany. But Hitler aspired to be still more. He claimed to be, essentially, 'Führer' of a movement – a revolutionary, historical development; a Führer who had, in the course of fulfilling his and Germany's destiny, 'drawn to himself' the supreme offices of state. To some the process was more mystical than constitutional or legitimate, and this was the mumbo-jumbo which the elite of the National Socialist movement absorbed and preached. To the SS – whose status had significantly advanced with the bloody eclipse of the top SA leadership – Hitler was to be venerated and unswervingly obeyed in all things, not as Supreme Commander, not as President, least of all as Chancellor. He was to have their devotion simply as Hitler, Führer, once-for-all-time phenomenon. Any man joining the SS was forsaking the sphere in which obligations were simply those of the loyal citizen and entering that in which ideology was paramount.[11]

Such propositions were dangerous: they divorced the concept of loyalty from that of office and attached it to a personality, an essentially political personality. They also promoted duality between the organs of state – to the bitter end having at least some continuity of tradition, some connection with legality – and the organs of the Party (primarily the SS under its Reichsführer, Heinrich Himmler), which evolved into an almost complete organizational shadow of the national administration; so that a Minister (increasingly) would find that his normal functions were being duplicated or usurped by some SS office, outwardly promoting 'liaison' or 'political coordination' or some such. But these theoretic propositions were remote from the awareness of a regimental officer, and when Hitler visited Goslar that September he inspected a guard of honour provided by Rommel's battalion, a guard of honour mounted on him as head of state. Rommel was on parade. The story is told that when he found that a line of SS men were planned to be drawn up in front of his battalion for the personal protection of the Führer, he threatened to march his men home unless the SS were removed, since he regarded their presence as insulting by its suggestion that his own troops were inadequate to protect Hitler. He won his point.[12]

* * *

A year later, in September 1935, Rommel gave up battalion command and was posted, once again as an Instructor, to the War Academy at Potsdam, where he served for the next three years. He had led his Jäger Battalion in Goslar in a highly personal way and every man recognized it. He encouraged his men to approach him with problems, with requests, and many remembered this long afterwards. He believed in the value of recreation, as well as duty, and was devoted to sport, whether skiing, riding or shooting. He had learned all about the internal combustion engine immediately after the war and enjoyed riding a motorcycle and taking his wife pillion on long tours. Rommel was nearly forty-four but he very certainly still was – and remained – the young commander who had done legendary deeds in France, in Rumania, in Italy, the young warrior who had been a byword for courage and energy. He had few doubts, however, about his abilities. 'As a young captain,' he said to Manfred once, 'I already knew how to lead an army.'[13]

Lucy, in the view of some of their friends, was the dominant personality of the family. She was more demonstratively patriotic than her husband, and she tended to see issues in terms of black and white. Rommel reckoned that in domestic disputes he had few victories – one of which was to prevent a piano being introduced into the house.[14] He deferred to her a good deal on the home front – a happy home front on which there was a lot of laughter.[15]

But heroes can be demanding fathers. Rommel, like most such, expected Manfred to develop those qualities, whether of physical daring or intellectual application, he had developed within himself. Manfred, however, showed independence of temperament from the start. He would not be pushed by his father beyond what he judged right, not for his father but for him. He would not seek to emulate his father's natural daring if he felt disinclined. Rebelling as a youth against conventional religion (Lucy's original Catholicism and Rommel's Evangelical faith had led to a somewhat disunited family in that respect), Manfred listened to his father's arguments for the existence of a God – a somewhat authoritarian God in the military manner – but stuck to his own views, which his father respected.[16] He was as sturdy as Erwin, in his own way; and Erwin Rommel's pride and affection shine through his letters until the end.

CHAPTER SEVEN

A Personal Assignment

AT THIS time the Reichswehr suffered a certain amount of egalitarian pressure from the Party,[1] and individuals reacted to it as they would in similar circumstances in any army and any society. There was a good deal of 'socialism' – sincere socialism – in the 'National Socialist' philosophy; the boasts of Röhm and his like that the National Revolution would sweep away the outmoded and reactionary habits of mind of the ruling circles of the Reichswehr found a good many echoes and did not expire with the virtual elimination of the SA as a political force.

In May 1934, for instance, immediately before the 'Night of the Long Knives', the War Minister, von Blomberg, had issued an order seeking to impose a greater degree of social democratization on the habits – and habits of mind – of officers. They must banish, the order stated, the concept of exclusivity, even in small things. At 'all ranks' social functions officers must not all sit together but must mix with others. The reaction to this sort of thing differed – no doubt it aroused contempt from some as meaningless populism, irritation in a number as being unnecessary in a good regiment, and applause from others as a progressive step in a still excessively traditionalist and caste-ridden community. But such directions, as well as the enormous expansion of the army from 1933 onwards, had naturally meant that the rigid principles and standards of the Seeckt regime were no longer significant. Curiously – because in some ways running against the spirit of the times – duelling, which had been prohibited under the republic, was legalized again in 1938. Duelling had not played any significant part in the army for a long time, and was infrequent. Its 'restoration', more a gesture than anything else, was probably a nod towards the neo-pagan ideals of Nazi 'warrior manliness' rather than a concession to the military customs of an earlier Prussia.

Rommel never joined the Nazi Party, even when it was legal for officers to do so, but he found this trend in policy perfectly congenial. Although himself a strict disciplinarian, punctilious in the proper observances of military life, he had a strong streak of no-nonsense egalitarianism in his character. By temperament he was a democrat. In early life he had actually been thought something of a 'socialist'. He liked very direct speech, no matter who was being addressed. He said what he thought, in soldiers' language. He remained to the end capable of speaking his mind to a general in terms which a sergeant-major would have recognized, and he was glad of the capability although it was not appreciated by all.

Nor was there now anything in Nazi attitudes towards the Reichswehr (at least as exemplified by the Führer himself) which was disturbing for a man of Rommel's tastes, temperament and character – a practical, vigorous and patriotic professional soldier. Hitler was, it seemed, promoting modernity, whether in attitudes to man-management or in his evident interest in weaponry. That was good. By his ostentatious friendliness, his show of confidence in the army, Hitler was proclaiming to Germany that to be a soldier was again a proud, honoured thing. That was very good. And Hitler had gone out of his way to make his obeisance to the ancient military traditions of Prussia. By his early dramatic gesture in the Garrison Church of Potsdam he had shown that his sort of modernity did not threaten but actually drew strength from the honourable past. That, too, was good. Nor did neo-paganism seem seriously to threaten. The Government had itself proclaimed that it regarded 'Christianity as the unshakeable foundation of public morality and ethics'.

There had been inevitable debate about which direction a newly stimulated Reichswehr, based on the Seeckt* foundation, should take, and some had argued for a small 'mobile' force of regulars together with a large 'defensive militia' based on cadres. Hitler rejected such a concept. A pupil of Ludendorff, he believed in the necessity for 'the big battalions', and his more secret ambitions could be satisfied by nothing less. But Hitler also believed that the army was a necessary school for youth. And Rommel's time at Potsdam brought him, for the first time, into professional contact with the National Socialist Party. In February 1937, while at the War Academy, he was assigned as War Ministry Liaison Officer to the Hitler Youth organization.

* Von Seeckt died in December 1936.

In September 1936, in his second year at Potsdam, Rommel had met Hitler for the second time. At the Nuremberg Party rally he was attached to the Führer's military escort, and on one occasion was asked by Hitler to limit the number of cars accompanying his own on some recreational excursion. This he did, annoying a number of Nazi bigwigs who had aimed at a place in the cortège and earning Hitler's congratulations for his firmness.[2] Hitler had a notable, and of course popular, memory for names and faces, and when Colonel Rommel was nominated for attachment to the *Hitler Jugend* it is sure that Hitler recognized the name and the personality.

The *Hitler Jugend* at that time consisted of five and a half million boys, led by the twenty-nine-year-old enthusiast Baldur von Schirach. This youth organization was devised for sport, instruction in German culture, and a good deal of National Socialist indoctrination. To a youth still perhaps remembering depression, hunger, parental unemployment and urban violence it offered exercise, companionship, encouragement and fun. The roads of Germany, in those days, were often covered by cheerful, fit-looking columns of youthful, brown-shirted bicyclists, waving at passing motorists, led by a rider whose machine sported a small swastika-emblazoned pennant. The *Hitler Jugend* seemed to embody the resurrected and smiling self-confidence of a new and optimistic generation. The War Ministry proposed that the boys should, additionally, receive some military instruction and Rommel, whose reports from his superiors had all commented on the *rapport* he could achieve with the young, was regarded as a very suitable emissary from the Reichswehr.

Rommel's book *Infanterie greift an* was about to be published.* Since it was primarily autobiographical – and extremely exciting – it brought his name and fame to a considerable readership in Germany, probably including Hitler: the book, published by Voggennreiter of Potsdam, sold well and soon earned Rommel substantial royalties which, in his thrifty Swabian way, he asked the publisher to keep on deposit, paying him an annual sum, which suited him better from a taxation point of view. Rommel, therefore, met the boys of the *Hitler Jugend* and their leaders as one whose publicized exploits qualified him to speak with authority of the spirit of adventure and the spirit of service.

Rommel's liaison with the *Hitler Jugend* was not entirely happy, and he fell out with Baldur von Schirach. The reasons given for their

* In the same year as Guderian's influential *Achtung Panzer*.

disagreement differ. According to von Schirach, Rommel, who had proposed that young army lieutenants should be ordered to devote weekends to youth instruction, wished to instil too much military content into the prescribed curricula.[3] According to others, Rommel's objections to the regime he found were the exact opposite – he thought that there was excessive emphasis on sport and paramilitary training, and that education, in the widest sense, was suffering.[4] The Hitler Youth tended – a natural instinct – to resent treatment as schoolboys, to aspire to adulthood. Rommel, on the other hand, son of a headmaster and a dedicated mathematician, reckoned that military training could be imparted at any age. He told von Schirach that the latter knew nothing about the subject, had never been a soldier. While they were young, he said, boys needed their minds and characters to be formed by education. Whatever their respective points of view, Rommel and von Schirach quarrelled sufficiently for the attachment – which was only part-time, since Rommel throughout remained on the establishment of the War Academy – to cease in 1938.

These were remarkable, often intoxicating, and in retrospect ominous years for Germany. The necessary instrument by which Adolf Hitler attained such overriding power in Germany was the state of emergency. It was, from the standpoint of Hitler and his trusted associates, essential to maintain a certain atmosphere of crisis, requiring extraordinary measures, measures which would arouse revulsion in 'normal' times. At first this had not been difficult – the crisis (albeit largely caused by the Nazis themselves) had been real. It was necessary to keep it alive. Hitler had been granted power to rule by decree for four years from March 1933, a period he had no intention whatever of not renewing; but the pace of revolutionary change needed to be kept up, and needed to be stimulated by fear of enemies both without and within. The initial 'state of emergency', indeed, continued until the final disaster.

As far as the general body of the Reichswehr was concerned the 'emergency' meant that Germany must, as soon as possible, be in a position to 'defend itself' against those dangers of which Hitler talked often, although (to the general public) in vague terms. In certain privileged and confidential circles he sometimes spoke in words which foretold actual aggressive and expansionary possibilities,[5] and sometimes in phrases which dismayed the more cautious or principled among his listeners. In November 1937, for instance, he talked to the War Minister (von Blomberg), the Foreign Minister (von Neurath), the Army Commander-in-Chief (von Fritsch), the leader of the Navy (Raeder) and the Luftwaffe (Goering) in terms which foretold an overtly expansionist policy for Germany, a policy which must involve the rapid absorption of both Austria and Czecho-Slovakia and could clearly involve confrontation with Britain and France. The risks appalled his listeners and Hitler shrugged the forecast off as not yet seriously intended, as dependent on other developments. But more generally he only uncovered each step which he proposed in foreign affairs shortly before he took it, almost as if he were guided by some infallible instinct or mystic impulse. Each step, it then seemed, was successful and was accomplished without bloodshed. Each step, he explained to the German people and the German army, was consistent with Germany's historic rights; and no more. And each step involved an army increasingly conscious of its recovered importance and developing strength.

Conscription had been restored in March 1935. One year later, in March 1936, German troops marched into the Rhineland, a part of Germany demilitarized under the provisions of Versailles, and of obvious and symbolic significance. This was in breach of the treaties of both Versailles and Locarno. Hitler, however, argued that a recently concluded Franco–Russian treaty of mutual assistance itself breached the understandings of Locarno and justified the action, but to most Germans this, whether casuistry or legalism, was unimportant compared to the natural justice involved.

In March again, in 1938, after a further two years, Hitler by a combination of intimidation and intrigue procured an Austrian invitation for German troops to occupy that country, and for a peaceful merger, an *Anschluss*, to take place between Austria and Germany. Austria's situation had indeed been parlous in the aftermath of 1918, with the great Imperial capital of Vienna shorn of most of the Empire which had sustained it; and union with Germany was popular with large numbers of Austrians. The Austrian Nazi Party was strong, and German troops drove peaceably into the country amid scenes of considerable acclamation.

None of this was repellent to ordinary Germans, and it was certainly not repellent to Erwin Rommel. Within Germany there was little serious thought or fear of war, while abroad, up to and including the Austrian *Anschluss*, Hitler's policy could be convincingly represented as a natural process of rectifying some, at least, of the injustices of Versailles, to the material advantage of the peoples of Europe and without threat to their existence in peace. These moves had been accomplished without physical opposition or the slightest shedding of blood. When the cinema audiences of Europe watched German troops driving or riding or marching into the demilitarized Rhineland or the streets of Vienna they saw and heard cheering, ecstatic crowds, throwing flowers. And close on the heels of the troops they saw, standing impassive in a pennanted car, arm outstretched in the Nazi salute, the inscrutable figure of Adolf Hitler.

One month before the *Anschluss* Hitler had effected a reorganization of the High Command of the Armed Forces, a step which was to have grim consequences. On 4 February 1938 he had announced his intention of exercising personal command over all services. The *Wehrmacht* office of the War Ministry would henceforth become the High Command of the *Wehrmacht* (OKW), its head – with Ministerial rank – directly responsible to Hitler and exercising all the responsibilities of the previous War Minister. Von Blomberg (victim of an unfortunate scandal affecting his wife) departed. Henceforth the head of OKW, General Keitel, became the Supreme Commander's most intimate strategic adviser; the army's authority, within what was now a dictatorship, was correspondingly diminished. The commands-in-chief of the Army (OKH) Navy (OKM) and Air Force (OKL) were, at least in theory, to be subordinate to the impartial and elevated oversight of OKW; and from the first OKW was a tool of Adolf Hitler's will and prejudice rather than in any way a fount of independent assessment and advice. In theory the new system would ensure a coordinated inter-service approach to strategic priorities. In practice it merely emasculated the professionals.

The German General Staff viewed the reorganization with scepticism. Cautiously welcoming the support Hitler had instantly provided for rearmament and expansion, they had been concerned at the decision to re-enter the Rhineland in defiance of Versailles, in spite of Hitler's arguments. Would the Western powers object with violence, even act? Nothing happened: there was, certainly in Britain and even to some extent in France, a recognition that the settlement of 1919 needed revision, that it was time for reconsideration, for healing. The Austrian *Anschluss* of early 1938, tactlessly (it seemed) effected while numbers of fortunate foreigners were spending skiing holidays in Austria, ruffled feathers somewhat more, but the feathers soon subsided. So many Austrians seemed so happy; and the economic plight of pre-*Anschluss* Austria had been so manifestly awful. Again the Reichswehr, initially nervous, responded gratefully to the peaceful simplicity with which the operation was conducted, the military column commanders taking the lead in festooning armoured vehicles (large numbers of which, disappointingly, broke down en route) with decorative greenery to show the celebratory rather than warlike spirit of the whole business.[6]

Czecho-Slovakia was different.

The western *Sudetenland* regions of the state of Czecho-Slovakia, a state created by the Treaty of Versailles, were inhabited by Germans. These, protesting (with little justification) that they were treated by the Czech Government with hostile discrimination, were militantly organized by local National Socialist sympathizers and clamoured for incorporation into the German Reich. The Austrian *Anschluss* – much of Czecho-Slovakia had belonged to the pre-1918 Austro-Hungarian Empire – served to reinforce this clamour; and the presence of German troops in Austria, on Czecho-Slovakia's south-western frontier, exposed the Bohemian west of the country, in case of war, to threat of concentric attack. During 1938 Hitler's bombast over the plight of the *Sudeten* German Czechs increased in stridency, and he ordered military planning and preparation. In fact the Sudetenland issue was largely a pretext for a policy on which Hitler had decided, for different motives, well before even the Austrian *Anschluss*. He had indicated in his confidential session with von Blomberg and others in the previous November that the absorption or emasculation of both Austria and Czecho-Slovakia was central to his plans.

But Czecho-Slovakia had strong ties and understandings with France, and in Germany the French army was regarded as by far the most formidable force in Europe. And the Czech army was itself strong – thirty-four divisions, with a possible mobilized strength of forty-five, against which at that time Germany could only oppose fifty-five; the Czech army, furthermore, was supported by a considerable armaments industry and a fortified line of frontier defences. Any view of possible war with Czecho-Slovakia must posit the intervention of France – and, quite conceivably, Britain; the outcome would materially affect the entire balance of power in Europe.

To the German General Staff an aggressive policy towards Czecho-Slovakia was a high-risk policy indeed; and militarily they felt the army to be still weak – still in the course of an immensely rapid expansion and still, despite Seeckt's careful planning and husbandry, crucially short of experienced, trained men in the senior ranks, as well as of modern equipment in quantity sufficient for the greatly increased number of divisions now being formed. The period of greatest vulnerability had been between 1935 and 1937, but so dangerous did the military leaders, and especially the Chief of the General Staff, General Ludwig Beck, think Hitler's Czech ambitions that they prepared resistance.

Beck – he was at this time the moving spirit – first tried, without success, to organize a *démarche* in July 1938 by those generals primarily involved in the contingency planning of operations against Czecho-Slovakia.[7] Led by the Commander-in-Chief of the Army, General von Brauchitsch,* they would jointly approach Hitler and tell him that his proposals would end in disaster for Germany. Hard facts about the army's unpreparedness would be adduced.

This came to nothing. The necessary degree of unity and leadership was absent. Hitler had already determined on operations – he had made up his mind in May. Close to despair about the future, Beck applied in August to retire. Hitler had just informed his Generals that they were afraid of shadows. France and Britain would not fight for Czecho-Slovakia.

Beck was succeeded by a Bavarian Catholic, Franz Halder. Halder inherited from Beck more than his office – he inherited a plot.

A number of generals, led by the Commander of the Berlin Military District, General von Witzleben, had resolved to seize the person of Adolf Hitler immediately he gave orders for the execution of 'Operation Green', by which Czecho-Slovakia was to be invaded. They were convinced, like Beck, that Hitler's policies were leading Germany to ruin. The plot, of course, had a necessary prerequisite for success – it must be able to be presented as Saving Germany from War. And for this to apply, war – war with Britain and France – must actually threaten. The German people, to whom so far the Führer had apparently brought the benefits of conquest without the hazards or losses of campaign, had little enthusiasm for war itself. The grimness of the years between 1914 and 1918 was still very real to every German over thirty years of age: and deliverance from war would bring relief, a relief which, with luck, would overshadow the grief felt at the Führer's failure and departure. Beck had allegedly said that if he were brought certain proof that England would fight if Czecho-Slovakia were attacked he would put an end to the regime.[8] Whether or not this 'plot' amounted to as much as was later claimed, the agreements secured by Hitler from Chamberlain and Daladier at Munich certainly cut the ground from under the plotters' feet.[9]

* Von Brauchitsch's attitude to the Nazis was at best equivocal, at worst sycophantic.

A moment had perhaps been missed – war against Czecho-Slovakia would not have been popular in Germany.

For now Czecho-Slovakia's western provinces were, by international concordat, ceded to the Reich; and now, like the entry into Austria, the bloodless occupation of the Sudetenland by the German army took place amidst scenes of ecstatic enthusiasm. This, Germans said, was not action against enemies – these so-called Czechs were Germans. The only enemies – outwitted by the Führer's resolution – were those who might have persuaded the Czecho-Slovak Government to obduracy for their own reasons of hostility towards Germany. Thus men and women felt throughout the Reich, and thus, very probably, felt Erwin Rommel. He had for several years had profound and grateful admiration for Adolf Hitler. Now, in September 1938, Hitler's star was high indeed. The generals, the fainthearts, who (unknown, of course, to the general public) had urged caution upon the Führer were, in Hitler's eyes, discredited, their advice manifestly inept or corrupt. Goebbels, Reich Minister of Propaganda, described them as a group of reactionaries. Hitler's political instincts had been proven right and Germany recognized that fact. Most of the German army recognized it also; and Hitler's principal opponents within the High Command had already made way for others. Von Fritsch, Commander-in-Chief before Brauchitsch, and a bitter enemy of the regime, had been accused by an SS-suborned perjurer of homosexual offences and left;* Beck, sick at heart, had resigned. These men, for varying reasons, had, they believed, seen the writing on the wall, and its message in the longer term was grim for Germany.

In October 1938 Colonel Rommel, seconded for special duty from the Potsdam Academy, was ordered to command the Führer's escort battalion for the occupation of the Sudetenland. He had been personally selected. 'A crystal clear character,' wrote the Commandant at Potsdam of him, 'selfless, unassuming and modest . . . popular with his comrades and highly respected by his subordinates.'[10] The assignment again brought him, during the short time it lasted, to the personal notice of Adolf Hitler.

It was to be his last attachment while on the staff at Potsdam. On 10 November 1938 the Rommels moved from Prussia to Austria. Rommel had been appointed Commandant of the War Academy at

* Insisting upon court martial, he was acquitted on the day Hitler drove into Vienna.

Wiener Neustadt, south of Vienna. And on the same day there occurred in many towns and cities of Germany an event which displayed to the world another ineradicable facet of the National Socialist regime.

The history of anti-Semitism in Germany, as in most European countries, is long and disagreeable. It culminated, during the Second World War, in a crime of almost unimaginable proportions.

When Hitler first came to power the National Socialist movement's ideological hostility to the Jews was given an increasingly ferocious character, finding its first constitutional expression in the Nuremberg Laws, enacted in September 1935, the same month that Rommel had begun duty at Potsdam; already, in 1934, Jews had been expelled from the army. These laws had effectively conferred second-class citizenship on Jews, who lost such basic rights as unfettered choice of employment (since a virtual 'quota' of Jews was assigned to each calling and profession) and equality of treatment in a number of other areas. The phrasing of these laws was, and was intended to be, discriminatory and humiliating; but the Nuremberg Laws themselves marked only a stage in a process which had started with anti-Jewish rhetoric and had already been stained by sporadic, albeit not officially organized, violence. Jewish-owned shop windows in Göttingen, for instance, were all broken in March 1933, in the first flush of Hitler's advent to power.

The process intensified and was accompanied, within Germany, by inadequate protest. Courageous individuals helped and supported Jews, thereby incurring the disapproval of the Party, a disapproval which became increasingly dangerous as the climate worsened, and mortally dangerous with the advent of war. Some churchmen were outspoken. But there was little organized expression of revulsion, let alone counter-action. For later generations to understand this dispiriting truth it is necessary to recall the general atmosphere of opinion in Europe, by no means only in Germany.

Jews had, of course, been marked folk since medieval times. Their religion had identified them, at least superficially, as irreconcilable enemies of Christianity, and prayers for their conversion were standard in the liturgy of most Christian communions. Jews, again from medieval times, had been associated with loans and banking, with the successful management of money. They had, therefore, attracted that envy and dislike which attaches to the materially successful alien – for the Jews, because of religion, customs and necessary intermarriage had remained alien. In Germany in the 1920s, in the aftermath of financial collapse and the great inflation, Jews were widely regarded as having profited while non-Jewish Germans lost all. A new myth was created, deriving easily from more ancient myths: the Jews had 'done well' out of the Fatherland's defeat.

Then there was the cultural question. Jews were especially prominent and influential in the arts, in literature, in what would now be called the 'media', and were accused of favouring their own kind and their own kind's interests and opinions. To the extent that there was substance in this belief it doubtless derived from that artistic talent and sensibility found in many of the Jewish race: but it assisted another myth – a myth of whole sections of German life (and particularly important sections) being 'dominated' by Jews. This domination was often stigmatized as destructive and cynical – it was ironic that one Nazi description of Jews was of a people incapable of creative activity[11] – this in a nation which had produced Mendelssohn and Heine! The Jews, in Nazi mythology, dominated to their own advantage German finance, culture and communication.

There was also, and very importantly, the question of patriotism, of loyalty to the Fatherland. The Jews – it was an occasional comment heard in most European countries, certainly in England and France as well as Germany – were 'a nation within a nation', inherently suspect in loyalty because owing a prior and obligatory allegiance to Jewry. This was an ancient as opposed to a specifically National Socialist dictum, but in Germany it had been inflated by the traumatic events of the collapse in 1918. Certain German Jews had been associated in the public mind with the movement for peace, and it became customary to associate Jewry with the (guilty) men of November 1918, an association of most dubious historical validity.

In Imperial Germany, as elsewhere, there had certainly been prejudice. Nevertheless Jews had served and served gallantly in the Imperial army, had gained commissions, and had been as assimilated and as patriotic as in any European nation. Ancient mistrust, however, the legend of a 'world Jewish domination conspiracy', had its roots in the 1880s, and after 1918 it waxed. Now it waxed grossly. Disgracefully, Jewish names were erased from many German war memorials – action ostensibly prompted by 'local feeling' – in 1935. And since Karl Marx was a Jew and a number of Jews, notably Trotsky, had been prominent in the Russian Bolshevik Revolution and in the later ruling Communist hierarchy of the Soviet Union (and prominent in some of the revolutionary movements in Germany itself), it was inevitable that 'Jewish' was an adjective often attached to 'Bolshevik'. That it was also impartially attached to, for example, 'international financiers' by Nazi propagandists – even in the same sentence – disturbed the credulity of an insufficient number of Germans.

To this blend of economic, religious, cultural, historical and political myth Hitler and his acolytes brought a theory of race. The greatness of Germany should be built on racial purity; Jewish blood must diminish that purity. Miscegenation could only lead to racial degeneration. Jews were, by definition, an 'enemy within', and the state of emergency which Hitler maintained in Germany was intended to deal, *inter alia*, with enemies within.

By no means all such 'enemies' were Jews – the Jewish population of Germany only numbered some half-million, of whom 175,000 emigrated in the years between the Nazi assumption of power and the outbreak of the Second World War. The concentration camps – places of strict-regime internment for political opponents of the state – began to be established in 1933 and by 1939 numbered six, but they included prisoners of every sort of political as opposed to criminal category, as well as such irreconcilables as Jehovah's Witnesses. Nevertheless a significant number of the concentration camp population – an overall total of 21,000 in September 1939[12] – consisted of Jews; Jews who had been convicted of 'anti-social behaviour' or some other prohibited activity. The concentration camps became the sole responsibility of the Reichsführer SS, Heinrich Himmler; they were run by a department of the SS. The Gestapo, the secret state police who investigated the sort of activity which could lead to a concentration camp, were also subsumed by the SS and answered to Himmler. These formidable instruments of the dictatorship, therefore, were Party instruments first and foremost; and at no time were they subject to effective judicial control.

Round the Jews a formidable and ever-increasing fabric of legalized discrimination, Party rhetoric, political hostility and simple prejudice was woven. And because of the diverse threads which had come together in this fabric it was by no means only the Nazi ideologue who came to accept the situation without much misgiving, even with tolerance. This was certainly true in the Reichswehr. So hostile (to Hitler) an officer as Fritsch had earlier dumped together 'Pacifists, Jews, and democrats' as people who wanted to destroy Germany[13] (Fritsch was bitterly opposed to the Weimar Republic). The prejudices of Fritsch were probably typical of the earlier, non-theoretic prejudices of many in Imperial Germany. Even among

Germans too intelligent to regard Nazi racial theories with anything but contempt and too historically sophisticated to accept myth as reality, there was a 'Jewish problem'. And although ordinary Germans were often repelled by the worst of the Party-inspired propaganda which permeated Germany in the 1930s, by the shop windows placarded with '*Juden Unerwunscht*', by the clusters of uniformed Party stalwarts mocking or bullying a Jew, by the disgusting and public humiliations of the defenceless, they did little about it; it might be dangerous to do more than little about it; and in many cases, without question, accompanying the shrug of the shoulders was a spoken or unspoken 'They've been asking for it, after all!'

Then came 10 November 1938, the day Rommel assumed command of the Academy at Wiener Neustadt, housed in the huge Maria-Theresa Academy building.

A German diplomat in Paris, Freiherr vom Rath, had been murdered by an adolescent Jew on 7 November. The Party newspaper trumpeted that the German people would be able 'to draw their own conclusions from this new outrage'. The murder had to be regarded as part of the 'Jewish conspiracy', and Goebbels made a ranting speech to the Party faithful in Munich two days later. His message was clear. Why should patriotic Germans tolerate this 'terrorism'? Goebbels carefully warned his listeners against appearing as instigators of demonstrations, but left them in little doubt of what he trusted would 'spontaneously' happen. On the following day, the day the Rommels started a new life at Wiener Neustadt, seven thousand Jewish shops were destroyed in various parts of Germany, and in a country which since the eighteenth century had taken justified pride in its religious toleration synagogues were burned throughout the Reich. Jewish communities were subjected to a huge corporate fine, their insurance compensation for damage was confiscated, and over ninety Jews were killed in sporadic violence orchestrated by the SA. Thirty-five thousand Jews were temporarily taken into custody in concentration camps, to be later released, cowed and fearful. The broken glass from Jewish shop windows led the sardonic Berliners to name the occasion the '*Reichskristallnacht*'.

The immediate – and undoubtedly intended – consequence was an increased emigration of German Jews; Hitler's own view of the 'Jewish question' at this time probably envisaged a massive departure from the Reich of the Jewish population as the simplest way of 'cleansing' Germany of Jews, making the Fatherland *Judenrein*. The reaction abroad was predictably hostile – this seemed yet another example of that brutish instigation or tolerance of violence which had from the start characterized the Nazi regime. The Jewish emigration very naturally transplanted to the countries reached – notably Britain and America – a large Jewish influx irreconcilably opposed to Nazi Germany; thus lending bizarre credibility to the propaganda of Goebbels that hostility in the West to Germany was primarily an expression of Jewish influence, exerted in the Jewish interest.

The majority of Germans strongly disapproved of these outrages – the violence, the destruction, the temporary anarchy in the streets, the damage to the name of Germany. Revulsion within Germany was, nevertheless, sadly muted.[14] This disorder was, no doubt, people said, regrettable – typical of the over-excited hooligans the Party pretended, mendaciously, were now under control; but it was certainly not a reason for revolution. The anti-Jewish rhetoric of the Nazis, people muttered, was extreme and distasteful, and within the still-boisterous (but now politically weakened) SA there were a lot of thugs who were given too much rope; but there *was* a 'Jewish problem', the problem of a powerful and influential 'nation within a nation', and at times there would, regrettably, be popular excesses in consequence. The habit of mind of the anti-Nazi traditionalists, too often tainted by indefensible prejudice, showed at that time little promise of the sort of resistance to the prevailing climate of opinion which many ultimately adopted. Fritsch, for instance, wrote that after the war it had been necessary to win three battles – against the working class, against the Catholic Church and against the Jews. He wrote this, furthermore, less than a month after the *Reichskristallnacht*.[15] While Brauchitsch, two months later, was writing in a training directive of the necessity for German officers to be unsurpassed in the purity of their National Socialist outlook.[16]

Rommel had no personal anti-Jewish feelings. In the Goslar days Manfred, accompanying his father on a walk, had innocently indicated the Battalion Medical Officer's large, curved nose and asked if he was a Jew. He received a swift reprimand.[17] Rommel had Jewish acquaintances, his temperament was tolerant, he thought of Jews when he thought of them at all as fellow-humans who he wished were Christians, wished would see the light.[18] He was too intelligent to place credence in the conspiracy theories which the Party peddled. Furthermore – the syndrome was frequent and long persisted – the Führer himself, idealistic, immaculate, was far removed from such excesses as took place, was above the more violent, the loutish behaviour of some of his adherents, probably knew little of it. The

Führer's mind was occupied by plans for the restored honour of Germany, and for improvement and greater justice among her people; by such current measures as industrial protection against dismissal, paid holidays, greater social equality within industry; by the need for a higher level of cooperation between all classes.

These things were happening and were congenial to Rommel; and it was with these things that the name of Adolf Hitler was associated. The rest was largely aberration. In the minds of many decent and patriotic Germans, despite some ominous evidence, Hitler's achievements were surely such that he deserved from them the benefit of any doubt.

And yet even to Rommel there was a 'Jewish problem'. He had attended Nazi indoctrination courses. He noted with approval, in December 1938, Hitler's speech within the War Ministry when he spoke of the necessity for the modern soldier to be 'political', to be 'ready to fight for new policies', to understand his part as a warrior for the new German philosophy of life. Hitler believed with passion in what would now be called 'motivation': and Rommel, who understood soldiers, respected that belief. When Rommel, also in that December, visited Switzerland to lecture to Swiss army audiences on his recollections of the war he reported that 'the younger [Swiss] officers . . . expressed their sympathies with our new Germany. Individuals among them spoke with remarkable understanding of our Jewish problem.'[19] This 'Jewish problem' was, to such as Rommel, only that concerning the divided loyalty which the Jew must have, the supposed reservations the Jew must entertain about giving his whole heart to his German homeland; the resentment aroused by Jewish clannishness and financial power in some Gentile communities. But the formulation was ominous; problems demand solutions.

Rommel, however, was remarkably – naively – immune from racial conspiracy theory and from understanding of the extent to which it affected the mind of his admired Führer. Long afterwards, in 1943, when circumstances had led him often to Hitler's intimate circle, Hitler's table, Rommel volunteered a suggestion. Germany's official attitude towards Jews had led to a bad press abroad, was misunderstood. It would be, surely, a shrewd move to appoint a Jewish Gauleiter if one suitable could be found. The suggestion was met by a stunned silence into which Hitler snapped, 'You've understood nothing of what I want!' After Rommel had left the room Hitler remarked bleakly, 'Doesn't he realize the Jews are the cause of the war?'[20]

* * *

On 10 March 1939 Hitler presented an ultimatum to the Government of Czecho-Slovakia, a country now deprived of its western, *Sudeten* German, provinces and populations and under threat of the secession of the eastern, Slovak, regions. Hitler proposed what would, in effect, be the dismemberment of the Czecho-Slovak state. Slovakia, already autonomous since November with German encouragement, was to be independent. Over Bohemia and Moravia a German protectorate would be established.

If the ultimatum were not accepted Czecho-Slovakia would be subjected to merciless military attack, beginning with aerial bombardment. The Czech President Hacha (his predecessor, Beneš, had resigned in the aftermath of Munich) acquiesced. He had few options. The Munich arrangements had deprived the Czechs of a fortified frontier and of important armament production capacity, and had indicated pretty clearly that if Czecho-Slovakia were to fight Germany she would, at least initially, do so alone. She was anyway beset by other resentments beside those of Germany. Poland had occupied the Teschen area, which it claimed, after the Munich pact, and had further, unsatisfied, demands. There was a Hungarian minority along Czecho-Slovakia's southern border which had led to certain territorial adjustments. And in the east of the country there was a sizeable Ukrainian population – with the Soviet Union, feared by every East European nation, across the frontier. To many the protection of Germany, however brutally imposed, made sense. On 15 March Hitler drove into Prague.

But to most people in Western Europe the occupation by the Germans of Bohemia and Moravia seemed the final piece of evidence that German ambitions were as insatiable as they were intolerable, and that Hitler's territorial plans had finally passed from what could, without difficulty, be described as a phase of rectification – rectification of the unfair provisions of Versailles, rectification of frontiers imposed by victors on resentful peoples artificially separated from their fellow nationals – to one of naked and unprincipled aggression. The Czechs had made clear six months ago that they disputed the ambitions of Germany. Hitler, furthermore, had then spoken of having no more territorial ambitions in Europe. He had persuaded the governments of Britain and France that this was so and they had acquiesced in arrangements based upon that premise, and upon certain frontier revisions credibly consistent with the principle of popular self-determination which the Allies had preached – and then too often failed to practise – in 1919. Now Hitler had, it seemed, contemptuously flung this principle and the Munich agreement itself in the face of Europe. Nobody suggested that Czechs were German. Hitler, arguing from a clearly spurious invitation to the Germans to 'protect' a neighbour, and from the equally clearly German-contrived secession of part of that neighbour's country, had simply walked in under a rascally cloak of legality. There was no effective counter-action from the world: but within the minds of men there was a decisive shift in perceptions. A shift, it might be said, from a post-war to a pre-war psychology.

To the majority in Germany – but by no means to all – the Czech business seemed wholly different. To them it was perfectly natural that the peoples of Bohemia and the rest should desire protection and should seek it from Germany. Racially these folk might not be German, but they, or most of them, had spent centuries as part of the Holy Roman Empire, that empire known for much of history as the German Empire, ruled by Hohenstaufens or Habsburgs or whatever. Bohemia was the cradle of many families famous in the Imperial past. Near Prague, at the battle of the White Mountain in 1619, and in consequence of a dynastic quarrel superimposed on a religious struggle, the Thirty Years' War had started – a very German affair. These lands were woven into the tapestry of German history and if they were now again gathered under the Führer's protective hand, all the luckier they: the Western powers, Britain and France, must surely appreciate that a strong and reinvigorated Germany was the best guarantor of a Central Europe secure against both chaos and the Soviet Union. To the minority who already felt despair at the direction in which Hitler was leading Germany the Czech business might have precipitated an anti-Hitler coup – Halder, now Chief of the General Staff, claimed that he would have supported such had war actually been declared – but the claim appears as irrelevant as it is insubstantial.[21]

For Hitler's personal drive into Czecho-Slovakia and to Prague, Rommel was detached from his duties at Wiener Neustadt and placed in command of the Führer's escort. Question arose about the prudence of Hitler driving into the Czech capital – there was a hitch concerning the SS troops who were assigned to his personal protection. Hitler asked Rommel's advice, to be told 'Get into an open car and drive, unescorted, to the Hradcany Castle!' Rommel sensed that the gesture – undoubtedly courageous – would command admiration, even if grudging. It was what Rommel himself would have done, and to a man he now revered he could only give advice to do as he, personally, would do. Hitler took the advice. Rommel always appreciated Hitler's courage, and the gesture further strengthened the instinctive bond between the two men. Then Rommel returned to his agreeable bungalow at Wiener Neustadt where he was entirely happy gardening, touring in the Austrian Alps and, increasingly, practising photography, at which he became proficient. The storm clouds had drifted away for a little.

Poland – which had for its own reasons joined with Germany in attempting to bully the Czechs during the preceding months, and which at this stage had a sharply anti-French Government[22] – showed itself warily content with the Czecho-Slovak outcome, despite the inevitable consequence that a German army could henceforth stand along the southern Polish frontier. Poland at that time unrealistically still envisaged an Eastern European combination of states in which she would act as leader, and which would by its spread and strength deter further expansionist moves by either Germany or the Soviet Union.

Such a combination needed an external ally or sponsor, and Poland found one in the United Kingdom. The British – and in particular the British Prime Minister, Neville Chamberlain, who had taken the leading role in promoting the Munich agreement and who now felt personally betrayed by Hitler – were ready to sign a bilateral Polish–British security agreement. Britain now guaranteed the territorial integrity of Poland, and to most people in Western Europe – certainly in Britain – this was a clear and deserved signal to the German Führer: thus far and no farther. All parties in Britain supported it. 'God helping,' said Churchill, then out of government on the Conservative benches of the House of Commons, 'we can do no other.' The sentiment was general, and in due course – by a reversal of recent policy – France followed suit and also gave a pledge to Poland.

This new-found determination on the part of Britain and France was injected into a different situation, insofar as German cohesion was concerned. Whereas war with Czecho-Slovakia in 1938 or 1939, the war which never took place, would have been unpopular in Germany (not least with the General Staff, fearful of such a war's ramifications), Poland was a different matter.

'The Polish question' had always aroused a degree of unanimity in Germany. It had drawn the arch-conservative Seeckt to regard Poland's existence as providing common ground between Germany and Bolshevik Russia. It had, with the territorial concessions deriving from Versailles, produced German minorities in the western Polish districts. It had, most emotively of all, divided East Prussia, cradle of Teutonic legend, from the rest of the Reich by driving a wedge of artificially created Poland – the 'Polish corridor' – to the Baltic at Danzig. And Versailles had fathered the concept of Danzig as a 'free city' – populated largely by Germans, it belonged neither to Poland nor Germany, a troubling symbol to both. In fact what might be called the Seeckt attitude to Poland had been modified a good deal, especially among thinking German soldiers, by an apparent increase in the strength of the Soviet Union – if the Soviet Union were ever to be a significant threat on Germany's eastern flank the Polish question took on a somewhat different colouration. Nevertheless to most Germans, and by no means only to adherents of Hitler or enthusiasts for the regime, the 'Polish question' represented unfinished business, demanding resolution.

Hitler – in view of his subsequent record, ominously – had negotiated a pact with Poland in 1934 and his first years of power had been accompanied by somewhat improved relations. There was a certain ideological parallelism – Jews in Poland, as in Germany, were subject to special disabilities, their commercial activities curtailed in an historic attempt to protect the peasantry from them, their participation in professional life limited. The Polish Government had then watched, with anxiety but not without profit, the events of 1938 and the dismemberment of Czecho-Slovakia.

Until the spring of 1939 German military attitudes towards Poland had been entirely defensive,[23] and certainly Poland, regarded since 1919 as the Germanophobe tool of the French, appeared to the German General Staff chiefly as threatening a possible war on two fronts in alliance with France – the traditional nightmare of Berlin. Polish military expenditure had sharply increased during the 1930s, and to others besides Germany Poland's ambitions to play a major role on the Eastern European stage seemed at times overwhelming. Some circles in Poland, indeed, appeared to dream of expansion, to imagine recovery of the era when the Polish–Lithuanian lands had extended from the Baltic to the Black Sea and had included much of Byelo-Russia. And in fact France progressively distanced herself from Poland during 1938. The Polish Foreign Minister, Beck, was intelligent but mistrusted, and France certainly did not wish her relations with Germany to be controlled from Warsaw.

In early 1939, therefore, Poland was largely friendless – until the

British guarantee. The German–Polish frontier, Hitler ordered, was to be fortified, and work went on rapidly during the summer. This particular measure seemed inconsistent with anything but a defensive mentality, in spite of the obvious fact that the British–Polish pact and French–British guarantee had thrown down something of a challenge to Hitler.

Hitler had, in fact, decided on war with Poland in March 1939, and had so told the compliant and equivocal von Brauchitsch, Commander-in-Chief of the Army.[24] He had been infuriated by the British guarantee and had hoped to inhibit its effects by signing an Alliance with Italy in May – France and Britain would now have to reckon not only with Germany but with Italy in the Mediterranean if they adopted a hostile attitude on Poland's behalf. He had not been deflected from his determination to crush Poland. Nevertheless when Hitler spoke to the assembled Army Group Commanders and their Chiefs of Staff at the Obersalzberg on 21 August 1939 his hearers were, in many cases, startled. They had recognized that Hitler was determined to bring increasing psychological pressure on Poland: his utterances had made that very clear. Plans were ready, preparations were complete. But if this really was to be war the public position of Britain and France seemed likely to expose Germany once again to a war on two fronts.

Hitler, however, spent time in his address on countering those fears. He told the assembled generals that the hour had come. Britain and France were, and knew that they were, very backward at that moment in air and anti-aircraft capability; they could not possibly assist Poland except by an offensive in the west, for which it was certain they would have no stomach in view of the bloodletting of the last war; and the leaders of neither France nor Britain would seriously contemplate war for Poland.[25] His audience was by no means convinced that the Führer's intentions represented, as he claimed, a comparatively low-risk policy; but they listened in silence, audible astonishment only breaking out when Hitler told them that he had dealt with the chimera of a two-front war; that a pact was about to be signed with the Soviet Union – an extraordinary reversal of policy which astounded the world and was everywhere taken as a clear indication that Germany had decided to draw the sword.

Many Germans, however, still reckoned that, as at Munich, there would at the last moment be some sort of settlement reached. The issues between Germany and Poland were hard but not intractable; and this Obersalzberg conference, unlikely to remain secret, might, some told themselves, itself constitute a move in that game of poker, that technique of alternating menace and conciliation, at which Hitler had hitherto shown himself such a master. Franco–British undertakings to Poland were likely, the Führer implied, to be the bluff of irresolute players.

Matters this time were crucially different. On 25 August a formal Anglo–Polish alliance in case of war was signed in London, which seemed inconsistent with Hitler's expressed optimism. That notwithstanding, after some characteristic attempts by the Führer at instant diplomacy by surprise and a last-minute and disruptive postponement, a signal was flashed to all affected formations of the Wehrmacht on the afternoon of 31 August: *Fall Weiss** D day 1 September: H hour 0445. If there had been bluff, Hitler was calling it.

Rommel had been briefed in Berlin on 22 August, the day after the Obersalzberg conference, on new personal responsibilities, a new assignment. He was forty-seven years old. He was to be promoted to the rank of Major-General, a promotion retrospective to 1 June. His mobilization appointment was to be commandant of the Führer's special field headquarters, the field headquarters of the Supreme Commander of the German Armed Forces. In this capacity, on 4 September 1939, Rommel crossed the frontier into Poland. At eight o'clock that morning he had reported personally to Hitler – the War Diary recorded simply, 'Report to the Führer of Commander-in-Chief Army Group North, Colonel-General von Bock, and Commandant of the Führer-Headquarters, Major-General Rommel.'[26] Rommel had joined the *Generalitat*. The Second World War had begun.

* The invasion of Poland.

PART 3

1939–1940

CHAPTER EIGHT

Command from the Front

ROMMEL'S FIRST command as a Major-General, in terms of the number of men under his immediate authority, was actually smaller than many of those he had held as a Lieutenant, whether on the Western Front after 1914, in the Rumanian campaign or while fighting in the Italian Alps. The Führer Headquarters train, with the usual complement of signallers, clerks, orderlies and so forth was protected by an escort, '*Begleit*' battalion, 380 strong, including four anti-tank guns and twelve anti-aircraft guns for close protection, the whole under Rommel's authority. Its officers and soldiers were largely, like Rommel himself, drawn from the staffs of military schools and training establishments, and the whole *Sicherungsbataillon* numbered twenty-five officers and six hundred men.

The train itself consisted of twelve or more coaches, drawn by two engines, followed and preceded on the rails by an armoured anti-aircraft wagon. There were, among the coaches, Hitler's extensive personal quarters, and equally extensive guest, dining and personal staff accommodation. Hitler used the train as a base – his personal adjutants and visiting dignitaries slept on it but Rommel never. The train was also a command post; it had communications facilities so that Hitler could discuss with and get information from Armed Forces Headquarters in Berlin, from Ministries, and from subordinate commanders at the front. There was also a large conference compartment with map table where Hitler would daily survey the war fronts, surveys which Rommel usually attended. Each day started at nine o'clock with a personal report on the general war situation to the Führer in the conference compartment. From it he could direct the business of the embattled Reich. Moving according to separate orders, meeting the train as directed, was a column of troops and vehicles to be at Hitler's disposal and to see to his security, for the local situation was often unclear and the countryside bestrewn with Polish soldiers.

Hitler spent his days touring from the train as base, moving in a small convoy of heavy six-wheeled armour-plated Mercedes vehicles with armoured car escort. He made few attempts to interfere with the conduct of operations in Poland, but he took an enormous interest in the battle details of this first great campaign in history in which the mobile arm largely consisted of tanks and tank-led columns. He noted the destruction of Polish unarmoured forces caught on the roads in the open and was astonished at first to learn that this had been caused not by the Luftwaffe but by the bold handling of the German Panzer troops, pushing deep and fast, ignoring threats to flanks, ignoring the untidiness of the 'front', simply conquering by shock, speed and mobile firepower. And Rommel, accompanying Hitler on these journeys as he did, also worked his eyes, his ears and his mind hard.

This warfare, Rommel realized, was new, in that the technical means of mobility now produced more opportunity than had favoured the striding, climbing infantrymen of his old *Gebirgsbataillon*, while it gave to the mobile arm new personal protection, in the form of armoured-plating, which the mounted cavalryman had lacked. But to Rommel there was nothing novel in the principle of what he saw in Poland, astonishing though it seemed to a world still accustomed to think of war in terms of the entrenched confrontation of twenty years earlier, the unbroken trench lines, the earth churned by massed artillery. This 'novelty' was, in its own idiom, war as Rommel had always practised it. He had always believed in shock, surprise, concentration. He had always believed in the effectiveness of pushing forward, almost regardless of risk, into the enemy's flank and rear. He had always believed – as an inexperienced young Lieutenant he had instinctively believed – that war is boldness and movement. In Poland he saw, exultant, modern demonstration of the validity of his own military principles and – in a larger than purely tactical sphere – the validity of the lessons he had consistently preached.

In Poland Rommel saw something else which impressed him. He noted Adolf Hitler's grasp of and fascination with the details of the campaign; but he also observed Hitler's own courage, and it warmed him. Hitler's personal convoys ranged far: he visited most parts of the front, he drove through country allegedly infested by Polish snipers, to command posts and troops engaged by Polish artillery.

Hitler had little concern with his own danger, was marked as one who, as Rommel wrote to his wife, seemed to enjoy being under fire. Hitler had been decorated in the earlier war for exemplary courage, with the Iron Cross both Second and First Class, and was in fact an extremely brave man. He always looked back on his service in that war of his youth with justified pride and even with nostalgia. It had been the most formative influence of his life, his battle experience, his *Fronterlebnis*. He always reckoned – and the belief could lead him on occasion into dangerous illusion – that he understood better than any General the feelings, the sufferings, the morale of the ordinary front-line soldier. He was, he said on 1 September 1939, nothing more than the first soldier of the Reich. There was bombast in this, and later Rommel would come to realize, also, Hitler's repulsive indifference to the actual sufferings of his soldiers; but the bombast had a golden thread of genuine feeling.

Hitler had always enjoyed reminiscence of the front in that earlier war, including reminiscence with former enemies, with British officers who had shared with him the terrible experiences of First Ypres, for instance, of the *Kindermord** in 1914.[1] Indeed it is possible that the name given to Hitler's headquarters train, '*Amerika*', derived from the Führer's memory of that time. The name may have had different (albeit surely puzzling) origins – Goering's similar train was named '*Asia*' – but Hitler liked taking a personal, almost a romantic, hand in codenames (as when he named the great 1941 invasion of Russia, culmination of his ambition, '*Barbarossa*' after the gifted, crusading Emperor Friedrich Hohenstaufen). And 'Amerika' happened to be the name of a small hamlet near Gheluvelt, south-east of Ypres. Near Amerika, in 1914, Lance-Corporal Adolf Hitler, a *Meldeganger* (messenger), leading a remarkably dangerous life carrying reports and orders from one command post to another within the 16th Bavarian Infantry Regiment, won the Iron Cross for the first time. It may not be too fanciful[2] to suppose that the name stayed vividly in Hitler's mind and that when he again went to war, not as a Lance-Corporal runner between command posts but as a supreme commander with a specially designed command post of his own, he named this wartime amenity for the Belgian cluster of houses he had once known in the bloody confusion of 1914.

* 'The slaughter of the children' – so called in Germany because of the new divisions hastily assembled to extend the German right wing and largely drawn from the very young student-volunteers.

Rommel now saw Hitler every day, accompanied him almost everywhere, attended the morning conferences (was even, at times, encouraged to speak, to offer a view), and he developed for the Führer an ever-increasing admiration; it was born of observation.

Hitler, he thought, showed complete mastery of the war situation. His nerve was strong: the perpetual anxiety of all during the Polish campaign was that Britain and France would use the opportunity to attack the *Westwall*, which was, of course, denuded of troops; but Hitler consoled the General Staff, and told them it wouldn't happen. Theirs was a very natural anxiety. The French army already numbered over ninety divisions, including a considerable mass of tanks, while the Germans, having attacked Poland with forty-three divisions, including six armoured and four motorized, had only some twelve regular divisions in the west, and no armour. There were, additionally, some thirty-five newly-formed and second-line German divisions, but the nominal total in the west of forty-seven – and effectively only a proportion of that – were by any count heavily outnumbered by those of France to which the British, it had to be presumed, would make in due course some small contribution.

Britain and France had indeed declared war but, Hitler said, they had no intention of waging it for Poland. General Gamelin, the French *generalissimo* of the Western Allies, had promised the Poles a western offensive with the bulk of the French forces after fifteen days of war, but Hitler was confident that such projects would amount to nothing, and he was right. In Hitler's – and in all German – eyes the Polish campaign had to be a local affair. Poland's western friends would make speeches but little else. The General Staff nervously hoped, and Hitler assured them. He spoke, indeed, as if with the Polish question decisively settled the Western Allies would again find some basis for peace. Rommel, listening, wrote home when the Polish campaign was only four days old that he thought the whole war would soon end.

The campaign was, indeed, moving fast and decisively, and such thoughts did not seem bizarre. The first night in Poland was spent by 'Amerika' at Topalno (the Polish rail network connected with the east-running German lines from Stettin and Berlin). 'Amerika' ranged widely, parked sometimes for several days in a siding near some country town or village while Hitler toured, with his small column, a sector of the front, every move meticulously recorded in the War Diary; or flew by light aircraft, Rommel frequently with him. Already by 6 September, after two days, he was at Graudenz

VIII Poland. Area of operations 1939

in the corridor, sixty miles south of Danzig. Four days later, on 10 September, he was at Kielce, due south of Warsaw. On 17 September, by secret agreement with Germany, Soviet forces crossed the frontier into eastern Poland. On the nineteenth Hitler, Rommel with him, drove into Danzig and broadcast to the Reich from that historic German city. On 26 September Rommel flew to Berlin and arranged new quarters for the *Führerhauptquartier* in the Reich Chancellery. At five o'clock that afternoon he was on duty to meet Hitler at the Stettiner Bahnhof in the capital; and on the twenty-eighth the whole Headquarters Battalion paraded to receive new Colours presented to them by the Führer.[3] The Germans never made the mistake of dispensing with the trappings of war. The campaign was virtually over. The weather had been glorious.

Warsaw capitulated on the twenty-seventh, and Rommel flew there early on 5 October for the two-and-a-half-hour-long German victory parade in a capital city he found shattered, filthy and depressing. Together with the great men of the conquering army, von Brauchitsch, Milch, von Rundstedt, Blaskowitz and von Cochenhausen, he was on duty to receive Hitler at the Warsaw airfield at eleven o'clock that morning.[4] The parade started at noon. The Polish affair had lasted three weeks.

Rommel's part in that campaign was essentially one of privileged spectator. He saw, for the first time, war at the top, he was able to follow with the same immediacy as his master the moves of the German armies and the progress of the great grim chess-game which was played over Poland. He was able to deduce clearly and satisfyingly from it. The Polish campaign was a triumph.

It was a triumph well before the Russian invasion sealed the fate of the unfortunate Poles. The Poles had deployed their army – an army with obsolete equipment, in most cases, when compared to the German enemy – well forward, to defend their extensive borders and conduct a gradual withdrawal on a broad front. They had to cover the German frontier to their west and the Slovakian frontier to their south; and could, of course, be immediately threatened in the north by a German advance from East Prussia. They were, in fact, menaced by envelopment from the logic of geography, as well as by a superior enemy able to select his points of attack. In such circumstances – and to Rommel it was a vivid example of a principle with which he was very familiar and had personally illustrated on smaller fields – he who decides to defend everything defends nothing.

The only strategic hope for Poland lay in some sort of offensive from the west against Germany, a fragile hope indeed; but the only operational hope lay in the winning of time, in an attempt to prevent the main body of Polish forces being entrapped in the west of the country, west of the river Vistula.

Instead, the Polish plan was based on a delaying but essentially linear defence for the eight weeks they reckoned (optimistically) they would be fighting alone. For this the Polish forces were quite inadequate. The deep outflanking German moves in north and south might have been checked if the task had been given the right priority – natural obstacles exist – while the mass of the Polish army might have been held well to the east, clear of the envelopment; but forward defence on a broad front was doomed.

To the Germans the validity of such operational arguments was self-evident. A Pole could with justice have retorted that with a Western guarantee, accompanied by what looked like indefinite Western military inertia, and in the face of a German attack launched with the connivance of the Soviet Union and under promise of Soviet military intervention, there was nothing that Poland could do anyway, and no Polish operational plan had the smallest chance of anything approaching success. True, in the event; the Soviet–German Treaty of August, in its secret clauses, had provided for what was, in effect, the fourth partition of Poland by her rapacious neighbours. Perhaps one sombre lesson, re-emphasized, was that Poland was indefensible against either neighbour except in alliance with the other, an alliance which, for entirely comprehensible reasons, the Poland of 1939 had rejected as inevitably leading to Soviet infiltration, atrocity and servitude, as surely as did defeat in the field.

To the Germans, of course, Polish maldeployment implied something else, something which, however bizarre it may seem in retrospect, was a reality to many Germans at the time. Poland – the instinct went back to 1918 – was seen as an aggressive danger on Germany's eastern flank, an ally of France, always restless and always ready to exploit Germany's weakness and to attack.[5] Traditional Polish military deployment reflected, in this German view, Polish unease at the unjust territorial acquisitions made by the Poles at Germany's expense; but reflected, also, Polish visions of a Poland expanding into East Prussia or Upper Silesia, German lands. Such visions would revive immediately the main German strength was drawn westward by French activity; and then Poland would be ready.

Whether these ambitions played the part in Polish strategic thinking attributed by Germans may be uncertain and was certainly irrelevant. By the end of September Poland's military power had been, at least for a time, extinguished.

Above all, for Germany, the Polish campaign established the soundness, the comparative skill and the high morale of the new German army. Leadership was first class. Better armed though he might be, the German soldier also demonstrated personal superiority on the battlefield, a combination shown at all levels and by all ranks of energy, initiative, self-reliance and discipline. The triumph could only have been achieved (by an army which had expanded on the near-incredible scale of the Reichswehr in five years) as a consequence of exceptionally intelligent leadership; and could never have been attained without a bold, far-sighted tactical and operational philosophy – the first '*Blitzkrieg*' – nor without that cadre of very high-quality commissioned and non-commissioned officers the Reichswehr had represented. In Poland von Seeckt was richly justified.

To what, if any, extent did Rommel at that time guess anything of the fate in store for huge numbers of the Polish nation? Poland was effectively dismembered – this fact was public, and was generally accepted as just. Districts previously German were again incorporated into Germany. The eastern districts were annexed by the Soviet Union, and from them mass deportations – and mass executions of irreconcilable elements, or those so supposed – took place in horrific circumstances. The central rump of Poland became the 'Government-General', in effect a German-administered colony (although in November the Germans formally declared that an 'independent Poland' was to be established), and in this area the SS were given responsibility and a free hand.

This free hand had been anticipated and prepared before the campaign had started. To each of the five German armies invading Poland was attached an SS *Einsatzgruppe*, sufficiently strong for an *Einsatzkommando* to be assigned to each corps. These were responsible for counter-espionage and political control in the occupied areas, and they were answerable, ultimately, to the Reichsführer SS, Heinrich Himmler. Since, however, these security units accompanied the army their activities had to be communicated to and coordinated with army commanders. A brief historical summary, written in July 1940 by Reinhard Heydrich, head of the Security

Directorate, the *Reichssicherheithauptamt*, dealt with Poland and implied much. It noted that while cooperation with the troops and various army staffs was in general good, the more senior officers had adopted a fundamentally different approach to suppression of 'enemies of the state', and that this had led to friction and counter-orders at variance with those of Himmler and, indeed, of Wehrmacht High Command (OKW).

It had not been possible, the summary ran, to iron out difficulties by personal explanation and contact because the directives governing police activity were especially far-reaching, involving the liquidation of 'numerous Polish leading circles, running into thousands'. It was remarked that orders of this kind could not be divulged to ordinary army headquarters and staffs, and in consequence the actions of police and SS had appeared brutal and unauthorized, whereas they had in fact been ordained.[6]

Wehrmacht commanders had indeed been directed not to interfere with the authorized actions of the SS; and it was clear, at least afterwards, that 'authorization' had covered widespread murder, and went far beyond even ferocious anti-guerrilla activity – Polish partisan operations soon began and lent credible cover to atrocity parading as counter-terrorism. Nazi paranoia played its part in the general brutality. The mass shooting of Jews as, by definition, 'enemies of the state' was in many cases enjoined by verbal order on particular *Einsatzkommandos*. German-occupied Poland was to be the fief of the SS, and the process began as soon as victory was assured, if not before.

It is highly unlikely that Rommel, who left Poland for good immediately after the Warsaw victory parade, knew anything of what was planned. These appalling policies were put into effect with a certain amount of discretion, were certainly not discussed in detail at military conferences, and were smothered by euphemism: 'anti-partisan activity', 'counter-terrorist operations'. That the Poles would be treated pretty roughly was very generally known, and very generally accepted – all able-bodied men, Rommel had noted while the campaign was still under way, were being rounded up and put to hard labour under German supervision. This, to him, was wholly undisturbing. Rommel was a hard man. The Poles had been irreconcilable nuisances to Germany. There had been a 'Polish question'. The Wehrmacht had answered it.

Nor was there any disposition within the Wehrmacht to be tender with partisans indulging in guerrilla activity. Colonel Wagner, of the

General Staff, already involved with Beck in the tentative anti-Nazi conspiracies germinating periodically within the General Staff building, wrote on the second day of war about the guerrilla warfare that had already broken out in Poland: 'We are issuing fierce orders which I have drafted myself. Nothing like the death sentence! There's no other way in occupied territories.'[7] Such draconian sentiments are not uncommon in war and may even be inseparable from its successful conduct, unpalatable though this may be in easier, less challenging times.

But Rommel's sort of harshness, frequently expressed in most armies towards a beaten enemy who has recently spilled the blood of comrades, and even the sort of ruthlessness expressed by Wagner's words, were compatible with correctitude, albeit severe. They were not incompatible with Rommel's frequently displayed chivalry of temperament, and were a long way from the mass persecution and mass murder to be undertaken by the SS. It was only later – a good deal later – that Rommel began to learn something of what German victory on the Eastern Front had from the first portended. His friend General Blaskowitz, Commander of Eighth Army in the Polish campaign, asked him at a later date in the war why, did Rommel suppose, was he, Colonel-General Blaskowitz, not a Field Marshal? Blaskowitz supplied the answer: that he had refused to condone the activities of the SS in Poland.[8] Blaskowitz, indeed, had written a report severely condemnatory of SS excesses and the effect of witnessing them on the discipline of the troops, a report shown to Hitler who received it with a contemptuous word about attitudes prevalent in the higher ranks of the army. 'One does not make war,' Hitler observed, 'with the methods of the Salvation Army!'[9]

Rommel had been given several days' leave at Wiener Neustadt after the end of the Polish campaign, returning to Warsaw to stand at Hitler's tribune on 5 October, as the troops stepped past to the thrilling strains of *Preussens Gloria*. When *en suite* with Hitler these days Rommel lunched and dined at his table, sometimes sitting next to him. The Führer's headquarters, whether in the field or at home, always represented a court, with all the attributes of a court, including considerable jealousy among courtiers; and Rommel soon found that the obvious rapport he had established with Hitler attracted a certain degree of mistrust.

For himself, Rommel admired Hitler unreservedly. He had always been grateful for what, as he reckoned, Hitler had so far done for Germany. He had been comforted by the immense restoration of national morale Hitler had brought about. He had marvelled at what seemed the diplomatic skill by which Hitler had achieved so many patriotic ends with (until now) so little international opposition; and he had turned his attention away from the excesses, the blemishes, as things typical of some of the undesirables who had always attached themselves to Hitler but utterly untypical of the Führer himself. With Hitler, Rommel's relations were now easy, friendly and gratifying. He had, he recorded during the campaign, frequent chats with him, and even a talk of about two hours on military problems. After the campaign was over Rommel suspected a certain animosity towards him in Colonel Schmundt, Hitler's devoted senior Wehrmacht Adjutant (this is surprising, and was no doubt temporary, as Schmundt was later a trusted friend and a useful conduit of information to Hitler in less auspicious times). But with Hitler himself the sky seemed unclouded.

Rommel now felt confident of his future. In November he was forty-eight. He had, very clearly, made his mark with the Supreme Commander himself. He had once again, although as an onlooker rather than a participant, tasted battle. He had instantly recognized that all his experience, all his reflections, all his training and, surely, all his qualities were as apt for the war of 1939 as they had been for the war of 1914. Still formally on the strength of the War Academy, he told Lucy that he didn't reckon he'd be left as such much longer – the Führer's headquarters were, naturally, 'stood down' at the conclusion of the Polish business, but Rommel was kept in Berlin, ready once again to move with 'Amerika' if a new and active campaign opened.

Rommel was now consulted about his own career and said bluntly that he hoped for command of an armoured division. His background as an infantry officer was raised by the personnel authorities as an impediment: what did Rommel know of tanks or armoured tactics? Rommel himself had few doubts. He realized, as a soldier, that the handling of armoured, mobile forces owed little to specialized knowledge of technology (although he could always absorb technical knowledge, if required, with great facility). It owed everything to an understanding of war – the shifting but essentially unchanging principles of war. And Rommel reckoned that he understood very well the principles of war. Hitler is said to have intervened on his behalf – it is very probable, and Rommel certainly believed it.[10] In February 1940 Rommel was appointed to command the 7th Panzer Division, then stationed at Bad Godesberg in the west and soon deployed in the nearby valley of the Ahr. He took leave of Hitler and was given an inscribed copy of *Mein Kampf* as a present.

When it became clear to Hitler that the Western Allies were not disposed to accept the German conquest of Poland, to make peace, the next move on the chessboard of Europe was reviewed. Hitler had in one respect at least miscalculated. He had always believed that when faced with the ultimate reality of German power and German resolve Britain would choose the path of friendship. In this he was acutely disappointed and remained disappointed. Hitler found British objections to his European ambitions, his Polish policy, incomprehensible. They did not threaten British interests, to which he would, he professed, always remain sensitive. In the years immediately preceding war, however, Hitler had grudgingly conceded the fact of British opposition, and had marvelled at the extent to which it appeared accompanied by British weakness and infirmity of purpose – Chamberlain's Britain did not seem the Britain Hitler had taught himself to revere, Britain the ruthless imperialist power. Nevertheless, to a very late hour in his stormy life, Hitler hoped for and actually believed in the possibility of an axis between Britain and Germany.

But in the autumn of 1939, Hitler found that he was faced instead with Britain's irreconcilable hostility. This infuriated and often puzzled him. He spoke of the necessity for Britain to rediscover her common interest with Germany – the essential enemies were the United States and the Soviet Union, 'hemming in' the nations of Europe.[11] He had been infuriated by the British declaration of war on 3 September, a development he had been encouraged to suppose would not arise, despite British bluster. Now the Polish business had been satisfactorily concluded by the Wehrmacht, but military victory had not stilled the sympathies of the Western Allies with a shattered Poland. And strategic logic was failing to divide the Western Allies from each other – to divide France, with her Continental jealousy of Germany, from Britain, with her overseas possessions and maritime empire, towards which Hitler felt a certain envious benevolence. Undivided, the Western Allies were stronger than Germany; numbered more divisions of troops, more tanks; commanded naval forces to which Germany as yet had only a tiny counter; possessed empires spanning the world. It would be necessary, Hitler reluctantly appreciated, to deal Britain and France a sharp blow, perhaps a decisive blow, in spite of their advantageous position. It would be necessary to launch an offensive in the west in order to secure the Atlantic front. Ultimately, Hitler's plans and visions had a certain consistency. Such an offensive would be essentially a preliminary. Germany needed space, and that space could only be won in the east. Hitler had dreamed thus for twenty years.

An offensive in the west, however, appeared to the German General Staff a formidable task. A party within the High Command, and their friends elsewhere, still believed Germany might avoid disaster and lever the Nazis out of power if Hitler were denied further conquests; and they hoped that a *démarche* by senior generals refusing to undertake a western offensive might do the trick. Of this company – plotters of the future against Hitler's life as well as his rule – were such dedicated anti-Nazis as General Hans Oster of the Abwehr, the German Intelligence Service; Fabian von Schlabrendorff and Hans von Dohnanyi, lawyers; and others, able to see through the temporary intoxication of victory into the heart of National Socialism and to live by the lonely light of conscience. But in the main, as would happen in all armies and all societies, professional men simply acted professionally. When Rommel took over command of 7th Panzer Division in February 1940 plans for the western offensive were already far advanced.

Hitler had wanted an autumn offensive, and had only reluctantly accepted professional advice that weather conditions made this inauspicious. Such an offensive would require good going rather than rain and mud for tracked vehicle movement off roads, and also optimum flying conditions and, preferably, long hours of daylight. The Polish campaign, although recognized as having been conducted against a very differently equipped enemy from the forces of France, had demonstrated the speed and shock with which armoured forces, supported by close-support aircraft operating according to an integrated army plan, could achieve a decision; but this needed favourable conditions. In all armies, including the German, there was the haunting vision of another stalemate, another slogging match such as had characterized the Western Front from 1915 to 1918, and to Germany the vision was potentially disastrous. As ever, Germany needed to win a quick war. A protracted, expensive war of attrition was not a reasonable option.

But Hitler was contemptuous of the operational plans submitted to him by the General Staff for the offensive – *Fall Gelb*, 'Operation Yellow'. He had, after the victories in Poland, decided that he had both the military wisdom and the constitutional authority to override the High Command of the Army (OKH) even on military matters. His 'decision' to launch an autumn offensive in the west – and to violate the neutrality of Holland, Belgium and Luxembourg, arguably a primarily political factor – had been communicated without prior consultation and found expression in a Wehrmacht High Command (OKW) directive of 9 October. That this had been revised, withdrawn, amended and so forth in the face of professional protest did not alter the fact that in issuing it Hitler – and his tool, OKW – had shown that he intended to act untrammelled even in purely military, indeed detailed military, matters. Hitler – and the trait had ever-increasing effect on the destiny of Germany, of the German army and, one day, of Erwin Rommel – intended to dictate, and only to dictate. It was his way.

Hitler – described in this context by the shrewd Erich von Manstein* as 'utterly unscrupulous, highly intelligent and possessed of indomitable will'[12] – had defensible reasons for impatience with the professionally produced plans he was receiving. It was not only – although it certainly was to a considerable extent – that he distrusted the inadequate political motivation, the lukewarmness towards National Socialism he knew existed in the General Staff; it was also his sense that they were old in mind if not in years, unimaginative, unoriginal. He told his personal staff that he would like to clear out the entire High Command building in Bendlerstrasse in Berlin and fill it with young optimists instead of pessimistic reactionaries.[13] It was consistent with this mood that in February 1940, the month Rommel assumed the command for which he had longed, Hitler heard with considerable interest from his principal Wehrmacht Adjutant, Schmundt, of a visit the latter had paid to Army Group A Headquarters at Coblenz. Schmundt had had a discussion with the Army Group Chief of Staff, von Manstein, about the coming operations. A fortnight later Hitler sent for Manstein personally, regardless of protocol. The consequence was a radical change of plan and the adoption of a new concept.

The previous concept had not, in fact, been the brainchild of the General Staff (who had actually favoured a defensive strategy in the west). It had been largely adumbrated, *faute de mieux*, by Hitler himself, and amounted to a repetition of the Schlieffen wheel of 1914, a great enveloping movement through Belgium and Holland

* At this stage of the war Chief of Staff to Army Group A (von Rundstedt) in the west. Later Field Marshal.

round the Allied left flank. But the General Staff, hoping that weather and, perhaps, political factors might delay and conceivably abort the whole business, had certainly not proposed an improved version. Some of them, indeed, believed – or pretended to believe – that the best German strategy would be to await an Allied initiative and then succeed with a counter-move, the Allies having incurred the odium of violating Belgian neutrality and the casualties associated in conventional thinking with the launch of an offensive. The idea was not compelling, if only because it was perfectly well known that Gamelin had not the slightest intention of attacking the *Westwall* before 1941.

Reluctantly, the General Staff had worked on *Fall Gelb*. Hitler had spent the winter with the infuriated feelings of a man whose achievements and concepts are likely to be frustrated by unwilling or half-hearted executives, who is denied the stimulus of enthusiastic, like minds. And now there was a new concept.

The essence of this new concept lay in its object, which radically differed from its predecessor. The original *Fall Gelb* object had been to defeat the largest possible elements of the French and Allied armies and simultaneously to gain as much territory as possible in Holland, Belgium and northern France as a base for successful air and sea operations against Britain, and as a broad protective zone for the Ruhr. The proposed aim of the new concept was, quite simply, 'to force an issue on land'. Von Manstein, entirely supported by his commander, von Rundstedt, had been fighting for this fundamental shift of emphasis since October. The best strategist on either side in the Second World War, he regarded the official plan as a half-measure which would negate Germany's greatest asset – the offensive power, now tested, of the army and its ability to reach a decision with strategic consequences in a short, fast-moving campaign. A new plan – '*Sichelschnitt*' – was adopted.

Because its object was so ambitious its detail was utterly different. It aimed at nothing less than total victory. This was to be attained not by a repetition of Schlieffen's outflanking movement, of the days of 'keep the right wing strong', but by a shift of weight southward, so that the main offensive strength – to be exerted by Rundstedt's Army Group A – would strike the Allied left-centre, in the area of the Ardennes. It was correctly assumed that the Allied left wing would have advanced into Belgium immediately a German offensive opened, and it was planned that the French and British so advancing would meet the strong but not predominantly armoured forces of

General von Bock's Army Group B on the German right. Meanwhile Rundstedt, with the majority of the Panzer divisions, would have broken through near the hinge of the Allied advance, and north of the Maginot line; would have thus instantly threatened the inner flank and rear of the Allied left wing, and cut it off near the stem of the advance; would, by a remorseless western advance, drive a wedge between that Allied left, including virtually the entire British Expeditionary Force, and the bulk of the French army including the garrisons of the Maginot line; would, in fact, split the Allied front in two. Thereafter greatly superior strength could be brought against the main body of the French army south of the Somme.

The new concept was agreed by Hitler and embodied in an Operation Order issued on 20 February. Its success would largely depend on the performance of the seven Panzer divisions of Army Group A. These seven divisions, out of a German total of ten at that time, constituted the *Schwerpunkt* of *Sichelschnitt*, the cutting edge of the sickle. The German armour overall was heavily outnumbered by that of the French army; nor were French tanks technically inferior. The battle would be decided by the skill and energy of the leaders of the Panzer divisions, and by the courage and stamina of their troops. The initial advance of Army Group A's armoured spearheads would be through country thought difficult for armour, defensible, afforested, seamed by a great river, the Meuse, beyond which lay the last, western ridges of the Ardennes; but beyond the Ardennes were the open plains through which ran the Sambre, the Aisne, the Somme. And beyond that was the English Channel.

It was, by every reference to history and to temperament, exactly Erwin Rommel's sort of battle.

Rommel thus assumed command on 15 February 1940, when the operational plan to which his division was committed had just been radically changed, and changed in a way he would have entirely approved had he been privy to the high-level arguments and consultations which had produced this development.

Rommel's first activity was to get to know his division and to become familiar with the characteristics of its equipment, notably its tanks. 7th Panzer Division consisted of a Panzer regiment (25th) of three tank battalions – a total of 218 tanks – and an armoured reconnaissance battalion (armoured cars); two rifle regiments, each of three battalions; a motorcycle battalion and an engineer battalion;

and a divisional artillery with one field regiment (nine batteries, 36 guns) and an anti-tank battalion of seventy-five anti-tank guns. The division had been converted from a light (that is, motorized) division to a fully armoured formation during the winter. Rommel's tanks were of mixed type; over half were of Czech manufacture and comparatively lightly armoured. The other half were the excellent German Panzer IIIs and IVs. The armoured reconnaissance battalion (37th) had new three-axled cars mounting a 38-millimetre gun and was a formidable fighting as well as scouting unit.

The principal visionary force behind the creation of the German armoured troops had been Heinz Guderian, who was now commanding a corps in Army Group A. Guderian had for years been a prophet, a preacher of the virtues and potential of the Panzer arm, and like most prophets had not always earned honour in his own country. He had had to contend with the obstinacy of those who refused to believe that armoured troops generally and tanks in particular could achieve, technically, what was claimed for them: and he had had to convince his brother officers and the General Staff that fast-moving formations could be controlled in a fluid battle, that communications could be developed adequate to the task. Guderian had had experience of communications, including the rudimentary wireless communications of the day, as early as the opening phase of the war in 1914, and he consistently preached the *essential* nature of command and control by radio, as well as its corollary (as he saw it) that the commander of mobile troops must be far forward and command 'from the saddle'. In this, as well as in his corresponding belief in movement – the engine of the tank, he said, was as much a weapon as its gun – he was a fellow-spirit of Rommel.

Guderian had, as Chief of Staff of the Panzer Truppe Command when the new arm was formed in 1934, fought a hard and often lonely battle; but he had, by and large, won it, and the demonstration of victory had seemed to come in Poland. In Guderian's book *Achtung Panzer*, published in 1937, he argued that everything hinged on the ability to move faster than had hitherto been envisaged; to move faster, keep moving and by sheer impetus make it hard or impossible for the enemy to build successive defensive positions. To the extent that such thinking was now dominant in the German army in the west, Erwin Rommel had come home.

The extent of such dominance was, however, still limited. The 'battle' for Guderian's – and, as they became, Rommel's – ideas was by no means easily won. As late as 1944 Rommel was to write, in retrospect:

> There was a particular clique that still fought bitterly against any modernization of methods and clung fast to the axiom that the infantry must be regarded as the most important constituent of any army. This may be true for Germany's Eastern Army*... but it will not be true in the future when the tank will be the centre of all tactical thinking.[14]

Certainly – and naturally – Rommel from the first day of his assumption of command of 7th Panzer Division was a vigorous and articulate exponent of the ideas which Guderian had promoted with such insistency.

The core of these ideas was the proposition that an armoured, mobile force of *all arms*, sufficiently strong and with its logistic support sufficiently well organized, could dominate a campaign and achieve decisive operational – and, ultimately, strategic – results, provided that it was handled with sufficient skill and boldness. Flanks would be exposed: the remedy was not to halt and redeploy to protect them, but to drive ever more furiously, paralysing the enemy's reactive capacity by the threat to his own rear areas and communications. The 'front' would be ragged, a patchwork, a confusion of intermingled enemy and friendly troops rather than a definable line; the secret was not to attempt any 'tidying' of the battlefield, but to impose the victor's pattern upon it by the successful momentum of his own advance, supported by air power handled in close cooperation with the ground forces.

This was exactly what, at the tactical level and on his feet, Rommel had always both practised and taught. Now, exultant, he commanded armour, and his theories of war could, if all went well, be put into effect at greater speed and on a grander scale.

There was, of course, antithesis between this core idea and that thinking, excoriated by Rommel, which 'stuck to established methods and precedents'.[15] But although Rommel, particularly after later suffering multiple frustrations, tended to identify this 'traditional' thinking with the German General Staff and particularly its chief, Franz Halder, there was little if any antithesis between Guderian's and

* Which, at the time Rommel was writing, had been engaged for the previous three years in Russia.

Rommel's ideas and the actual operational plans which distinguished both the Western Front in 1940 and the Eastern Front in 1941. It has been argued[16] that the 'pure' ideas of Guderian concentrated upon the use of mobile forces in creating paralysis by their threat to the enemy's headquarters and communications, while more traditional manoeuvres aiming at encirclement and annihilation constituted a dilution of the pure milk of armoured theory. Certainly a great aim of mobile operations must be to induce paralysis in an enemy, but it is a mistake to suppose that this will only or best be achieved by attacks upon his headquarters and communications, or that the capacity for such attacks is or was the most significant consequence of the development of armour. On the contrary, paralysis of enemy communications was achieved by artillery hurricane bombardment in the great German breakthrough of March 1918; and by Rommel's days in high command this was already more attainable by air action.

What should be aimed for, as Rommel perceived very plainly and at every level, was paralysis of the enemy's *will*, of his capacity for clear thought and measured response, not simply or even primarily by physical destruction of his communications and threat to his headquarters (although, certainly, such can play a part) but – and essentially – by the threat from actual and rapid manoeuvres: threat of encirclement, threat of annihilation. It was these threats, posed by the armoured forces with their speed and shock, which would really induce paralysis of the will, and lead to victory. And these threats, these manoeuvres, were consistent with the traditional lore of a General Staff steeped in history. Antithesis was between method, pace and detail; not fundamental concept.

All this was academic, and Rommel always rebelled against the academic – 'This unnecessary academic nonsense,' as he called it.[17] He was supremely objective and concrete in his thinking. Every instinct told him that if a force of all arms such as he now commanded could win a tactical battle and break through the crust of enemy defence, it could thereafter exploit success and keep going in a way unenvisaged in previous eras; and this exploitation would create a momentum which, if supported, could have great operational and even strategic results. Manoeuvre, in other words, was restored to warfare. He had seen the results in Poland, and soon, perhaps, he would help to produce them himself, against the greatest army in Europe.

The plan for Rundstedt's Army Group A, with its forty-five divisions, was basically simple. Responsibility for the main break-in and exploitation was entrusted to General List's Twelfth Army, with two armoured corps coordinated by General von Kleist; these constituting Panzer Group von Kleist, leading Twelfth Army. The southernmost of Kleist's corps was commanded by Guderian. North of Twelfth Army General von Kluge's Fourth Army was to advance, led by General Hoth's armoured corps, which would move with two divisions on different axes, on the right 5th Panzer under General von Hartlieb, and on the left Rommel's 7th Panzer. Rommel, therefore, would be commanding one of two armoured divisions covering the right or northern flank of Panzer Group von Kleist, itself considered to be delivering the *Schwerpunkt*. Rundstedt's seven armoured divisions, advancing on a forty-mile front through the Ardennes, constituted the tip of the Army Group spear, whatever their placing or subordination.

Each divisional commander was given a centre line, roads he might and might not use, boundaries and objectives. Each was given the sector in which his division would be expected to reach and cross the Meuse – the principal obstacle. Each realized that there could be serious traffic congestion – the country was well-wooded, and although by no means terrain impossible for armour, there were many areas where columns would inevitably close on roads to pass what were, in effect, defiles, with all the traffic problems and vulnerability which that could produce. And after the Ardennes were passed, and the Meuse crossed, every divisional commander knew that a new situation might evolve, new opportunities open, unforeseen dangers threaten. The elder Moltke had said 'No plan survives contact,' and the German plan was the reverse of rigid. For the first phase – the planned phase – 7th Panzer Division was to drive from its 'peace' location in the Eifel, cross the Belgian frontier near the point where Germany, Belgium and the Grand Duchy of Luxembourg met, and force the Meuse in the sector immediately north of Dinant – a direct distance from frontier to river of about seventy miles. Thereafter the general intention of Hoth's corps – and of Fourth Army – was to drive west, to drive deep. Rommel needed, at this stage, no more.

In the months of March and April, before embarking on this remarkable adventure, Rommel learned more of his own troops, did a great deal of thinking and imposed his personality on his division. It was a very distinctive personality – brisk, incisive, intolerant of slackness or infirmity of purpose, inventive, questioning, essentially business-like, enormously energetic. Rommel believed with passion in physical fitness, going for early-morning runs himself and intolerant of flabbiness or inertia in his subordinates. He had always been impatient of others, whatever their rank, who failed to meet his own standards, and with the authority of a divisional commander his impatience was fortified. He dismissed one of his battalion commanders within three weeks of assuming command and noted with some satisfaction that the dismissal would send salutary shock waves through the division. Rommel was by instinct a humane man, but his kindness never extended to toleration of inadequacy.

Insofar as politics impinged, his admiration and affection for Hitler were fortified at every turn; and after the first phases of the Norwegian Campaign, when German forces had successfully invaded the country (as well as occupying Denmark unopposed) and were in process of driving the British intervention forces back to the Norwegian coast, Rommel wrote exultantly of Hitler's genius for both military and political leadership. He was sent, as junior Staff Officer and aide, a thirty-six-year-old National Socialist, Karl-August Hanke, an official of Goebbels's Propaganda Ministry; Rommel put him into a tank to teach him something about being a soldier, but eventually came to respect him. Rommel had no illusions about the atmosphere Hanke would find among the officers, or the lack of enthusiasm many of them felt for the political leadership: he did not share it.[18]

Any doubts 7th Panzer Division had about the competence of this incoming infantryman to understand armour were soon dispelled. The division was put through as intensive a final training period as facilities and the constraints of security allowed – incessant firing practice, movement and traffic control drills, the coordination of all arms in the tactical battle, and ever more training in communication. Rommel gave considerable thought to his own method of command in the forthcoming campaign. Like all contemporary German commanders of mobile troops, he believed (and had always believed, in whatever incarnation) in commanding from the front. The opportunities of battle present themselves fleetingly, and can only be seen by the eye and seized by the mind of one at the actual critical point. But to command a large and complex formation of all arms while simultaneously placing oneself at such a critical point or points requires a well-thought-out technique.

Rommel's technique – modified but essentially the same when in due course he commanded not a division but a Panzer army – was to spend much of his time in a specially converted tank, or armoured vehicle, in which he could range any quarter of the battlefield and from which he could communicate efficiently with subordinate commanders or with his own headquarters; and to hold regular conversations with his senior operations officer at divisional (or later corps or Army) headquarters. Often he would take to a light observation aircraft if he thought it would afford him a better visual picture or save time in travel (he himself was an efficient pilot). He trusted his staff to 'read the battle' and deduce exactly what should be done if, as would inevitably happen, they knew of circumstances of which he, because of his absence far to the front, did not. If his staff – it happened more than once – actually countermanded an order or operation because of this superior knowledge he would, when he was apprised of the whole situation, wholly support them. He needed and generally got a staff with a mind, with expert professional judgement, and with the moral courage to act independently, taking responsibility thereafter. This was in the highest tradition of the German General Staff, and Rommel, although no member of the General Staff himself, admired it and used it. He often had words of scorn or criticism for the 'Great' General Staff, the superior planning authorities of OKH in Berlin whom he regarded as too often detached, timorous and excessively academic – and as also excessively dominated by artillerymen, without fundamental sense or feel for combat 'at the point of the spear'; but for the '*Truppenstab*', the General Staff officers at the headquarters of field formations like his own division, he had high regard provided he was satisfied that they served him and not some other remote authority, and provided he was assured of their toughness and ability.

No detail within his division was too small for Rommel's attention. He held conferences for all officers, down to a very junior level. He devised novel methods of movement for use if circumstances demanded, such as the '*Flachenmarsch*', a march across country of the entire division in a box formation. He made very clear that he would always expect subordinates to show initiative, not to await detailed orders but to conform to a general plan and to use their heads and their energy and their judgement; it was and must be the German way. If Rommel, with his extraordinary battle instinct, his *Fingerspitzengefuhl*, wished to interfere in the minor or major tactics of a subordinate unit, interfere he would: but he wished no man to wait for his approval.

Above all, Rommel impressed on all the lessons he had learned twenty-five years before – in an encounter battle, open fire with everything available, no matter the theoretic suitability of the weaponry. Open fire, blanket the enemy, and use speed and shock to confound him. In an advance ignore threats to flanks or rear – march fast, march deep, throw the enemy off balance. Do not be hamstrung by convention in producing effects to distract an enemy – set his buildings on fire to produce smoke if you need smoke, persuade him to cease resistance by subterfuge if you can get away with it, surprise, deceive, never be afraid of the original or the unorthodox: be your own man: command from the front.

CHAPTER NINE

'To the Last Breath of Man and Beast'

For general Erwin Rommel as master of battlefield manoeuvre, the Second World War began on 10 May 1940. Such absolute security had been imposed within the Wehrmacht that most officers only received the order to move at or after midday on the ninth. Numerous map exercises and war games had, however, been played – to such an extent that during the campaign itself at least one staff was able to use exercise orders with amended dates to cover an episode of actual battle during the earlier phases. But the date and the hour were kept secret, and the secrecy assured surprise. During the next five days Rommel's conduct and characteristics in command as a general officer set the pattern for almost the whole of the rest of his career. To those who knew him and his reputation of old it was a wholly predictable pattern.

Rommel's 7th Panzer Divisional centre line ran from the divisional concentration area in the Eifel across the Belgian frontier and thence by St Vith and Vielsalm to cross the river Ourthe at Hotton. From Hotton the centre line ran via Marche and Ciney to the Meuse at Dinant. The advance was through well-wooded country with many narrow defiles and steep hills. It was appreciated that the Belgians would have prepared demolitions and blocks at every point which could hold up a column, and it was clear that traffic congestion could be produced with considerable ease, and that the German divisions, largely roadbound and formed of many thousands of vehicles, would be exceptionally vulnerable to air attack.

The heroes of the approach march were the traffic police and control authorities; and the Luftwaffe. The first, at every level, were fast and ingenious at devising and marking changed routes where

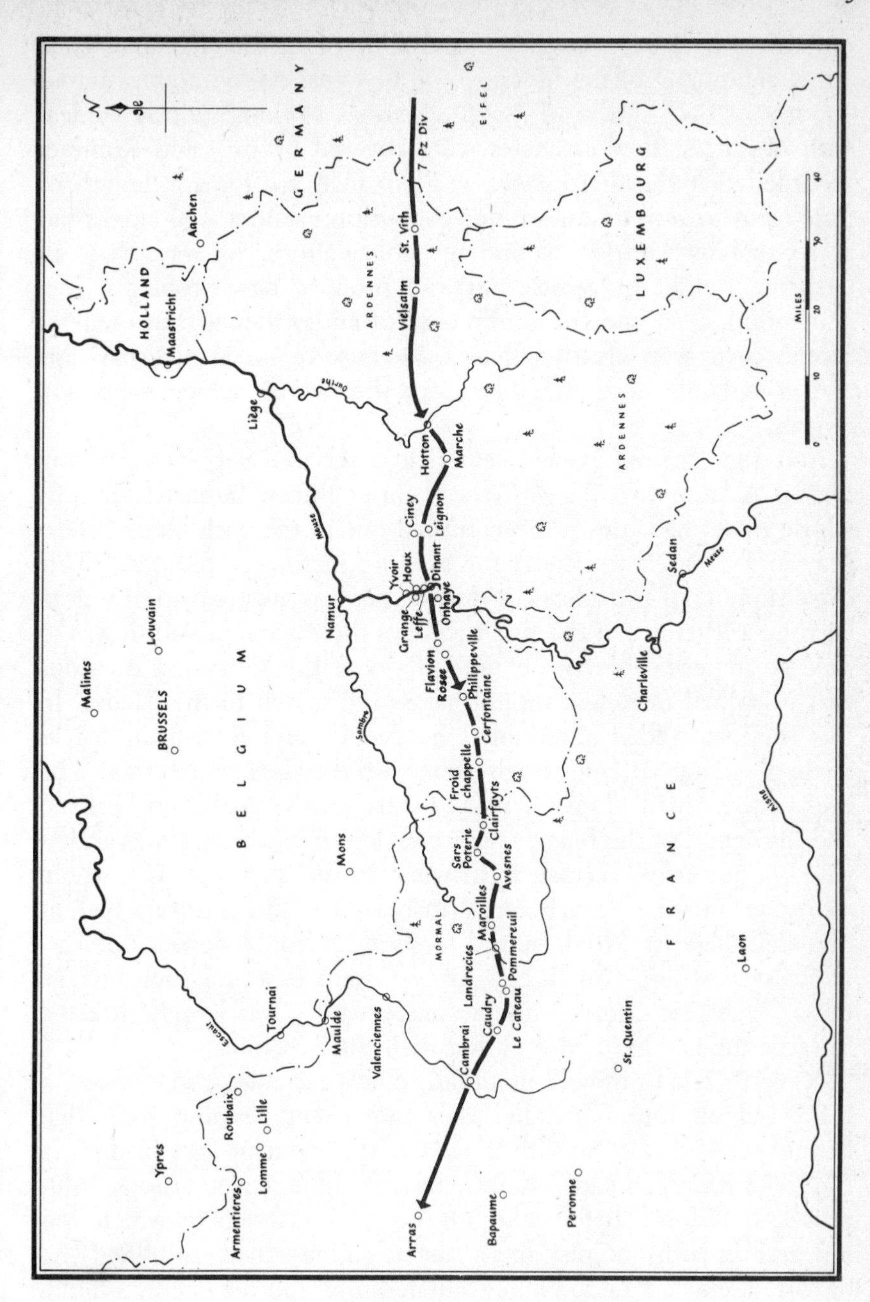

IX Western Front, 1940. Area of operations, Army Group A. Penetration and Advance of 7th Panzer Division

deviations became necessary – and they often did. Rommel later wrote that many of the blocks could be bypassed by moving across country or over side roads, with all troops working quickly to deal with obstacles. Few obstacles were covered by fire, and Rommel recorded that the division was seldom held up 'for any length of time'. But to reroute an armoured division round a well-sited road crater or blown bridge, to find and police alternative ways that can bear the weight of vehicle and not produce new problems, is a demanding task, and the traffic control authorities within General Hoth's corps were clearly efficient. There were, inevitably, snarl-ups and complaints here and there, but the overall achievement was impressive.

And the Luftwaffe gave remarkable service. Supporting Army Group A were two *FliegerKorps* – about fifteen hundred aircraft, whose pilots had mostly been turned out in the early hours of 10 May and ordered to report for briefing in fifteen minutes. Their close support of the advance when it met serious opposition was to become a legend. On the first two days their prime function was to hold so impermeable an air umbrella over the advancing divisions that no Allied intruders should harass the march to the Meuse. In this they succeeded admirably; the French and British air forces made no serious attempt to interfere with the German advance. The eyes of the Allied High Command were on the north, on Holland and the centre of the Belgian–German frontier, where, it was anticipated, a powerful German right wing, in the spirit of 1914, would soon confront the French and British armies. These armies had, as Manstein had predicted, moved forward, untroubled, to a line east of Brussels where von Bock's Army Group B would soon engage them. Allied air effort – anyway inadequate – was largely directed towards the northern wing rather than the Ardennes.

Nevertheless Rommel's first hours contained at least some frustration, and he found his division's movement running somewhat behind schedule. In St Vith, however, three out of four bridges in the town had been successfully seized by 'undercover' troops, infiltrated in disguise before the advance – a manoeuvre which was repeated at many points, and a *ruse de guerre* which, amplified and distorted, created a good many agitated myths in the enemy's mind. But Rommel's advance was not unopposed. The Belgian Chasseurs Ardennais – charged simply with imposing delay upon an enemy – often fought bravely and checked the German advance. By the middle of the second day, however, 7th Panzer Division was across the river Ourthe at Hotton: forty miles. Twenty-four hours later Rommel was through Ciney and Leignon: fifty-eight miles. He was well ahead of his right-hand neighbour, 5th Panzer Division. Conscious that in Rommel he had a thruster whose pace and success deserved reinforcement, the Corps Commander, Hoth, made an additional Panzer regiment – 31st – available to Rommel from the neighbouring division. By the standards of the day Rommel, with only one Panzer regiment (albeit of three tank battalions instead of the established two), might find himself short of tanks.

Hoth's was a gratifying gesture. Rommel's own 25th Panzer Regiment, with 218 tanks, certainly had as many as the division's allocated roads and terrain could accommodate, as yet; but the effect of this reinforcement was to widen Rommel's frontage of responsibility, for 31st Panzer Regiment (Colonel Werner) was some way ahead of the rest of 5th Panzer Division to Rommel's north, and he now, therefore, had an additional thrust line and an additional Panzer regiment with tanks and armoured cars moving west along it. His task was to reach the Meuse valley, cross the river, secure a bridgehead, build a bridge (on the reasonable assumption that the enemy would have blown the bridges) and then put armour and heavy equipment across. Like all commanders in such circumstances, Rommel had hoped to advance with such speed that he would 'bounce' a bridge before demolition, and he nearly did so. But not quite, and when the leading German troops, Werner's armoured cars and the motorcycle battalion of 7th Panzer, reached the Meuse valley on the afternoon of Sunday 12 May they found the bridges blown. Sixty-five miles. By last light the east bank and much of Dinant were in Rommel's hands.

Rommel's motorcycle battalion now undertook a bold step. North of Dinant, halfway between that town and Yvoir, where Werner's armoured cars had reached the valley, the river divides around a small island in mid stream opposite the village of Houx on the east bank. Running from that bank to the island was an old stone weir, offering precarious footholds, and over the weir to the island – which might or might not be tenanted by the enemy – dismounted men of the motorcycle battalion now cautiously edged their way. None fell into the Meuse; the island was unheld; and beyond its far bank the Germans found a crossable lock gate, connecting with the west bank of the river. The first troops of 7th Panzer Division were now across the Meuse, and shortly afterwards were reinforced by several companies of Rommel's 7th Rifle Regiment.

These companies were subjected to Belgian machine-gun fire from positions on the river, and a good deal of shelling. Most of the French – purely by chance, Rommel had struck a French inter-corps boundary – were deployed on higher ground, back from the river, and the crossing had been accomplished to some extent unseen; but the enemy now appreciated what was happening, the Germans began to take considerable casualties, and soon neither reinforcement nor casualty evacuation across the river were practicable. The situation when Rommel explored it in the small hours of Monday, 13 May was tense. From this point he personally set the pace of events.

Rommel's divisional artillery were already deployed to bring down fire anywhere in the Meuse valley or on the high ground beyond, and his artillery observers were with the forward troops at the crossing places. Dinant and the valley were under French artillery fire; French anti-tank fire from the west bank had knocked out a number of German tanks approaching the river. Attempts to get reinforcements across in rubber boats were failing – enemy fire, whether Belgian or French, now had the Meuse enfiladed from well-concealed positions on the west bank. Casualties among the Germans on the west bank were undoubtedly mounting, and in the pale light of the early May morning the enemy's artillery fire was also being accurately directed on to any German movement on the east bank and the approaches to the river. Shell fire several times landed near Rommel himself.

Rommel directed the firing of several houses in the Meuse valley in order to provide a pall of smoke and obscure enemy observation. He then drove south, where he realized that in the southern sector of Dinant enemy fire was too heavy for further crossing. He drove back to his headquarters, where the Army Commander (Kluge) and the Corps Commander (Hoth) met him, told them how things were going, and drove back to the Meuse north of Dinant. Rommel left his command vehicle a few hundred yards east of the river, and raced forward on foot (being, as he noted, 'bombed on the way by our own aircraft'), aiming to reach the Meuse at another weir, at Leffe. The valley was dominated at this moment by enemy fire.

Rommel had ordered some tanks of the Panzer regiment, and two field guns, to meet him at Leffe weir. The commander of one of them takes up the tale:

> I look left and see General Rommel and his Adjutant, Major Schraepler. I halt my tank and the General climbs into the turret, with me, squeezing in on my left, Major Schraepler

> behind. We head off toward Dinant through artillery and machine-gun fire.
>
> After about five hundred metres I see three Frenchmen in a trench to the right, halt my tank and the French put their hands up. They'd taken their time about it so I draw my pistol to shoot them and make sure of our own skins since, half out of the turret, we offered a tempting target. We decided not to, however, and my tank rolled on.[1]

Shot and shell were flying, and Schraepler was hit in the upper arm. Eventually the tank raced into Dinant and Rommel with his wounded Adjutant climbed down. The three Frenchmen had fired at them after their lives had been spared, and had been shot by the companion tank-commander.

Rommel now directed the nearest tanks to drive, line ahead, northward up the Meuse road along the river, turrets traversed left, firing incessantly at any possible enemy position on the west bank. He reached the northern crossing point and, in his own words, 'took over personal command of 2nd Battalion, 7th Rifle Regiment'. Under cover of the tank fire which was now being poured indiscriminately across the river the rubber boats were relaunched and crossing restarted. Rommel went across in one of the first boats.

He found the position on the west bank better than had appeared when viewed across the (impassable) Meuse – the rifle companies had dug in, and were, where possible, expanding the shallow bridgehead; but they had with them no anti-tank weapons, and when the cry went up that enemy tanks were attacking, Rommel himself directed the response. Small-arms fire, his frequent and typical reaction – to check, harass and deter enemy tanks, however invulnerable to it – was effective. His personal intervention steadied the troops in the bridgehead and his vigorous handling of an admittedly tentative French counterattack gave them confidence. The motorcycle battalion from the Houx crossing over the weir were now moving away from the valley towards the village of Grange. The defence was being punctured but was still intact and potentially formidable.

Rommel now recrossed the river to the east bank and drove north to the northern (6th) Rifle Regiment area. Here not only infantry but anti-tank guns had been ferried to the west bank, and bridge-building had already started. Rommel – allegedly jumping into the river to lend a hand with bridging – gave direct orders to the engineer company commander: he was to switch effort to a large-type bridge

– sixteen tons. This was necessary to bear tanks, and Rommel's urgent desire was to get tanks across. He showed, in later recounting all this, little sense that a heavier bridge might have been preplanned; Rommel sometimes acknowledged personal inadequacy of forethought, but seldom to excess.

Rommel was soon able to bring his signals vehicle, driven by the later commissioned Sergeant Hansel, to the west bank.[2] Meanwhile there was heavy French shelling of the crossing places and approaches, and the motorcycle battalion in Grange had been strongly counterattacked by French troops. Rommel recrossed the river yet again, determined to organize the ferrying of tanks during the night. Darkness was falling. Rommel seemed to have been everywhere – 'like a whirlwind', one of his subordinates remembered. And men asked each other, 'Is Rommel immune?'[3]

Three miles west of Dinant, and on Rommel's centre line, lay the village of Onhaye; and immediately north of Onhaye was a long east–west wood. Onhaye had featured in a map exercise which Rommel had undertaken with his subordinates at Bad Godesberg, weeks before. It had been agreed that once the Meuse was crossed Onhaye would be seized, as opening the gate for the next major advance, and that the seizure of Onhaye would best involve a move round its north side, near the wood, followed by the cutting of the road west of the place.

At nine o'clock on the morning of Tuesday 14 May Rommel heard that 7th Rifle Regiment (Colonel von Bismarck), despite its casualties in the Meuse valley and the still tenuous lifeline across the river behind it, was close to Onhaye and was planning immediately to send a company, as anticipated, round its northern side. Rommel, back on the west bank, had by now about thirty tanks across the river. He decided to climb aboard one of them and lead 25th Panzer Regiment (Colonel Rothenburg), when it arrived, to the wood north of Onhaye, whence it could deploy in any direction, whether to defeat a French counter-move towards the valley, to clear Onhaye, or to rejoin the centre line west of it. He also told Rothenburg to allocate five tanks to Bismarck to assist the men of the rifle regiment engaged in outflanking Onhaye to the north. Rommel moved with the remaining tanks, watching the riflemen and their allocated five tanks to his left front, and aiming at the south-west corner of Onhaye wood where he hoped to assemble such tanks as were west of the Meuse.

As he reached this point Rommel's own tank was hit twice. French artillery and anti-tank fire had opened up from another wood west of Onhaye, and Rommel's tank slithered down a steep slope and was canted at an angle which made it impossible to traverse the turret. Rommel, face pouring with blood from a splinter, abandoned the tank, taking the crew with him, and clambered up to the wood itself, which his signals command vehicle had reached and where it, too, had been hit and was immobilized. He ordered the tanks to move east into cover, through the wood. The setback was temporary, and later that evening Rothenburg's tanks drove the French from the neighbourhood of Onhaye. Bridging was now complete, and 7th Panzer Division had won its bridgehead west of the Meuse. It was not alone (although the leading troops of the division had the honour of first crossing), for by the end of 14 May, far to the south, all three divisions of Guderian's corps in Panzer Group Kleist were also across the river in the area of Sedan. Rundstedt's *Schwerpunkt* was under way.

Rommel gave orders for the following day. The division, led by Rothenburg's 25th Panzer Regiment, would drive west, avoiding villages where possible, cross the railway line east of Philippeville (which, unbeknown to the Wehrmacht, had been set by the French Army Commander as a fall-back line for his troops retreating from the Meuse valley) and seize ground around the village of Cerfontaine, twenty-five miles west of the Meuse. The order was to keep moving – enemy positions, tanks, columns, were to be engaged by German tanks on the move. The momentum of the advance was at all costs to be maintained. Flanks were to be covered, where necessary, by predicted artillery fire, or by tank rounds loosed into likely enemy concealed positions without delay or investigation. Rommel would accompany the Panzer regiment, and there would be no stop or pause between the assembly area around Onhaye and the Cerfontaine objective.

Rommel himself wrote that the secret of the forcing of the Meuse was what he called 'tight combat control' and the ability of the divisional commander to give personal orders to the forward regimental commanders. Radio and necessary encoding would have spelled unacceptable delay. Rommel, in effect, directed the tip of the spear personally, by verbal order, again and again. He personally organized the resumption of the assault crossing, with troops undoubtedly shaken by their early losses and under heavy enemy fire. He personally stimulated the bridging; and it might well be remarked that he could have planned it better and have needed to interfere later all the less. He – a company commander again – instructed the troops on the west bank how to drive off a French tank attack when they had no anti-tank weapons. He led – being wounded in the process – the leading armoured elements of 25th Panzer Regiment to his ordained assembly area for the next advance, when only about one sixth of their number were yet across the Meuse; and he led them back again to cover when the French reaction, shortlived though it was, spelled trouble.

It was, perhaps, fitting that for these essentially front-line achievements Rommel was awarded a clasp to his Iron Cross (2nd Class) for his deeds on 13 May and another to his Iron Cross (1st Class) for those on 15 May. It may be said that each of these actions was, essentially, the business of a subordinate, or a subordinate's subordinate; that the divisional commander's duty was with map and radio, reading the battle, preparing the next step, and not with his vehicle, and his binoculars, and his boots and his lungs (Rommel was hoarse from shouting in those days) in the midst of shot and shell. It may be said that his actual effect was to interfere, to override. It may be asked what he did that the responsible lieutenant or captain or colonel would not have done, if sufficiently energetic and resourceful; and the Wehrmacht did not lack energetic and resourceful officers.

It may be so said and so asked. Rommel would have replied, as he did of differing levels of command and from the experience of many campaigns and two wars, that the pace of *offensive* battle is set by the decision of the commander, and that there is one critical point only where that decision can be taken – the point of contact. No man has ever been more alive to the importance of opportunism in battle. No man has ever been more conscious of time, of the fleeting nature of opportunity, of the rapidity with which the commander must act or react. Rommel had an instinctive feeling for where and when a chance of battle would present itself, for where his own dynamic presence would make the difference – perhaps a difference only measurable in minutes, but minutes of inestimable value – between a seized or a lost opportunity. Battle can turn on the action of a tiny handful of subordinate, perhaps very junior, commanders or soldiers. Their initiatives or failures, cloaked in the bland language of official military jargon, translate to arrows pushed forward or lines retracted on staff maps, but their decisions, generally unrecorded, and often made by instinct or in fear, have been crucial. In Rommel's command Rommel himself, trim, compact, dynamic, shouting like a sergeant-major where necessary, leaving others behind with the speed of his mind and his movements – Rommel himself, as often as not, was by that subordinate's side. That he sometimes made mistakes in so acting is certain; but that tactical victory sometimes turned on the voice and gesture of this man is beyond doubt.

On 15 May Rommel's advance gathered pace. On that day, following a general withdrawal order given to the French troops further south, Panzer Group Kleist was through the last Ardennes and Meuse defences and was driving west: the German frontage of advance in the sector of Army Group A was now sixty miles. On that day, far to the north, a German attack penetrated the front of the First French Army, deployed on the right of the British Expeditionary Force under General Viscount Gort, which had advanced as anticipated into Belgium to the line of the river Dyle, east of Brussels; and on that day, too, the first British withdrawal was ordered, to conform to the French withdrawal in the south and in initial, still half-incredulous, recognition of the fact that the Allied front had been ripped open. To the Germans, or most of them, the extent of their own achievement was still unclear, if not incredible. On Rommel's front there still lay ahead the allegedly impenetrable Maginot line. Rommel's division was still in Belgium and beyond the Belgian–French frontier lay a thin extension of the Maginot line proper – a shallow belt of pillboxes and anti-tank obstacles. This was, for a short while, a more formidable defensive position in the minds of the advancing Germans than they would find it to be on the ground.

Rommel had ruled on the map, copied on the maps of his subordinates, a straight thrust line – from Rosée, a village several miles west of Onhaye, to the church in the village of Froidchapelle, four miles west of Cerfontaine and still in Belgium, still short of the Maginot line. He requested fire or directed troops to objectives by reference to this thrust line and the sectors into which he divided it; and the personally devised system pleased him. After a brief encounter with some French armour at Flavion – heavy tanks, capable of outgunning any German tank by a combination of larger gun with armour plate of nearly twice the German thickness – Rommel directed the division westward, riding as usual with the leaders.

Unlike in the Ardennes, the 'going' off roads was now good. Routes were canalized by woods or villages but there was little to impede deployment or cross-country movement. The weather was excellent, and a quota of dive bomber support had been allocated to 7th Panzer. The division soon breached – an exaggerated term in the circumstances – the French so-called 'stop-line' east of Philippeville. Rommel's technique was everywhere and all the time the same – to keep moving, to scatter fire from traversed tank turrets at possible enemy positions on the edges of woods, in farmsteads, in villages; to maintain momentum, to intimidate, to shock, to paralyse. North of Philippeville, his leading tanks opened fire at extreme range on a force of French tanks observed south of that place, but on the whole nothing held up the advance.

Prisoners were now giving themselves up in huge numbers. A tank commander recorded: 'General Rommel orders me to take the lead and in a village square, where I have a feeling there must be Frenchmen, I shoot into houses with my pistol and yell "*Soldats françcais, venez!*" and on that command the doors of every house open and a great crowd of Frenchmen, perhaps several companies, flock into the square with their hands up.'[4]

Large numbers of French vehicles, including tanks, were overrun without opposition: tanks were ordered to accompany the German column while unarmoured vehicles and dismounted men were ordered to march eastward along Rommel's centre line. Everywhere the enemy seemed shattered, bemused, looking only for a way of escape from a situation of incomprehensible disaster. The scenes of chaos and disorganization among them were to be repeated and repeated again in the days ahead. Stories were soon rife of French traffic police and troops helping to guide and direct the German columns, of French soldiers seeming willingly to cooperate in the conquest of their own country; such stories certainly ran like fire in corn through the Allied armies, and helped speed an atmosphere of hopelessness in the face of an irresistible enemy, of disaffection, uselessness, betrayal.

Extreme deductions from those days would be unjust, and Rommel never personally made them. It was true that morale in many parts of the French army was fragile, a fact clear even to friendly observers during the preceding winter,[5] and enthusiasm for the war was scant. It was also true that the Allied plan of deployment, with its reliance upon a linear defence (hinged on the Ardennes) and a paucity of reserves, was defective; true, too, that French armoured theory was reactionary rather than progressive, with armoured units sprinkled over basically infantry formations to a large extent (French tanks were in many cases superior to German in tank-versus-tank combat, a form of combat which Rommel anyway shunned) and without sufficient emphasis on mobile divisions of all arms.

But, all that said – and, of course, it was fundamental – troops suddenly struck by wholly unexpected catastrophe, with the fabric of military command and organization disintegrating, with everything familiar, orderly, disciplined in their military society dissolving – such troops, not unnaturally, will react simply as individual human beings, seeking safety where they can find it, conforming to the will and the orders of whoever appears to have self-confidence. Uncertainty is relieved by authority – any authority. Fear is temporarily stilled by realization that not every enemy soldier is about to kill – that the enemy soldier may be pointing the way, however brusquely, to shelter, rest, food even. The consequent relief can induce, at the least, cooperation; and that cooperation will be accorded the more promptly when large numbers of comrades are doing the same thing. Rommel reflected a good deal about how to deal with troops who have broken, lost cohesion, run away. He believed that it was hopeless to try to rally such men by draconian measures – their willpower, their self-respect, their discipline and thus their courage and coherence had all gone and would need orderly restoration.[6] Instead, he wrote, it was best to shepherd them into specific areas (the shepherding being itself a consolation) and then take a little time in the restoration of order and morale. But such measures were quite beyond the French command in 1940; and such measures require, somehow, somewhere, that a front be held.

On 16 May Rommel's task was to advance across the French frontier and through the Maginot line extension. He had received a message ordering him to stay at his divisional headquarters, and during the morning he was visited by the Army Commander, Kluge, to whom he explained his intentions.

They were understandably deliberate, and illustrate that Rommel, dashing though he was, was a man who planned battle carefully if he thought that the enemy's circumstances made that necessary. Sometimes he was to miscalculate those circumstances, but in this case he supposed that the Maginot line might be a hard nut to crack.

His thrust line was to run across the frontier at Sivry and thence via Clairfayts to Avesnes. He deployed every gun of his divisional artillery and ordered 25th Panzer Regiment to advance in extended order. When the main French positions had been reached the two rifle regiments – their movements coordinated by a brigade headquarters established within the division for such a purpose – would pass through the Panzer regiment, covered by its fire, and breach the 'fortifications'. Thereafter the Panzer regiment would again take up the lead and advance to Avesnes.

Rommel rode in the same tank as the Panzer regimental commander, Rothenburg. The first part of the plan went well. The tanks bypassed Clairfayts, the assault engineers having dealt with the first French 'fortifications' – a concrete pillbox – one hundred yards beyond the frontier, and with a steel anti-tank hedgehog on the Clairfayts–Avesnes road. French artillery and machine-gun fire now opened up, and there was a certain amount of anti-tank fire; but Rommel's instinct told him that if the Panzer regiment, now in the lead, kept moving they would probably be able simply to motor through. He gave orders that there should, therefore, be no pause, no waiting for the rifle regiments to dismount or deploy according to the plan. Rommel now reckoned, truly, even if somewhat belatedly, that the 'Maginot line' was an inconsiderable obstacle, and that if the tanks of 7th Panzer Division drove ahead at maximum speed, firing on the move to left and right at every spot which might conceal a Frenchman or a French weapon, they would reach clear country beyond.

And so it proved. Without stopping, Rothenburg's Panzers, their guns using a good deal of tracer for heightened effect, sprayed the area which was supposed to constitute a main French position, and carried on towards Avesnes. Engineers disposed of concrete bunkers with demolition charges. Rommel led the division north of his thrust line to Sars-Poterie, through which ran a larger and straighter road to Avesnes. Sars-Poterie was packed with French troops, horses, vehicles – as was, by now, every village, every farm – all in what seemed a state of total demoralization. There was no resistance. Avesnes itself, a sizeable place, was crowded with French. Rommel directed the Panzer regiment round the south of it and soon reached the main road running west from Avesnes to Landrecies.

Rommel was now riding by moonlight behind the leading Panzer company. Avesnes had been shelled by German artillery and every road and track out of it was crowded with refugees, carts, cars, all mingled with the fleeing forces of the French army. The chaos was indescribable. Rommel reckoned that his prisoners must amount to the equivalent of two French divisions, and gave orders to improvise a prisoner-of-war cage in a field beside the road.

Since Rommel was with the leading troops his command was

greatly extended and radio communication was proving difficult. There was now a gap in the Panzer regiment, with one battalion of tanks still the far side of Avesnes, and with what appeared to be a French tank battalion moving into the town, temporarily cutting 7th Panzer Division's centre line. Rommel could get no response from his own divisional headquarters, by now far behind him. He was sending signals from 25th Panzer Regiment command tank but was unable to hear acknowledgement.

He could, of course, only communicate with superior headquarters through the rear link at his own divisional command post, with which he was out of touch. He had reached his objective beyond Avesnes – the division strung out behind him over many miles of country between Avesnes and Philippeville – and was untroubled by the possible reactions of the enemy. He was, however, keen to press on, with anything which he had under his hand (which at that moment consisted only of the leading battalion of the Panzer regiment and the motorcycle battalion), to Landrecies, during the darkness if need be, and to seize a crossing over the Sambre. This went some way beyond corps orders, which had directed 7th Panzer on Avesnes. Frustrated in his attempt to get approval for this onward move, Rommel nevertheless gave orders in the small hours of 17 May: on to Landrecies. He himself moved with the leading Panzer battalion, setting forth at 4 a.m. He wrote afterwards that he was firmly convinced that behind the leading Panzer elements were the rest of the division, who would in time catch up and take part in whatever action lay ahead.

In fact 7th Panzer Division was extended well beyond the power of anyone to control it, so rapid had been the advance and so dense was the confusion of refugee and other traffic. No tanks had been resupplied during the few hours of darkness, and since they had, by order, been using their armament liberally as they smashed through the occasional French defences, ammunition was short. Just as Rommel set out towards Landrecies, he learned that the French tanks in Avesnes had been dealt with – he had sent his National Socialist Adjutant, Hanke, back in a Panzer IV to help clear up the situation, and Hanke (a considerable feat) had done so. The two Panzer battalions were now reunited, and 25th Panzer Regiment, unsupported by much else, drove west in the light of dawn. At Landrecies the bridge across the Sambre was unblown.

Landrecies – scene of a famous encounter in 1914, with the right wing of the Schlieffen wheel swinging down through the Forêt de

Mormal after first meeting the British Expeditionary Force at Mons twenty-five miles to the north, and meeting them again as they withdrew towards Le Cateau – was packed with French troops in huge numbers, anxious to surrender; with French military traffic; with refugees, confusion, hubbub and terror. Hanke, always at hand to discharge a wide variety of functions at the personal order of the divisional commander, was told to take control of one large mass of Frenchmen still in their barracks, to parade them, disarm them and start marching them eastward.[7] Rommel drove on. He clearly felt that nothing could stop him, that 7th Panzer Division was irresistible, could win the war single-handed.

There was, as there has to be, an element of competition in all this. Rommel could not at that moment know exactly or even vaguely what point his southern neighbours, Panzer Group Kleist, Reinhardt's and Guderian's Corps, had reached, but it is a fair presumption that he hoped to be ahead of them – just as he was certainly ahead of his right-hand neighbour, 5th Panzer, within Hoth's corps. Rommel was like a racehorse which needs challenge to achieve real pace. In fact the Kleist Group were, on 17 May, about level with him, twenty miles to the south – the breadth, now, of the German penetration; and on that day Rommel told his division his ultimate objectives. 'General Rommel,' an observer noted, 'collected his Panzer commanders and gave his orders in his own classic way – "Thrust line, Le Cateau–Arras–Amiens–Rouen–Le Havre"'[8] – an apparent fantasy which came true or almost true within four weeks.[9]

Rommel drove on for two hours, through Pommereuil (where large numbers of Frenchmen surrendered) to some high ground just east of the town of Le Cateau. By now only the leading Panzer battalion was with him and he decided that he had best try to bring up rather more of his command – and, indeed, discover where they were. Ordering the Panzer troops to form an *Igelstellung*, a 'hedgehog' of all-round defence, at Le Cateau, he drove back down his own centre line, taking with him one Panzer III for personal protection. There were reports of enemy tanks now in Landrecies.

Rommel saw French troops bivouacked everywhere. He passed Landrecies and drove eastward through Maroilles on the road back to Avesnes. The only 7th Divisional troops he met were one rifle company; of the rest no sign. At Maroilles he found a column of French traffic moving from the north and, telling the ever-available Hanke to jump on the leading French vehicle and direct the driver to change course, Rommel diverted the column like a traffic policeman, eastward towards Avesnes, then decided to escort it thither and led it into Avesnes, where he ordered the troops to parade and lay down their arms. Of such incidentals was Rommel's day composed on 17 May 1940.

It was now afternoon, and at about four o'clock 7th Panzer Division headquarters and the rest of the division began to arrive. Rommel, selecting off the map, pushed them into various locations between the frontier and Le Cateau, both east and west of the Sambre.

At midnight he received new orders. 7th Panzer Division would continue the advance to Cambrai, a further fifteen miles from Le Cateau. Another bridge across the Sambre had been taken north-east of Landrecies, and 5th Panzer Division were routed towards it and were catching up.

Rommel has been criticized for certain episodes during the campaign of 1940 in which he had, by his own admission, no accurate knowledge of where his command was, and thus had little ability to react in a balanced way to any new situation. The first such episode was on 16 and 17 May after the Meuse crossing, and it is certainly true that the speed with which he personally drove forward the spearhead of the division, and his insistence on accompanying that spearhead, meant that coordination and resupply became near-impossible amidst the chaotic traffic conditions of the day. In all armies in those May days movement was laborious and frustrating; and in the case of Germany's enemies extremely dangerous as well, so complete was the Luftwaffe's command of the air. It is perfectly true, therefore, that Rommel's ability to deal with, for instance, a coordinated counterattack from a flank, perhaps demanding rapid redeployment and regrouping, the switching of artillery to new fronts and new tasks, would not have been easy, for he had thrown away the reins and was riding with whip and spur alone. But his instincts – which were extraordinarily good, and were again and again justified by events – told him that it didn't matter; that what mattered was to keep up momentum and to face new dangers and new situations only if they arose. (His instincts were soon to be tested sharply. A jolt was impending, but it was still several days away.) It speaks nothing but credit to German military appraisal that it was for his division's exploits on those two days that Rommel was awarded the Knight's Cross of the Iron Cross.* The citation for the victories on

* The highest degree of the Iron Cross, open to further honourable embellishments of oakleaves, swords and diamonds.

16 and 17 May, 'of decisive significance for the whole operation', referred to the divisional commander's personal courage, 'regardless of danger'.[10]

Rommel was in fact troubled by a very different aspect of the advance. Nobody could enjoy the spectacle of French refugees trudging, terrified, to escape the battle and the Luftwaffe, but war was war, and the sooner it ended in German victory the sooner everybody would be able to return home. Rommel was always convinced that the only sensible outcome must be reconciliation and lasting friendship between France and Germany, and he recorded the shooting of a French officer with an unease which comes clearly through the perfunctory description, referring to the obviously fanatical hatred which led the Frenchman three times to refuse an order to get into Colonel Rothenburg's tank, after which there was nothing to be done but to shoot him.[11]

Rommel, then and later, had a deserved reputation for chivalry, for not only treating prisoners and enemies decently, but feeling decently towards them. Prisoners were sometimes killed by their captors in the Second World War, by all armies (on the Eastern Front later, indiscriminately and with every circumstance of atrocity). But this was never condoned, let alone ordered, by Rommel, and was utterly repugnant to his mind. That one brave French Colonel was clearly determined to court and face death rather than obey a legitimate order was a matter which troubled him and which he did not forget.

Early on the morning of 18 May Rommel retraced his steps and headed westward again, having ordered the second Panzer battalion to join the first in its position east of Le Cateau. He caught up with them at Pommereuil, through which he had driven the previous day. He found that the town had now been reoccupied by the French and that, since he had last been with them, the leading Panzer battalion had been attacked by French heavy armour and subjected to a good deal of shelling. In a confused situation he led the second battalion on a southern detour via Ors, ultimately to join Rothenburg and the leading battalion. By three o'clock in the afternoon, and at long last, 25th Panzer Regiment was reunited and replenished; and the rearward centre line, through Landrecies, was clear. Rommel set the leading Panzer battalion on the march towards Cambrai.

The country between Le Cateau and Cambrai is that which saw General Smith-Dorrien's Corps of the British Expeditionary

Force deploy to meet overwhelming numbers of the German army advancing from the north in August 1914. It is open and unbroken, the only impediments to movement being a few small villages and the little town of Caudry. Rommel ordered the Panzer regiment to act as they had done when advancing two days earlier – that is, to move on a broad front across country to the area north of Cambrai, scattering fire into the northern outskirts of the place and cutting all roads into it. Once again it worked, and worked perfectly. With Cambrai soon in the hands of the Wehrmacht, Rommel was ordered to rest his division for at least two days. Its achievements had been remarkable.

Since crossing the German frontier on 10 May, eight days earlier, Rommel's division had advanced about 175 miles; had forced one of western Europe's most formidable river barriers, the Meuse; and in the last two days alone had sent on the road eastward about ten thousand French prisoners of war. It had destroyed over a hundred tanks, thirty armoured cars and twenty-seven guns, and had lost, in those two days, only thirty-five men killed and fifty-nine wounded.

In spite of the authorities' concern that he and his division needed a little longer for recuperation, Rommel persuaded them that it would be wrong to give the enemy the slightest avoidable remission. 'Pursuit,' ran the traditional Prussian maxim, 'should be to the last breath of man and beast'; and to Rommel this campaign was now surely a pursuit, a phase of war in which enormous efforts are needed, in which great fatigue should if necessary be accepted, in order to pre-empt harder penalties later, in order to reap the harvest. *Sichelschnitt* was already cutting through the ripe corn. One hour after midnight on 19–20 May Rommel, again riding by moonlight, and again with the tank spearhead of the advance, set off towards Arras.

CHAPTER TEN

The Ghost Division

IT WAS by now pretty clear to the Allied High Command that the German 'penetration' was in fact a massive offensive, thrusting on a comparatively narrow front but with weight and strength behind it: and it was clear to the British, north of the penetration, that withdrawal in that sector could only be west or north-west. On the evening of 19 May, nine days after the original German attack, Gort discussed for the first time with his subordinate commanders the awful possibility of evacuation.

The British Expeditionary Force was now deployed on the line of the river Escaut, its right at Maulde, a small place twenty miles south-east of Lille; and, behind their right shoulders, only thirty miles away, Rommel was approaching Arras. Gort had a garrison in Arras, and two divisions – 5th and 50th – north of that place in General Headquarters reserve. His garrisons and bridge guards were largely drawn from three 'labour divisions' sent from England in the preceding months to complete their organization, equipment and training in France: divisions largely without support weapons, artillery or communications, but whose men in many cases and places, fighting in the only way open to them – with their rifles, against a well-equipped and largely mechanized foe – often performed valiantly. Extending Gort's east-facing line and facing south to protect his right flank were troops of the French First Army. Gort's force, as well as all French north of the penetration and the entire Belgian army (deployed on Gort's left flank), were now simultaneously threatened from north-east, east and south. And the southern threat – Rundstedt's Army Group A – had now reached the mouth of the Somme, decisively carving the Allies in two, and cutting Gort's line of communication with his base at and around Le Mans.

For his part Rundstedt reckoned that his troops had almost completed their decisive part in the first phase of the great operation. The Kleist Group had done all that was expected of it and more. Guderian's corps would reach the English Channel on 21 May and Rundstedt's eyes were already on the south, on the bulk of the French army now, it was supposed, deploying south and west of the Somme and the Seine. In the north Army Group A was ready to press the beleaguered British and the French First Army, to widen the gap between the Allied wings and to harass what must now be a defeated enemy: but Rundstedt was content that in that quarter his principal part should now be as anvil to the hammer of von Bock's Army Group B which was advancing westward through Belgium to deal the Allied left wing a *coup de grâce*. The German advance had been made with dazzling rapidity; the view that threat to flanks should not be allowed to reduce momentum had been triumphantly vindicated.

Hoth's corps was ordered to move round Arras and then operate north-west towards Bethune. There were considerable French forces in and around Lille and the British were still in front of it, on the Escaut. Hoth's move would threaten them with immediate envelopment, and cut off any westward withdrawal before Bock. 7th Panzer Division would advance on the left, having swung round the south of Arras, with 5th Panzer Division advancing north on its right, east of Arras. On Rommel's left would be the SS *Totenkopf* (Death's Head) Division. The British garrison of Arras would be isolated, and any delaying screens would be swept away.

Then, in the afternoon of 21 May, Army Group A received disturbing reports. 7th Panzer Division were being attacked from the north, by five British divisions and a considerable mass of armour. The reaction all had feared had come at last.

Rommel was, as usual, up with the leading 25th Panzer Regiment, moving west and then north round the southern edge of Arras. He was – as often – irritated by what he thought was the sluggish movement of some of his following units, in this case 6th Rifle Regiment, and at three o'clock in the afternoon, to hasten them, he drove back along the route he had taken, via Ficheux, a village five miles south of Arras. He found his rifle battalions and led them north towards the head of the divisional column, which had reached the village of Wailly. As he rejoined the leading troops half a mile east of Wailly they came under fire from the north. One of the German howitzer batteries was already in position firing rapid on enemy tanks attack-

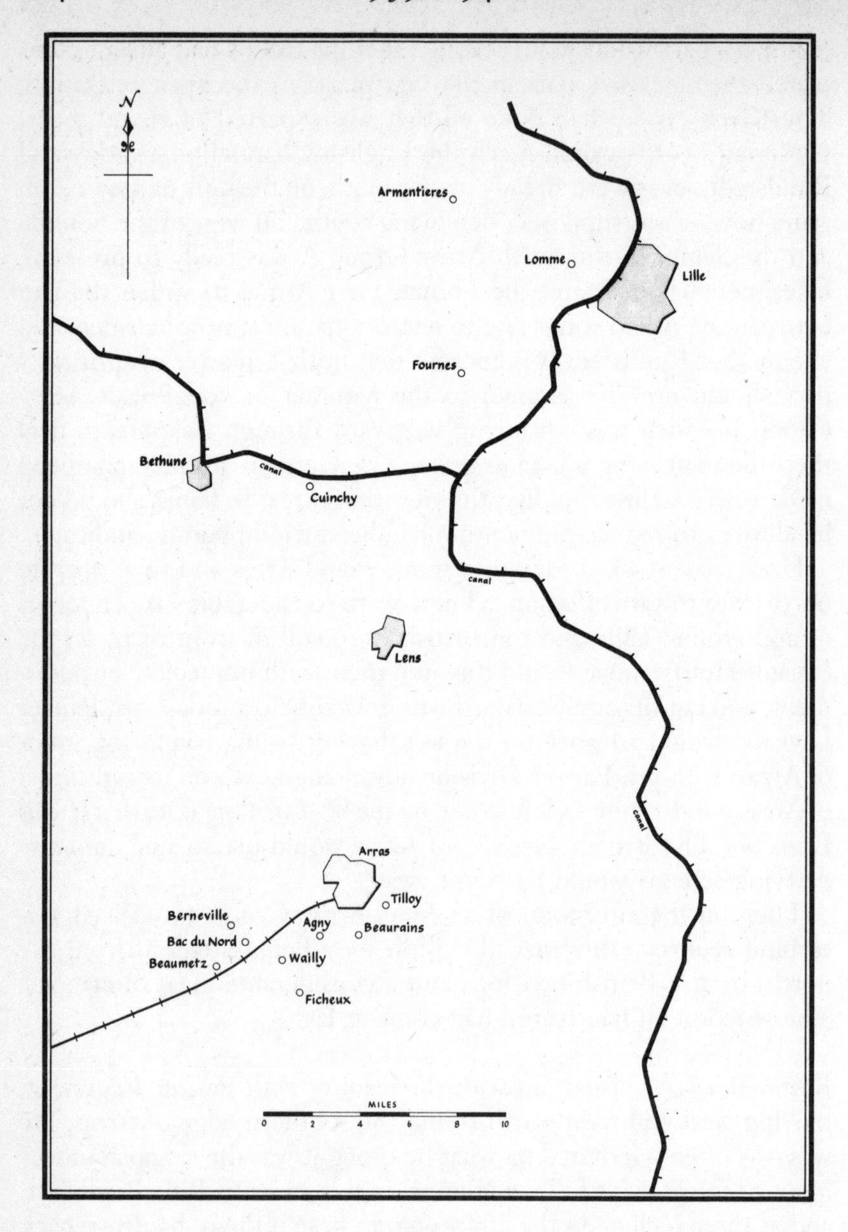

X Area of operations, 7th Panzer Division, Arras–Lille, 21–28 May 1940

ing southward from Arras.[1] Rommel jumped from his vehicle and took to his feet. Infantry were by now everywhere down in position and Rommel, moving behind the line of his anti-tank guns engaging the enemy, reached Wailly. The enemy tank fire had created chaos and confusion among the troops in the village and he tried to re-establish order.[2]

Reunited with his armoured car, Rommel now drove west from Wailly to a higher point where he could see something of what was happening. He found that enemy tanks were advancing from the west, having crossed the railway line that runs from Arras south-west through Beaumetz; and that more enemy tanks were moving south-east, from the direction of Berneville, Bac du Nord. Rommel ran from gun to gun along his anti-tank line, determined to put down enough fire to persuade the heavily armoured British Matilda tanks to halt, and overruling gun commanders' objections that the range was too great. After a little the enemy tanks were either halted, turned or set ablaze. Rommel's artillery commander, pleased with the scoring of his guns, wrote that this was 'the finest success his battalion had enjoyed in the campaign so far'.[3]

Meanwhile Rommel's 6th Rifle Regiment, which had deployed east of where he was himself, in the area of Tilloy, Beaurains and Agny, was also being attacked. Rommel wrote that very powerful armoured forces had attacked southward from Arras and had inflicted heavy losses. This attack, too, was halted by a line of anti-tank guns, as well as the mighty 88-millimetre anti-aircraft gun already used in the anti-tank mode, a weapon soon to be dreaded by the Wehrmacht's enemies. 25th Panzer, which had by now advanced northward leaving the rifle battalions facing the enemy tanks south-west of Arras, was turned round by Rommel at seven o'clock in the evening and directed to move south-east and to attack the right flank of the attacking British. This led to a tank-versus-tank contest in which the Panzer regiment lost nine Panzer IIIs and IVs and a number of light tanks, destroying seven Matildas in the process; a British win on points.

What had happened was that the British Commander-in-Chief, Gort, had been exhorted on 20 May by his government to 'move southward on Amiens, attacking all enemy forces encountered, and to take station on the left of the French army'. A glance at the map shows that to obey such an instruction the British army would have had to break contact with von Bock's Army Group B attacking it from the east, waved goodbye to the Belgian army on its left and the residue of the French First Army on its right, and moved some sixty miles across the front (at that time) of the westward-moving armoured divisions of von Rundstedt's Army Group A. Fortunately for Britain Gort made no concessions to so ludicrous an instruction. He was, however, very aware that the only hope of imposing delay or difficulty on the German advance would be by a flank attack against the penetration, and on 21 May he undertook to set in hand an operation from the north with three divisions, divisions relieved in the east-facing British front by Belgian and French troops for that purpose. This operation was to be coordinated with a major French offensive from the south and could not start before 26 May. In the event it never took place.

Gort, however, had reckoned that if anything effective was to be done it should be done early rather than late. He had, therefore – before an inter-Allied conference at Ypres which had approved the aforesaid major counterattack – already ordered a limited southward offensive, to begin as soon as possible, that was to say on 21 May. For this operation Gort had the two divisions, 5th and 50th, of the GHQ reserve; but he also had the only tanks (apart from light tanks of the reconnaissance regiments) in the British Expeditionary Force, the Matildas of the 1st Army Tank Brigade – seventy-two in number, of which only sixteen carried the two-pounder anti-tank gun while the remainder, albeit heavily armoured, were armed with machine guns only.

Although two British infantry divisions were allotted to this operation they were under strength, with only two brigades each. Of the four brigades available two had been committed to the defence of Arras itself. That left two – and of these one was held in reserve. The Arras counterattack by 'two divisions', therefore, was made by one brigade of three battalions; and in that brigade, again, one battalion was held in reserve and only two advanced. The thrust-point of two divisions was reduced, as far as infantry were concerned, to the assaulting companies of two battalions.

'As far as infantry were concerned'; but, of course, infantry alone would have had no sort of effect. The impact of the Arras counterattack lay in the movement of armour, the first British armour Army Group A had seen. And although only sixteen gun-tanks took part, the appearance of a respectable number of Matildas (achieving a very respectable score indeed in Panzer IIIs and IVs knocked out) was sufficient to send reports flashing upwards through the command echelons of Army Group A that a serious and massive armoured offensive had been mounted against the German northern flank.

The effect in tactical terms was soon over. Rommel had acted with characteristic energy, personally organizing the firing line and by his presence, his example, his shouted commands, reassuring men who had clearly been badly shaken by this sudden and surprising development, this interruption to a triumphant progress. The sense of inevitable victory had already penetrated the German army, and it is a sense which can be vulnerable to disconcerting developments: 7th Panzer Division lost nearly four hundred men that day, at least ninety of them killed – four times the losses of the entire break-in and race across northern France. Rommel lost his aide-de-camp, Lieutenant Most, killed at his side – his previous Adjutant, Major Schraepler, who had been wounded while crossing Leffe weir, now rejoined for duty.*

Rommel, although his impressions of enemy numbers were hugely exaggerated (a facet of every battle and every army), was undisturbed once the counterattack was defeated. He had certainly been caught off balance; a prudent commander would have been ready to defend his rifle regiments at all times against enemy armour striking their flank. That commander would, however, have been less likely than Rommel to cover the distances achieved by 7th Panzer Division; and the appearance of Allied armour, boldly handled, was sufficiently unusual to justify a measure of risk. Hoth's corps was now directed to continue northward, and Rommel wrote home on 23 May that the war in France might be over in a fortnight. It was not a bad forecast.

Although the tactical and local effects of the Arras counterattack were spent by nightfall of 21 May, the psychological effects were more significant. Great decisions were now being made.

On 24 May the northward operations of Army Group A were halted by orders from Hitler's headquarters. This – the 'Halt order' – has often been historically dissected and attributed to many motives, but it seems certain that the nerves of the superior Wehrmacht commanders, including the Army Commander, Kluge, and the Army Group Commander, Rundstedt, were affected by the shock of the operations round Arras and that their anxiety communicated itself to Hitler. Hitler had watched the westward progress of the

* The dangers to which those who served Rommel in a personal capacity were exposed call to mind the story of the peppery eighteenth-century British general who growled at his ADC, when the latter sharply reined to one side to avoid a cannonball, 'Do you flinch, sir? Do you not know that the King of Prussia had three aides killed beside him at Rossbach?' to receive the smooth reply, 'I had not thought that you, sir, could afford so many.'

Wehrmacht with awe and a good deal of astonishment. He readily accepted Rundstedt's proposal that the Panzer divisions should be husbanded for future southward operations, and Rundstedt's view that it would be a waste of effort, a blunting of the tip of the spear, to use it incautiously against the now-surrounded British in the north. The British had shown that they could sting. They could be dealt with at leisure, and to a large extent left to Army Group B and the Luftwaffe. Hitler sanctioned the 'Halt order' with some relief, despite the judgement of OKH that an opportunity was being missed. To the thrusting commanders of the Panzer troops, notably Guderian, the order was disastrous. Rommel – unusually – recorded no particular disappointment. He was sure the war was as good as won, he looked forward to the next phase, a northward attack across the La Bassée canal and completion of the encirclement of Lille, and he was glad to give his division what turned into two days' rest.

On 24 May the Anglo–French counterattack from north and south against Army Group A was abandoned, as to the French movement from the south; and on the following day it was abandoned as to the British movement from the north. It had long been fantasy, as Gort recognized, although he had loyally set preparations in hand under the direction of his III Corps. Equally illusory had been various instructions to Gort from the British War Cabinet (Churchill, who had become Prime Minister that month, had been conferring in Paris), such as one to 'attack south-west towards Bapaume and Cambrai . . . with eight divisions, the Belgian cavalry on the right of the British'. Gort ignored such absurdities. It was becoming very plain to him what had to be done.

The British Expeditionary Force was now facing Army Group A on the line of the La Bassée canal to its south. To the east the garrison of Arras was withdrawn during the night of 23–24 May – it had been at the tip of a long salient since Hoth's corps had swung round the place on the twenty-first, and German troops were already moving up to the canal; while facing east the British were confronting Army Group B on the line of the old 'frontier positions' in front of Lille, positions they had abandoned to advance into Belgium only two weeks previously. Gort, when he withdrew from Arras, knew that there was little the British were going to be able henceforth to do except defend, withdraw, survive and – conceivably – reach England to fight again.

On 25 May, the Germans of Army Group B drove back the Belgian army in a major attack, leaving a gap on the British left; and captured

German plans showed Gort that von Bock was about to commit two German corps towards Ypres, further separating the Belgians from the British. It was this, as much as his accurate reading of the general situation, which led Gort to abandon all thoughts of the southern counterattack, an operation already discarded by the French.

On 26 May further German attacks in the north drove the Belgians yet further from the British left. On that day Hitler rescinded the 'Halt order'. Army Group A's leading divisions were now again to operate northward, to strike into the deep flank of the British army, to cross the La Bassée canal and cut off that army from the sea. Also on 26 May, in the evening, the British Government ordered the evacuation of the army from Dunkirk. That afternoon Rommel's 'National Socialist aide', the brave Lieutenant Hanke, had obeyed a personal order of the Führer and decorated Rommel with the insignia of the Knight's Cross of the Iron Cross.

During the evening of 26 May Rommel's 7th Rifle Regiment managed to push some troops across the La Bassée canal in the area of Cuinchy, and the bridgehead was widened until two rifle battalions were deployed on the northern bank.

Rommel, as ever, took personal command. In the early morning of 27 May he found British snipers active at the crossing point. He also found that the rifle battalions had not established a deep or defensible bridgehead, nor managed to get anti-tank weapons across. It was, in some sort, a repetition of the situation Rommel had initially found at Dinant a fortnight previously. As on that occasion he ordered the building of a bridge sufficiently strong to take tanks. As at Dinant he himself directed fire – in this case from a tank and some light anti-aircraft guns – against places thought to hold enemy snipers, ultimately silencing them; standing in full view of the enemy on a railway embankment, pointing out targets, careless of danger. And, not unlike at Dinant, he directed fire at every house and bush within hundreds of yards of the bridgehead, sweeping the ground with shot and shell, ordering a Panzer IV (again, with an echo of Dinant and the Meuse crossing) to move along the south bank engaging British opportunity targets, including Matilda tanks, north of the canal. Soon he had field guns, anti-tank guns, 88-millimetre pieces and tanks across to the north bank. The 5th Panzer Brigade, with its two regiments and four tank battalions, (an organic part of Rommel's neighbour, 5th Panzer Division), was transferred to Rommel's command at midday.

Thenceforth Rommel's northward movement met a good deal of opposition and went slower than he hoped. The hours of daylight – and there are always too few of them in war – slipped away. Frustratingly, Rommel found that his ability to convey personal orders to the substantial mass of armour he now commanded was inhibited by the difficulty of cross-country movement in the dark. His own 25th Panzer Regiment was now at Fournes, north of La Bassée and only five miles from the western outskirts of Lille. Rommel's objective was to attack northward and reach Lomme, in the north-western suburbs of Lille itself. A blocking position at Lomme would cut the principal remaining road westward out of Lille, the road to Armentières. To his west the divisions of Panzer Group Kleist were now moving north, taking the Channel ports as they could.

Colonel Rothenburg asked whether Rommel would be accompanying the attack, and received an uncharacteristic 'no'. Rommel remembered the way 25th Panzer had been isolated and surrounded without supplies east of Le Cateau, and was determined on this occasion to make sure that the Panzer regiment would be supplied, and reinforced as soon as possible by the rest of the division. 5th Panzer Brigade was several miles to the south-east, radio contact proved impossible, and when night fell Rommel's seven tank battalions were disunited, but in various positions west, south-west and south of the city of Lille. He had now, as he intended and supposed, almost cut off the – mostly French – Allied troops in Lille.

Rothenburg reported 25th Panzer in a hedgehog position at Lomme just before two o'clock in the morning of 28 May. He had crossed the La Bassée canal against opposition and advanced in what he described as heavy and costly fighting – a remarkable tribute to the French and to Gort's south-facing forces which by now consisted, in all, of one Regular and two Territorial divisions, exhausted but holding a fifty-mile front from Lille to St Omer and along the Aa canal to the sea. On this day the Belgian army, on the left or north of the BEF, surrendered. They had been pressed very hard, and the British decision to evacuate, taken on the previous day, had not been communicated to them. A further and futile fight to buy more time for the evacuation of others was unappealing.

To the Germans, however, the weakness of their enemy was less apparent. Rommel set out in the darkness to lead Rothenburg's supply column, together with the divisional reconnaissance battalion, through to Lomme, and reached 25th Panzer Regiment's 'hedgehog' just before dawn. Rothenburg reported that he had clashed with 'enemy tanks and a strong motorized force' and shot them out of his way. With the dawn remnants of the French divisions in Lille made a number of attempts to break out to the west, using tanks and under cover of artillery support. Rommel's first actions after daylight on 28 May were to organize a strong east-facing defence, to keep the besieged Allied garrison of Lille hemmed in.* That night the British divisions which had faced Rommel on the La Bassée canal and then withdrawn northward were ordered back to the line of the Yser. The British were now manning a perimeter, intended to cover their last remaining port, Dunkirk; and 7th Panzer Division were given what proved to be six days of much needed rest.

The Allied evacuation from Dunkirk was finished by 3 June. The remarkable total of 337,000 troops (two thirds British and one third French) had been embarked.

In the campaign so far Rommel's division had taken just under seven thousand prisoners of war, had captured a considerable number of tanks and claimed to have knocked out over three hundred, including eighteen French 'heavies'. A copy of 7th Panzer Division's report was sent to Hitler; and certainly from this time forward Rommel's reputation for a certain egotism and self-advertisement increased among some of the Wehrmacht's senior officers. He had a reputation, too, for possessing Hitler's favour, and so far that reputation was deserved. Rommel wrote in glowing terms of Hitler's visit to his victorious troops on 3 June, and recorded that he was invited to accompany Hitler for the rest of that day – the only divisional commander who did so.

Rommel was supremely unfussed by what he would have regarded as the jealousy of others; and he was perfectly aware that his genuine regard for the Führer was not shared by all. Rommel, of course, felt considerable personal gratitude to Hitler for, as he was sure had happened, giving him his chance to win his spurs as a general, a commander in mobile operations – more mobile and more adventurous than Hitler himself had anticipated. 'We were all very worried about you!'[4] Hitler had remarked on his visit, referring to the thrusting exploits of the '*Gespensterdivision*', the 'Ghost Division', as 7th Panzer had been named by commentators both friendly and hostile,

* The gallant French defenders of Lille eventually surrendered with the honours of war, marching past the German commander, with their arms, on 1 June.

a formation so elusive, so hard to pin down whether by an enemy or on staff maps, so unpredictable.

'We were all very worried about you.' The words conveyed affection as well as admiration. Both were reciprocated. Rommel realized that Hitler was an essentially intuitive rather than reasoning creature, that he felt himself inspired, that his judgements appeared to him as prophetic revelations as much as statesmanlike pronouncements. And while to many in the higher ranks of the Wehrmacht these characteristics aroused mistrust and alarm, to Rommel, who had felt the Führer's personal magnetism, Hitler's prophetic and intuitive quality had to be judged by events. And events were now, more than ever, flattering to Hitler's judgement. A biographer has written that Hitler's ideas aimed to 'burst the bounds of tradition and lead to some sort of biological Utopia'.[5] Rommel had neither knowledge of nor taste for such visions. To him Hitler was his Führer and Commander-in-Chief, who had restored the morale of the German people and their army, and who had presided over the fortunes of both in a brilliantly successful campaign waged in the course of a just war, with every hope of culminating in an enduring peace. It was as simple as that.

For the campaign had so far indeed been astonishing, a three-week wonder. The great French army, the victors of 1918, had been either trapped in Belgium or sent scurrying beyond Somme and Seine, having suffered stupendous losses, clearly broken in morale. The British army had been hustled to the sea, driven from the Continent with its tail between its legs. The Wehrmacht had avenged Versailles, avenged the humiliations of Compiègne. There was another act to be played out before the curtain fell, but nobody doubted that it would be played successfully. And none of this would have happened, Germans said with either enthusiasm or reluctance, depending on their disposition, except for the will, the genius, the courage of Adolf Hitler. Rommel, with most of the German people, acknowledged it with gratitude.

Like many of his contemporaries in the Wehrmacht, Rommel felt little animosity towards the French, and even less towards the British. French bystanders now waved cheerfully to his troops,[6] and he was sure – all sense and history dictated it – that a good peace demanded friendship and cooperation between Germany and France; while it seemed evident that, faced with such a peace, with such a reconciliation, Britain would have no motive in continuing a pointless war.

Rommel had no knowledge of Hitler's wider intentions. In fact

Hitler had made clear to his senior generals as early as May 1939 that the object of German policy in the east had little or nothing to do with the Polish Corridor, with Danzig. Danzig was a pretext; the object was to win *Lebensraum* and food supplies by conquest. And in this victorious June of 1940 Hitler was already telling his confidants of his next and most far-reaching plans.[7]

Meanwhile Hitler was riding high. His triumphs not only dazzled the loyal and essentially apolitical like Erwin Rommel – in June 1940 they dazzled most of the German people, and indeed the world. Neutral nations tended to discover within themselves a certain coolness towards Britain while Britain appeared recalcitrant. Within Germany those circles which had been repelled by Nazi rhetoric and practice found themselves diminished and isolated: reactionary irreconcilables, or so it seemed, churlishly incapable of recognizing genius and success. Some among those irreconcilables were men and women with judgements based on absolutes of right and wrong, uninfluenced by the tides of victory. Dietrich Bonhoeffer actually described Hitler as Anti-Christ in 1940, at the moment of his highest triumph,[8] and Helmuth von Moltke wrote to Peter Yorck von Wartenburg on 17 June: 'We must today reckon with having to live through a triumph of evil.'[9] All three were later to die horribly but heroically by the executioner's hand. But for most honest, patriotic Germans the sun was shining and the night was far away: Rommel was certainly of their number.

It now remained to settle matters militarily with the French. The remaining French divisions, about forty in all, were deployed on the general line of the Somme and the Aisne. On 27 and 28 May two minor offensives had taken place from the south against the German forces arrayed between the Ardennes and the sea in a corridor now being filled by the marching infantry divisions. The first of these attempts, on the twenty-seventh, had been made by two French Colonial divisions, towards Amiens. The second, on the twenty-eighth, was made by a French armoured group (based on the French 4th Armoured Division, led by Charles de Gaulle). Neither sortie had made more than a temporary impression on the Wehrmacht.

In the first attack, however, the two French divisions had been supported by the light and cruiser tanks of the British 1st Armoured Division, which had only begun to land in France a few days before. The chain of command above it was confused – it had landed at Cherbourg and there could, therefore, be no question of its uniting with the main body of the British Expeditionary Force north of the

German penetration. On 27 May, under orders of the French Seventh Army, the tanks of this division had found themselves attacking alone, rather than supporting infantry (for which their tanks were anyway unsuitable, but the task required a combined arms attack), and attacking without artillery support against well deployed German infantry and anti-tank guns. By the end of the day they had lost sixty-five tanks: and by the end of 29 May both offensive attempts had expired and the French were everywhere facing north on the defensive.

1st Armoured was not the only British division south of the German penetration. 51st Division, largely composed of Highland troops, both Regular and Territorial, had been serving under French command on the Saar front of the Maginot line, sent there to gain experience as had been the agreed British practice during the period of inactivity before 10 May. As *Sichelschnitt* gained momentum the division was withdrawn, to join the BEF. By then, of course, such a move was impossible.

51st Division had then been moved to the west of the French Tenth Army front, south of the Somme, and on 4 June the divisional commander, General Fortune, was allotted two French divisions besides his own and ordered by the French IX Corps commander to attack some of the German bridgeheads across the Somme, south of Abbeville. The attacks met a certain amount of success and many of the division's objectives were taken.

Next day, however, the German divisions facing the Somme passed to the offensive. On 7 June 51st Division, deployed on the river Bresle twelve miles south of the Somme, had the disagreeable sensation, familiar to the rest of the BEF, of German mobile forces threatening to drive deep past their right flank. For, twelve miles south-east of Abbeville, General Hoth was already starting yet another great operational march.

There had been a regrouping of forces. Hoth's corps, still with Rommel's 7th and von Hartlieb's 5th Panzer Divisions, was now one of three Panzer corps in von Bock's Army Group B. After Dunkirk had marked the end of fighting in the north, Army Group B had deployed along the Somme, on the right of the German south-facing forces, while on its left, facing the Aisne, was von Rundstedt's Army Group A. On 5 June, at 4.30 a.m., Hoth's right-hand division, 7th Panzer, crossed the Somme canal between Abbeville and Amiens, seizing two railway bridges which engineers quickly began converting to take road traffic. The first vehicle across, by his personal order, was the divisional commander's armoured command vehicle. Rommel was now across the Somme.

The first day's fighting, the break-in to a potentially strong French position, showed Rommel at his more deliberate. Undoubtedly he was conscious that in the preceding weeks, successful though he and his division had been and strongly though he would have defended his personal initiatives, his tireless command 'from the point of the spear', there had been times when the division's coordination, balance, and general handiness – to say nothing of its internal resupply – had suffered from the feverish pace at which he had led it. This, Rommel was convinced then and later, had been abundantly justified: but there might be situations in which a more controlled advance, while not so dramatic in terms of mileage achieved by the spearhead, might produce a steadier rate of progress, no less impressive overall.

Rommel accordingly directed the division with as much aplomb as hitherto, but keeping a strong grip on his whole command, flanks and rear as well as front. He had prepared his attack with a considerable artillery and machine-gun concentration on the bridges, and was rewarded by their seizure intact. He had made the canal crossing with two battalions of 6th Rifle Regiment, and when his first bridge could carry traffic he brought 25th Panzer Regiment across with orders to move to a hill north of Quesnoy for an attack on it; while one Panzer battalion was to harass the village of Hangest, clearly held strongly (as was Quesnoy) by the French. Throughout the day French shelling was heavy and French infantry positions and anti-tank gun nests were manned with determination. A new system had been hurriedly decreed by the French command, and although undermanned and inadequately prepared it was sound. There was emphasis on depth rather than linear defence, on strongpoints based on woods and villages with all-round anti-tank defence (the 75-millimetre artillery piece was ordered to be used in the anti-tank mode), with gaps accepted between strongpoints. Rommel's command vehicle itself was plastered with fire from Hangest, one such strongpoint. It was perfectly clear that 7th Panzer was not dealing with a beaten enemy; or not yet. The weight of French artillery fire was affecting German *élan* and morale; and Rommel noted the fact.

In mid-afternoon Rommel gave out verbal orders for deliberate attack, for what he hoped would be breakthrough. At four o'clock 25th Panzer Regiment would attack Quesnoy, moving round it from the north and smothering it with intensive fire. Close behind and

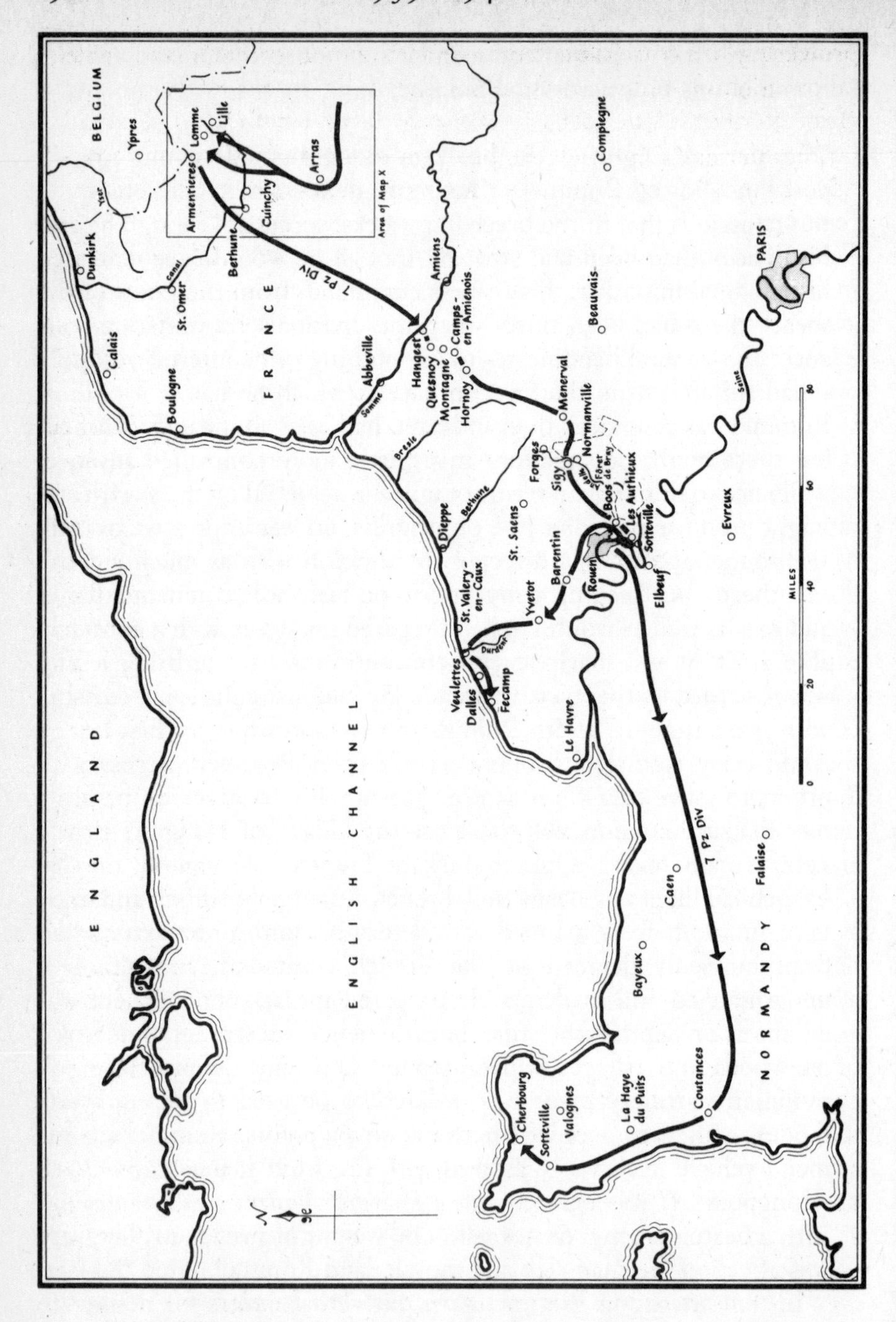

XI Western Front, 1940. The advance of 7th Panzer Division to the English Channel and into Normandy

protecting the rear would move the armoured reconnaissance battalion. The battalions of 7th Rifle Regiment would follow up and clear the village. Once clear of Quesnoy the division's advance would be resumed on a thrust line Montagne–Camps-en-Amienois/Hornoy. Rommel himself moved immediately in rear of the Panzer regiment.

Rommel was pleased with the upshot. He congratulated himself that movement and coordination had been 'as on an exercise'. Communications had worked admirably. There may have been in this a tiny, half-conscious implication that earlier, perhaps avoidable, perhaps even Rommel-induced, shortcomings had been overcome.

The French had fought fiercely (although a number of the prisoners taken were drunk). 7th Panzer was stopped by corps order at Montagne – divebombers had been allocated to support the attack, and to move south of Montagne risked confusion and assault by the Luftwaffe. Rommel's advance, following a successful assault crossing of the Somme canal and against a brave enemy well-deployed and supported by plenty of artillery, had been less sensational than some of his earlier exploits, but it had prised open the French Somme front and the ball was now at his feet. There had been strong French reactions, including counterattack by tanks, frustrated by the fire of the 88s. More and more these pieces featured well forward in Rommel's order of march. Towed and unprotected, they were of course vulnerable, and next morning Rommel found a number had been knocked out by French artillery fire.

At nine o'clock that morning – 6 June – Rommel gave out orders at the headquarters of 25th Panzer Regiment. In the next two days he put into effect something he had rehearsed before *Sichelschnitt* started; a *Flachenmarsch* (literally 'area march'). The division was formed up on a two-thousand-yard front – occupying a depth of twelve miles. This huge rectangular 'box' then moved across country, ordered to avoid villages and main roads, firing on the move to front or flank where places or woods might harbour enemy and be designated strongpoints, ready to react to a new situation or to fight an unexpected battle with no element of the division at any great distance from the rest.[10] The advance began at ten o'clock.

A *Flachenmarsch* was clearly only a possible manoeuvre in open country, and the area between the Somme and the Seine offers a good deal of such. Many of Rommel's vehicles – as in all armoured divisions in all armies – were, of course, wheeled and with an indifferent cross-country performance if any. These were ordered to use the tracks, hedge-gaps and so forth made by the Panzers.

On 6 June the division travelled by *Flachenmarsch* some thirteen miles; and by the same method on 7 June about sixteen. Such distances do not compare to the dramatic advances which Rommel had achieved on certain days in May, and would again. They do not appear as great or rapid operational movements, but it must be remembered that the whole division and not simply the head was moving, and moving through country in which French positions were known to be sited in depth and organized for all-round defence. By five-thirty in the afternoon of 7 June Rommel had his division in the area of Ménerval, over forty miles beyond the Somme, with reconnaissance troops pushed forward to the narrow and shallow river Andelle and having cut the Paris–Dieppe road in the Forêt de Bray.

The eyes of Hoth's corps were now on the Seine. On 8 June Rommel discovered a ford over the Andelle at Sigy and started to push troops across. Shortly afterwards reconnaissance troops found an unblown bridge at Normanville, a small village south of Sigy, and Rommel switched the divisional centre line to that place and set out south-west in the general direction of Rouen.

Rommel planned to send a force to a crossroads a few miles east of Rouen from which a powerful fire demonstration could be opened towards Rouen itself. Covered by this, and (he hoped) deceiving the city's defenders into expecting a frontal assault from the east (Rouen lies on both banks in a great loop of the Seine), he intended to move south and south-west to the river valley and seize one or more bridges in the area of Elbeuf, a town on the next loop immediately south of Rouen. 7th Panzer would then be south of the river.

It was not to be; and since a new task was about to be given to Rommel's division the fact that operations on 8 June went less happily than he hoped worked ultimately to the German advantage. The country where the Andelle runs into the Seine is wooded and broken with a large number of small villages; certainly not ground for a *Flachenmarsch*. Rommel sent one Panzer company with a strong force of field guns and 88s towards Rouen, moving by the narrow roads, skirting the woods. On the way this column ran into parties of British moving south across their front. These were elements of the 1st Armoured Division which, after its doomed counter-offensive attempt near Abbeville, had been ordered by the new French commander-in-chief, General Weygand* himself, to face east and hold the Andelle, an equally doomed hope. Now, together with some hastily-scratched-together battalions from the base organization, they were doing their best, but had, on 8 June, been ordered to make their way south of the Seine, thus crossing Rommel's path.

Some alarms and encounters took place, and the progress of Rommel's Rouen crossroads task force was correspondingly slow. Enemy prisoners of war were appearing on all sides, confused, fearful, necessitating organization and causing delay. It was nightfall before the task force reached its destination, and even later before the artillery were able to deploy for Rommel's planned demonstration towards Rouen. Meanwhile he set the remainder of 25th Panzer Regiment on the march towards the Seine at Sotteville, driving by Boos and Les Authieux and himself riding immediately behind the Panzers. The French inhabitants of the country, where they appeared, were confused and terrified. The friendly waves of other places were absent. A woman took Rommel's arm and asked if he were British; she was shaken by his reply.

At two o'clock in the morning of 9 June, Rommel was with the head of the division in the valley of the Seine, on the north bank of the river. He had sent his motorcycle battalion westward along the valley in the darkness, with orders to seize the bridges at Elbeuf. His radio communications with the rest of his command had failed – a factor caused by night and the close country of the Seine valley. He reckoned that within two hours it would be dawn, and he had no desire that dawn should discover 7th Panzer Division strung out along the Seine, hemmed in the valley. He needed a bridge, to get troops across on to the high ground south of the river. Failing that, he would need to pull them back well north of the river – anywhere except trapped in the valley itself, with French artillery ranged upon it. Rommel, anxious and impatient, set off for Elbeuf to find out what was happening. He was uneasily aware that he had little idea where most of his division by now were.

In Elbeuf Rommel found total confusion, the streets crowded with men and vehicles. All that was clear was that the storming parties of the motorcycle battalion ordered to seize the bridges had not yet made the attempt. Rommel, on foot and furious, gave orders for immediate action. There had been a crush of French military and

* Weygand, said to have been named by Foch as the man whom France should summon in an hour of need, had relieved the shattered Gamelin.

civil traffic moving through the town and it was clear to Rommel that in this extraordinary situation the officer commanding had shown insufficient energy and ruthlessness. He himself got the assault moving just before three o'clock in the morning – to be rewarded a few minutes later by the blowing of both bridges by the enemy, and by the sound of explosions up and down river which indicated similar French action to east and west. In Hoth's sector, at least, the Germans had been checked on the Seine.

Rommel decided that the division must tidy up the area through which, well-scattered, they had driven in darkness. There could be no crossing of the Seine yet. Next day, 10 June, 5th Panzer Division took Rouen and Rommel received fresh orders.

The British 51st Division, under General Fortune, after its attempts to counterattack under French command near Abbeville, and its subsequent withdrawal to the Bresle river, had been retained by the French IX Corps on that river until 8 June. By that day Rommel was already approaching Rouen, many miles behind 51st Division's right shoulder. And on that day Fortune was ordered to withdraw his division south of the Seine – a marching distance of fifty miles, planned to take four days.

The progress of Army Group B clearly made this movement impossible. Instead 51st Division and the remaining French divisions of General Ihler's IX Corps withdrew westward parallel to the coast with orders to reach Le Havre. Fortune sent two brigades – one of his own and one from the improvised base and line of communication 'Beauman Division' – to Le Havre itself, and ordered the remainder to withdraw towards them, with intermediate positions on the small rivers Bethune and Durdent. The main threat, however, came now not from the front but from flank and rear. The Bethune runs into the Channel at Dieppe, and 51st Division reached it: but the Durdent meets the sea twenty-five miles on, near Veulettes, a small place about six miles west of St Valéry-en-Caux and still forty miles from Le Havre. At nightfall on 10 June Fortune learned that Veulettes was in German hands. Some of the exhausted divisions of IX Corps, French and 51st Highland alike, moved towards St Valéry where the small harbour offered some slender hope of embarkation and rescue.

For the Franco–British withdrawal towards Le Havre had been anticipated, and Hoth had been ordered to turn away from the Seine and drive north to cut it off. 7th Panzer was to move on Le Havre itself, and by 7.30 in the morning of 10 June Rommel, with a somewhat widely dispersed division behind him, had driven into Barentin, ten miles north-west of Rouen. He had already assembled 25th Panzer a few miles south of that area and had ordered the reconnaissance battalion to reach Yvetot, twelve miles further north, and to explore towards the sea as soon as possible. 'Establish yourself,' Rommel had told the battalion commander, pointing out a distant objective on his map, 'until I move with the tanks. Don't look to left or right, only forward all the time. If you get into difficulties let me know.'[11] Two hours later he had brought 25th Panzer up to Yvetot; and then he learned that a 'strong enemy force' was moving west along the axis of the main road from St Saens.

This, of course, was part of that general Allied retreat westward from the Somme of which 51st Division was by now the northernmost element. Rommel himself organized and brought up anti-aircraft guns – light, as well as 88-millimetre – deployed facing east and covering Yvetot. He had then driven the Panzer regiment and the reconnaissance battalion northward as fast as possible, driving two abreast on the road itself where necessary, and on a track parallel to the road where practicable. There were by now enemy vehicles on every east–west running road, vehicles carrying French and a few British troops aiming to reach Fécamp (ten miles further west) and the possibility of evacuation. Everywhere vehicles were being destroyed, and from all directions scattered parties of prisoners were coming in. Rommel pressed on and pressed his leading troops on. He, with his command vehicles, reached the sea at a little place called Dalles, about halfway between Fécamp and Veulettes and some ten miles west of St Valéry-en-Caux. Fortune's information was accurate. The trap was closed. No Allied troops would reach Le Havre. No Allied troops would reach Fécamp.

Amid general confusion, but in a mood of immense elation shared with every man in his division, Rommel drove west to Fécamp. The sight of the sea, the sense of being at last at the rim of Europe, of reaching one sector of the campaign's ultimate culmination – this was intoxicating indeed. Rommel, since forcing the Somme canal, had been on the move almost without pause, making little distinction between day and night. He sometimes observed, almost with surprise, that a particular operation or move had been laborious or gone undesirably slowly because of darkness. He sometimes remarked, with weary resignation, that night had brought particular difficulties of radio communication (as it always did). He brushed these problems away. As always he drove himself and he drove his command, regardless of light or circumstance, while there was still strength in bodies, ammunition available, and fuel still in vehicle tanks.

Rommel knew that after the last four weeks the men of his division were extremely tired, that their reactions were not as sharp as they had first been, that there was a certain sense of achievement, of merited relaxation expected. He would have none of this. He himself was full of the joy of victory, and he was as tired as any, but his temper was sharp and the slightest hint of slackness or irresolution roused him as it always had and always would. As he drove east again, after a brief look at the southern outskirts of Fécamp, he took station behind the leading three tanks of one of the Panzer battalions, now ordered to turn and move along the coast towards St Valéry. A lone French anti-tank gun opened fire and hit the leading tank. The tank commander abandoned tank and the next two tanks, failing to return fire, drove off the road, leaving Rommel's command vehicle in the direct line of fire. Several rounds were fired at it but missed. Rommel, dismounting, organized return fire from the two undamaged tanks, which silenced the French gun. He next told the leading tank commander exactly what he thought of his performance and that of his command, then drove on into Veulettes.[12]

By 11 June St Valéry was packed with French and British vehicles and soldiers, and a Franco–British defensive position had been organized on the high ground west of the little town. Rommel edged 25th Panzer Regiment forward where it could be done, and eventually managed to get some tanks to a lodgement on ground from which they could see and fire at the port itself, thus halting any attempts at embarkation. That afternoon Rommel sent an envoy under white flag to summon the garrison to surrender. The summons was refused. Intensive fire was then opened by the Panzers and 7th Division's artillery, and Rommel's infantry of the rifle regiments were deployed on the hills above the town. Still no surrender.

Fire on every visible target in St Valéry was kept up all night, while the Panzers were withdrawn and replenished. Early next day, 12 June, Rommel brought his armour back into position, hemming the town and inching into it where possible. He brought his own command vehicle into a house on the north-west outskirts and, dismounting, moved cautiously on foot with some of 25th Panzer into the town itself.

St Valéry was a mass of fires. Eventually Rommel and the leading tanks reached the harbour, where earlier that morning, from the heights, a transport had been spied attempting to take on men, and

Rommel had personally directed fire at it, first (unsuccessfully) from an 88-millimetre and then, with success, by artillery. Now there were no transports and no hope of them. Prisoners of war were coming in on all sides, and a French general, approaching Rommel, identified himself as General Ihler, commander of IX Corps. He had already ordered Fortune to surrender, an order the latter had for some time refused to obey; but with French troops surrendering on all sides, with an earlier counterattack by his Highlanders masked by French soldiers carrying white flags, Fortune now had no alternative to ordering a ceasefire. North and east of the Seine the war was over; at least for four years. In Fécamp that evening, Rommel's divisional band gave a concert on the promenade.

There was one last brief chapter to be written before Rommel's part in the campaign of summer 1940 was ended. On 13 June, the day after the fall of St Valéry, the British General Brooke arrived at Cherbourg. Commander of II Corps, he had been evacuated with his troops from Dunkirk and was now directed by the British War Cabinet to return to France and organize a new British Expeditionary Force. There were still a considerable number of British troops in France, south of the Seine, troops from the base and the lines of communication who had not yet been overrun by the Wehrmacht. Another division – 52nd – had been embarked in England for France and its leading brigade had already arrived at Cherbourg and been placed under French orders. A Canadian division was planned. More would follow.

On 14 June Brooke consulted with the French generalissimo, Weygand, and with General Georges, Commander-in-Chief of the group of armies facing the German onslaught. There was discussion of an allegedly agreed (between the Allies) strategy of holding a 'redoubt' in Brittany – a defensive line across the neck of the Brittany peninsula. Brooke remarked that at least fifteen divisions would be required for this, and asked where they were. He, at that moment, had one. The French reply told him all. The Brittany plan, Weygand said, was absurd. An armistice was absolutely necessary and Weygand had (although he did not say so) already advised the French Government accordingly.

Brooke was grimly certain of what needed to be done. That evening he spoke by telephone to the Chief of the Imperial General Staff in London, and to Churchill himself. There must be no delay, no question of reinforcement, and no British troops left under

French command. There must, as swiftly as possible, be another evacuation; for the most part from the ports of St Nazaire and Cherbourg. With a good deal of difficulty Brooke carried his point. He sailed from St Nazaire on 18 June, and at four o'clock in the afternoon of the same day the last British ship left the harbour of Cherbourg. By that time German artillery was already ranged on the heights around the town. On the previous day the French Head of Government, Marshal Pétain, recalled to save France in its hour of trial, had broadcast to the French nation in terms which made perfectly clear that further serious fighting was inconceivable: the broadcast was widely regarded in France as a call for an armistice.

The Germans had crossed the Seine without difficulty or opposition. German infantry divisions were now marching south towards the Loire; and on 17 June 7th Panzer Division, brought down from its success at St Valéry, was ordered across the Seine, to drive west as fast as possible and then to swing north and take the great port of Cherbourg. A prime strategic object would be to cut off from evacuation any British (or French) troops minded to escape to England.

Rommel, besides his own division, now had under command a motorized brigade commanded by General von Senger, another very independently-minded commander. On 17 June Rommel's division had little fighting, but covered a distance of 150 miles – driving through daylight and darkness, a remarkable achievement. More French troops appeared and were ordered to lay down their arms. Only towards the end of that exhausting and joyous drive, when the head of the division had reached Coutances and was swinging north towards Cherbourg itself, did it come under fire – heavy artillery and machine-gun fire, north of La Haye du Puits. Rommel's men had run into part of the covering force deployed to enable evacuation from Cherbourg to proceed. It was already midnight.

It was clear that if there was an armistice – and Rommel had had no official word of it – the defending troops were unaware. Rommel sent word that unless they surrendered by eight o'clock that morning – 18 June – he would attack. By eight o'clock the defensive positions had been abandoned and the advance towards Cherbourg continued.

Rommel had no intention of taking chances, of losing men by motoring into enemy fire on the assumption that a war was over which was still in progress. On the road north the head of the division met a roadblock, and French fire. Soon, and typically, the divisional commander was returning fire from his own armoured command vehicle and ordering the leading platoon into action.

Rommel was now leading the division with 6th Rifle Regiment (Colonel von Unger). He gave orders for a considerable artillery barrage to be fired on what he assessed were the French positions covering Cherbourg, and he was soon to be glad of it. As the Germans advanced, uncertain of what would meet them, during 18 June they encountered a considerable weight of shelling from some of the Cherbourg forts. The town was ringed by forts, and within the forts were emplaced guns. Rommel's plan was to reach the coast west of Cherbourg itself, to maintain artillery fire on the heights around it, and to push 7th Rifle Regiment from the west into the town. He decided that this should not be done until next day – the division was still strung back over the two hundred miles of road which lay between himself and the Seine, and he needed his artillery, his tanks and both rifle regiments as well as the Senger Brigade for what might be a major battle.

By nightfall on 18 June Rommel had his artillery (with the exception of the slow-moving heavy batteries) in position where he wanted it, covering the approaches to Cherbourg; and most of his division were up and in hand. He took a few hours' sleep at divisional headquarters, established in the Château de Sotteville, which happily turned out to be the residence of the Commandant of the Port and fortress of Cherbourg, and to contain in a drawer plans of the Cherbourg forts. Early on 19 June, Rommel made his way forward to join 7th Rifle Regiment, now moving towards the western outskirts of Cherbourg. French artillery fire was still coming from at least one of the forts, and Rommel, from 7th Panzer Regiment command post, enjoyed himself personally directing against it the fire of one of the German guns, and silencing it. He was still anxious that there should be no unnecessary casualties arising from complacency, and soundly rated a machine-gun platoon (and its commander) which he found lying about and doing nothing, idle and vulnerable, as he moved forward on foot to join 7th Regiment. He knew very well how easy it is for troops to relax at the hour of victory and to suffer thereby.

But the end was very near, and at midday a negotiating process started. Two French civilians, persons of some responsibility, had appeared, and Rommel had told them to seek the French commander with a message to surrender by quarter past one in the afternoon. The attack would be resumed if there were no answer.

There was no answer.

At exactly that hour, therefore, German divebombers attacked and Rommel's artillery resumed firing, with especially heavy fire directed at the dockyards, where numerous fires were started. First into the town was the Senger Brigade, which had moved round to the east and attacked from that quarter. Soon resistance ceased. A formal surrender document was signed at five o'clock, with Rommel expressing satisfaction that there had been no blood shed by the civilian population. Particular care was taken that there should be no indiscipline or looting.

Germany now regarded itself as virtually at peace with France, the campaign over, terms to be negotiated. An armistice was concluded on 22 June. Rommel's 7th Panzer Division had lost 682 men killed, 1646 wounded and 296 missing. His score of enemy prisoners and equipment taken was enormous. As always in the aftermath of fighting he busied himself with correspondence, with condolence, with recommendations for decorations – done at considerable length and with exactitude and a personal touch. Rommel was as generous in commendation as he was unsparing in criticism. The highest awards were not scattered liberally in the Wehrmacht, but both Rothenburg and Bismarck were awarded the Knight's Cross, as were several others.

Rommel now had a great name. His exploits and that of the *Gespensterdivision* – for the name stuck – were the stuff of excellent publicity, and Rommel did not find it disagreeable. He had kept a good personal photographic record of the campaign, and he took painstaking interest in the accounts of his division's part in it which soon appeared in print, commenting scrupulously on points of detail (and style), making sure that justice was, as he saw it, done but deploring any vulgarity of advertisement. The verb 'to Rommel' had been coined, although Rommel firmly said that it had certainly never found currency in his division![13] But he took great and justified pride in his command.

His methods continued to be contentious. All German commanders of mobile troops believed in leading from the front – commanding 'from the saddle', as Rommel himself said, in the manner of Seydlitz or Ziethen, in the spirit of the great cavalry commanders of history. But Rommel's technique raised eyebrows on two especial counts. First – and unimportantly albeit significantly because so consonant with his character – Rommel ignored regulations and irritated his superiors thereby. As an oft-quoted example, the German system of marking routes was laid down, and individual variants on it were forbidden: such private enterprise could cause confusion and route marking, above all things, demanded clarity. Rommel, disregarding this, ordered 7th Panzer's route to be invariably marked 'DG7' – *Durchgang* 7. This was irregular and it rankled. It is not always easy for staffs, loyal to a commander but also loyal as they must be to a system, to serve such a man. Rommel was an individualist, his opponents said an egomaniac and ungenerous about others (as some were to say of Montgomery), a commander with absolute faith in himself and his guiding star, of iron will. Such men can awake the indignation of superiors, especially superiors with more pedestrian minds. General Halder, Chief of the German General Staff and the epitome of such, was to refer to Rommel, on one of his later exploits regardless of authority, as 'This General gone raving mad!'

The second criticism of Rommel's methods has already been often noted in this narrative. He was and remained absolutely convinced that in mobile operations a commander must influence decision at the critical point. In an advance this is at the tip of the spear. Again and again Rommel was to be found with the leading tank, the leading platoon – certainly with the leading company commander. A division is a huge organization, with different components to be brought into harmonious play for a well-orchestrated battle, a well-balanced operation. An orchestra needs a conductor; needs him on a rostrum and not in the orchestra pit. Furthermore the initiative of subordinate commanders (by which Rommel set explicit store) could be stunted rather than encouraged by the sense that the General himself would at any moment take personal command of some tactical battle, perhaps on the smallest of scales, rather than do his own job. 'Don't keep a dog and bark yourself' – the injunctions against interference with subordinates are legion and normally justified.

Rommel recognized this. He would have rationalized it – he acted instinctively, but it was thoughtfully schooled instinct, and he always examined his own actions retrospectively with a rational eye – by saying that, in a fluid battle situation, near superhuman energy is required at particular places and particular moments; and that the function of the commander is to supply that energy where it is even temporarily defective, regardless of formalities. Again and again we find Rommel, visible, vulnerable, obvious to all with his gleaming boots and red patches at the throat, standing on a bank shouting, a target for any enemy marksman; managing to galvanize the builders of a bridge into building it more quickly; directing fire like a corporal and somehow encouraging men who were waiting, perhaps cowering, to move, react, shoot; guiding a Panzer column to a different route and position because he, alone, has spotted the sense of it; even leading an essential supply column to where it is needed. Again and again we get the inescapable feeling that, but for Rommel, matters would have gone a little more slowly, a little less decisively, even a little awry; that, but for Rommel, a chance would have been missed. Rommel was the least formal, the least procedurally scrupulous of commanders, because he believed that war – that extraordinary activity – is a reckless, untidy business, and that the habits of mind of the methodical manager are alien to what is required. Furthermore – and it was an integral part of his character, demonstrated again and again both in youth and middle age – Rommel could on occasion be prudent where prudence was demanded, could plan as carefully as he could lead with every appearance of recklessness. His soldierly judgement of what qualities a particular situation demanded was generally (although not always) impeccable.

But by instinct his place was at the tip of the spear. Rommel, throughout his career, was criticized for ignoring the complex mechanics of a military command, whether an army or a division; and for neglecting its constraints, particularly the need for forethought in the organization of supply and the dissemination of information. His critics blamed him for failing to use his staff correctly, for relying excessively on verbal orders to individuals, for taking insufficient account of the inadequacy of radio communications – and the distances travelled during the French campaign often left such communication unreliable, although German radios were technically first-class. Rommel's critics blamed him, in short, for failure of system coordination, for failure to use the machine which lay to hand, for frequent failure himself to know where and how placed the parts of his own command were. During the advance to Cambrai, for instance, divisional headquarters had little knowledge of where Rommel was or into what scrapes he had got himself; and they let Hoth's headquarters know of their unease. The senior General Staff officer, the Ia, Major Heidkamper, wrote a memorandum on the difficulties the staff found with Rommel's methods and submitted it (bravely) to Rommel just after St Valéry.

This infuriated Rommel. His own view was that the divisional staff had been laggardly and timid in not anticipating his requirements and the requirements of the Panzers in particular. Failure to bring up supplies to 25th Panzer Regiment's 'hedgehog' near Le Cateau had been a failure of staff forethought and staff initiative, not the consequence of an over-bold commander's leadership. Rommel knew, however, that his corps commander, Hoth, himself had certain reservations about the Rommel technique, much though he valued 7th Panzer's performance; and Rommel had a talk with him and made peace with his own Ia. It is easy to sympathize with the latter, and to blame Rommel's impetuosity and determinedly personal style for the administrative difficulties which sometimes – then and later – beset his command. But Rommel reckoned that if he had had a longer period of training and preparation with the division his own staff, as well as others, would have come to understand his methods and his military philosophy; and to conform. Hoth's own report on Rommel, written on 7 July, was generous. It referred to his '*Fronterführing*', his sense for the decisive point of battle, and said that General Rommel 'has explored new paths in the command of Panzer divisions'.[14]

Rommel was not an easy man to serve, and the irritation and sometimes confusion his rapidity and energy could generate were the reverse side of the triumphs his genius equally often procured. He could only command in his own way, in the way he had learned from his own experience against the French in the Argonne, the Rumanians on Mount Cosna, the Italians at Matajur; and had now proved to himself again as a General, on the Meuse, at Le Cateau, at Arras. He had to be himself. He could only play the commander as hero. And whatever the impact of his methods on his superiors, his colleagues, sometimes his own staff, there was no doubt among the men of his division. In response to their formal congratulations to their commander on the award of the Knight's Cross Rommel published an order, expressing his thanks and reminding them of their achievements: '*Dinant – Avesnes – Le Cateau – Cambrai – Arras – Lille – Somme – Rouen – Fécamp – St Valéry: werden fur alle Soldaten der Division stolze Erinnerungen Zeitlebens bleiben**.'[15] He gauged their pride correctly and would forever, in their minds, be associated with it. On his later triumphs he would receive signals from 7th Panzer Division and members of it assuring him that the old 'Rommel spirit' was still alive. To them he was always, simply, 'Rommel'. His trim figure, rapid movements, penetrating look, ready smile when amused, sharp decisive voice with its Swabian accent – all these had printed an unforgettable image on the mind of every soldier in his command.

* '. . . will remain proud memories for all soldiers of the division throughout their lives.'

PART 4

1941–1943

CHAPTER ELEVEN

Sunflowers in Africa

ROMMEL'S STAR shone brightly now. It had been agreed with the Führer's headquarters that he should send back for Hitler's own information a map showing 7th Panzer Division's progress through France, and this he had done, by hand of officer – an ordained act to inform the Supreme Commander of progress of a typical Panzer division at the front, or an eagerly-accepted chance to push Rommel's personal fortunes, as his critics would certainly aver. The name of Rommel was becoming widely known. Propaganda Minister Josef Goebbels admired him and undoubtedly took pleasure in publicizing the exploits of this unpretentious yet martial figure with his Ghost Division: a soldiers' soldier. Goebbels's diaries abound with adulatory references to Rommel, almost until the end – an exemplary character, an outstanding soldier, a commander who put others in the shade. The commendations of Goebbels may not add much to Rommel's stature in professional terms, but they witness to the personal magnetism and charisma he exuded, and of which very dissimilar acquaintances have almost invariably spoken and written.

For Rommel, as for many in Germany and in the Wehrmacht, peace was now on the point of breaking out, a just peace, a peace to be enjoyed. He did not know, and at that time few knew, that Hitler had already decided war must now begin against the Soviet Union as early as possible. Nor did he know that Hitler, facing with a good deal of reluctance and disappointment the reality that Britain was determined not to make peace, was already speaking of the possibility of Germany becoming ready to resume the responsibilities of a colonial empire.

Much publicized was Hitler's intention, announced amidst many crocodile tears at the blind folly of the British in spurning his friendship, to invade England; a plan which Hitler was persuaded without much difficulty would be impracticable after the air battles of summer 1940, and which – also without much difficulty – he decided to abandon altogether shortly afterwards while retaining it, as he reasonably supposed, in the form of an advertised option, a threat and distraction to the enemy. Rommel always thought it should have been attempted. His division was involved in the preparations while they continued, and had the invasion – Operation *Seelowe* (Sealion) – actually happened Rommel would have undertaken yet another armoured thrust: this time initially from Rye to Hawkhurst, fifteen miles inland to the north-west.[1]

Hitler, in fact, was dreaming dreams that autumn and winter of 1940, some long-cherished and welcome, some a good deal less so. Dreams of conquest and occupation of England faded, and had perhaps always been entertained with reluctance. The unfading dreams were of a swift victory over the Soviet Union before too late in the following year: of an army 'thinned out' after triumphs in both west and east to a level of some sixty divisions to secure the eastern empire by September 1941; of emphasis, thereafter, on naval and air forces. As far as the army was concerned there was, however, another, even more ambitious, alternative thought – that a German army based in Russia might be made ready for a further offensive operation from the Caucasus into Iran and Iraq, together with an advance from Libya (Italy consenting) into Egypt, and from Bulgaria (Turkey consenting) into Syria. All this was adumbrated in or before the first months of 1941,[2] and part of it would one day find an echo in Rommel's own strategic thinking – actually, indeed, found expression in a draft 'Plan Orient' in June 1941. For Hitler, of course, it was part of his increasingly grandiose visions, and consistent with his idea of a united Europe ultimately challenging America in the economic sphere, having become mistress of a great abundance of raw materials and sources of energy.

Rommel knew nothing of all this, although he might well have found such visions spell-binding – Rommel was a political *naïf*, and Hitler was adept at binding his loyal listeners with spells. What Rommel and the world did know was that on 10 June 1940, a week after most of the British Expeditionary Force had reached England, and with the Wehrmacht already across the Seine and closing in to deal the death blow to the French army, Italy declared war on the Western Allies, Britain and France. No development had a greater ultimate effect on the destiny of Erwin Rommel. Meanwhile he remained for the rest of 1940 in France, training, preparing contingency plans for dealing with unoccupied France if there was serious trouble between Frenchmen in the south and the German authorities; chafing a little.

Mussolini had joined Hitler in time to appear at the victors' table just before an armistice was arranged with the French. His ambitions, however, ranged more widely, and it was in Africa that the Duce proposed to expand the new Roman Empire. He had already conquered Abyssinia in a campaign of 1936 and now had an army of quarter of a million men poised there, in Eritrea and in Italian Somaliland, to threaten the tiny British garrisons of Kenya, British Somaliland and the Sudan. In Libya the Italians had deployed fourteen divisions under the command of Marshal Graziani, and it was Mussolini's intention to launch this force on an invasion of Egypt, a country under treaty with Britain wherein was stationed a small British army; a country, because of the Suez Canal, of strategic importance.

On 13 September 1940 the Italians began moving towards the Libyan–Egyptian frontier. Having advanced to Sidi Barani, fifty miles inside Egypt, they then assumed the defensive, started to construct fortifications on a twenty-mile frontage, and manned them in considerable strength. Here, on 9 December, the British and Imperial Western Desert Force under General Richard O'Connor attacked, an attack launched against an enemy of over five times the attackers' strength. Within three days nearly forty thousand Italian prisoners, seventy-three tanks, 237 guns and a thousand vehicles were in British hands.

The Italians were now pursued throughout January 1941 along the North African coast, and their successive fortified places of Bardia, Tobruk and Derna were taken – Tobruk with its harbour being an especial prize and potential advanced base for further operations. O'Connor then drove his mobile forces across the Cyrenaica 'bulge', south of Benghazi. At Beda Fomm on the Gulf of Sirte O'Connor's troops blocked the Italian southward escape route and on 7 February took the surrender of the Italian Tenth Army. Some ten Italian divisions had been destroyed, with the loss of 130,000 prisoners, five hundred tanks of poor quality and over eight hundred guns. British and Imperial losses had been less than two thousand. On 8 February the British occupied El Agheila, near the border between Cyrenaica and Tripolitania.

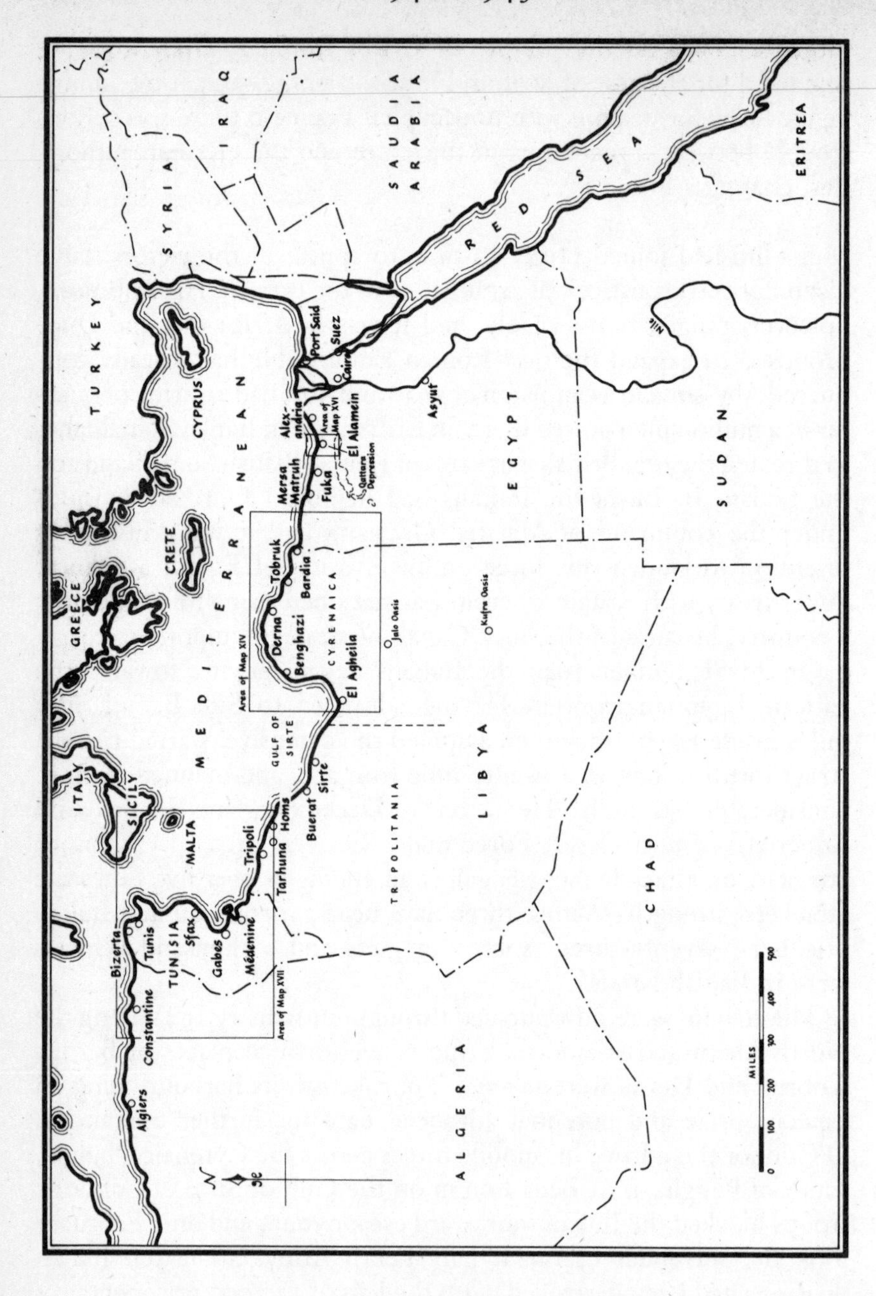

XII North Africa. Area of operations 1941–43

They advanced no further. O'Connor had been ordered to stop; it had been decided to send British help to Greece, and it could only come from North Africa. Greece had been attacked by Italy in October 1940 (operating from Albania, which the Italians had invaded on Easter Day in 1939), and the Italian forces had been checked by the Greeks with considerable spirit.

In Eritrea the Italian army was attacked in January by British and Indian forces under General Platt, invading from the Sudan; while in southern Abyssinia a British, British Colonial and South African force under General Cunningham advanced northward, having subdued the garrisons of Italian Somaliland. The two expeditions ultimately joined hands; the Abyssinian capital, Addis Ababa, was to fall to Cunningham on 6 April; and by the end of June 1941 Mussolini's empire in East Africa would be no more.

In February 1941, therefore, the Italians were already in an unhappy military situation throughout the eastern Mediterranean and Indian Ocean theatres. They had been checked in Greece; they were in the process of being defeated in Eritrea, Italian Somaliland and Abyssinia; and in Libya they were about to lose an entire army and had been driven back into Tripolitania itself. This was the general situation when Rommel, on 6 February, was summoned to Berlin.

Rommel reported in the morning to the Commander-in-Chief of the Army, Field Marshal von Brauchitsch. In the afternoon of the same day he reported to Hitler personally. He had been selected by name to take command of a small German force of two divisions – one Panzer and one light* – which it had been decided to send to help the Italians in North Africa. This was Operation *Sonnenblume* (Sunflower), by no means wholly welcome to the Italians, who recognized its inevitability if there were to be a chance of warding the British off Italian possessions but who foresaw loss of influence and prestige in a theatre of war which geography had made their own. Hitler was sensitive to this, and directed very specifically that German troops must treat their allies as equals. An earlier German offer (of 3rd Panzer Division) had been first accepted and then declined by the Italians in the preceding November. Mussolini was reluctant to share with Germany the coming triumph.

After a few hectic days of briefing and kit assembly Rommel flew on 11 February to Rome and saw the Chief of Staff, General Guz-

* Light divisions were mechanized, but not normally equipped with tanks. Establishments varied.

zoni, of the Italian Armed Forces Headquarters, the Commando Supremo, which had only been in existence since November. He then flew to Sicily to meet the German Air Force commander, General Geissler, responsible for operations in the central Mediterranean.

In Sicily Rommel spoke to Geissler about immediate air operations in North Africa. It was wholly characteristic of Rommel – and curiously comparable to the actions of one who would one day be his chief British adversary in similar circumstances – that although he had not yet reached the North African theatre of operations, nor assumed command of any troops there, he immediately began to give orders (or, at least, urgent operational requests) to Geissler. The news from Cyrenaica was bad: O'Connor had completed his masterpiece and, for all the Germans at that moment (11 February) knew, was poised to advance without delay on Tripoli. O'Connor had, indeed, pressed to be allowed to do exactly that, and was sure to the end of his life that he could have thus pre-empted the whole North African campaign. However that may be – and there are imponderables in the speculation – he would have almost certainly pre-empted the arrival of Rommel; but the Greek adventure, soon to become the Greek fiasco, made it impossible.

Meanwhile Rommel, in Sicily, contemplated a map of North Africa where there seemed to be nothing between the British, who had taken the port of Benghazi and were on the Gulf of Sirte, and Tripoli itself. He asked Geissler that the Luftwaffe should, that night, bomb Benghazi and should next morning attack any British columns moving south between Benghazi and the Tripolitanian border. Geissler explained that the Italians had particularly asked him not to bomb Benghazi – many Italian officers and officials owned property there.

Rommel was accompanied by Hitler's principal Army Adjutant, Colonel Schmundt, with whom his relations were now, and would remain, good. On hearing what he regarded as Geissler's unacceptable response Rommel acted as he often would in future – he used Schmundt as a conduit to Hitler. Schmundt explained to the Führer by telephone Rommel's concern and his wishes, and received Hitler's consent. Geissler was ordered to go ahead.

Both the request and Hitler's agreement to it were, from the point of view of the German–Italian alliance, improper. Rommel had demonstrated two constant characteristics: readiness to assume personal responsibility as early and as decisively as possible for any situation which affected him, regardless of formal correctitude; and readiness to appeal to Hitler himself. Both characteristics were weapons he used, often with startling success; and both would one day break in his hand.

Next morning, 12 February, Rommel arrived at Castel Benito airfield, Tripoli.

The German decision to intervene in North Africa derived from a number of factors: and although the strategic decisions of the British in the Mediterranean theatre were driven by very different factors there was a certain correlation, a mirror-image effect, between the policies of the two sides.

The British had a base in Egypt and certain responsibilities towards that country as well as a mandate in Palestine and an agreement with Iraq. Egypt, and the Suez Canal zone, were an essential centre of British Imperial communications both by sea and air; particularly with India. Egypt had long been the hub of British political influence in the Levant and the whole region. The Middle East was the source of most of Britain's oil. An enemy power established in the Middle East would not only threaten the supply of oil but would in effect cut communications within the British Empire. And if the Suez Canal – and the Mediterranean route to it – were unusable by Britain because dominated by enemy air and maritime forces, based on possession of the North African littoral, as well as the Balkans, then the consequential strain on British shipping would strike Britain at its most vulnerable point. Every journey to India or South-East Asia – or, for that matter, to Egypt itself – would be made round the Cape of Good Hope, and Britain's ability to survive a battle of shipping attrition in the Atlantic would be weakened accordingly. Ultimately, of course, any hopes Britain nurtured of taking the offensive against Germany would also turn vitally on shipping, but such hopes were remote in early 1941. Also remote, as yet, was war against Japan; but when this came at the end of the year all these factors stood out in even sharper relief. The Italian offensive against Egypt had therefore threatened what was arguably a vital British strategic interest, and O'Connor's desert victory had signified more than the comparatively small forces engaged might have indicated. Furthermore, an Italian army had been destroyed; and to a Britain which had suffered the defeat of its only Expeditionary Force in France seven months earlier that was a great deal.

To the Germans, and particularly to Hitler, there was argument for a gesture of support to the Italian ally, both in the Balkans and in North Africa. The Italians had played no useful part in the French campaign, but if they could be kept reasonably effective they had the capacity to tie down a significant number of British forces; and the place for this was in North Africa, where the British would, for valid reasons, be particularly sensitive about their own position. When the possibility of a small German expedition to help the Italians in Libya had been mooted in the autumn of 1940 a leading Panzer expert, General von Thoma, had been sent to look at the situation and report. Von Thoma had accurately described the difficulties which would attend any such venture and it had been left pending, with an offer first accepted and then rejected. With the alarming collapse of the Italians before the British offensive in January and February 1941 the idea had been swiftly resurrected.

But German – some German – reasoning went beyond the desire to give limited help to Italy and, in an equally limited way, harass the British far from home. There was also a major strategic issue. The German Navy regarded British sea communications as, ultimately, the vital factor of the war in the west, and their chief, Admiral Raeder, was in no doubt that the most effective measure against those communications would be a successful campaign against Britain in the Mediterranean. This argument failed to find instant favour with Hitler, but it periodically resurfaced. It was the mirror image of the British concern for shipping.

Other factors were reflected, as the campaigns of 1941 and 1942 took their course. Just as the Germans perceived a vital interest in keeping Italy in the war – not least because of the strain on German manpower which would be caused by replacing Italian forces in Greece and the Balkans, a consideration which later emerged powerfully – so Britain, for the same reason, would come to see a strong motive in knocking Italy out. Just as the Germans wished a secure European southern flank so that the main campaign, always regarded as that to be waged one day against the Soviet Union, could proceed without distraction, so the British (and later, with reservations, the Americans) saw a Mediterranean front, which might ultimately take the war to Italy itself, as an important element in multiplying the directions in which the Wehrmacht had to operate. Finally – and the matter remained contentious, especially for Rommel – some Germans (Rommel naturally and persistently, Hitler sporadically, OKW and OKH for a short while only) dreamed of an ambitious and triumphant German campaign – 'Plan Orient', the 'Great Plan'

– which would take the Wehrmacht through Egypt, across the Syrian desert and into Persia, threatening the Soviet Caucasus from the south and in the process denying Middle Eastern oil to Britain. The British were to suffer periodic nightmares about a German army, successful against the Russians in the Caucasus, erupting into Persia and threatening the back door of the British Middle Eastern position and the oil it covered.

These dreams and nightmares, mirror images of each other, brightened or faded with events. Most lay well in the future, as indeed did the German invasion of Russia itself, when Rommel first set foot in Africa on 12 February 1941. Initially the Germans saw their intervention as strictly defensive, against what seemed as if it might be an imminent and overwhelming British offensive. They perceived *Sonnenblume* as a rescue operation.

Rommel's information about the enemy was both grave and defective. He knew that Tobruk, Benghazi and the whole of Cyrenaica were in British hands. He knew that the British had effectively destroyed an Italian field army and were standing, presumably poised to advance at any day or hour, on the borders of Tripolitania. He was given an enemy order of battle which mistakenly showed the British as having at the front two army corps – an armoured corps and an ANZAC corps. He did not know, and the Italians did not know, that O'Connor had been replaced first by General Wilson and then by General Neame, and was back in command of British troops in Egypt. He did not know that the British 7th Armoured Division, the armoured core of O'Connor's victories, had also returned to Egypt for re-equipment. He did not know that it had been replaced in Cyrenaica by the British 2nd Armoured Division, fresh from England, new to the desert; nor that one of this division's two armoured brigades had been removed and sent to Greece, while the brigade remaining in Cyrenaica (the only British tank force there) had two regiments of light tanks and a regiment of heavier cruisers of which only twenty-three were fit. He did not know that the only other enemy division in Cyrenaica, 9th Australian, had had two of its three brigades recently replaced by new and largely untrained formations, and was critically short of transport, so that it chiefly consisted of non-motorized infantry. He certainly did not know that the British had been ordered to advance no further, but simply to 'make good' Cyrenaica, as a largely static 'Cyrenaica command': nor did he know that British orders were that the forward troops should be prepared

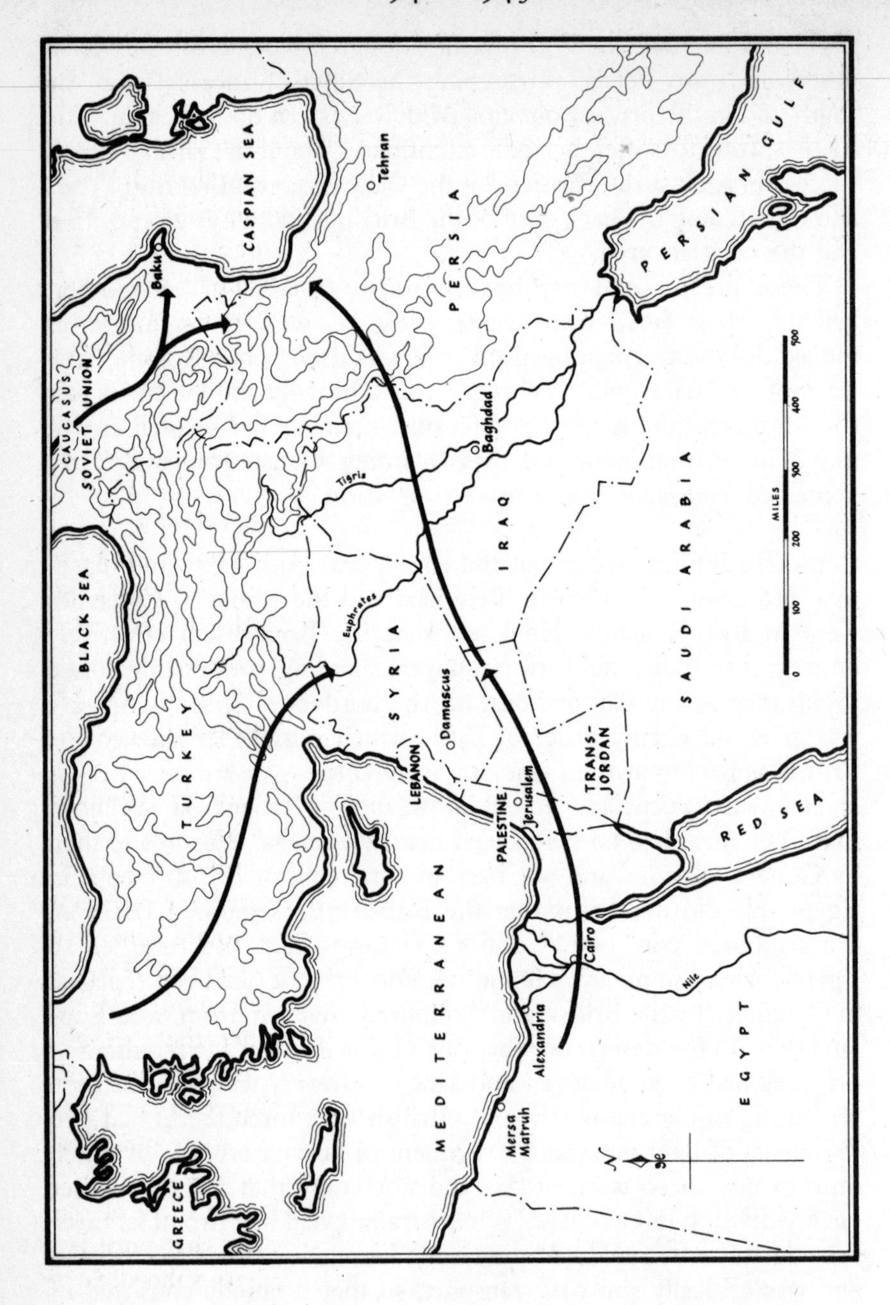

XIII 'Plan Orient'

to withdraw if attacked, although attack was not expected. Above all he did not know of the British decision to withdraw forces from North Africa in order to send them to Greece.

Rommel, assuming that he would soon be dealing with a British offensive in locally overwhelming force against demoralized Italians, reckoned that the immediate salvation of Tripolitania would lie with the Luftwaffe. On the ground, Italian motorized infantry formations were now to be placed under Rommel's command, and he was determined that both they and the Germans, as they arrived, should be deployed well forward, near the frontier of Cyrenaica, making a brave show as far to the east as possible. He had been allotted only two German divisions – 5th Light Division and 15th Panzer Division – of which the latter (an infantry division until the previous November) would not arrive until May. Initially, therefore, he would have one German division, whose reconnaissance battalion started to disembark at Tripoli on 14 February; such Italians as the Italian authorities could be persuaded to confide to his command – primarily the Brescia and Pavia Divisions, at present mostly in the Tripoli area; the Italian Ariete Division, with sixty light tanks of obsolete pattern; and Luftwaffe sorties as assigned. Rommel himself was to be subordinate to the Italian theatre commander-in-chief, General Gariboldi, although he had 'right of appeal' to Berlin.

Operationally, however, Rommel had already made up his mind about the immediate posture which needed adopting – as bold a façade as could be shown forward as early as possible. For this he needed troops not at Tripoli but on the Gulf of Sirte. He reported to Gariboldi at 1 p.m. on the day of arrival and found him of a different opinion: the best strategy would be defensive, near Tripoli itself. Rommel set out to fly eastward and inspect the country. What he saw convinced him of the rightness of his instinct: there should be strength built up on the Gulf of Sirte. When he reported back to Gariboldi that evening he had already determined to carry his point. He had also determined to take over personal command at the front as soon as there was anything there to command.

Because Rommel so stamped his personality on the story of the North African campaign it is easy to ignore the fact that until his last two weeks in Africa, in Tunisia, he was a military subordinate, first as a corps, then a Panzer Group, then a Panzer Army commander. In all cases he had a superior officer in the theatre, and until February 1943 that superior officer was Italian. Libya, an Italian colonial possession, was the responsibility of an Italian General, initially Gariboldi, himself responsible to the Italian Government, the Duce, through the Italian High Command in Rome, the Commando Supremo. Rommel was under his command and thus, at one remove, under the authority of Commando Supremo.

This subordination was very real for two practical reasons. First, at all times, the majority of Rommel's troops were Italian. He was given – to different degrees at different times, and sometimes wholly – operational command of some or all of the Italian divisions in North Africa, and they invariably outnumbered the Germans; at no time did he have more than five German divisions (of which one, at the high point of the count, was in fact a large parachute brigade) under his command. Most of his troops were Italian and their Italian commanders, *pace* Rommel, with whom relations varied, had a national responsibility to their Italian Commander-in-Chief.

The second practical point was logistic. Rommel's supplies – in virtually every commodity since little was locally produced – came from Italy, across the Mediterranean. They originated in Italy or in Germany (or German-occupied Central Europe) and were transported south by rail and shipped from Italian ports in Italian ships. Human reinforcements generally took the same route. Rommel – through his formal superiors in the first instance – submitted his demands. The Italian authorities did what they could – at least allegedly – to satisfy them.

Over this second point, therefore, Rommel had no direct authority, and neither did the German Government. The latter could discuss – and remonstrate – with the Italian Government if, as happened incessantly, supply was inadequate; and Rommel, like most men in similar situations, grumbled that in his view the German authorities were too feeble, too diplomatic, in their representations. In one way, however, the Germans could assist directly. The transport of supplies from Italy was vulnerable to interdiction by British air and sea action, much of it based on the island of Malta. Counter-action demanded both air and maritime effort. Here the Germans made a massive contribution, by moving air forces to the Mediterranean from other fronts and by actually redeploying submarines from the Atlantic in 1942. The operations of these forces had to be coordinated with the Italians, but it lay with the Germans to contribute, and they did: often to a near-dominant degree. All of this lay well outside Rommel's sphere of authority, but its effectiveness crucially affected his ability to discharge his task.

This task, however – and the point bears making again – was

itself a subordinate task; in theory, and in all propriety, Commando Supremo approved operations, through the Italian Commander-in-Chief. Yet Rommel (because German troops were committed) had a right of appeal to the German High Command – to OKH, to OKW, to Hitler himself as Supreme Commander of the German Wehrmacht. In theory, again, this was arguably only in respect of his German troops; but the personal primacy Rommel soon attained in the campaign, the prestige of the German General Staff, and the dominant position of Germany in the Axis meant that for major decisions affecting the conduct of operations in North Africa, Rommel on occasion looked to and was governed by the decisions not of Commando Supremo but of the German Führer and his creature, OKW, in this essentially Italian theatre of war.

Coalition warfare is never simple. The situation of a military commander, subordinate to an allied superior of a different nationality but with right of access to his own Government, often provokes conflict, accusation and counter-accusation, divided loyalties. The position of Gort in 1940 offers an example. There were plenty of others in the conduct of war by the Anglo-American alliance, which adopted a combined Chiefs of Staff organization. This was empowered to coordinate Allied strategy at the highest military-political level and to issue agreed directives to subordinate Allied commands or agencies, to whom or to which national resources were allocated. There were upsets, but there was at least machinery for discussing and if possible correcting them. The Axis powers had little of such. In the Mediterranean the higher direction – the grand strategy – depended largely on communications between Hitler and Mussolini, each of whom was surrounded by excessively subservient advisers. There was no regular method of collusion between OKW and Commando Supremo, no regular exchanges of ideas, no regular meeting of minds. There were periodic sessions at which Marshal Cavallero (for instance), the head of Commando Supremo, would be exposed to the geopolitical ramblings of the Führer, but these could hardly be dignified by the name of strategic consultations. There was much correspondence. There was a German General, von Rintelen, accredited to Commando Supremo (and also German Military Attaché in Rome), through whom much passed and to whom Rommel's expostulations were addressed or repeated; but machinery for agreement on and enforcement of priorities, for giving practical effect to generalities, was largely defective. The Italians, furthermore, were determined to wage not an integrated but a parallel war, a difficult concept; indeed their declaration of war in 1940 had astounded Hitler, and their attack on Greece was entirely concealed from the Germans.[3]

Directives to Rommel, therefore, and to Rommel's Italian superior, were given within an intrinsically unsatisfactory inter-allied framework. Rommel had to work a flawed system. When all was going well – again, the phenomenon is familiar and not confined to what became Panzerarmee Afrika – he found it periodically convenient to be able to plead the exigencies or directions of one authority against the inconvenient demands of the other. When all was going less than well the inherent anomalies were more obtrusive.

The first German troops, 5th Light Division's* 3rd Reconnaissance Battalion, were paraded in Tripoli immediately they arrived, and with their disciplined, formidable appearance made an immediate and powerful impression on the population. Within hours they were on their way eastward, towards the front: and within forty-eight hours were deployed and in contact with the enemy 280 miles east of Tripoli. A myth spread among their enemies that great care had been taken with German equipment, uniform and acclimatization. Nothing could be less true; they were unused to the climate and the conditions, their uniforms were unsuitable. They had everything to learn, but they learned extraordinarily fast. The basic training of the German soldier, the adaptability of the formation staffs and the energy with which all were directed ensured it.

The next weeks were hectic for Rommel. He flew almost every day between Tripoli, where German troops were disembarking and there was essential business to be done, and the Sirte front. He allowed absolutely no delays at Tripoli – ships were immediately unloaded by the troops, working if need be through the night, until stores and vehicles were clear and a column moving eastward towards the desert and the frontier. In hastily established workshops he arranged the construction of dummy tanks made of plywood and canvas – he believed that every measure of deception should be employed to persuade the enemy of greater German–Italian strength than actually existed. And he had already succeeded in persuading Gariboldi to move his Italian formations eastward, to Rommel's personal command; the leading Italian division began to move to pos-

* 5th Division was drawn from elements of 3rd Panzer Division in Germany and included its Panzer Regiment 5. Later it was retitled 21st Panzer Division, and so remained.

itions west of Buerat on 14 February, the day 3rd Reconnaissance Battalion disembarked at Tripoli. Days passed, and still the British on the frontier did not attack; Rommel was able to write to Schmundt (knowing that it would reach Hitler) that everything was going better, day by day.[4] On 19 February a new formation title for the German troops was promulgated, one that was to go down in history: the Deutsches Afrika Korps.

The first contact recorded by Rommel's staff in North Africa between the British and the Deutsches Afrika Korps was on 24 February when a British armoured car reconnaissance patrol of the King's Dragoon Guards, one officer and two soldiers, were taken prisoner. Despite the uncertain nature of Rommel's intelligence hitherto he, and his admirable intelligence staff (his 'Ic', manned by at the most two officers), became increasingly convinced that the situation facing them differed radically from that which they had expected and of which they had been warned. Some time later Rommel remarked to Lieutenant Behrendt, working with Ic, 'It's given to me to feel where the enemy is weak,'[5] and that feeling was already influencing him. It was clear to him that something had happened to the British, although it was not yet clear exactly what. Whatever it was probably spelled opportunity.

Weeks passed. Rommel wrote to Lucy that he was on excellent terms with 'our allies' and that a visit to an Italian division had made a very good impression on him. He had enjoyed being asked by an Italian officer where he had been awarded the *Pour le Mérite* and replying 'Longarone!' German forces continued to arrive, and 5th Light Division (General Streich) began its concentration at the front. Its 5th Panzer Regiment, consisting of 120 tanks, half of them light tanks and half Panzer IIIs and IVs, disembarked at Tripoli on 11 March and moved east. Two days later Rommel finally moved his headquarters forward to Sirte. Henceforth he would seldom be far from the front. There were reports of Free French activity against the Italian desert garrisons far to the south, of activity from Chad, and Rommel sent a small motorized force south under Lieutenant-Colonel Graf von Schwerin to watch and ward. The threat was as yet minimal, but an Italian garrison at Kufra had capitulated. Von Schwerin's force was ultimately recalled, and rejoined Rommel on 3 April.

Still the British did not attack. Rommel's intelligence reports from Berlin attributed to the British in Cyrenaica a second armoured division (whereas in fact, of course, one was by now in Egypt and half the other was in Greece). Rommel, however, was less concerned with the enemy's order of battle than with what he sensed as British hesitancy and sluggishness. He deployed his dummy tanks, and ordered the Brescia Division (General Zambon) to take over defensive positions at Mugtaa, to free 5th Light Division for mobile operations. Mugtaa was a strong position, a defile, difficult to attack from the east and, because of the terrain, hard to outflank. Such positions were infrequent in North Africa and where they existed held considerable operational significance. A corresponding position thirty miles on the far side of El Agheila was Mersa El Brega, hard to attack from either direction. Rommel appreciated that when the German–Italian army opened an offensive the first objective, the door to burst open, must be Mersa El Brega; and there was water-bearing land around it. Meanwhile a small British garrison was deployed at El Agheila.

In the four weeks he had so far experienced of Africa, Rommel was beginning to come to terms with its salient features for a soldier – the enormous distances; the incidence of blinding sandstorms; the vulnerability of machines and equipment to weather and sand; the hazards of blocked air filters, sand-choked working parts of armaments; the primacy and rarity of water; the enormous strain on vehicles from wear and tear; the scorching days and chilling nights. He loved it. He was a 'natural' player in the great game of desert war. On 19 March he flew to Berlin.

Rommel's first return to Berlin from Africa was marked by the award of the Oakleaves to his Knight's Cross. It was also marked by a discouraging douche of cold water on his African venture. 'I was not very happy,' he wrote afterwards, 'at the efforts of Field Marshal von Brauchitsch and Colonel-General Halder to keep down the numbers of troops sent to Africa and leave the future of this theatre of war to chance. The momentary British weakness in North Africa should have been exploited with the utmost energy.' Rommel could write with hindsight – on 19 March neither he nor OKH knew much about the British weakness, momentary or otherwise: nor could they foresee the development of the Greek situation. But Rommel would have declared that factual knowledge, intelligence, might still be patchy but that his 'feel' for the British, their sureness of touch or lack of it, their boldness or its absence – that this 'feel' was strengthening day by day. The British, he believed, were vulnerable to a strong, decisive blow, and after that who knew what might come? When Rommel felt like that, even when (perhaps especially when) his superiors did not share the feeling, events generally moved fast. He was told that 15th Panzer Division would not join his command until late May, and that he could plan to attack in the region of Agedabia thereafter, possibly exploiting as far as Benghazi. His first priority, however, must be the secure defence of Tripolitania.

Rommel contented himself with pointing out that to move on Benghazi must imply the occupation of Cyrenaica; one could hold on one side or other of the Cyrenaica 'bulge' but not halfway, exposed to outflanking at will across the chord of the arc. He flew back to Africa, and had the satisfaction of 3rd Reconnaissance Battalion, first in Africa, taking El Agheila, with its airfield and water points, on 24 March. The British did not contest and withdrew. Rommel had ordered this before flying to Berlin, and he now considered the future.

The British had withdrawn to Mersa El Brega. The longer they were left there undisturbed the stronger their position would be. Rommel did not believe that time was working for him; he had been instructed to await the arrival of 15th Panzer Division, but that would mean a further two months before battle, two months in which Mersa El Brega would be developed and mined. Whereas now –

Rommel decided to attack Mersa El Brega. He argued to himself that the main attack in May, explicitly sanctioned at least by OKH, was to be directed on Agedabia, possibly Benghazi. Mersa El Brega had always constituted an essential first step. Like Mugtaa, Mersa El Brega was a defile and to win it would give Rommel a highly defensible east-facing position during the next phase of campaign; if, in fact, it proved to be defensive.

On 31 March 5th Light Division advanced on Mersa El Brega. The immediate response of the British was stubborn. Rommel himself, at last in his element again, found a way to outflank, driving through sandhills north of the coast road, and by the evening had moved a machine-gun battalion round and taken the Mersa El Brega defile. A good many British vehicles were captured, and next day air reports showed a widespread British withdrawal. It was, Rommel wrote later, a chance he could not resist, and on 2 April he launched his entire available forces on an adventure which resulted, in a very few days, in the expulsion of the British from Cyrenaica. What started as a preliminary, a probe, had become by whatever name a major offensive and an act of implicit disobedience which Rommel counted on justifying by its success.

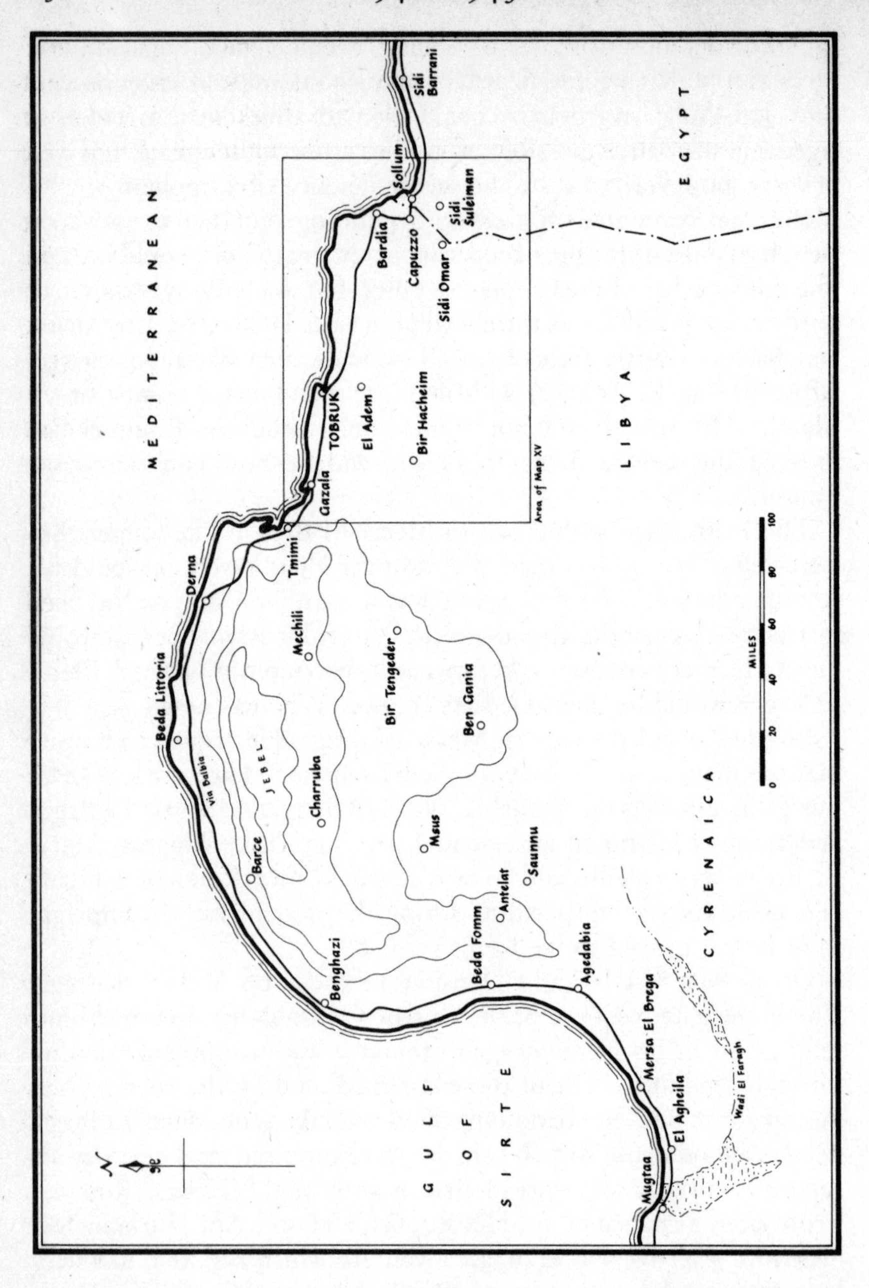

XIV Cyrenaica. Area of operations, German–Italian forces, March–April 1941, December 1941–January 1942, November 1942

Rommel, unbeknown to him, was greatly helped by British orders and by a confused and inappropriate British operational concept. General Wavell, the Commander-in-Chief, Middle East, had given somewhat detailed orders to General Neame, commanding in Cyrenaica. If the Germans attacked the British must be prepared to withdraw, fighting a delaying action as far as Benghazi – and could abandon Benghazi too, if necessary, in order to avoid being cut off. The concept, permissive and operationally unsound because of the terrain of Cyrenaica – mirror image, indeed, of the OKH instruction to Rommel which he had criticized and was now disregarding – accounted in part for the hesitancy which Rommel now sensed in the British Command. The other part of British hesitancy derived from inexperience. Rommel's 5th Light Division were also inexperienced in desert fighting, but the basic training, discipline and battle drills of the German army gave them more than an edge over their opponents; and unlike those opponents they were led by a genius.

There was also a certain British complacency. They had been told – and ULTRA* had encouraged Wavell to the conclusion – that the German–Italian enemy could not possibly be ready for any sort of offensive for some time. This was a reasonable conclusion, especially to one privy to OKH instructions to Rommel. It took, however, no account of the personality of the commander of the Afrika Korps, nor of his readiness to vary his instructions from above as he went along.

Rommel's plan, with the 'door' at Mersa El Brega broken down, was to move on Agedabia, forty miles from Mersa El Brega, and thereafter to divide his force. One wing, on the left, led by Lieutenant-Colonel von Wechmar's 3rd Reconnaissance Battalion, was to advance on Benghazi along the coast. A right wing, Schwerin with part of 5th Light Division and a reconnaissance battalion from the Ariete Division (General Baldassare), was to strike across the chord of the Cyrenaica arc by Ben Gania and Bir Tengedir, aiming to reach the coast at Derna and cut off the enemy escape route, the coast road running from Benghazi through Derna to Tobruk. Between these two wings a strong armoured force, consisting of the balance of 5th Light Division and formed round Colonel Olbrich's 5th Panzer Regiment, was to drive across Cyrenaica by Msus, fifty miles north-west of Ben Gania, and then make for Mechili. Moving

* The all-important and highly secret British reading of German high-grade cypher traffic.

with von Wechmar's left-hand column was the Italian Brescia Division, and with the Olbrich column the main body of the Ariete. In the event the reconnaissance battalion on the left wing was checked after reaching Benghazi, moved due east, and joined the centre and right wings in the vicinity of Mechili.

The operation began on 2 April and Rommel, as usual, spent his hours throughout the next week goading each column to further efforts and greater speed. He moved his headquarters forward to Agedabia, and was determined to test for himself the terrain and the conditions under which his columns were to advance. He had been told that the right-hand column's route, via Ben Gania, was almost impassable in its early stages; he explored it, and found that to be nonsense. He was told, on the second day of the advance, that fuel requirements and the turn-round time of replenishment vehicles would mean 5th Light Division halting for four days. Rommel's reaction was vigorous. Such a wait, caused by shortage of fuel-carrying vehicles and distance from replenishment point, was wholly unacceptable. Naturally there had not yet been time or opportunity to build up dumps forward, but the immediate expedient must be to increase load-carrying capacity by improvisation. He ordered the division to dump every load from every vehicle and use them for one function – fuel. This meant a 5th Division temporarily (for one day) immobilized, and with normal loads, commodities, men, stranded for that period: but thereafter the division should have enough fuel to complete the operation. 'That,' Rommel observed to his staff, 'saves blood and wins us Cyrenaica.'[6]

General Gariboldi, Rommel's titular superior, had already strongly objected to what he sensed was a subordinate determined to ignore his instructions and the limits placed on the action of the German and Italian forces under his command. To Gariboldi, who had wished to hold Tripolitania only and form a front not too far east of Tripoli, the prospect of what was now looking like a general offensive was intolerably rash. Rommel had at once appealed by signal to Berlin and to his considerable satisfaction had received authority to act, in the circumstances, as he felt best. It was an early example of his natural and generally successful exploitation of having two masters.

The advance, although Rommel perpetually fretted to his staff over its slowness, continued. Rommel spent a good deal of time in his light Storch aircraft (he was a competent pilot), and on occasion he would drop a message on a column: 'If you do not move on at once,' it might read, 'I shall come down! Rommel.'[7] Schwerin's advance guard from the Ariete reached Bir Tengedir at 9 o'clock on the evening of 4 April, and on the previous day the left-hand column had entered Benghazi.

Rommel was lucky in his first North African contact with British armour. The British 3rd Armoured Brigade, the only remaining brigade of 2nd Armoured Division, had lost a few tanks at Mersa El Brega, and had had large numbers break down. Like the Germans they were widely dispersed over the desert north and east of Agedabia. Battle, where it took place, resolved into the actions of small bodies of tanks and men, and in the complacent words of Rommel's intelligence staff, 'under uncertain leadership, without proper coordination, receiving constantly changing orders and objectives', the British fared poorly in the face of an enemy serving under 'Rommel's energetic coordination and leadership from the front'.[8] By the morning of 6 April Rommel's information from air reconnaissance confirmed his optimistic assessments from the start. The British were trying to get back to Tobruk as fast as they could. At half past seven that morning a mass of vehicles were reported evacuating Cyrenaica eastward.

The progress of Rommel's columns was however uneven and confused. It was typical of this sort of operation in the largely featureless and enormous desert or Cyrenaican Jebel that parties of friend or foe were often near-inextricably interwoven, and that there was little semblance of tidiness in such movements as were ordered or such combat as took place. Thus, on the British side, the remnants of 2nd Armoured Division were ordered on 6 April (by General O'Connor, who had been sent forward from Egypt by Wavell to advise Neame and, at an appropriate moment, take over command from him) to concentrate at Mechili. The armoured brigade had withdrawn after the earlier encounters first to Msus and then northward to Charruba. The brigade commander decided that he had insufficient fuel to reach Mechili and had better move north to the coast road with the remaining elements of his brigade, and try to reach Derna. Towards Derna, however, along the same road, were moving 9th Australian Division, from Benghazi. The congestion on the coast road was immense: the Jebel flank was tenanted by little but 2nd Armoured Division Headquarters, and 3rd Indian Brigade, which consisted of infantry riding in unarmoured trucks without supporting arms or anti-tank weapons of any kind. And across the Jebel were driving towards Mechili von Wechmar's reconnaissance battalion from Benghazi via Charruba, Olbrich's 5th Panzer Regiment and the Ariete via Msus, and Schwerin's force of Germans and Italians via Ben Gania and Tengedir.

Had the British known exactly what was advancing against or round them it would have seemed formidable enough; but the tally of their enemies, as usually happens, was magnified by report and rumour, the Germans discovering from prisoners that at least a Panzer Corps was imagined to have moved to the attack in Cyrenaica.[9] On the German side there was also a considerable amount of confusion, uncertainty of where anyone or everyone was, change of plan and order, periodic infuriating delay. Thus Rommel took personal command of the advance towards Mechili on the afternoon of 5 April at Ben Gania, having earlier and throughout the previous day flown over the Jebel and seen what he was tolerably – and in the event correctly – sure were German–Italian columns rolling eastward, on course. He had already set in hand what he hoped would be the convergence of the Schwerin column and the Olbrich column at Mechili. At Ben Gania he heard from air reports that Mechili appeared clear of enemy, and sent an order to Schwerin's right-hand column: 'Mechili clear of enemy. Make for it. Drive fast. Rommel.'

Yet later in the day, after flying to the head of the column and then flying back to see where the centre column – Olbrich – had reached, Rommel was told that Mechili was not clear of enemy, but was in fact 'strongly held'.[10] It was now dark and Rommel set off from Ben Gania north-eastward to place himself at the head of the right-hand column – now consisting of most of 5th Light Division except the Panzer regiment, and moving under its commander, General Streich, together with the Ariete. Rommel believed that if he could, as often in the past, bypass opposition at Mechili he could cut off an enemy seeking to escape north and east from there, and push on to cut the coast road itself.

It did not work out like that. Next day, 6 April, when Rommel hoped to have enough forces round Mechili from the right-hand and centre columns, progress had been disappointing. He had ordered the troops from the right-hand column to be in position to move on Mechili from south-south-east and east but most of them only arrived that evening, while the centre column (Olbrich and 5th Panzer Regiment) had run into trouble from weather and fuel replenishment delay, and had not left Msus towards Mechili until the same evening. Then, at 2 a.m. on 7 April, the right-hand column also reported such shortage of fuel that the accompanying Italian artillery could not be deployed for the attack on Mechili.

An hour later Rommel had assembled every can of petrol – thirty-five, he wrote afterwards – he could lay his hands on and drove off to find the right-hand column's artillery, refuel them and deploy them. This took time, and it was not until later in the morning of 7 April that (having encountered a small party of British vehicles, driven at them with determination and watched them hurry away) he had seen to the fuelling of the Italian artillery and deployed the right-hand column for its operation against Mechili. There was no sign yet of Olbrich's Panzers. Time, as ever in battle, was slipping away.

Rommel – it was now the afternoon of 7 April – flew westward in his Storch, patrolling the Jebel, searching for 5th Panzer Regiment. He narrowly escaped landing in the middle of one column which proved to be British, and landed by another, a mixed force of German and Italians, who had lost time through being deceived by a mirage. Returning to his *Führungsstaffel*, or tactical headquarters, Rommel found that there was still no sign or report of Olbrich and his tanks.

He took off again in his Storch. Light would soon be failing, and upon the arrival of Olbrich with the main Panzer strength of the Afrika Korps depended, as Rommel later wrote, 'the decision in eastern Cyrenaica'. Eventually he found them, just before dark, well to the north of the direct route to Mechili which was their intended thrust line. Rommel, furious, ordered them to get on as fast as they could in the darkness, and despite that darkness flew back and found his own headquarters.

Next day, 8 April, troops from the right-hand column opened their attack on Mechili, and soon Rommel – in the air again – saw enemy vehicles streaming westward out of Mechili. These would soon run into Olbrich's Panzers, and Rommel flew west to find them, without success. Returning to the area of Mechili he found it taken by the original right-hand column; British attempts to break out eastward had been blocked. Soon 5th Panzer Regiment at last appeared from the west. A small force which Rommel had sent north to reach the coast road had successfully blocked it but needed reinforcement. A considerable number of prisoners were being taken, moving east from Benghazi. Rommel set reinforcement in hand from Schwerin's column and himself drove to Derna, reaching the coast by six o'clock in the evening of 8 April. Von Wechmar's reconnaissance troops had, when checked north of Benghazi, turned east by Charruba and joined in the battle at Mechili, while the Brescia Division marched along the coast road.

The operation had been in progress for six days. The British had evacuated Cyrenaica. On 9 April Rommel found 5th Light Division at Mechili planning a two-day halt for essential maintenance, and sharply enlightened them. They were to drive through the night to Tmimi* and be at Gazala by daybreak next morning, ready for attack on Tobruk.

To read Rommel's own account of these extraordinary and often wholly disorganized events is to receive an unusual impression of a commanding general; and Rommel, by now, had not only the Afrika Korps of one German division with some additional elements, but three Italian divisions in his force. He can be tracked either with his *Gefechtsstaffel*, his small fighting command group of three vehicles, or flying over the troops or the battlefield in his Storch, always hurrying in light or darkness from one place to another, hunting for a column which had taken too long or mistaken its direction, goading another for its dilatoriness in refuelling, finding another detachment actually out of fuel and improvising arrangements so that at least some vital vehicles and weapons could move a few more vital miles: invariably, as he always had done, galvanizing every part of his command, every subordinate, by his sudden presence, his peremptory directions, his sharp tongue, his vitality, his energy. Rommel's visitations were memorable, his face and posture familiar to every man in the Afrika Korps, with his blue-grey eyes, lines of humour running from their corners, his clenched fists and arms slightly bent at the elbow when he addressed a group, his emphatic facial expressions, his remarkable eye for detail, his abrupt, precise, soldierly manner. Memorable too was his voice, sharp, incisive; and the ferocious language with which – not always fairly, but the stuff of legend nevertheless – he castigated shortcomings, particularly of senior officers. Rommel could show much charm but his tongue was rough. Yet everywhere he went he taught.

No staff college, no war academy, no General Staff prescribed such activities (or turns of phrase) as those proper to a higher commander, a general. Rommel, it is clear, often had but the haziest notion of where in the Jebel or the desert large parts of his command were – he acknowledged it himself, and although German communications were good for that age radio communication in North Africa was often rendered defective by conditions, and plotting the

* Rommel found it an awkward word, and generally pronounced it 'Tmini'.

whereabouts of a dispersed command in a mobile campaign was always a good deal more difficult than telephone and map exercises can generally indicate. But surely, the reader of Rommel may ask himself, he could and should have relied more upon his staff, upon giving directions and planning ahead rather than insisting upon so much hectic personal intervention, almost as exhausting to read about as it assuredly was to suffer or to make? And were fuel shortages, so often infuriatingly almost critical, not failures of system and prevision? Failures of Rommel?

On the latter point, Rommel has often been described as having neither interest nor grasp of logistics, especially in Africa, where the supply line from base forward has been likened to a length of elastic, liable to snap or jerk an army backward if over-extended. Similarly, in the actual army area, any force was essentially limited in its range of operations by the facts of fuel replenishment: the availability of the fuel itself in the theatre, the resources available to bring it to forward base or dump, the load-carrying vehicles, the distances to be covered, the efficiency of internal system, the turn-round times. Other commodities, notably ammunition and water, were also, of course, crucial to battle and to life, but without sufficient fuel troops and vehicles could not move or manoeuvre, and in the desert of all places battle could not possibly be won.

It is absurd to presume Rommel ignorant or careless of this matter. It dominated much of his thinking. He was a highly professional soldier, and the suggestion that to Rommel logistics were of secondary rather than central importance is wholly incredible. Furthermore he knew as well as any ally, any subordinate or any historian that success in North Africa would be dependent on supply – the fact lay at the root of his agonies when they came in the following year, and his lamentations after the event dwell upon it. Immediately after his arrival in Tripoli and first activation of the Afrika Korps Rommel had, with his quartermaster branch, initiated the movement forward of supplies in small ships to Sirte. His later preoccupation with Tobruk was essentially logistics-driven. Certainly he often complained that 'our petrol stocks were badly depleted', and that his tactics were going to be dictated by that fact: but it was to a large extent outside his control. Rommel had to base plans on forecasts and promises, and on occasion these – for reasons he entirely understood – miscarried; while his own plans were sometimes no doubt optimistic. Fuel and water, ports and communications, transportation and supply – these matters were never far from his mind. It could not be otherwise, and the matter recurs again and again in telling Rommel's story.

But did he do enough about it? Whether Rommel, by earlier intervention in planning, could have obviated some of the fuel crises which arose during the first campaign in Cyrenaica is questionable. He intervened once, directly and forcefully, after the event, as already described. Fuel replenishment thereafter, with the transport resources available, was always going to be difficult, a marginal calculation; essentially a calculation for the staff. Rommel should be blamed if, at some point, he was explicitly warned that unless he took certain steps part or all of his command would be grounded for lack of fuel, and chose to disregard the warning.

This seems unlikely, but it is possible. After major replenishment, as recounted, near Agedabia, vehicles had to travel on average 150 miles to reach the area of Mechili, and the range of armoured vehicles on full tanks, although it differed type by type, meant that replenishment would be an urgent requirement before the operation was over. Rommel certainly knew this – it was a fact impossible to ignore. Did he choose to be blind to it? Choose to believe that if quartermaster staffs tried a little harder the problem could be resolved?

Probably. It *was* resolved, after a brief lapse of time. It was only time – but Rommel, of course, grudged time – which was necessary for the replenishment of 5th Panzer Regiment, of the Italian artillery vehicles. Rommel always believed that if every subordinate and every staff officer possessed the same sense of urgency as himself time – somehow – could be saved and the difficult become less intractable. Furthermore his staff were new to him and he to them. Just as 7th Panzer Division had taken time to adjust in 1940 to their new commander's methods and expectations so did the Afrika Korps.

At the somewhat higher level of logistic management Rommel believed that staffs were inherently cautious, fearful of underprovision and of failing the command at a crucial point. He always believed in exploiting an adversary's weakness or defeat, in pursuit after victory 'to the last breath of man and beast'. He had learned it as a lieutenant in Rumania, in Italy. He had demonstrated it as a Divisional Commander in France. He was always suspicious of the advice of quartermasters, writing that they complained at every difficulty instead of developing their own (frequently non-existent) powers of improvisation. Instead of accepting the estimates of the quartermaster branch, which a commander often weakly did, Rommel wrote that he, himself, should form his own clear picture of the potentialities of the supply organization and base demands thereon. There would be grumbling but the result would be better. And in the same context he wrote that standards set by precedent are based on something less than average performance and that one should not submit to them. The commander, Rommel believed, must be a critical and informed listener to advice, and make his own judgement. That is not the picture of a man ignorant or neglectful of administrative factors.

Rommel, however, in his firm attitude to logisticians, in his belief that over-insurance is the enemy of enterprise, undoubtedly went well beyond prudence on some occasions. The Italian – and German – High Commands were often to seek to rein him in, pointing to inescapable logistic factors – the overall supply situation, the shipping situation, the length of his line of communication if he advanced as he proposed – and as often Rommel was to make his own, more optimistic, assessments of what might be possible; and be justified. Such risk-taking was deliberate. It stemmed, always, from Rommel's sense of opportunity, not from ignorance or neglect. Mistakes he may have made, like all men, but unconcerned with logistics he most certainly was not.

Rommel may not have gained a reputation for painstaking personal absorption in the detail of supply within the North African theatre before campaign, like a Marlborough before the march to the Danube, a Napoleon before the triumphant progress to Ulm, to Austerlitz. He certainly sometimes ran out of fuel – and, often by his own initiative, repaired the situation; and he more often ran the risk of running out of fuel. It is likely, however, in the sort of fluid situations, the sort of mobile manoeuvres at which he was master, that the man who never risks running out of fuel is inclined to risk nothing; and he who risks nothing is seldom crowned with the laurels of victory.

But what of the criticisms of Rommel – criticisms which would have been equally apposite on the Hilsen Ridge, at Dinant, at Avesnes, at Le Cateau – that he spent his time in perpetual motion, commanding many subordinate units in turn, out of touch with too much of his force for too long, intervening and personally stimulating and setting the personal pace to a degree inappropriate in a lieutenant-general with more than four divisions spread over many thousands of square miles? The charge would be as true as on previous occasions – and as false. Rommel could have commanded in a completely different way; but he would not have been Rommel. He would not have been what a future member of his staff, an exceptionally thoughtful General Staff-trained officer, called 'The Seydlitz of the Panzer Corps, perhaps the most daring and thrustful commander in German military history.'[11]

By now, throughout the Afrika Korps, men were speaking of Rommel's extraordinary resourcefulness, his almost superhuman capacity for work, his hardiness and lack of concern for personal comfort, his disregard of danger; and, despite his reputation for driving them remorselessly, despite his sharp tongue on occasion, despite his exacting standards, they were also speaking of his concern for them, of his fellow-feeling with them, of his utter lack of pretension or pomposity, of the fact that he was manifestly moved, and only moved, by the interests or the sufferings of his troops. He would, they said, do anything for them, and they knew it.[12] Above all, they had come to know that with him they would win. He had a '*sechsten Sinn fur Lagen*', a sixth sense for tactical situations. He was wholly decisive, and when he rapped out orders in his own voice by radio, as he often did – restricting transmissions to not more than twelve words, if necessary followed by further brief transmissions of the same sort – there was never the smallest doubt of what he wanted. He taught wherever he went, and at every level; his eye was sharp, his military instinct as sound in schooling a platoon commander as a general. He was, it was said, everyone's instructor, '*Der Lehrmeister Aller*'.[13] In the desert he knew where he was by instinct, his *Orientierungsinn* was impeccable.[14] A less interventionist commander, a less hard-driving commander, a more prudent commander, might have exposed the troops to fewer risks and fewer sufferings; but such a one would have been unlikely now, after a mere nine days and with remarkably light casualties, to have driven the British from Cyrenaica and be standing with his soldiers before the fortress of Tobruk. There were criticisms, subsequent angry recollections by some of his subordinate commanders of the changes of plan, the confusions.[15] But so far *Sonnenblume* had scored an intoxicating success.

'The Erwin Rommel of those early days is instantly recognizable as the man who half-hypnotized an enemy army twenty-five years later.' Rommel wearing the Iron Cross and *Pour le Mérite*, at the end of the First World War.

'An energetic Battalion Commander, chasing his officers up steep, snow-covered slopes, leading them on the ski-run down and leading them up again before they could collect themselves.' Rommel in the Harz Mountains.

General Hans von Seeckt, 'the dedicated and inscrutable genius who guided the Reichswehr'.

'Rommel heard that 7th Rifle Regiment was close to Onhaye.' Rommel with Colonel von Bismarck.

'Rommel rode in the same tank as the Panzer Regimental Commander, Colonel Rothenburg.' France, 1940.

'His Corps Commander, General Hoth, had certain reservations about the Rommel technique.'

'A French General, General Ihler, had already ordered General Fortune to surrender, an order the latter had for some time refused to obey.' Rommel at St Valéry with Ihler and Fortune, June 1940.

'His friend General Blaskowitz asked why, did Rommel suppose, was he, Blaskowitz, not a Field Marshal. He had refused to condone the activities of the SS in Poland.'

‘Rommel stamped his personality on the story of the North African campaign.’

‘He allowed absolutely no delays at Tripoli.’ Rommel watches German armour being unloaded.

ABOVE: Wavell. Rommel carried his essays on generalship with him on campaign.

LEFT: 'The first German troops were paraded in Tripoli immediately.' Rommel with General Gariboldi, February 1941.

BELOW LEFT: Left to right: Field Marshals Keitel and von Brauchitsch, Adolf Hitler, General Franz Halder, in the *Führerhauptquartier*.

BELOW: 'He would take his Chief of Staff, Gause, with him, against the principles of command in every army.' Rommel on a desert flight.

'Rommel and Bayerlein have criticized the British Command in the early phases of Crusader.'

'There were many moments when Rommel's bacon was saved by the sober moral courage of his staff, or by Cruewell.'

CHAPTER TWELVE

'The Seydlitz of the Panzer Corps'

ROMMEL HAD reason to be satisfied with his opening campaign so far. For a little it seemed to him that nothing could stop him, and he proclaimed on 10 April: 'Objective – the Suez Canal.'[1] The British had been driven from Cyrenaica, and the Cyrenaica airfields, together with the port of Benghazi, were in German hands – a benefit of which, Rommel soon complained, the Italians were slow to take advantage, using Tripoli to excess and Benghazi too little and thus avoidably lengthening the supply link to the desert army.* A considerable number of British vehicles and a mass of prisoners had been taken, including three generals, Gambier-Parry (commander of 2nd Armoured Division), Neame and the unfortunate O'Connor, who had not yet assumed responsibility and who had the reputation with the Germans, because of his exploits in the previous winter, of being the one British general whose watchword was not 'Safety first'.[2]

Beyond all this a great many documents had been captured at Mechili, from which Rommel's intelligence staff were able to piece together a much more accurate picture of British organization, strength and personalities than anything they had inherited from the Italians or received from OKH.

Above all, however, the port of Tobruk seemed within Rommel's grasp, and Tobruk offered the possibility of an advanced base for those operations which would always be vulnerable and laborious with the place in enemy hands. To some extent Tobruk obsessed Rommel – a wholly understandable obsession, as it was a considerable

* But see discussion in Chapter 16.

strategic prize. His first two main attacks on Tobruk took place on 14 and 30 April, and both failed entirely.

Rommel was, to some extent, intoxicated by the speed with which the advance from the Gulf of Sirte had gone, by the suddenness with which his still tiny forces had ejected the enemy from an enormous area of North Africa. He felt that fortune was smiling – as it was – and he persuaded himself that speed and determination, as so often in times past and in times to come, would carry his troops even into a major defended fortress. He persuaded himself that he could take Tobruk on the bounce, so to speak; that the enemy was on the run. It was speed he had to thank for victory, Rommel wrote,[3] but where prepared positions have to be assaulted speed is seldom enough.

Rommel had assumed that the defenders would have insufficient time to recover their balance and their nerve. In this he was wholly mistaken. Tobruk, the detail of whose defences was at first unknown to the Germans, was a hard nut to crack, particularly so because it was held by 9th Australian Division under General Morshead, which had withdrawn thither after evacuating Cyrenaica and was to be reinforced by a fourth brigade, moving by sea from Egypt. Rommel had already noted Australian prisoners – 'immensely big and powerful men', he wrote, 'without question . . . an elite formation of the British Empire as battle showed'. In Tobruk he found the Australians in position, defending stubbornly and showing no sort of inclination to yield. Their artillery fire was heavy and accurate on every occasion; their strongpoints were manned and fought with tenacity. The defences of Tobruk, prepared by the Italians, consisted of two concentric perimeters, separated by some two thousand yards from each other. In each perimeter were a number of defensive nests, concealed dugouts and fire positions connected by communication trenches, eight feet deep, and with emplacements designed for mortars, machine guns and anti-tank guns. Outside the outer perimeter was an anti-tank ditch. Each of the defensive positions was planned for about forty men, and was roughly circular with a diameter of some ninety yards. There was, of course, extensive wiring. Tobruk was formidable. It could only be taken – unless the spirit of its defenders totally failed – by method and by siege.

Rommel appreciated this, especially when, somewhat belatedly, Italian maps of the defences reached him. Meanwhile he planned to envelop Tobruk swiftly and attack it from several directions. The Brescia Division and the Trento Division (General de Stefanis) were to make maximum dust and demonstration from the west while 5th

Light Division was to envelop the place from south and south-east. The Ariete Division was ordered to move to El Adem, south of Tobruk, ready to assist from that flank. Meanwhile the new 15th Panzer Division (General von Prittwitz) had started to arrive, and Prittwitz was immediately given by Rommel command of some units already in the area of Tobruk, including 3rd Reconnaissance Battalion which on 10 April took El Adem. On 11 April the 'envelopment' was complete, and the tanks of 5th Panzer Regiment were deploying that afternoon south of Tobruk.

On the previous day von Prittwitz had been killed; he had driven into undiscovered British defences and was found dead in a trench. The 5th Division attack was in the hands of General Streich, who had little faith in the plan. Streich regarded Rommel as an opportunist, impatient and egocentric. The two men had met in France, when Streich had commanded a Panzer regiment in Rommel's neighbouring 5th Panzer Division in Hoth's corps and had formed a sceptical opinion of 7th Division's leader. Now he undoubtedly felt – and the feeling had reason behind it – that his men knew too little about the fortifications they were to attack, and that the odds were heavily in favour of defenders, entrenched, covering the whole field of attack with vision and bullets as well as observed artillery fire. Streich objected again when a similar operation was mooted some time later.

Rommel's first attempt on Tobruk, beginning at 4.30 a.m. on 14 April, was made with the machine-gun battalion and Panzer regiment of 5th Light Division from the south. It failed completely, with considerable losses. Initially Rommel reckoned that all was going well and drove forward himself at dawn, but a few hours later Streich and his Panzer regiment commander, Olbrich, reported at Afrika Korps headquarters that their tanks had become separated from the infantry and had been unable to move forward or maintain their positions due to the heavy enemy anti-tank fire from the flanks of their penetration. Furthermore – a disagreeable development – British Blenheim bombers were pounding the attackers. For a while the Royal Air Force had got the edge.

Rommel was angry. He reckoned that the coordination of tanks and infantry, the combined arms training, was defective. His opinion of Streich did not improve, and he was deeply depressed by some of the instances he found of Italian behaviour – as his Adjutant, Major Schraepler, wrote to Lucy, they did not come forward at all, or they ran at the first shot. One participant observed restrainedly that the attackers suffered from the perhaps decisive disadvantage of inadequate knowledge of the defences,[4] a disadvantage apparent to Rommel, whose judgement it nevertheless was that the benefits of intelligence would be offset by the penalties of delay. The judgement proved wrong.

Rommel made a further attempt two weeks later on a commanding point in enemy hands eight miles west of Tobruk, Ras el Madamer, and ultimately took it, not without difficulty, thus eliminating on the perimeter a place from which the enemy could observe and harass his supply convoys from the west. But before this happened he had a visitor from Berlin.

General Paulus, head of the Operations Branch at OKH, arrived on 27 April to consult and assess the situation. Rommel had astonished his superiors with the speed and depth of his advance. He was known to have talked to his staff about reaching the Suez Canal. He had, alleged OKH as relayed by Paulus, so dismayed the British that they had withdrawn prematurely from Greece, thus defeating a German design whereby they would have been trapped there. (This last was totally untrue. Evacuation of Greece by British forces was agreed on 19 April between Wavell and the Greek Government; it had nothing to do with Rommel's progress and everything to do with the threatening situation within Greece itself.) Rommel replied, with irritation, that he had known nothing of any German plan to keep the British trapped in Greece. Anyway, he said, the Greek adventure was a mistaken dissipation of force from the German point of view – better far to have concentrated more forces in North Africa and aimed thereby to deny the Mediterranean to the British than squander them in the Balkans. And, as he wrote later, better far to attack Malta, a key to sea communication between Italy and Africa, than Crete, invaded by the Germans in May. Thus might well have written O'Connor!

Paulus was a contemporary of Rommel in age, a courteous, cultivated and intelligent man, an extremely hard-working staff officer but with a less than dominant personality. He had authority to sanction any major move proposed by Rommel or refuse it;* and he sanctioned a further attempt on Tobruk, an attempt which took place (Paulus present) three days after his arrival. The attempt again failed. Rommel had originally intended to attack from the west with whatever had by now arrived of 15th Panzer Division, a manoeuvre

* Another noteworthy German intrusion into an Italian domain.

to which Streich had already raised objections and about which he was pessimistic; the attack was in fact limited to the operation against Ras el Madamer and cost Rommel a good many men. He claimed it cleared his supply line of enemy observation, but it certainly did not offer opportunity to develop that movement on the main Tobruk position Rommel had hoped would follow.

But Paulus observed Rommel in action, and when he reported back to Berlin after a fortnight in the desert he described a headstrong, tempestuous commander of the Afrika Korps, whose wilfulness could yet involve OKH, against its better judgement, in reinforcing Africa to the detriment of the enormous operation in the east on which all minds at OKH, and especially that of Paulus, had for some time been bent. Paulus noted that Rommel complained about Italian failure to use Benghazi to better effect, but disregarded the fact that the sea route to Benghazi was longer than that to Tripoli and thus more vulnerable to British interference, while the port capacity of Tripoli was greater. Rommel, of course, would have pressed methods of increasing Benghazi's capacity and would have pointed to the balance of advantage, of risk, in undertaking a rather longer voyage for a dramatic improvement of supply to the field army, but to Paulus the rational case was that of the Italians against Rommel.

The comments on Rommel are by no means surprising. There is no particular need to invoke the story that Paulus contemplated recommending Rommel's replacement by himself, but was warned by his beautiful and ambitious Rumanian wife that North Africa would not be the place to make a reputation.[5] Paulus was very different from Rommel as a type. A staff officer with little experience of command, he was by general consent a man to whom the letter of orders had absolute authority. Rommel's readiness to back his own judgement, to act first and argue afterwards, to exploit his distance from Berlin and his nominal subordination to Rome, to hope that victory would be his justification and his advocate – such attitudes were anathema to the meticulous, efficient Paulus.

Paulus was also appalled by the conditions under which the troops around Tobruk were fighting, and formed the view that Rommel would do well to withdraw to Gazala, thus shortening his supply line and improving matters for the soldiers. Rommel, sharing his soldiers' food and hardships at all times, thought differently. It would be a further twenty-one months before Paulus presided over the destruction, the grinding into death or captivity, of the entire German Sixth

Army at Stalingrad, having scrupulously observed Hitler's murderous order that the army must not attempt to break out through the Russian encirclement, that it must stand its indefensible ground. A brother officer of both men observed long afterwards that a Rommel, at Stalingrad, would have signalled, thrillingly, to Hitler: 'Am passing to the offensive, attacking,' and, with every man gathered, would have broken out just the same.[6] Whether this can be sustained is speculation – Rommel later, and tragically, observed a *Führerbefehl* as rigidly as Paulus, for a few precious and irrecoverable hours at Alamein; but the observation bears witness to the essential difference between the impressions made by the two men.

Meanwhile, however, Paulus was unimpressed; and the 1941 attacks on Tobruk undoubtedly show Rommel at his worst – leading troops into battle hastily and without preparation or coordination, sacrificing method to speed in a way which the situation condemned. To some the magic touch, the *Fingerspitzengefuhl*, the Rommel instinct for victory was temporarily absent. There had been much chaos – less important, perhaps, than the essential momentum – on the advance across Cyrenaica, and now there were angry mutters from some commanders. The casualties at Tobruk reinforced them. 'Those were my men he sacrificed,' said Schwerin with understandable bitterness, years afterwards.[7] The accusation of Rommel being a gambler was made, and here and there it stuck. A good many men died.

Rommel now had to accept that for a little Tobruk must be invested rather than stormed. Meanwhile on the Egyptian frontier, beyond which the British had withdrawn their main forces, he aimed to establish as strong an outpost line as he could manage – 'outpost' because he regarded the frontier defences and complex based on Sollum, Bardia, Sidi Suleiman and Fort Capuzzo as essentially what used to be called a 'line of circumvallation': the positions a besieger established to prevent a relieving army marching to lift a siege, relieve a garrison. Rommel's eyes were on Tobruk. He certainly did not ignore the frontier but his attention – perhaps excessively, as matters turned out – was focused on Tobruk and, while he was ready to deal with any opportunity, he regarded himself as the besieger of Tobruk. Nor was this unreasonable. To take Tobruk and use its port would have been of considerable value to the German–Italian army. Rommel's forces on the frontier necessitated a long and difficult supply chain from Cyrenaica, moving round Tobruk; and the daily replenishment requirement of Rommel's forward forces was 1500 tons.* Tobruk remained a thorn in the side, a '*Dorn im Fleisch*', of the Afrika Korps.[8] Later Rommel persuaded the Italians to improve east–west communications round Tobruk by a bypass road, the *Achsenstrasse*, constructed in a creditably brief three months in the autumn of 1941.

For Rommel it was a matter of priorities: how much did he need to keep near the frontier to tackle the British if they attacked to relieve Tobruk, as they surely must, and how much could he devote to another attack on the place? And when might that be? And when would a British attempt be made? He had followed up the British as far as the Egyptian border. Bardia had been taken by 3rd Reconnaissance Regiment on 12 April while Rommel was still wholly absorbed with Tobruk – he did not visit it for a further week. Sollum and Capuzzo were seized at the same time. At all times, too, Rommel appreciated that he must keep a mobile force in hand to counter any British attempt to swing round the southern, desert, flank, whether to encircle his forces on the frontier or to drive straight for Tobruk or even behind and west of it, a deep thrust menacing Bir Hacheim and Gazala.

Fighting near the frontier would be greatly influenced by terrain. In that area the coast road is hemmed by a steep escarpment, offering only a few points at which vehicles can climb to the desert plateau or descend from it to the coast. These points became passes of considerable significance, and the pass nearest to Rommel's forward positions was at Halfaya, five miles south of Sollum. Possession of the Halfaya pass gave or denied the ability to move between desert and coast road.

Rommel planned to move an Italian division, the Trento, to the frontier, and to deploy another, the Brescia, on the east flank of Tobruk as part of the investing front around the fortress. Unlike his enemy he had a considerable force of non-motorized infantry, Italian infantry; and the only use for dismounted infantry in the desert, with its enormous open spaces and huge fields of fire, was in places where terrain and conditions gave some practicability and significance to static positional warfare. The frontier positions, including the Halfaya pass, offered one such; the static investment of Tobruk (until a major attack could be mounted) another. Non-motorized infantry

* The *total* import bill into the theatre had reached seventy thousand tons monthly; more than Tripoli could handle.

divisions, however, not only suffered the limitation that they were of little use in manoeuvre operations, having no ability to fight except when dismounted and dug-in (and then at very short effective ranges, vulnerable to stand-off tank fire and observed artillery fire) – they were also a potential liability to a commander in that transport had to be provided to move them: stranded in the desert they were hostages with little hope of survival. Because they had no mobility, their effective deployment, in such places as suited their characteristics, demanded sufficient numbers, and Rommel decided to ask Commando Supremo for two more Italian divisions. He hoped to minimize the extent to which his mobile forces, above all the Afrika Korps, would ever be committed to any sort of positional warfare or pre-ordained manoeuvres.

These redeployments took time, and Rommel did not believe he had a great deal of time. The British, he assumed, would appreciate how over-extended he was, attempting both to hem Tobruk and fend off any relieving force. He felt particularly short of German troops and asked for the build-up of 15th Panzer Division to be expedited. The position at Sollum was not fully taken over by Italian infantry even by mid-May. On 14 May Rommel's intelligence staff heard on British radio nets, down to subordinate levels, a codeword: 'Fritz'. It was not clear to them from this what was about to happen, but the following day the expected British attack was launched and a connection was made, profitable for the future.[9]

Rommel's battle intelligence in North Africa, his knowledge of his enemy, derived from a number of sources.

First, of course, was direct observation by the Luftwaffe and by his own troops, patrols, reconnaissances. Nothing could replace these, and initially the Luftwaffe, in particular, had things much their own way. Later the British Desert Air Force won mastery and the British knew more of Rommel from air reconnaissance than he of them. His reconnaissance patrols, watching the deep desert flank, were necessary, to watch and ward; but he seldom got the measure of the British raiding forces, in particular the Long Range Desert Group, of which German intelligence staffs were very aware, and whose strength and exploits they sometimes magnified, even (in December 1941) faultily deducing from a captured personal letter (addressed to a British officer at his parent regiment and battalion) that the entire 22nd Guards Brigade might have been converted to Long Range Desert Group duties![10]

Second were captured documents, of which many were recovered, sifted and analysed after the taking of Mechili and the overrunning of the British 2nd Armoured Division's headquarters. Any headquarters, and most units, held considerable stocks of classified documents from which a picture of strengths, weaknesses and conditions in the enemy camp could be constructed. Nor were only official documents useful. A number of officers kept diaries – actually forbidden in the British as in the German army – and a number of these were in time captured, and gave useful information about the views – and the spirits – prevalent in the enemy camp, both in the field and at home.

Third was the interrogation of prisoners. The Germans admired the British order whereby a soldier was taught that if taken prisoner only his army number, age, rank, name and home could be given to his interrogating captor. On both sides, of course, attempts were made with differing degrees of success to get more than that, and the usual techniques of interrogation, the varyings of flattery, hectoring, insinuation and inducement were tried. Rommel – like Montgomery later – enjoyed meeting and talking to enemy prisoners, particularly although not solely those of distinction or senior rank. He would chide them with any breach of the rules of the game – for the Desert War was, to some extent, a grim game, and it had rules. When he saw (later in the campaign) Brigadier Stirling, the captured deputy commander of the British 4th Armoured Brigade, he rallied him, telling him that the first days of captivity were inevitably the worst, but that the Germans esteemed brave soldiers. All the more deplorable, said Rommel, to see (as he had seen) photographs of wounded Italian soldiers at Tobruk, mutilated by their enemies after capture – 'the behaviour of beasts, not men'. Stirling reacted vigorously. No British soldiers would have done such a thing – possibly it was the work of Abyssinian auxiliaries. To which Rommel countered with expressions of regret that the British had employed such people against whites, '*gegen Weisse in dem Kampf*' – a not uncommon European reaction at the time. The conversation ended pleasantly, however, with Stirling asking the interpreter (a German Naval Captain) to tell Rommel that he admired him; and Rommel, with a handshake and a smile, hoping that Stirling's captivity would not last too long, and remarking that there was surely room in the world for both the British and the Germans without need for fighting – a sentiment (according to the German war reporter who publicized the interview) with which Stirling heartily agreed.[11] Such exchanges yielded little information, but they conveyed, to an extent, the atmosphere of the Desert War.

Again, in September 1942, at a critical point in the campaign, the New Zealand Brigadier Clifton talked to his captors about personalities, about Churchill's recent visit to Egypt, and about higher command changes, including the advent of General Alexander. Clifton told the Germans they had missed their opportunity – a few weeks earlier they might, he said, have reached Cairo and Alexandria.[12] Such conversations – natural and not very difficult to prompt in the immediate shock of captivity – were invariably interesting to Rommel. He recorded his admiration for Clifton, a brave and inveterate escaper. Like every distinguished commander Rommel was absorbed by study of the mind and psychology of his opponents, and trivia or prisoners' opinions relayed to him by his intelligence staff all helped compose the picture. His own conduct to prisoners was invariably chivalrous, and those who saw him in person could seldom avoid being powerfully impressed. Such exchanges between captor and captive were irresistible to both sides. Two months after Clifton's capture General Montgomery entertained to dinner General von Thoma, the captive commander of the Afrika Korps under Rommel, and formed from him some thoroughly useful as well as gratifying impressions.[13]

Occasionally, of course, some nugget of information from a prisoner might suffice actually to affect decisions and the course of battle – as when, during the battle of Alamein in October 1942, German interrogators reported that, in conversation in a British prisoner-of-war cage, a prisoner 'who had lived a long time in Paris'* was giving as his opinion that the next British main thrust was coming in the north of the front rather than the south. Rommel was impressed by this. It clearly corresponded to his own instinct, although corroborative evidence may have been non-existent or scanty: and that evening (28 October 1942) 21st Panzer Division was moved to the north of the Alamein front – a movement already started despite some German misgivings because of the paucity of armour in the south but now pressed forward with more assurance and, as it turned out, wholly justified.[14]

But by far the most productive and timely battle intelligence available to Rommel came from interception of British signals traffic. In 1941 British radio security was poor, and Rommel's staff had little

* Why this strengthened his credibility is obscure, but it did.

difficulty in piecing together from it a picture of enemy strength and positions which led Rommel to boast that he reckoned he knew more than the opposing commander about where British units and formations were. The science and interpretation of radio interception, on both sides, was an evolving art, and each side had periodic reason for self-congratulation, but, at least in the earlier days of the desert war, German battle intelligence from this source was notably superior to British.[15] The Germans noted that British codes were not changed sufficiently frequently (this was remedied later in the war); that lack of radio discipline often made the relationship between stations evident – a rebuke, for instance, would instantly establish relativity; that 'disguised speech', whereby a word would be addressed obliquely was often clumsily obvious – Rommel's Ic observed that the British must be naïve indeed if, rather than saying 'London', they imagined they had baffled interception by referring to 'capital of England'! British reports in clear (uncoded), for example of Luftwaffe attack, were often picked up and used to correct the German pilots' targeting. Rommel looked at the intercept reports, produced by the highly skilled Oberleutnant Seebohm, every evening, and valued them highly.

Knowledge by both sides that the enemy were inevitably and invariably on the line could sometimes serve different and humanitarian purposes. A severely wounded British lieutenant, taken prisoner in Cyrenaica, was anxious that his wife (who was in Cairo) should if possible learn that he had survived before the procedures of the Red Cross ran their course; the Germans passed a message, naming him, in clear, and had it acknowledged.[16]

Between reconnaissance patrols far out in the desert, watching each other and watching the enormous southern flank of both armies, direct radio communication sometimes took place. A patrol was occasionally captured, and a message from the enemy would be received enquiring about the whereabouts of Lieutenant or Corporal so-and-so; an answer would be sent, sometimes leading to an exchange – of prisoners or (at least as requested) of soldiers for cigarettes! Such traffic was consistent with a war in which observation patrols could do little useful in open desert during the hours of darkness, and which – at some periods and in some areas – was marked by an explicit agreement to take no hostile action between specified times.[17] Rommel, when he learned of these irregularities, was wholly content with them. He was a soldier to whom hatred or even churlishness towards an enemy was anathema. In a sense war was to him a sport, foolish perhaps, grim perhaps, serious certainly, often tragic in its personal consequences, but a sport in which he, Rommel, was above all things a professional. His own reflections would be one day produced under the admirable title *Krieg ohne Hass*. War without hatred.

On 15 May the British on the frontier attacked.

The first move was made by the 7th Armoured and 22nd Guards Brigades, formations at once identified by Rommel's staff. German suspicion had been aroused by an unusual and absolute enemy wireless silence. Rommel learned that the British had taken Halfaya pass and inflicted considerable casualties, moving above the escarpment on Capuzzo and Sollum. The British High Command knew from ULTRA of the impending arrival with Rommel of 15th Panzer Division (much of it, in fact, already in the desert; commander, General Freiherr von Esebeck) and wished to pre-empt, to knock Rommel off balance and worsen his position before he was further reinforced.

Rommel on the same day sent more troops forward – a Panzer battalion with some 88s – but in a confused situation the British, finding the German reaction stronger than anticipated, withdrew, leaving a garrison at Halfaya. Rommel was somewhat perplexed but he had no intention of leaving Halfaya in British hands, and on 27 May he organized a three-pronged attack and recaptured the pass. These actions, albeit small-scale, exemplify a facet of Rommel's – indeed to a large extent of German – command technique, already noted. He used two 'groups', one commanded by Colonel Herff and the other – the reinforcement on 15 May – under Lieutenant-Colonel Kramer. Rommel had, at that time, 5th Light Division and 15th Panzer Division (still arriving), but he tended to use his divisions as, in some sort, military supermarkets from which he selected and threw together units and arms into a group composed for a particular operation and, certainly as far as their enemies could observe, working perfectly happily together as to communication and the cooperation of all arms. Rommel often complained, especially in the early days, about the training of his formations, but the flexibility displayed by the Afrika Korps in this respect was impressive. It spoke, above all, of admirable signals and command systems. And Rommel, in a letter to Lucy on 23 May, after visiting the frontier positions and before launching his counter-move against Halfaya, wrote that he had come back from the frontier impressed. Command 'up there' was good. The machine, clearly, was beginning to be more instantly responsive to Rommel's will and Rommel's way.

Rommel was also, at this time, engaged in perfecting certain tactical techniques, simple in essence but demanding exposition. Above all he developed the practice of drawing, or attempting to draw, enemy armour on to his own anti-tank guns while aiming to keep his armour for manoeuvre against the enemy's more vulnerable targets – supply columns, dismounted infantry, headquarters. This involved the bold use of anti-tank guns, many of them towed (although Rommel was setting his excellent workshops to improvising gun-mountings on chassis, often captured chassis). In the desert the side with the most powerful gun had a huge advantage, as there were few obstacles to vision or to engagement at maximum range. The German 88-millimetre anti-aircraft gun, by now regarded by both sides and feared by the British as a desert *dominatrix*, was the largest anti-tank gun in the game, but those of smaller calibre, 75s and 50s, were also used to good effect. Rommel taught the virtue of pushing out screens of guns and then manoeuvring against enemy armour in such a way as to lead British tanks upon a line of guns in position; and of coordinating all sorts of anti-tank weapon so that each could open fire at optimum range. He had also formed a reserve of artillery under Afrika Korps command – the Bottcher Group.

Rommel, however, was feeling uneasy, and suffering from a certain volatility of temperament, no doubt exacerbated by the climate of Africa and the disagreeable physical conditions under which he and his troops were living. He was rebuked by the Commander-in-Chief, von Brauchitsch, for the lack of restraint in his reports – exaggeratedly up or down, von Brauchitsch wrote, rather than temperate and balanced (the message of Paulus was clearly penetrating).[18] Rommel, angry at the criticism, assumed that a contributory reason for OKH's displeasure was that he was also sending his reports to OKW – a procedural confusion, as he wrote to Lucy. He undoubtedly adjusted his reports towards the achieving of particular effects, exaggerating shortages and difficulties, for instance, to stimulate effort. (This device had unpredicted consequences, in that his messages were read by ULTRA, leading London to assess his situation as worse than it was and chide the British commanders for failing to seize their opportunities; Berlin was probably more sceptical.)[19]

Rommel felt a lack of appreciation of his efforts in Berlin, a failure to realize the enormous stride across the map of Africa made by the German–Italian army, a failure to realize the great strategic benefits which could follow a successful Mediterranean campaign. He knew that he could do little to improve his overall supply situation because decisions on maritime and shipping priorities were made elsewhere – he could only state his requirements and base his plans on hopes and promises: but when the Italians told him the limitations of Tripoli's or Benghazi's port capacity he could point to their failure to build or improve certain roads on the line of supply which would have materially eased the requirement by easing the journey forward. And he periodically fulminated against the Italian Navy's reluctance to give trans-Mediterranean supply protection the priority it deserved – a reluctance he sometimes ascribed to treachery, to hostility to the regime and the German alliance. He thought Berlin was too easy, too diplomatic with Rome. Rommel, at the beginning of June 1941, was feeling that every circumstance and most men were conspiring against him or failing him. Anybody who has spent the first months of any year in that part of Africa knows the feeling.

Rommel knew that the British must shortly attack again, and he felt his troops thin on the ground and vulnerable. 15th Panzer Division was now pretty well complete but he appreciated that the British, too, had been reinforced, perhaps rejuvenated. He replaced the commander of 5th Light Division, Streich, with General von Ravenstein, of whom he thought highly, another holder of the *Pour le Mérite*, won in 1918; and was soon writing of the division's transformation under the 'intelligent and firm leadership of its new commander', reporting in the same letter (to Brauchitsch) that he was now satisfied with the quality of the officers of the Afrika Korps.[20] There had been something of a purge. Rommel had dealt savagely with a tank battalion commander who had broken down emotionally during the last attack at Tobruk; and he had changed the commander, Colonel Olbrich, of 5th Panzer Regiment. Anybody who challenged Rommel's judgement – as Streich had done in the Tobruk assault; or apparently shown insufficient drive and initiative – the verdict on Olbrich at Mechili; or cracked in battle itself – such officers were broken swiftly and remorselessly.

This was often thought unjust, and probably often was. Streich, for instance, was regarded as an excellent divisional commander by his officers, and his remonstrances with Rommel over the first assaults on Tobruk manifestly had substance. Afterwards he claimed that Rommel told him he was too concerned for his troops, to which he had returned the admirable response that in his eyes no compliment to a divisional commander could be finer. Rommel, Streich said, was speechless at that – and, perhaps, humbled; for Rommel, too, was careful of his soldiers' lives. His vigour in attack, his pertinacity, stemmed generally from the judgement that this would be more economic in the longer run, brutal though it might appear; the judgement of a Marlborough at the Schellenberg. Sometimes the judgement was wrong.[21]

But in spite of his ruthlessness, his difficult moods, his rough words, his apparent reluctance to admit error, his intolerance, to the vast majority of the Afrika Korps he was unique, he was 'Rommel'. He was harsh but totally unpretentious, brusque but wholly without pomposity. Men immediately felt his charm, his warmth of heart. He drove them, but he drove himself hardest. He loved his 'Afrikaners', would do anything for his soldiers and they for him. When he won a victory he always knew how to find the right words with which to thank them. It was their sufferings and their sufferings alone which sometimes nearly crushed his heart.[22] Larpent, British Judge Advocate General in the Peninsular War, tells the story of how Wellington, learning of wounded soldiers in a certain village in northern Spain lying in the streets while officers had cover, ordered his horse and rode twenty miles through darkness to the village in question, not leaving it until he had turned the commanding officer and every other officer out of his billet and seen the wounded soldiers housed.[23] In similar fashion Rommel once visited a dressing station and found wounded soldiers lying on the sand while officers had been placed on beds or boards. The staff there felt his tongue and the unfortunate officers were changed over with the slightly more fortunate men. Such was Wellington. Such was Rommel.[24]

On 14 June Rommel's interception service reported that a codeword 'Peter' had been passed on British nets to subordinate units. This was, must be, a repetition of 'Fritz'; a warning of an imminent major operation. Troops on the frontier were alerted.

The British called it 'Battleaxe'. The Germans called it the '*Sollumschlacht*', and Rommel's intelligence staff and radio interception service regarded it as the high point of their achievement in the Desert War. Certainly they plotted the movements and deduced the intentions of the enemy with a good deal of accuracy, materially assisted by the capture of a list of British codenames and call signs at midday on the first day of battle, 15 June. This was particularly gratifying since British radio security in the previous weeks had been noticeably more effective, and German amendment of the enemy order of battle had been, they themselves reckoned, inadequate.[25]

At the highest level of command the British were under political pressure. With considerable risk to the Royal Navy the British Government had sent a convoy to Egypt through the Mediterranean – Italian maritime and maritime air activity meant at that time that the only safe route to Egypt for the British was round the Cape of Good Hope. The convoy had carried, among other reinforcements of men and equipment, 240 tanks. It had reached Alexandria on 12 May, and the British 7th Armoured Division, withdrawn from the desert and virtually without equipment since February, had been re-equipped. The Commander-in-Chief, Wavell, thus had a refurbished armoured division: and had moved to Egypt, after a triumphant campaign in Abyssinia, the 4th Indian Division with one Indian brigade, to which was added the 22nd Guards Brigade. These two divisions were now combined in XIII Corps, as the original Desert Army had been designated, with a new commander, General Beresford-Peirse.

Wavell was being powerfully urged by London. The British were sore at their somewhat ignominious expulsion from Cyrenaica two months previously, and the Government was conscious of having taken serious risks to reinforce North Africa; Wavell ordered Beresford-Peirse to carry out an offensive. XIII Corps was to attack the Germans in the frontier region, secure the Halfaya pass and advance thereafter to relieve Tobruk. If successful, Beresford-Peirse was ordered to exploit, after Tobruk, to the line Derna–Mechili, the inner border of Cyrenaica. The British knew from ULTRA of Paulus's visit a few weeks previously. They knew that Rommel's wings had been clipped, that he had been instructed to stand on the defensive.

Rommel had strengthened his gun numbers by taking into service a good many Italian pieces he had found in the frontier area, abandoned the previous December: he had them repaired in his workshops and in some cases mounted on chassis. There were on the frontier a total of forty-six anti-tank guns, including thirteen 88s. A large number of these were deployed on the Hafid ridge, seven miles west of Capuzzo, and others around the defensive positions at Halfaya.

The British attack began at four o'clock in the morning. Rommel had infantry garrisons in the fortified places of Sollum, Capuzzo and at the Halfaya pass; he held 15th Panzer Division – as a division new to the desert and with its Panzer grenadiers (who had arrived at

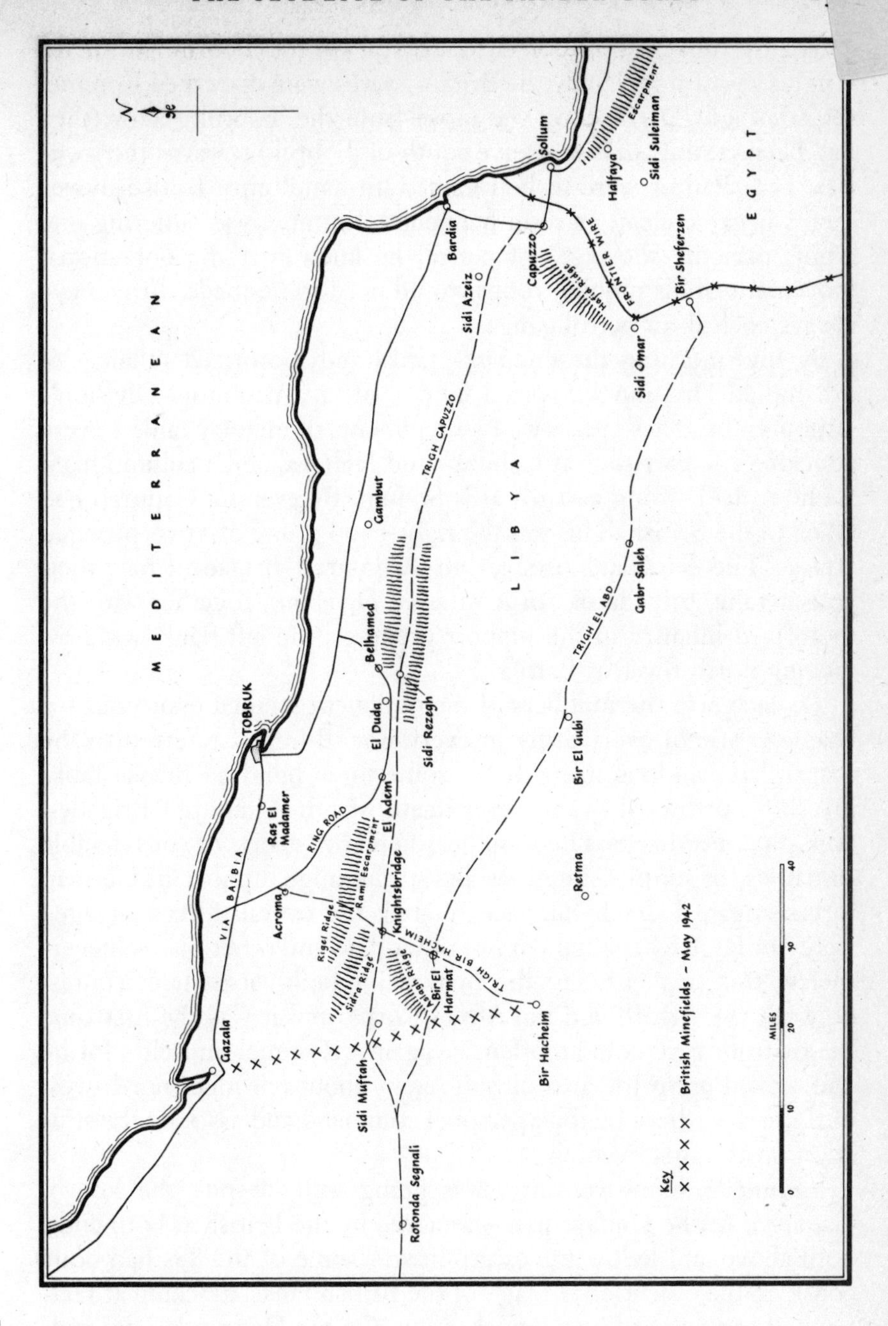

XV Area of operations, 'Battleaxe' (summer 1941), 'Crusader' (autumn–winter 1941) and Gazala (summer 1942)

Derna by Junkers transport aircraft) still on the Tobruk perimeter – in reserve. Immediately the British moves were discerned Rommel set 5th Light Division on the move from the Tobruk sector (they had been resting and in reserve south of Tobruk) towards the frontier. 15th Panzer were to be retained 'in hand' until British movements became clear, as soon happened. Rommel was suffering one of his periodic shortages of petrol: he knew that the operational movement of his mobile troops would need to be made with an eye always cocked on petrol gauges.

By mid-morning the enemy – tanks and motorized infantry of 4th Indian Division supported by one of 7th Armoured Division's armoured brigades, 4th, with two regiments of infantry tanks – were attacking the garrisons at Capuzzo and Halfaya, and a column from Capuzzo had swung east towards Sollum. By evening Capuzzo had fallen to the British. This was the right-hand prong of a two-pronged attack. The left-hand prong, 7th Armoured Brigade (the other, cruiser tank, brigade of 7th Armoured Division), together with the motorized infantry of the support group on the left flank, was now moving north towards Bardia.

To each side the numbers of the opponent seemed enormous – a characteristic of every army in every war. Rommel referred to the British 'tremendous strength', to some three hundred British tanks pressing northward – an overestimate of 7th Armoured Brigade's tank state. For his part he soon heard enemy reports of considerable casualties, of serious situations and of the huge number of German forces engaged. Each side exaggerated; the overall forces engaged were not large by the standard of other days and other places; nevertheless this was virtually the first major tank-versus-tank contest between the British and the Afrika Korps, and it was the first time the British carried out an offensive against Rommel himself – for he had moved up to the area of Sidi Azeiz, about ten miles north-west of Capuzzo, where he took personal command and assessed the situation on the first evening.

Rommel's defensive battle was going well, despite the loss of Capuzzo. At the Halfaya pass – attacked by the British at both ends, both above and below the escarpment – some of the 88s had done considerable execution. In fact, of the British tanks attacking at Halfaya, all but one had been knocked out. On the Hafid ridge, towards which the British 7th Armoured Brigade, the left-hand prong (assessed by the Germans as surely intended to be the main attack, the *Schwerpunkt*) had advanced on the first morning, Rommel's anti-tank guns had also scored highly, their greater effective range in open desert proving of battle-winning quality. Rommel was learning of enemy losses to his anti-tank guns: he was able to chart the progress of the British prongs of attack with a fair degree of accuracy through radio interception, while air reconnaissance flew regularly: and his own mobile forces were to some extent still in hand although the Panzer regiment of 15th Panzer Division had been committed, had lost a significant number of tanks in tank-versus-tank combat around Capuzzo and was at less than half strength by evening. But although Rommel did not know it his enemy, Beresford-Peirse, would have only thirty-nine tanks – twenty-two cruisers and seventeen infantry tanks – by mid-morning on 16 June, the second day of battle, having started Battleaxe with two hundred.

Rommel decided that the moment was right for a counter-move, an 'operation'. At dawn on 16 June 15th Panzer Division was to attack south, on either side of Capuzzo, striking, as Rommel saw it, both the right- and the left-hand British prongs of attack and holding them. 15th Panzer would probably be outnumbered, but its function was to pin the British frontally. Meanwhile 5th Light Division was to attack southward on the western flank, to march by Sidi Omar and swing eastward towards Sidi Suleiman. This, Rommel hoped, would cut across the stem of the British advance. He directed 5th Light Division, after turning east, on the Halfaya pass as its objective. Rommel had no expectation of the operation being easy. He scribbled a line to Lucy at half-past-two in the morning – it was going to be a hard fight.

It was a hard fight, and it lasted throughout 16 June. Both on 15th Panzer's frontal attack and 5th Light's right hook there were considerable tank-versus-tank engagements and, on both sides, losses; but after a vigorous encounter between 5th Light and the cruiser tanks of 7th Armoured Brigade west of Sidi Omar Rommel reckoned that the mobile battle was going his way. Intercepts seemed to show the British bewildered. He ordered 15th Panzer now to disengage, to move west and join 5th Light for a concentrated armoured thrust towards Sidi Suleiman and thence Halfaya. The movement started at 4.30 a.m. on 17 June. The Afrika Korps reached Sidi Suleiman at six in the morning, and Halfaya at four in the afternoon.

By then the British had evaded Rommel's right hook operation and had withdrawn, with a total loss of ninety-one tanks. The Germans' loss of tanks destroyed was twelve. They had deduced –

correctly – from interception that the enemy Commander-in-Chief himself had visited the front, and it was soon apparent that the visit had led to abandonment of the British offensive. Battleaxe was over.

The three-day battle, Rommel wrote to Lucy on 18 June, had ended in complete victory. His jubilation was amply shared by the Afrika Korps. Despite the local superiority which the Royal Air Force had attained over the Luftwaffe, he felt – and his men felt – that with forces outnumbered overall they had shown themselves the masters; and they also felt, with justification, that their equipment was superior to their enemy's. Rommel paid generous tribute to his Italian infantry, some of whom had fought extremely well, particularly at Halfaya where the German–Italian garrison, under the heroic Major Bach,* had put up a formidable defence. He paid tribute to the improved tactical handling within the Afrika Korps, particularly the coordination of tanks and anti-tank guns, their trained and rehearsed method of leapfrogging forward during an attack.

It is also possible to detect in Rommel himself an increased assurance in his handling of his small army. Hitherto he had led – and emphatically led rather than organized or orchestrated – a division in France in what had rapidly become a pursuit. He had done, or attempted to do, much the same in Cyrenaica, and it had taken him to Tobruk. He had failed there – hoping to fight one sort of battle, he had found himself confronted with another. He had undoubtedly learned: it is a fallacy to suppose that any commander, however naturally gifted or inspired, does not need to learn, to practise, to experience in the doing, the problems of command at a superior level. In Battleaxe, the *Sollumschlacht*, Rommel's handling was more assured, his timing more carefully considered than before.

Rommel hoped that his demonstration of success would convince his superiors, and especially Commando Supremo, of the possibilities of the North African theatre and lead to greater emphasis on its support. He did not know, when Battleaxe finished, that within five days a completely new chapter in the war would begin, a chapter which would henceforth dominate the strategic situation for Germany and lead the German nation to ultimate catastrophe.

On 22 June 1941 Germany invaded the Soviet Union, and the German army began the long march into Russia which would take it in

* A German Evangelical pastor, as well as an outstanding soldier, and known to his men as 'Vater Bach'.

less than four months to the gates of Moscow, to the envelopment of Leningrad and to the Sea of Azov. This had long been Hitler's intention. He had made clear, at least to intimates, that the Polish campaign was only a step in a process which would yield land and living-space for an enlarged German people. This living-space lay in the vast expanses and food-producing potential of Russia. Hitler – like Stalin – had regarded the German–Soviet Pact of 1939 as an expedient, inevitably temporary: and he breached it first. Unlike many of his predecessors in charge of German foreign policy Hitler did not believe that Germany's destiny lay in friendship to the east. To the east were lands to be conquered, exploited and settled; lands inhabited by peoples of inferior racial characteristics, born for various degrees of servitude.

There was also, of course, an ideological dimension to all this. The Soviet Union, home and fount of Bolshevism, represented for Hitler an evil to be extirpated: not, as others would have argued, because of its savagery, its irreligion, its destructive violence towards all which had gone before, its commitment to the doctrine of the universal class struggle; but because, to Hitler, it represented internationalism, it represented in some way Jewry. Hitler had always equated international Communism with international Jewry, drawing as evidence on Marx himself and on the number of Jews distinguished in early Bolshevik history. It suited him to meld his enemies and his aversions into one identifiable target, regardless of consistency or fact.

The anti-Communist facet of the German invasion of the Soviet Union at once gave a completely different character to the war. Firstly, Communist Parties throughout the world now discovered that the war against Germany was a war on behalf of the proletariat, whose vanguard, the Communist Party of the Soviet Union, was physically threatened. In Western countries still undefeated by Germany such as Britain, or in neutral countries such as the United States, Communists had hitherto opposed the war as a capitalist-imperialist venture without relevance to the class struggle. Now the war was a crusade, albeit one whose conduct and aims would need careful watching (and guidance from Moscow).

Secondly, in a Europe largely occupied by the Wehrmacht and in a not unfriendly relationship with Germany, reluctant or otherwise, there was born among some a certain idealistic response. This campaign might, indeed, be a crusade – an anti-Bolshevik crusade. Volunteers came forward from non-German nations to join the

Wehrmacht, and in spite of its harsh discipline and their own sufferings sometimes retained for it a memory of extraordinary affection.[26] Waverers in opinion, repelled by some of what they heard of the Nazis but respectful of German achievements, felt that in this struggle if in no other matter the German cause was in some sort their own. Men of property, perhaps more fearful of Communism than of anything else, in some cases felt likewise. And in Central and Eastern Europe resentment of Germany – a resentment amounting in many cases to irreconcilable hatred, especially in Poland and Bohemia – was to some extent offset (increasingly as the years passed and the tide of the German invasion ebbed) by terror of enforced Communism and of the Red Army. Even in the stable democracies of the British Empire and the United States the German onslaught on the Soviet Union (and the valiant Soviet resistance) gave a peculiar colouration to the war, defined political attitudes, often induced unease and made most things less simple than they had been hitherto.

Above all, of course, the invasion – Operation 'Barbarossa' – transformed the strategic situation. The chessboard and the pieces upon it now appeared completely different. For the Germans, the Eastern Front soon became in most important senses 'the war' – the Eastern Front, and, from the west, the aerial bombardment of the Reich which was soon to come, increasing in scale until much of urban and industrial Germany would lie in ruins.

The Wehrmacht invaded Russia with 145 divisions,* thirty of them either Panzer or mechanized, a total of over three million men. The bulk of the German armies consisted of marching infantry, supported by horse-drawn guns and transport (over 600,000 horses took part in the invasion, and the ordinary German soldier marched into Russia on his feet – and ultimately, if he survived, marched out again), but the mobile forces soon achieved astonishing successes. Against all this the size and feats of the Afrika Korps appeared puny indeed.

A considerable degree of surprise had been achieved, despite the mutual mistrust which had been shared by the two opposed dictators: Stalin had persuaded himself that Germany would not march against the Soviet Union until Britain had been beaten or at least made some sort of peace; and to some extent the exaltation of Rommel's

* Including an OKH reserve of twenty-eight divisions, but not including a large number of Finnish and Rumanian formations.

successes played their minor part in disinformation, giving an impression of disproportionate German absorption with the Middle East.

The Russians possessed much larger forces than their enemies, and many more tanks, some of high quality: but the dynamic of the German attack produced extraordinary victories, huge quantities of prisoners, and the conquest of vast regions of the western Soviet Union, including the fertile Ukraine. Not until winter did the tide turn. Hitler had been convinced (a conviction shared by many Soviet citizens) that Stalin's regime was so odious and so incompetent that, as the Führer expressed it, 'One kick at the door and the whole rotten edifice will collapse.'

It didn't. But this was war on an immense scale, and after the launch of Barbarossa there was no doubt where German military priorities would lie. Initially, even in OKH, there was optimism. The sedate Halder himself – who had been mainly responsible for the operational plan, a three-pronged advance, with each prong operating on a narrow front and aiming to encircle and destroy the forces opposing – was sure after a fortnight that the campaign would soon be victoriously ended. And in North Africa Rommel, distant from these remarkable triumphs, supposed that with an eastern campaign concluded Germany would have the chance and perhaps the inclination to wage a more strategically imaginative war against Britain.

Rommel contracted a bout of jaundice in August. Once clear of it he planned to meet Lucy in Rome in the autumn and take a short leave; and he sometimes at this time wrote that he would not be sorry to see service in another theatre. Nor is it difficult to guess where. Every evening in North Africa Rommel's staff briefed him on the latest information from Russia. And every evening his Ic listened to the British BBC news bulletin, translated it and brought it to Rommel. He was supremely optimistic, writing to Lucy on the day following the news of Barbarossa that in view of the superiority of the German forces 'the Russian affair' would soon end in victory.

Rommel was fortunate in not serving on the Eastern Front. Some German military reputations were made there, and his might have been one of them – certainly, and increasingly as the campaign went on, the ability to manoeuvre, to defend wide fronts with small forces by intelligent anticipation, movement, concentration, these things were paramount, and Rommel was a master of them. But however his personal star would have risen or fallen in Russia he would not have changed the ultimate course of the campaign, and would to some extent have been tarred by its eventual failure: whereas in Africa – although his campaigns there were also ultimately to fail – the failure was not felt so profoundly by the German people. The Mediterranean was remote from them, an Italian concern. The Eastern Front was beyond the horizon, but it was their own horizon, and the distance to it shrank as time passed. Rommel, therefore, retained in German eyes a certain aura of romantic success untarnished by depression, fear, and whispers of unbelievable horrors. He remained beyond others a symbol of uncomplicated, victorious decency: a hero. This may not have been particularly just to many of his comrades, but it was so.

Rommel was also fortunate in another and important respect. He had left Poland immediately after the conclusion of the Polish campaign in 1939, and although there were tales told of early and subsequent 'excesses' by organs of the Nazi Party, especially of course by the SS who held sway in the directly German-governed area of Poland, Rommel had no first-hand experiences of them; and for one whose admiration for Hitler was fervent and personal it was easy to be persuaded that such things were aberrations, failures of discipline, errors to be corrected. The systematic brutality practised in Poland towards the Jewish population – not only, but especially, the Jewish population – had started instantly, and had intensified throughout 1940 and the first half of 1941. Polish Jews were deported from their homes into city ghettos, were taken to forced labour camps and set to work under conditions of slavery, were, in large numbers, murdered for trivial 'offences', and were everywhere beaten and humiliated whether in custody or under the casual eye of the German authorities. The same practices had spread to greater or lesser degree wherever the German armies had marched, and throughout the Balkans.

With Operation Barbarossa, however, a new and dreadful chapter opened. On the Eastern Front there were large Jewish communities – the Jewish population of the Soviet Union numbered two million – and Nazi policy towards the Jews now operated in a vastly enlarged area of opportunity and challenge. Hitherto, unpardonable though the policy and its practice had been, no Jewish community in Europe had lost, murdered, more than 3 per cent of its citizens. These crimes have given historical colouration to the period, and justly so: but throughout much of Western Europe, and even within Germany, the numbers affected meant that the sufferings of the Jews had had only limited public impact.

The scale of crime was now hugely increased, and although the Eastern Front was remote from Germany, there were three million German soldiers serving there, and the word, after a while, began to get round. As the German armies advanced their reception by the Russian populations was often warm. Especially in such areas as the traditionally Catholic west Ukraine, which had suffered horribly (as had the whole Ukraine) from Stalin's rule, the Wehrmacht were welcomed as liberators, and in most parts of Russia German soldiers found a surprising friendliness; a confirmation, it seemed, of what they had been told of the unpopularity and brutality of the regime. The Russians are a large-hearted people, and the initial behaviour of the German front-line troops, disciplined and reared in a tradition of inherent decency, was generally good.

This atmosphere, however, could not last. It could not survive Hitler's doctrinaire enactments. The Russians were *Untermenschen*, and they were to be treated as such. The rule of the occupying forces was to be harsh and masterful. Hitler envisaged a fragmented and colonized society.

And the Jews were to be exterminated. This was never expressed in written orders but so it was to be. Until June 1941 massive 'patriation' of Jews was envisaged as the final solution of the 'Jewish problem'. Not for long thereafter.

Special SS Groups and Commandos, as in Poland, were established to implement a gruesome policy. In Russia, as in most of Eastern Europe, Jews had often been unpopular, victims of periodic mob hostility, pogroms. Nazi policy was thus not always anathema to the rest of the population; with dreadful frequency Jews were murdered actually before the Germans arrived, murdered by communities imbued with historic race hatred, and aware that in this they would be acting in harmony with the policy of the new conquerors.[27] Then came organized massacres, behind the fighting fronts, in areas often remote from the front-line troops, under the supervision of the SS whose responsibility the business was. Helped, all too often, by local police or auxiliaries, the SS drove Jews into trenches or pits, and machine-gunned them. Jews were driven into synagogues and the buildings fired. Jews were killed in public places, watched sometimes with curiosity, even with enjoyment, by civilians, families, off-duty soldiers. The killings started immediately Barbarossa started, and continued until there was no more to do.

Nor were Jews the only victims. Russian civilians in areas where partisan activity took place or was suspected against the German occupying forces were also dealt with without mercy. And there was another and deplorable aspect of war on the Eastern Front – the treatment of prisoners. The Soviets themselves had, from the first day, behaved with outrageous cruelty to captured German soldiers. The Germans had found mutilated corpses, clear evidence of torture in a great many cases, and this had engendered hatred, a sense of war against infinite numbers of merciless barbarians. In consequence Russian prisoners, at least in some instances, were also treated abominably, a treatment naturally exacerbated by the contempt Nazi ideology induced. Added to this, the treatment of Russian prisoners after transfer to Germany itself was often utterly inhumane. In many cases they were used virtually as slaves. Slaves may have decent as well as harsh masters, so conditions differed, but very generally they were half-starved, overworked and handled like unvalued animals. On the Soviet side German prisoners of war were lucky to survive alive and sane. Not a large proportion did, of the great numbers ultimately taken.

This appalling war, setting standards of inhumanity unseen in Europe for centuries, could not but infect the German army. Only a minority were involved in the actual perpetration of atrocities, but a good many knew something of it and in too many cases witnessed it, in spite of the fact that systematic murder was the function of the SS. Nor was it always systematic. There were instances of the wholesale murder of Jews by SS *Einsatzcommandos* even when the order, for some reason, had been to spare them. They were killed, one participant said, because they were hated.[28]

Official policy, however, was brutal enough. Only a minority of German generals had their hands soiled by transmission or implementation of orders from above which contravened every principle of human decency and military honour; but some did. The notorious *Kommissarbefehl* of 6 June 1941, which decreed that Communist commissars found with the Red Army were to be denied prisoner of war status and shot out of hand, was the subject of angry rejection by many; Rommel himself, when he heard of it, denounced it without equivocation, just as he totally ignored a later order that captured Jewish enemy soldiers were to be treated not as prisoners of war but as Jews. Yet this poison was now running in the arteries of the German army and state, and for every German soldier it was, at the least, an odious whisper in the background of consciousness. Some officers bravely refused to accept these outrages and submitted formal reports and complaints, but for the majority it was possible either to close eyes and ears, or to shrug, perhaps with distaste, and turn away. Complete ignorance in Russia was impossible. An Army Group Centre (on the Eastern Front) report at the end of the year referred to the opposition within the officer corps to these atrocities against Jews, prisoners, commissars and others, and stated grimly, 'Everyone now knows what is going on!' Some orders were carried out unevenly, or were circumvented; but they were known.[29] To be spared personal contact with the Eastern Front Rommel was fortunate indeed.

Nor was all this confined to the savage conditions of Russia. Within Germany itself official policy was now one of deportation – 'resettlement' was the euphemism employed – to areas somewhere in the east. Hitler was determined that the homeland, the German heart of the Reich, was to be made *Judenrein*, and trains loaded with Jews began leaving Germany in October 1941. More trains, similarly loaded and for similar 'resettlement', left Austria, Bohemia, even Luxembourg. The Jews had been induced to board the transports by various deceptions, believing that they were being offered job opportunities, decent conditions in a new environment. Few ever returned. Initially they were deported to ghettos in various cities in Poland or elsewhere in the eastern regions under control of Germany. Later they were deported to special camps. It was not until January 1942 that decisions were taken which, when put into effect, meant that Jews from all the occupied countries, from all over Europe, were either to be worked to death or, ultimately, done to death.

These latter sickening happenings and policies were as yet unknown to the great majority of Germans, whether in uniform or not, and they were as yet unknown to Erwin Rommel. He would certainly have known that on the Eastern Front fearful measures were being taken against what was often stigmatized as equally fearful terrorism and atrocity – officers joining him from Russia would have told him of it. In Africa, however, there were no SS units. There appeared to be few Jews. There were no commissars, no Communists, no Russians. There were not even very many Nazis, although SS Captain Berndt was attached from Goebbels's Propaganda Ministry to look after public relations (a job he did with considerable skill, becoming, to Rommel's approval, a thoroughly useful soldier in the process). There were no rebellious or recalcitrant civilians except the Bedouin, whose relations with the Germans (unlike with the Italians) were generally good. For most Germans the Jewish deportations, when heard about, were shrugged off as a sort of social engineering – rough, no doubt, but these people presumably constituted a security risk, and wartime was wartime. And in wartime security, and secrecy, was rigorously imposed as a discipline and accepted as a duty. It was forbidden to mention or speculate on certain subjects, and this was one of them. It might help the enemy. It might betray the gallant German soldier at the front. And – this has to be said, odious though it is – among many there was probably a whisper that Jews, after all, were only Jews.

Thus the massacres being perpetrated in the east, whether casually or by calculation, were matters of rumour rather than knowledge; and where word got out, because of indiscreet talk or letters from soldiers at the front, there were stern injunctions against believing, let alone repeating, slander; injunctions which in Hitler's Germany it was wise to obey. In wartime, too, there was – and is in most societies – a readiness to believe authority, to believe, even with enthusiasm, that the authorities 'had their reasons', 'knew what they were doing', and a corresponding reluctance – again, as in most societies – to believe discreditable stories about the state, stories spread, as like as not, by agents of the enemy. Natural and disciplined patriotism underpinned evil and served it.

The degree of ignorance of the foul things being done in the name of Germany is difficult to comprehend in an utterly dissimilar era. Some total ignorance – innocence – there certainly was, although perhaps not to the extent which many have wished to imply. But there was also comparative ignorance, even among sophisticated people who knew something but not enough. The enormity of what was being done was almost beyond the mind's acceptance, particularly any normally dutiful mind, respectful of authority and inherently inclined to give authority the benefit of any moral doubt. And this was particularly so among those – the majority – who remembered the past, who were grateful to Hitler for delivery from much of that past, and who regarded the war as just and those who opposed the war or the Government as traitors, actual or potential.

But there was not ignorance but knowledge, albeit partial, among those already opposed to the regime, those totally sceptical of official information; and some of these were in contact with men of influence in state and army. With knowledge came torment, all the worse for the impatience it accompanied. Count Helmuth von Moltke wrote in Berlin on 21 October 1941:

> How can one bear one's share of guilt? In one part of Serbia two villages have been reduced to ashes, 1700 men and 240 women of the inhabitants have been executed. That is the punishment for an attack on three German soldiers ... more than a thousand men are being murdered for a certainty every day and thousands more are being habituated to murder. And all that is child's play compared to what is happening in Poland and Russia. How can I bear this ... don't I make myself into an accomplice? What shall I say when someone asks me 'and what did you do during this time?'
>
> ... If only I could be rid of the awful feeling that I have let myself be corrupted, that I no longer react keenly enough to such things, that they torment me without producing spontaneous reaction.[30]

Moltke, an exceptionally gallant and noble character, inspirer of the 'Kreisau Circle', was a bitter opponent of Nazism and was to die at the executioner's hands. He was certainly correct to write that day: 'child's play compared to what is happening in Poland and Russia'. Two weeks earlier, at Rovno in eastern Poland, seventeen thousand Jews were marched to pits outside the town, ordered to undress and then machine gunned – one massacre among many. Moltke vigorously attacked official orders persecuting the Jews, not without some effect here and there. To those still open to arguments based on legality (he was a distinguished lawyer) he was able, as he put it, sometimes to 'stop the spread of savagery in the military mind'; and to those whose minds were receptive to moral arguments he opposed an uncompromising morality. Yet it is clear that even Moltke, passionate and committed enemy of the regime though he was, had only limited knowledge of the outrages committed in Poland from the beginning; was appalled by news of atrocities in Russia (of whose extent even he was probably unaware), yet treated them as something different in kind rather than degree from what had gone before; undoubtedly hoped, initially, for German military success there, where friends and relatives were fighting; and concentrated upon the philosophic and political problems of a post-Hitler Germany.

But a post-Hitler Germany, it was already alarmingly likely, could only be brought about by armed force: by violence and revolt; by a total breach with the disciplined traditions of the German army. Not constitution-mongering but the preparation of rebellion was needed, and it was unnatural to expect it. Rommel, still unaware, represented soldierly virtue at its most talented and straightforward in the service of evil at its most vile. To understand this paradox demands effort to comprehend the influences in Germany which had shaped the manhood of Erwin Rommel's generation, and the way in which the genius of Adolf Hitler had exploited the most ignoble prejudices as well as the most martial qualities of his people.

CHAPTER THIRTEEN

Panzer Gruppe Afrika

ROMMEL ALWAYS carefully studied his enemy. By now he reckoned that he understood the British way of warfare. Later he was to write about it extensively, but that was in the light of many experiences still to come. Some of his impressions, however, he had no reason greatly to adjust, and when he received at this time an official OKH assessment of British leadership there was nothing in it which he felt disposed to contradict. Indeed his own reports – as well as those from Greece, Crete and France – had significantly contributed to the assessment.

The British, ran the OKH summary,[1] showed considerable courage and capacity for self-sacrifice among their junior leaders, although those same junior leaders were apt to display a lack of independent initiative, and somewhat stereotyped and over-systematic tactical methods. At the higher level of command, however, the British showed a 'lack of operational dexterity', and had not mastered the flexibility needed in the leadership of mechanized formations. A word used to describe the British command at the operational level was '*Schwerfallig*', which may be translated as 'clumsy', or 'sluggish'. There was demonstrated, in British actions, rigidity of mind and reluctance to change positions as swiftly and readily as situations demanded. There was also – the point was often noted – great fussiness and over-elaboration of detail in orders, which thus became inhibiting, inappropriate and excessively long, with subordinate leaders given little freedom of action or decision. Some of this no doubt derived from established system rather than personal defect, and was more applicable to some cases than others, but the OKH assessment was, at least in part, echoed by Rommel in his own reflections.

Rommel, however, had not only general British characteristics and tactics to consider; he had individual opponents. In particular he had a new team of opponents, for after 'Battleaxe' the British had changed their senior commanders in North Africa. Rommel, in retrospect, thought that this was done by the enemy too often for their own good. He reckoned that it took time to get to know the ways of the desert war, that everyone had to learn from his own experience and suffer, no doubt, the consequences of his own initial mistakes. Mistakes, furthermore, almost inevitably attended promotion – experienced skill at one level of command had to be freshly attained at the next. Wider responsibilities begat different techniques and new rules. The truth of this was shortly to be demonstrated by Rommel himself, who was about to make some of the most resounding mistakes of his career – and then, like a naturally skilled fencer, to recover from overreach or misjudgement by sheer agility, to avert disaster and ultimately turn the tables.

For Rommel was now commanding at a higher level. In Cyrenaica and during 'Battleaxe' he had been commanding the Afrika Korps, albeit with a certain authority over some Italian divisions. In August 1941 his command was designated 'Panzer Gruppe Afrika'. The Afrika Korps, under General Cruewell, now consisted of two Panzer divisions – 15th Panzer (General Neumann-Silkow) was complete, and 5th Light Division had been redesignated 21st Panzer. Also under Rommel's command was another German 'light' division, to be numbered as 90th, originally called the 'Afrika' Division (General Summermann) – formed largely from various units already in Africa but not embodied in the Panzer divisions.* Also to some extent fighting under Rommel, although without formal assignment by the Italian commander-in-chief in Libya, now General Bastico, were two Italian Corps – the XX Armoured Corps, under General Gambara, of two armoured or motorized divisions, Ariete (General Balotta) and Trieste (General Piazzoni); and the XXI Corps (General Navarrini), consisting of four non-motorized infantry divisions whose subordination to Rommel was less qualified. A further Italian division, Savona (General de Giorgis), was detached from Navarrini's command but was also part of Panzer Gruppe Afrika.

Rommel therefore now had a considerable force – effectively ten divisions, and three corps commands – and, as he wrote home happily, everybody else in that position (a Panzer Group) was a Colonel-

* It was formally named '90th Light' on 28 November.

General, and if things here went well he, too, would probably get that rank after the war. He had been, in fact, advanced on 1 July to be 'General of Panzer Troops'. More importantly, he was sent a staff appropriate to these enlarged responsibilities; a staff assembled in Germany and then despatched to North Africa as a coherent unit under the skilled direction of General Gause, an intelligent, placid, reflective East Prussian with a dry sense of humour who had been serving as a liaison officer with the Italians and who, on reporting to Halder in Berlin in July, had been frank about the impression sometimes made by Rommel. It was, Halder noted, one of 'morbid ambition', ambition as a disease.[2] Such was generally Halder's own view: but Gause was to serve Rommel faithfully and well, despite what he had characterized in the same interview as Rommel's 'brutal methods'.

The Germans traditionally believed that staffs should be formed as self-sufficient cells, with individuals understanding each other's methods and requirements, able to respond to battle and a commander's will in battle like a brain and a nervous system. This practice, this emphasis on the unique importance of a well-trained, integrated staff, preserved as a unit, enabled the German army to be served by exceptionally small numbers of staff officers – and thus by economic and inelaborate headquarters. The other principle which enabled the work to be done by small staffs is implicit in the OKH comments on their British opponents. The Germans rejected both the principle and the practice of over-detailed orders. They believed that because subordinate commanders possess more up-to-date and often more relevant information about the battle than their superiors it is a mistake for the latter to interfere with the detail of how an action should be conducted – unless, of course, the superior in question actually appears on the tactical battlefield (as Rommel so often did), or knows something which his subordinate cannot know. Such things apart, a subordinate should be given the simplest of instructions and objectives, and be set free to discharge his mission as appears to him best. The principle was at least as historic as the elder Moltke.

It has been said that large staffs are an invariable sign of bad armies. The staff which General Gause introduced to the new commander of Panzer Gruppe Afrika in August 1941 was of very high quality, and it was extremely small. For example there were only two officers in the operations branch (Ia) – Lieutenant-Colonel Westphal (an officer thought arrogant by some, but of high intelligence[3]), assisted by one

Oberleutnant; and three officers in the intelligence branch (Ic) – Major von Mellenthin, with two reserve lieutenants. The Quartermaster Department, responsible for the organization of all supplies, was headed by one General Staff Major, Schleusener (succeeded in December 1941 by Otto), supported by two further majors, one of them with General Staff training. Together with a personal '*Adjutantur*' of two majors, a major as Camp Commandant, and the heads of the Engineers, the reconnaissance troops, the medical, repair, and ordnance services, Rommel's military family numbered a mere twenty-one officers – including a civilian representative of the German Foreign Ministry, Freiherr von Neurath. In few armies would a force of three army corps, amounting to ten divisions and operating at the limits of a line of communication extending many hundreds of miles, have been controlled by so exiguous a team.

This staff had been designed by OKH as a '*Verbindungsstab*', a liaison staff to facilitate relationships between the Italian Command in North Africa and OKH, to deal with German matters within the theatre insofar as they affected the Italian Command, and to regulate supply areas in the rear areas as well as liaising with von Rintelen, the German military representative in Rome, about trans-Mediterranean supply. Halder had written to Cavallero explaining this, and placing the German rear areas (in Africa) commander and Rommel's *Oberquartermeister* under this staff's direction. Rommel knew nothing of all this until he received a signal from Rome.[4]

Rommel, however, did not at first welcome them, and this is wholly unsurprising. The tenor of Halder's communications to Cavallero and instructions to Gause had been such as to derogate from Rommel's authority and – worse – to introduce ambiguity and divided allegiance. Difficult relationships were inherent in the situation anyway, and Halder's mistrust of Rommel exacerbated them. Rommel was also undoubtedly irritated at times by a sense of his own non-membership of the General Staff, by the feeling that self-consciously highly-educated officers, steeped in theory and academic studies of war and its systems, might seek to 'take over' more responsibility for his command than he, the epitome of a personal and highly individualistic commander, would ever concede.

The feeling, at least in the instance of Panzer Gruppe Afrika, was unjustified. Rommel briskly made clear his own understanding – that this was a staff to assist him command, not one interposed with some sort of nebulous intermediary role – and the new arrivals wholly accepted the situation. General Gause and his officers proved an admirable adjunct to command for Rommel to use, outstanding in their own personal qualities and in the loyalty they gave Rommel; for, being without exception thoughtful and experienced officers, they recognized and appreciated the extraordinary genius of the man they were now required to serve, as well as the challenging task he faced. Soon Rommel was writing home of his satisfaction with Gause and his happiness at having, for the first time, the support of a trained staff which, he said, 'functions very well'.

But although Rommel very soon came to value his staff and esteem their dedication and efficiency, he was not yet prepared to adjust his personal methods to take advantage of the admirable tool of command they represented, nor use them (on occasion) for the best. Thus, he would take his Chief of Staff, Gause, with him on his frequent, indeed incessant, visits to the fighting front. This was against the principles of command in every army: the Chief of Staff was the commander's deputy in fact if not in name, and was empowered to act for him, to take decisions in his absence, even – if he knew more than his commander at some critical point in time – to countermand his commander's orders. If the Chief of Staff was accompanying the commander, if Rommel and Gause were on tour together, it meant that a heavy load of responsibility lay with juniors. They were thoroughly capable of bearing it, but it was wrong, and the wrong was Rommel's. The reverse side of the penny of his astounding tactical flair, his 'feel' for the front line and the critical point of battle, was, at times, a disregard for the more methodical approach to problems which higher command necessitates; and with this disregard went disregard for the staff systems and usages which generations of war academies have taught. Genius may sometimes break rules and evade the consequences. Rommel had genius. He broke rules. Sometimes, but not always, he evaded the consequences.

The rules applied now to a wider scale of game. Panzer Gruppe Afrika was confronting the threat of a new offensive, undertaken under new British leadership. Wavell had been moved to India after the failure of 'Battleaxe' and been succeeded by General Auchinleck, and the British Eighth Army had a new commander, General Cunningham. Rommel had esteemed Wavell, while critical of some of his moves. He carried a translated edition of Wavell's published essays on generalship with him on campaign. Of Auchinleck or Cunningham he knew little.

* * *

Rommel's intelligence staff were well aware that a British offensive was pending. They supposed that it would be an advance aiming to link the Allied forces on the Egyptian frontier with Tobruk, to relieve Tobruk. The OKH estimate was that the offensive would take place about the beginning of October, and thereafter the odds on timing would shorten day by day. Rommel himself decided to lead a major reconnaissance in force across the frontier in September, and Operation '*Sommernachtstraum*' saw him driving east with 21st Panzer Division in the area south of Sidi Omar, hoping to discover (and overrun) British forward dumps, or at least establish whether the British seemed geared for an offensive. The raid – which suffered loss from air attack and long-range shelling, and which overran nothing – produced a certain amount of information (from radio interception) about British deployment and was over, with Rommel withdrawn, by 16 September. His intelligence staff had been impressed by the speed and accuracy with which the British patrols had reported his progress, as well as by the flexibility of British artillery.

The general deduction from *Sommernachtstraum*, however, was that the British were not yet deployed for an offensive. This was welcome to Rommel. He had his eyes on Tobruk. If he could take Tobruk – which, given a sufficient concentration of strength, he was confident of doing – he would not only improve supply and give Panzer Gruppe Afrika a greatly increased and extended range of operations, he would also transform the operational situation. The British, New Zealand, South African and Indian divisions of his enemy, instead of being poised to relieve a beleaguered fortress, would be lined up facing a concentrated Rommel force of ten divisions, suffering no distractions. Tobruk obsessed Rommel, and the obsession was justified.

The timing of an assault on the place, however, was delayed by a particularly tenuous supply situation. Both reinforcements and materiel supplies had, in the previous few months, only satisfied a small percentage of the requirement. Between June and October 220,000 tons of Axis shipping on the convoys between Italy and Libya had been sunk by the enemy – over half of it by air attack and a large part of that by aircraft flying from Malta;[5] Italian naval signals and convoy details were, of course, being read by British Intelligence. Nevertheless Rommel's instinct was to press ahead, and he overrode the advice of OKH (advice strongly supported by the Italian High Command) that an attempt on Tobruk should be delayed until 1942. With a certain margin of risk he believed he could attack it by the end of November, and the operation was planned for the twenty-first. Rommel was convinced that the balance of forces and thus of time was moving against him: and he persuaded Berlin to concur. Tobruk, he was therefore ordered, should be taken.

Rommel visited Rome on the 14 November and explained his plans. He dismissed Cavallero's anxieties about a British offensive. On this instance, at least, Cavallero had right on his side.

In the middle of November, therefore, Rommel was standing ready, impatient to attack Tobruk. He was perfectly aware that at some time he might – indeed would – need to deal with a British attack on the frontier, but in spite of OKH estimates and Italian fears it had not yet come. When it did, it should be limited, and with luck containable locally – a captured British sabotage and raiding party at Derna had spoken of a possible offensive in the region of Sollum, combined with landings from the sea.* Rommel appreciated that the best method of dealing with such an offensive would be to take Tobruk first, in which case it might not even materialize. As to direct defence, Rommel had strengthened all the fortified places on the frontier very considerably. Bardia, Capuzzo, Sollum, Sidi Omar and Halfaya were strong points, and Rommel was satisfied that the first four were now in a state to be defended by the non-motorized Italian infantry of the Savona Division, and Halfaya by a German–Italian force. The headquarters of the Afrika Korps were in Bardia and the two Panzer divisions were not far away – 15th north of the coast road, near Rommel's own headquarters at Gambut, and 21st near Sidi Azeiz, some twenty miles west of Bardia. Rommel, therefore, felt reasonably balanced to deal with any British initiative. He had a strong defensive position forward, and two Panzer divisions within reach. He envisaged 21st Panzer taking any British offensive in the flank, if it happened and achieved initial success.

15th Panzer – and 90th Light Division, still known as 'Afrika Division' – were to take the lead part in the onslaught on Tobruk. Tobruk was ringed by Navarrini's four infantry divisions – Brescia (Zambon), Trento (now General Stampioni), Pavia (General Franceschini) and Bologna (General Gloria). Rommel, furthermore, held south of Tobruk the two mobile divisions of Gambara's XX Corps, the Ariete (armoured) at Bir El Gubi, and the Trieste (motor-

* In November an ill-conceived attempt, by a raiding party from the sea, to kill or capture Rommel was made at Beda Littorio. This was Rommel's quartermaster headquarters; he himself would never have pitched his tent so far from the front.

ized) fifty miles to the west at Bir Hacheim. His full operational authority over these divisions had to be obtained from the Italian High Command, and in the event his request for it was not made until 22 November. Rommel believed that he could carry on with his plan to attack Tobruk regardless of anything the British could do; and that he had the strength – just – provided he had the will, to win the Tobruk battle and thereafter, if it proved necessary, to deal with a British incursion.

If that incursion had started on or after 25 November he might have been proved right. None can say. Tobruk was held by the British 70th Division (General Scobie), which had relieved the Australians, and by a Polish brigade group (General Kopanski). It would never have constituted an easy objective but Rommel would have given it all he had in terms of firepower and, no doubt, personal dynamic. The matter is academic, because on 17 November German interception stations ominously reported complete wireless silence on all British radio nets. At dawn on 18 November the British Eighth Army began its advance across the frontier. Operation 'Crusader', as the British had named it, had begun.

Rommel's first mistake in Crusader was undoubtedly to refuse for too long to accept that it was happening and was serious. This was a personal error. Like Wellington in June 1815 Rommel had perfectly well-planned dispositions but was reluctant to move early for fear of premature overreaction. Unlike Wellington, however, his main motive was determination not to give up a projected attack – on Tobruk – until and unless it proved really necessary. Very soon it was indeed necessary, and first in the catalogue of German shortcomings in Crusader (there were plenty on both sides) was Rommel's reluctance to accept disagreeable facts and draw unpalatable conclusions. It was not easy. British concealment of their intentions had been admirable, and not until Crusader had been rolling for many hours did the Germans fully appreciate the scale of the enemy offensive. Rommel's staff paid regretful tribute to the skill with which Crusader had been prepared – the camouflage, the night movement, the discipline.[6] They did not, of course, know that ULTRA had disclosed to the enemy Rommel's intentions and timing, and that Crusader had been sprung as close to the German start as could be managed. Furthermore, for most of Crusader the Royal Air Force dominated their opponents, and air reconnaissance yielded more to Cunningham than to Rommel.

The British plan was sound in concept, less sound in actual execution. The British believed that the vital element to defeat must be the German armour. British tank numbers were greater than those of their enemy – about six hundred tanks of all kinds, including specifically infantry support tanks – against 380 available to Rommel, including 140 in the Ariete Division. In quality, however, the British believed, with justification, that the Germans had the edge. They also believed (but were less than adept at giving effect to their belief) that the best way to defeat German armour was to induce the German tanks to advance and thus to take them from positions of the defenders' choosing. These beliefs (here simplified) reflected an inadequate understanding of the skilled all-arms cooperation fundamental to the handling of a German Panzer division. The Germans, whenever possible, manoeuvred their tanks with, and often behind, a screen of their formidable anti-tank guns, and to deal with the latter artillery were chiefly required. It would be rare – not unknown but rare – that German tanks could be induced to charge a line of British tanks (or guns) in position, exposing themselves. This, however, or something like it, was what the British hoped would happen.

To make it happen, it would be necessary to place overwhelming anti-tank (and predominantly, with the British, that meant tank) strength on defensible ground which the Germans would need to take, to contest. The choosing of that ground in the desert, with its paucity of features, was not easy: it could only be deduced from the operational situation rather than from the terrain. The British knew that Rommel planned to attack Tobruk, and when. A British offensive calculated to provoke a German counter-attack in strength would therefore need to have, as objective, some feature vital to the German operation against Tobruk. Nothing short of that could with certainty provoke the required Rommel reaction.

The British plan of advance was two-pronged. On the right, XIII Corps (General Godwin-Austen) – the New Zealand Division (General Freyberg) and 4th Indian Division (General Messervy) supported by 1st Army Tank Brigade (Matilda and Valentine tanks, designed for infantry support) – were to cross the frontier and attack the garrisons of Sidi Omar, Capuzzo and thereafter Sollum and Bardia. On the left, XXX Corps (General Norrie) constituted the main armoured force of Eighth Army. XXX Corps consisted of 7th Armoured Division (General Gott), a division of three armoured brigades and a support group of infantry and anti-tank guns: of the division's armoured brigades, however, one, 4th Armoured, was retained under corps command. Also in XXX Corps was 1st South African Division (General Brink) and 22nd Guards Brigade.

The mission of XXX Corps was to advance along a centre line of the Trigh El Abd – a track running from the frontier at Bir Sheferzen in a north-west direction, and passing about thirty miles south of Tobruk. It was calculated that this thrust line would be perceived by Rommel as so menacing to the investment of Tobruk that he would react, and react with his armour. The object given to XXX Corps was 'to find and destroy the enemy armour', and it was hoped that the latter would accordingly present themselves for destruction. The initial objective was a point on the map called Gabr Saleh, thirty-five miles from the frontier. 4th Armoured Brigade, under corps command, was to advance on the inner flank with the dual function of warding the left flank of XIII Corps as it moved north to attack the frontier strongholds, and acting as corps armoured reserve.

The defect of this plan was that it absolutely depended on German reactions; and if these were to be as desired an initial objective more threatening to German plans than Gabr Saleh should have been selected – as Norrie wished, overruled by Cunningham. In the event the XXX Corps advance to Gabr Saleh evoked little response. The Germans were simply unaware of the scale of what was happening, and although British activity, some sort of advance, was detected and troops alerted, it was not until next morning, 19 November – over twenty-four hours after Crusader had started and twelve after Norrie's main body had reached Gabr Saleh – that Panzer Gruppe Afrika had information of 'six hundred fighting vehicles marching north-west'. Earlier reports from 21st Panzer (some thirty miles north-east of Gabr Saleh) had referred to 'strong enemy forces moving west and north' – presumably both XXX and XIII Corps – but the size of the British enterprise was not yet assessed. Consequently there had been little reaction.

It is arguable that from the German standpoint too sharp a reaction would have been premature. So far the British had moved large forces into a particular area of desert, and it was unclear what they might do next. Concentrated German Panzer forces could have waited, ready. The German Panzer forces, however, were not concentrated, and they were still comparatively far away. On the afternoon of 18 November, about eight hours after Cunningham's columns had crossed the frontier, Cruewell had visited Rommel and proposed a concentration of the two Panzer divisions well south of their present positions. Von Ravenstein, he said, wished to move 21st Panzer to (of all places) Gabr Saleh. He, Cruewell, agreed, and wished to move 15th Panzer up behind and supporting them. Rommel flatly disagreed, showing some irritation. He believed that such moves would represent unnecessary overreactions to British moves which were still uncertain and whose object and significance were obscure. This may or may not have been a decision beneficial to the Germans in its outcome – nobody can say what would have happened following a movement of 15th and 21st Panzer towards Gabr Saleh late on 18 November, and Cunningham could well have argued afterwards that it was exactly what he had wished to anticipate. But Rommel disallowed it, and his staff were sure his reason was that he still hoped this British incursion could be prevented from interfering with the assault on Tobruk.

Next morning, however, 19 November, Cruewell again visited Rommel. It was now clear from every report received by the Afrika Korps and Panzer Gruppe Afrika that the British operation was a major offensive. Rommel, reluctant but now convinced and therefore vigorous, agreed. The attempt on Tobruk must be postponed. The German armour must be given a free hand to deal with the British mobile forces.

These mobile forces, however, had by now themselves dispersed. 7th Armoured Division had been effectively concentrated at Gabr Saleh. Now, since the Germans had not reacted to its presence there, Cunningham had ordered Norrie to probe further towards Tobruk, in order to provoke the German moves his plan demanded. One armoured brigade – 22nd, with the new Crusader tank – had been sent towards Bir El Gubi on the morning of the nineteenth; and the other, 7th Armoured, had been directed on Sidi Rezegh. Sidi Rezegh, with an airfield, lay on a shelf between two steps on the escarpment running down to the coast road twenty miles east of Tobruk. It also lay on or overlooking the Trigh Capuzzo, the track roughly parallel to XXX Corps' centre line which ran from Fort Capuzzo to El Adem, ten miles south of Tobruk. Sidi Rezegh mattered, and its occupation by the British 7th Armoured Brigade and support group on 19 November mattered to Panzer Gruppe Afrika. By now, however, 22nd Armoured Brigade had charged the Italian Ariete Division at Bir El Gubi and suffered very severe casualties, while 4th Armoured Brigade – now at Gabr Saleh – had been vigorously attacked from the north.

Von Ravenstein, near Sidi Azeiz, had become ever more anxious. He had wished to move south-west earlier and been forbidden. Now, on the nineteenth, anticipating Cruewell's orders, he sent a strong group, built on his 5 Panzer Regiment, of 120 tanks, to Gabr Saleh. This group, under the Panzer Regiment commander, Colonel Stephan, also included field and anti-aircraft artillery, a machine-gun battalion and an infantry regiment. Ravenstein's orders to it were to attack south to the Trigh-El-Abd, and then turn east towards Sidi Omar with the object of cutting off and destroying an enemy force reckoned as two hundred tanks.[7]

Group Stephan encountered 4th Armoured Brigade in a savage tank versus tank battle during the afternoon of the nineteenth, with fighting going on from four o'clock until after dark. On hearing of this the British command supposed that it might well be that reaction to XXX Corps advance which it had always been hoped to evoke (as in a sense it was, although von Ravenstein's rather than Rommel's or Cruewell's). But the Germans were coming off best from this encounter, using their tanks and anti-tank guns in combination – the latter were few in number but devastating in quality: and on the next day, 20 November, the British ordered their 22nd Armoured Brigade, mauled by the Italians at Bir El Gubi, back on their tracks towards Gabr Saleh. Ravenstein, as perplexed as any about where everybody was and what they were trying to do, had now proposed that the Germans should, above all things, concentrate their armour at some central point and await developments.

Cruewell, however, who had received Rommel's order to 'destroy the enemy battle groups in the Bardia, Tobruk, Sidi Omar area' (in effect the whole arena of operations!) and been given 'a free hand' to do it, was taking somewhat inappropriate action. On 20 November he moved both 15th and 21st Panzer towards Sidi Azeiz and set both on sweeps between that place and Gabr Saleh. At Gabr Saleh 15th Panzer had an encounter with the (now reinforced) 4th Armoured Brigade and inflicted further casualties: but this was fortuitous rather than a deliberate effort by united German armour. In fact, on the evening of 20 November, the British 7th Armoured Division was partially concentrated, albeit not as planned, in that two of its three armoured brigades were united at Gabr Saleh with nearly two hundred tanks between them; while the Afrika Korps was also to some extent concentrated (in that both 15th Panzer and 21st Panzer were both somewhere in the area of desert between Gabr Saleh and Sidi

Omar, 21st Panzer having run out of fuel). Rommel, Bayerlein* and other German (and British) participants and authors have criticized the British command in the early phases of Crusader for failing to observe the principle of concentration of force, for failing to see the battle as a whole and for offering British armoured formations up to be defeated in detail by the local superiority of German forces. The events of 20 November hardly sustain these criticisms. Where the Germans came off best, as at Gabr Saleh, it was generally because of the superior hitting power of their tank or anti-tank guns, or because of superior tactical skill. The principles of war played remarkably little part.

On the evening of 20 November Rommel met Cruewell. Rommel had taken no very direct hand in the day's fighting – with uncharacteristic abstinence he had given his orders to Cruewell and left him alone to discharge his mission of 'destroying the British battle groups' all over the desert. But now Rommel had come to terms with the situation. This was a major British offensive. The British were set to relieve Tobruk, as he had always feared and been determined to preempt. This was not a diversion. For Rommel it was not a matter of sweeping the desert clear of British raiding forces in brisk, haphazard fashion as Cruewell had tried and failed to do. It was not – or not simply – a matter of the frontier defences, although the British XIII Corps was remorselessly grinding them to pieces and would, on the following day, attack Capuzzo, Sidi Omar and Bardia. Here was a threat to the entire existence of Rommel's army, as well as to its ability to keep Tobruk invested. At any moment there must come from Tobruk itself a sally to join up with the Eighth Army offensive. Rommel set both divisions of the Afrika Korps on the march towards Tobruk – on the march as rapidly as possible to Sidi Rezegh.

The support group of 7th Armoured Division and 7th Armoured Brigade were in position at Sidi Rezegh: and from that base the British plan was that they should attack northward, at dawn on 21 November, to join hands with a breakout from Tobruk by 70th Division.

Rommel had deployed 90th Light Division – not, at that time, part of the Afrika Korps – between Sidi Rezegh and Tobruk, and

* Colonel, later Lieutenant-General, Fritz Bayerlein. At that time Chief of Staff of the Afrika Korps.

when the tanks and men of 7th Armoured Division moved forward they suffered heavy casualties from 90th Light and particularly from its line of anti-tank guns in position. Rommel himself had moved to that sector and was directing a battle in which his forces were being attacked from both sides – from Sidi Rezegh and from Tobruk. His staff diarist recorded his doings by the minute, as ever: 'Commander-in-Chief personally directs thrust and counterthrust of fast-moving tanks, and against a renewed attempt at breakout by enemy tanks in the west–east direction deploys a battery of 88s whose value is yet again shown to be outstanding.'[8] But relief would only come with the arrival of the Afrika Korps, and Rommel signalled to Cruewell to use his divisions to prevent any linkage between the Tobruk garrison and XXX Corps. This implied tackling Sidi Rezegh; by the evening Cruewell's divisions were attacking Sidi Rezegh from the east, and the east end of the escarpment was in their hands.

Cruewell, however, gave new orders to his two Panzer divisions during that night, 21 November. One division was to move to Gambut, the other to Belhamed – places separated by eighteen miles, and Gambut far from Sidi Rezegh. He was presumably responding to what he felt was the main threat to the besiegers of Tobruk – the westward advance from the frontier of XIII Corps. He intended to withdraw both of his divisions eastward 'to obtain freedom of manoeuvre', and only left 21st Panzer at Belhamed on Rommel's order. The effect on 22 November was a disunited Afrika Korps, an unresolved situation at Sidi Rezegh and a British 7th Armoured Division which could soon be concentrated. 7th Armoured Brigade had been mauled at Sidi Rezegh, but also approaching the area was the remaining armour of the British XXX Corps, 4th and 22nd Brigades.

In the frontier area XIII Corps was continuing its attacks and would clearly soon start moving west towards Tobruk, a move which would begin, led by the New Zealand Division, very early on the following day. Meanwhile Rommel urgently needed to clear up the situation at Sidi Rezegh. He visited von Ravenstein and told him that the decisive armoured battle would be in the area of Sidi Rezegh, and with an excellent 21st Panzer attack from west and north on the afternoon of 22 November he forced the remnants of 7th Armoured Brigade and the support group to withdraw, after heavy casualties, by nightfall. The British reinforcing armoured brigades, 4th and 22nd, drew off southward, and the 7th Armoured Division Commander, Gott, hoped to concentrate his surviving armour – he still had some 150 tanks, of different marks – somewhere south of Sidi Rezegh. His tank strength outnumbered Rommel's, although this was not a factor of supreme significance when quality and the effectiveness of anti-tank guns was taken into account. But Gott's plans were gravely affected by a movement of 15th Panzer Division, which attacked the withdrawing British armoured brigades from the east, south of the Sidi Rezegh feature, doing much damage in the darkness and overrunning 4th Armoured Brigade's headquarters.

The next day, 23 November, was a Sunday. It was, in the Lutheran calendar, the annual German All Souls' Sunday – *Totensonntag*, the Sunday of the dead.

Rommel now held Sidi Rezegh. He had, he knew, inflicted considerable tank and human casualties on the British at Gabr Saleh, at Bir El Gubi, and now at Sidi Rezegh, both when he had defeated 7th Armoured Brigade's attempt to attack north from it towards Tobruk and when the Afrika Korps had recovered it and routed a part, perhaps a large part, of 7th Armoured Division. He reckoned that a bold move might finally dispose of XXX Corps. During the night of 22 November he ordered Cruewell, by signal, to take 15th Panzer Division, reinforced by the tanks of Ravenstein's Panzer Regiment 5, in a south-westerly drive from a point east of Sidi Rezegh on the morrow; to meet the tanks of the Ariete Division (not engaged since their encounter with 22nd Armoured Brigade at Bir El Gubi on 19 November); and by this manoeuvre to encircle and thereafter destroy the British armour south of Sidi Rezegh.

Cruewell, acting in the spirit rather than with detailed knowledge of Rommel's orders (since decoding Rommel's long signal would have caused delay), set off from his headquarters just before dawn to lead the Afrika Korps on this somewhat complex operation. At 6 o'clock on the morning of 23 November most of the staff of Afrika Korps headquarters were surprised and captured by the leading troops of the New Zealand Division, advancing west along the Trigh Capuzzo.

Cruewell had escaped, and was marching south with all the German armour towards the Ariete. His move, executed with characteristic aplomb, soon ran into huge supply columns of 7th Armoured Division and 5th South African Brigade – the 1st South African Division had been moving north-west on the left flank of XXX Corps with orders to 'mask' Bir El Gubi. Cruewell was tempted to break off his southern advance and exploit the chaos, but he persevered southward, believing that for a decisive operation he needed the additional strength of the Ariete. Then he planned, with Ariete and most of the Afrika Korps concentrated, to move north as a solid mass and destroy the residual forces of XXX Corps, hammering them to pieces on the anvil of Sidi Rezegh. This was the operation known to both sides as '*Totensonntag*'.

It was not an easy manoeuvre. It involved a cross-desert march of a great mass of vehicles, a rally with an armoured division, the Ariete, of another corps, and then a sharp change of direction northward, in some sort of agreed formation with both Ariete and Afrika Korps.

Rommel took no part in this. He was concerned at the signs of XIII Corps' predicted advance westward along the Trigh Capuzzo, towards what had to be the focal points of the battle, Tobruk and Sidi Rezegh; and he moved to the Trigh Capuzzo himself. Panzer Gruppe headquarters had also moved, during the night of 21 November, from Gambut (threatened by any westward movement of XIII Corps) to El Adem, twenty miles south of the Tobruk perimeter, and Rommel, as usual, was away with his *Gefechtsstaffel* (the armoured vehicle 'Mammoth', captured from the British at Mechili, was now used by Cruewell) watching what he appreciated might be the next critical point in the battle. He left to Cruewell the execution of *Totensonntag*.

Cruewell had joined forces with the Ariete Division and now, using tactics somewhat novel to the Afrika Korps with its usual methodical interplay of tanks, anti-tank guns and (where necessary) mounted infantry, lined up his command – tanks in the front ranks, mobile infantry and soft-skinned vehicles behind – and charged northward, the Ariete on the left of the line. He suffered heavy losses from the rump of the British 22nd Armoured Brigade which moved from west to east across the tail of his line of advance, but he did considerable execution among the South African Brigade deployed facing south and barring his path towards Sidi Rezegh. By the end of the day – 23 November – both the Afrika Korps and most of the armour of XXX Corps were lying in different sectors of desert within a few miles of each other, each counting extensive casualties. To the north Sidi Rezegh was held by 21st Panzer Division (less their tanks). Any British attempts to break out of Tobruk were being contained by 90th Light Division.

Towards Tobruk were moving elements of XIII Corps. Nevertheless on that night of 23 November Rommel was, in the words of his staff, jubilant.[9] There was little cause for it. The investing forces at Tobruk were still confronted by an unbroken garrison – the British

70th Division – and were threatened to the east by the march of the New Zealand Division of XIII Corps, while all over the desert to the south were the somewhat disorganized forces of XXX Corps and Cruewell's command. Less than a hundred German tanks were runners. Rommel's intelligence staff had, on the previous day, made an assessment of total British tank strength at 660 of all types, and they were well informed on British replacement and repair capacity.

Rommel, however, sure that *Totensonntag* had almost destroyed the enemy, had reduced such estimates so far as to make them irrelevant (it is true that he had been given an overestimate, and perhaps sensed it). He believed that the British were within inches of defeat, were in disarray which had to be exploited by rapid, vigorous action, personally led. Nobody was superior to Rommel in the exploiting of an enemy's disarray. He saw a chance to turn this battle – so far a murderous, costly but scattered and chaotic one – into a battle of decision. He gave out new orders during the night, back in his headquarters at El Adem.

Rommel believed that by a move eastward, personally leading the surviving mobile forces of Panzer Gruppe Afrika, he could swing round the southern flank of the British invading forces. He would interpose powerful strength between the British and Egypt and, using the garrisons still holding out on the frontier, Bardia and Halfaya, as pivots of manoeuvre, would surround and destroy Eighth Army. At 10.30 on the morning of 24 November, accompanied by his Chief of Staff, Gause, Rommel set off eastward at the head of 21st Panzer Division, followed by 15th Panzer Division. His thrust line was eastward down the Trigh El Abd, by Gabr Saleh to the frontier near Sheferzen, the same thrust line as that adopted six days previously by Norrie's XXX Corps in the reverse direction. This was what became known to both sides as 'the dash to the wire' – a thrust, Rommel ordered, to be made by all commanders without troubling about what happened on their flanks.*[10] Ravenstein gave 21st Panzer their verbal orders: 'the enemy was beaten and withdrawing south east'. Group Stephan would lead, and break through to the frontier south of Sollum.[11] 15th Panzer's war diary, too, referred to a beaten enemy.[12]

* If they had troubled somewhat more, they would have found two enormous British supply dumps established simultaneously with the initial advance on 18 November, one fifteen miles south-east of Bir El Gubi, the other a similar distance south-east of Gabr Saleh, on Rommel's southern flank. Their discovery by Rommel would have transformed the battle.

The 'dash to the wire' did not impress Rommel's staff with its wisdom. Without exception they admired their commander – his coolness, his vitality, his extraordinary sense for ground, situations, battle – and they liked him, too, for once they knew him Rommel could be a charming companion and relaxed with his military 'family' in a most agreeable manner. But they were clearsighted about the occasional lapses, the infrequent but potentially calamitous over-reach, of this extraordinary man; and on 24 November Rommel probably had few supporters among his staff.[13]

For the staff of Panzer Gruppe Afrika thought the underlying situation dangerous. A strong division – the New Zealanders – was advancing west, to the relief of Tobruk. Soon this would be the critical battle. The enemy's mobile force, scattered around the desert, might be somehow swept up or cut down by Rommel's scythe, but the blade of the scythe was less sharp than it had been, and British tank strength, always very capable of restoration, would soon make this operation dubious. Furthermore nobody really knew where any-body was, and much time – and, more importantly, fuel – could be spent combing the desert in hope. This was the negative side.

General Bastico, the Italian commander-in-chief and Rommel's titular superior, spent most of 25 November at Panzer Gruppe Afrika's headquarters and was no doubt close to despair. It seemed a wild goose chase; whereas if there were to be a concentrated Afrika Korps, positioned somewhere not too far from Tobruk and Sidi Rezegh, the opportunity might come to deal a decisive blow to XIII Corps.

Rommel would have none of this. His mind was on a final and decisive intervention, to cut off the entire British Imperial Army from Egypt. He believed that with the frontier again in his hands the British would race eastward in disorder to escape the trap.

The ensuing story of the 'dash to the wire' is anti-climactic. Rommel was away from his headquarters for three days. During those three days he tried, and failed, to relieve the German frontier garrisons which had withstood XIII Corps' assault: they were beleaguered before he arrived and they were still beleaguered when he left. The divisions of the Afrika Korps, despite admirable system, had more than ordinary difficulty in replenishment, the war diary of 15th Panzer recording, delicately, that supplying the division on 25 November presented difficulties not before encountered![14] Rommel did not bring to battle any sizeable part of XXX Corps. He did not produce a British withdrawal from Libya. He produced, it was true, a significant effect on the mind of the opposing Army Commander, Cunningham, who believed that Rommel's move spelt disaster for Eighth Army. His own proposed moves conformed to Rommel's estimates and intentions, and he was almost overrun, personally, by Rommel's whirlwind advance. He had asked the Commander-in-Chief, Auchinleck, to visit him in order to be persuaded that a retreat was inevitable, but Auchinleck was not so persuaded. Crusader continued.

Rommel received on 26 November a grim signal from his Ia at Panzer Gruppe headquarters, Lieutenant-Colonel Westphal. Westphal, on his own responsibility, had succeeded in contacting 21st Panzer Division and had ordered them to march back towards Tobruk, an act of considerable moral courage. He now managed to some extent to impress Rommel with the perils of the situation. The fuel position was grave. Air losses during Crusader were approximately even but the British had assembled sixteen squadrons of fighters, eight of medium bombers; they were, by now, dominant. The Panzer Gruppe was in real danger of immobilization and destruction in detail.

At least Rommel was not harassed by anxious probes from Berlin; Halder, somewhat ironically, was able to record in his diary that day that Rommel seemed master of the situation. But Rommel realized that his adventure had failed, with remarkably little to show for it. He persisted on the frontier for one more day, when 15th Panzer Division captured a New Zealand Brigade headquarters at Sidi Azeiz, but the game was over. For in the Tobruk sector the New Zealanders, in a skilful, silent night operation, had attacked and taken the Sidi Rezegh feature, while an offensive sortie by the Tobruk garrison had finally succeeded while Rommel was away on his foray. A narrow and vulnerable corridor now connected Tobruk with the relieving troops of Eighth Army. It was ironic that Rommel, whose obsession Tobruk had been, was absent at this dramatic moment.

The 'dash to the wire' was over, and Rommel's position was dangerous. During the escapade, of course, Rommel had been Rommel. General Bayerlein paints a picture:

> Rommel continued to drive from one unit to another, usually through the British lines ... On one occasion he went into a New Zealand hospital which was still occupied by the enemy. By this time no one really knew who was captor and who captive – except Rommel, who was in no doubt. He enquired if any-

thing was needed, promised the British medical supplies, and drove on, unhindered.[15]

Now Rommel had again to focus on the freshly, albeit tenuously, relieved Tobruk. To breach the corridor and again isolate Tobruk became the object of Rommel's next – and last – manoeuvre in what was afterwards referred to as the second battle of Sidi Rezegh.

For this last effort in the first phase of Crusader Rommel had the two Panzer divisions concentrated east of Sidi Rezegh on the Trigh Capuzzo. He himself flew at last to join his headquarters at El Adem, where his return was greeted with considerable relief. Rommel now decided that the Afrika Korps must surround and destroy the New Zealanders on Sidi Rezegh – New Zealanders who were now in hope of being reinforced by the armoured brigades of XXX Corps, moving from the south and south-east once again. These brigades – and Rommel's staff were well aware of it – had been strengthened by spare tanks and fresh men. Not least of Eighth Army's achievements was the way replacement soldiers and vehicles had been carefully husbanded to be used when required, as well as the way the British administrative system, including fuelling, had coped with the extraordinary vicissitudes of battle. In logistic provision the British were flexible and effective, in notable contrast to their tactical handling on too many occasions.

Cruewell, accepting the task, decided to drive the New Zealanders off Sidi Rezegh, attacking from the east, with 21st Panzer along the north of the objective towards Belhamed and 15th Panzer in the south, along the axis of the Trigh Capuzzo, towards El Duda. He gave orders accordingly, for a battle to start on 29 November.

Rommel disagreed. He believed that it would be ultimately counter-productive to reinforce Tobruk by driving the New Zealanders into it – two brigades of the New Zealand Division were in the area of Sidi Rezegh, and the British XIII Corps headquarters had actually itself moved into Tobruk, the better to coordinate the operations of 70th Division with its relievers. Rommel was determined, on the contrary, to isolate the defenders of Sidi Rezegh and destroy them by an encircling movement. He countermanded Cruewell's orders. As a consequence 15th Panzer was sent westward, south of the Sidi Rezegh feature, and ordered then to turn north and attack in the direction of El Duda. The movement took place successfully, and El Duda was ultimately taken (although then again retaken by the enemy). Rommel visited Cruewell on the afternoon of the twenty-ninth to insist that the operational object must be to destroy those elements of XIII Corps holding Sidi Rezegh, as well as to keep it isolated from the garrison of Tobruk.

By now, therefore, the New Zealanders were disrupted and near surrounded. An attack from the south on the Sidi Rezegh escarpment itself clinched the matter on the next day, 30 November. At quarter to eight in the evening Freyberg signalled to XIII Corps: 'The enemy has Sidi Rezegh.' The corridor to Tobruk was still – somehow – held open by the British but Rommel was close to achieving his last desperate objective. Freyberg withdrew the residue of his division eastward and then south. On 1 December the Afrika Korps listened to British radio from London: 'General Rommel has thrown his last strength into the battle to break through the British cordon to the west.' This was somewhat misleading. Rommel was not ringed by a British cordon, but was frustrated by a still unextinguished British corridor to Tobruk.

It was certainly true, however, that Rommel's last strength had been committed to the battle. It was also clear that it would not be enough. On 1 December Rommel's staff were able to show him a very accurate picture of the British forces remaining. Tobruk was invested – just – but what now mattered was the overall comparative strength remaining to each of the contestants. If that were forecast to be moving remorselessly in the enemy's favour Rommel knew that he could not successfully continue a campaign whose strategic object was the capture of Tobruk. It would, increasingly, cost him more to maintain the siege than it would cost the enemy to press and harry the besiegers. After an unsuccessful attempt again to take El Duda with 21st Panzer Division Rommel accepted that Panzer Gruppe Afrika could not hold on ringing Tobruk. The fortress had been relieved. He must withdraw westward, break clear.

Rommel managed to extricate both German and Italian divisions during the days between 4 and 8 December, and to man a new position (once fortified by the Italians in an earlier phase) running south from Gazala, sixty miles west of Tobruk. He was helped by the relative inactivity of enemy air forces, but he had decided that the condition of his own forces and his logistic situation meant that, for a while, he must give up Cyrenaica; that he could not hold east of the Cyrenaica 'bulge'; that the taking of Tobruk had for a time faded into mirage. It was a bitter decision and it was, not unnaturally, bitterly contested by the Italian High Command. These were Italian provinces, and were settled by considerable numbers of Italians. The blow to Italian prestige, too, would be deplorable. At a conference on 15 December attended not only by the Italian commander-in-chief, Bastico, and Gambara from XX Corps, but also by General Cavallero, the Italian Army Chief of Staff, there were angry words. The conference was also attended by Kesselring, a German Field Marshal who had transferred from army to Luftwaffe and been given by OKW responsibility for general supervision of German forces in the Mediterranean and Italian areas (albeit without direct operational authority over Panzer Gruppe Afrika). Kesselring, a natural optimist and an excellent diplomat, tended to take the Italian part.

But Rommel was adamant. Panzer Gruppe Afrika, all German–Italian forces, must be got back to a defensible area behind Mersa El Brega. They must go back to the point whence he had led them in heady triumph in March and April. They were not, he said, in a condition to deal with another British offensive, an offensive which would – like all offensives in Cyrenaica, by either side – strike across the chord of the arc and isolate the forward troops by driving, in this case, towards the Gulf of Sirte. The British possessed freshly augmented forces.

British losses, although Rommel did not know it, amounted to some 15 per cent of men engaged; whereas his own losses in German troops had been over 20 per cent, and over 40 per cent of the Italians. In major equipments Rommel was clearly a net loser – and his enemy were masters of the field, with consequent ability to recover and repair.[16] The British had already attacked at Gazala, showing that they were still minded to maintain the offensive, and although the attack had been repulsed by 15th Panzer Division the respite must be temporary. The Afrika Korps had only forty tanks left.

Rommel got his way. He wrote to Lucy on 20 December that those of his commanding officers who were not dead or wounded were ill and that there was no alternative to pulling out and going back. Once the decision was made he conducted the withdrawal with his usual speed and mastery. The German movement was harassed by uncoordinated British attempts in pursuit but on Christmas Eve Rommel evacuated Benghazi, and after the Afrika Korps had counter-attacked at Agedabia against a somewhat ineffectual British attempt to envelop it Rommel finished the year with a more secure position. In the Agedabia fight he had destroyed sixty British tanks and only lost fourteen, his supply position was greatly eased by his shortened line of communication, he had received substantial tank reinforcements (a transport convoy had reached Benghazi just before evacuation, and the Germans were always able to bring tanks into service much more quickly than the British, who needed to carry out extensive desert modifications), and he reckoned that, like his own forces, the enemy were tired.

Rommel himself was tired, but ever restlessly vigilant. He saw one of his surviving 88s exposed to British long-range artillery fire in a poorly selected covering position during the withdrawal, and drove up angrily to rebuke the commander, only to find that the gun was a dummy, an Italian telegraph pole skilfully camouflaged. Rommel grinned. 'Take it with you!' he said. 'We don't want the enemy to find out all our tricks before we improve on them!'[17] The men of the Panzer Gruppe knew that they were still in the game.

On the reverse side of the account Rommel's surviving garrisons on the frontier which had not already fallen to XIII Corps' onslaught were in a parlous condition, surrounded, hungry and with little hope of relief. With the approval of higher authority Bardia surrendered on 2 January 1942, and Halfaya – whose Italian commander had provided brave and outstanding leadership – on the seventeenth. Rommel had also lost Cyrenaica and its airfields. He noted details of the reorganization of Eighth Army of which his Ic told him on 29 December and meditated his next move.[18] The year 1941 had been an extraordinary one for Erwin Rommel, with remarkable alternations of triumph and setback. He wrote to Lucy on 31 December that his thoughts were more than ever with her and Manfred, who to him meant earthly happiness.

But, as in the immediate aftermath of 'Battleaxe', the entire strategic situation had been transformed during Rommel's withdrawal from Cyrenaica by events at the other end of the world. On 7 December the Japanese Air Force had attacked the United States Pacific Fleet at its base at Pearl Harbor and invaded the Philippines; and on the same day Japanese armies had invaded the British colony of Hong Kong and the British protected territories of Malaya. A few days later, and fatally, Germany had declared war on the United States. The Third Reich, with Italy and Japan as allies, was now in arms against the United States, the British Empire and the Soviet Union. In the long run, if it could be brought to a long run, Germany could not win.

Rommel's star did not shine brightly during Crusader. He was unwilling for too long to be persuaded that it was as serious a business as it was. His moves during the early days left a surprisingly large amount of decision to subordinates – not by any means necessarily a point of criticism but for the fact that his own interventions, at least initially, were hardly inspired, and so interventionist a higher commander was bound to be looked and listened to at every turn. His sense of what was of fundamental importance to the battle seemed, for a little, defective. He did not give an impression of putting first things first.

This was curious, because Rommel's obsession with Tobruk – with maintaining its investment by his army, with preventing a link-up between Eighth Army and their beleaguered Tobruk garrison, with ultimately taking the place itself – was wholly justified. Operationally, Cunningham's declared object had been the destruction of the German–Italian armoured forces, and operationally Rommel's orders and initiatives on at least three separate occasions were concerned with the elimination of the British armoured groups; but strategically Tobruk was the prize, and Tobruk, therefore, drew the flow of battle towards itself as to a magnet. At times Rommel gave the impression of refusing to acknowledge this fact – notably of course, when he made the assumption that the Tobruk battle was pretty well won and that, by a bold stroke – the 'dash to the wire' – he could win a great, mobile battle of encirclement against residual British forces. The assumption was ill-founded, and Rommel had every reason to know that it was ill-founded – his intelligence staff were perfectly well-informed and perfectly clear-eyed about the facts of the situation, and they did not justify a foray of the kind Rommel attempted. The opening sentence of each Panzer Divisional Commander's orders on that occasion, that the enemy had been beaten, simply was not true.

The impression therefore remains of a Rommel who, unusually, failed to maintain a steady sense of priorities, to maintain his aim. Frustrated to some extent by the course of battle, he needed – almost a psychological need – to believe that things were better than they were, and that a chance had at last arisen to move from the sort of battle he found unsatisfactory to the sort – a bold, personally led manoeuvre, stunning a battered enemy with its speed, ingenuity and surprise, a Matajur, a Meuse crossing – at which he shone. His staff, on the night of 23 November (after *Totensonntag*), reported him jubilant when there were insufficient grounds for jubilation. *Totensonntag* had caused very great British and South African casualties, and Sidi Rezegh was in German hands, with the sortie from Tobruk contained; but the fundamental *facts* at that point were not encouraging for Rommel. XIII Corps, led by the New Zealanders, was advancing towards Tobruk along the Trigh Capuzzo. British armour had suffered but was known to be being replaced and restored. Tobruk was still held by the enemy, its potential for offensive operations still presumably formidable.

Yet Rommel chose to believe that he could impose the order of a victor on a chaotic battle, as he often had in the past, by gathering everything that could move in his wake and by leading it on a triumphant march of decision. It did not work. It was not even likely to work, unless the enemy's nerve cracked (as, admittedly, it almost did). And it removed authority from the German command for several crucial days. There were many moments during Crusader when Rommel's bacon was saved – to the extent that it *was* saved – by the sober moral courage of his staff, notably Westphal, or by the steady tactical shrewdness of Cruewell.

This is not all. It is difficult to resist the impression that Rommel's command organization (or, at least, the way he used it) was defective. Rommel was Rommel, and whatever textbooks rightly preach there was never a way in which he could be prevented from assuming tactical command at a chosen point of contact – it was a facet of his genius, his resourcefulness, his eye and feel for battle situations, his speed, his boldness and his sharpness of wit. But criticism of Rommel in Crusader need not focus on his abiding readiness to throw himself into the tactical battle beyond the measure appropriate to an army commander. It should also be directed at the organization itself. No man can deal with more than a limited number of subordinates, and their responsibilities – like his – should be clearcut. Crusader does not show Rommel observing this elementary principle. Neither his arrangements nor his technique nor his judgement were at their best; and if he had learned by the end of the battle to use correctly an able and sophisticated staff he had certainly not done so before it began.

Rommel, clearly, had three main concerns, each of which required a responsible subordinate commander. He had the frontier defences, and – after they were largely vanquished – the battle on the frontier, Bardia, Sollum, Capuzzo, Sidi Omar. He had the investment of Tobruk and the defeat of British forces advancing to threaten the investors of Tobruk and link up with the garrison. And he had the obvious and to an extent simultaneous need – and opportunity – to deal with British mobile forces in open desert. Each of these concerns merited a major responsible command.

The first of these – the frontier battles and garrisons – might have been entrusted to the Italian commander of the Savona Division, perhaps suitably reinforced. The second – the investment of Tobruk and the prevention of British link-up – might have been in the hands of General Navarrini and the Italian XXI Corps: it was, after all, Navarrini's divisions which were ringing Tobruk. Rommel had previously spoken harsh words about the Italians, and a good deal of them had been merited earlier in the year: but in Crusader several enemy reports credited the Italian infantry with fighting exceptionally hard – on occasion harder than their German comrades.[19] Basic to the investment of Tobruk was responsibility for the area of Sidi Rezegh. Only twenty miles from the Tobruk defensive perimeter, it acted as a pivotal point for the British battle of relief and break-out, and although the reflections of hindsight are notoriously unjust the impression remains that its importance was ignored by the Germans for too long. It was, or should have been, Navarrini's outer keypoint in his lines of circumvallation, if he (or another) had been given responsibility for the investment of Tobruk.

The third concern was with the ability to defeat any enemy incursion with the united mobile strength of Panzer Gruppe Afrika, and here Rommel clearly had need of a commander. Equally clearly, the man for the job was the commander of the Afrika Korps, Cruewell. There were two mobile corps, XX Italian (Gambara) and the Afrika Korps (Cruewell), and however Rommel decided to use Gambara's headquarters (which he seems scarcely to have done) it was evident that the most effective mobile instrument in the desert was the Afrika Korps.

And Rommel used it as such – but fitfully, periodically giving orders to Cruewell's divisions. He differed from Cruewell (and he was often justified, but not always) on particular orders, as he had every right to do, but the sense remains of a Cruewell uncertain from hour to hour whether he was acting with independent power of decision and action or whether primarily as a conduit for the ideas of another. This dichotomy surely led to that impression of confusion and lack of direction within Panzer Gruppe Afrika which distinguishes some of the Crusader battle and which is so strikingly different from most of the Rommel record.

Whether the command system could have borne improvement or not, the response within it of individual commanders was often admirable. The attack on Sidi Rezegh on 22 November by von Ravenstein, the flank attack on the withdrawing British that night by

Neumann-Sylkow – these were masterly, but they were not particularly coordinated. They were not especial feathers in Rommel's – or even Cruewell's – cap although they were creditable to all levels of the German command.

But, of course, Panzer Gruppe Afrika still responded to Rommel's dynamism. The tactical handling he had taught and encouraged in the Afrika Korps was still superb. It was that, and the range and power of German anti-tank gunnery, rather than superior operational wisdom which gave his troops their periodic and transient victories in Crusader. When he appeared, and 'gripped' a local situation it was a thing apart from the intervention of most generals, German or otherwise. He had generated a myth and Crusader, whatever his mistakes, did not explode it. When he set out on the 'dash to the wire' a *frisson* rippled throughout the British command, for Rommel's name was already, like Prince Rupert's, one 'very terrible to his enemies'. And on 24 November, with Rommel approaching on his imprudent escapade, his opponent, Cunningham, decided that the British must evacuate Libya, must call off Crusader, were morally defeated. Cunningham was overruled by Auchinleck and soon relieved of his command, to be succeeded by General Ritchie. Like Rommel, Cunningham had prematurely imagined that Rommel had won.

CHAPTER FOURTEEN

'Rommel an der Spitze!'

IN EARLY January Rommel studied an intelligence appreciation of the state of the British army facing him and likely British intentions. As far as strength and composition went it was very accurate.

A fresh division, 1st Armoured, new to the desert, had taken over from the experienced 7th Armoured. There were numerous changes of units and commands. The official strategic assessment of enemy intentions was that the next British offensive would aim to invade Tripolitania, link up with French forces in Tunisia (assumed likely to take the anti-German side when conditions were favourable) and, having cleared the North African coast, use it as a base for invasive operations against southern Europe.[1]

Rommel was now receiving periodic intelligence from a new source, known to the Germans as '*die gute Quelle*', 'the good source'. This was Major Fellers, United States military attaché in Cairo, who was the unwitting conveyor of a good deal of information on British orders of battle, plans and appreciations. The Italians, in late summer 1941, had cracked the American diplomatic code, the 'Black Code', by which Fellers was signalling, and had shared the results with Berlin, while the Germans, working analytically and independently, cracked it themselves and thereafter were profiting by it with great speed, to Rommel's advantage.[2] Fellers, (described by the Germans as 'enthusiastic in his enquiries') had excellent contacts in the British army in Egypt and the British High Command. He was, at first as an especially friendly neutral and after December 1941 as the representative of an ally, given privileged briefings and comprehensive information on a discreet basis.

Fellers tended to be a pessimist about British prospects. He kept the Pentagon fully informed on British strength and deployment, and on how they saw the situation, as well as on their knowledge of German strength in North Africa and on their intentions. The process of this information was rapid and efficient, so that Rommel's intelligence on his enemy for the first few months of 1942 was particularly exact. It was, of course, too good to last. Through ULTRA the British read the OKH signals; and in due course *die gute Quelle* dried up, but not until the end of June, by which time Rommel had driven the enemy from Libya.

Rommel now again reckoned that for a short time fortune was with him. He had received not only the consignment of tanks to Benghazi but a shipload of fifty-four tanks and twenty armoured cars to Tripoli on 5 January. His intelligence staff told him that his armoured strength would be greater than that of the British on the frontier of Cyrenaica until the end of the month – they gave the British 150 tanks in 1st Armoured Division and pointed to the latter's inexperience, while Rommel now had a total of 117 German and seventy-nine Italian tanks. German tanks were brought into service in the desert quickly after arrival and were soon familiar to their crews, to some extent owing to the commonality of mechanical parts and parentage which the Germans established; and Rommel could, for a very short while, count on an actual and local fighting superiority in numbers. In terms of human quality he knew that the Afrika Korps, in spite of losses and hardships, felt unbeaten.

Intelligence suggested that the British would as soon as possible build up for a new offensive and Rommel's superiority would pass. The suggestion was well-based. The British certainly intended next to invade Tripolitania; and Crusader with its aftermath had given them forward airfields at Gazala, Mechili and Msus. And the British assumed that Rommel could not possibly attack for some time. He had been beaten in Crusader, he had been driven from Cyrenaica, he had withdrawn Panzer Gruppe Afrika to Mersa El Brega, he would need time.

This assumption was wrong. Rommel had recovered his spirits. He was always mercurial – surprisingly so in one whose Swabian temperament held so much stolidity mixed with the fire, with the panache. His reports had been criticized by Brauchitsch for their alternating ebullience and depression, and the same flavour of exaggeration can be found in his letters. He had been fiercely condemned by the Italians for his decision to quit Cyrenaica in December and had rebutted the criticism with equal fierceness – and rightly so, for

Rommel knew his army, knew that it needed a breathing space, which could only be provided by the long leap back to Mersa El Brega. But Rommel certainly felt sore at giving up the fruits of his sensational victory in the previous March and April, giving up the chance to take Tobruk and carry the war towards Egypt, to fight it on his own terms with an advanced base and port behind him. When his intelligence staff told him that for a short while the numerical odds were in his favour he rejoiced. He was reported by his interpreter as in 'a foul temper' on 17 January but immediately thereafter the sun came out.[3] Rommel had reservations about the supply situation, but he overcame them. He decided to attack.

He was, however, determined on absolute secrecy. He forbade information on his intentions being passed to OKH or to his immediate Italian superior, General Bastico – the Germans knew from experience, Rommel wrote in his diary, that Italian headquarters could keep nothing to themselves and that everything they wirelessed to Rome got round to British ears. He forbade all reconnaissance. Regrouping for the attack was to be entirely by night march. He decided not to tell the commander of the Afrika Korps of his plan until five days before the day fixed for operations to start, and he decided that divisional commanders need know nothing until two days before – and that all orders then must be verbal. Little was confided to paper – Rommel himself signed the single operation order* – a mere twenty-one paragraphs, each, on average, only seven lines of typescript.

Rommel believed that if he was to smash the enemy quickly on the western border of Cyrenaica the best prerequisite of success would be absolute surprise. Thereafter speed and shock, the Rommel formula, should once again bring victory. His immediate objective was Benghazi. His ultimate, undisclosed, object was to drive the British again from Cyrenaica, but for the time being he determined on a limited, spoiling attack to deprive the British of the initiative they had won in Crusader.

Rommel knew that the British armour opposing him, one armoured brigade, was inexperienced. He learned of its poor tank serviceability state, actually transmitted in uncoded speech. He listened with enjoyment to the expostulations of General Messervy, commander of the 1st Armoured Division, about the relief of 7th Armoured Division's support group, now back on the Egyptian fron-

* The Panzer Gruppe was on 22 January elevated to the status of 'Panzerarmee'.

tier on what Messervy described (to XIII Corps) as 'a joyride which in my opinion will not help to win the war'.[4] Messervy's unease was, Rommel reckoned, well justified. Rommel had, for a little, the upper hand on the Cyrenaica–Tripolitania border and he meant to use it. He was opposed by a raw armoured brigade which had anyway been dispersed for much-needed training in January; by two weak brigades, totalling only four battalions, on the immediate front; and by a temporarily weakened 4th Indian Division on the coast road – 'weakened' because his intelligence staff reported one of its brigades withdrawn to Barce, east of Benghazi, for an intensive programme of courses and training. The stage was set.

Rommel planned to move with two wings. On the left 'Gruppe Marcks', with 90th Light Division (General Veith) and some tanks from 21st Panzer Division, would advance up the Via Balbia, the coast road; while on the right the Afrika Korps would advance north-eastward on the line of the Wadi El Faregh. Rommel proposed to place himself at the head of Gruppe Marcks. He set the starting day at 21 January, the time six-thirty in the evening, as the light began to fail. He wrote to Lucy that afternoon of his complete faith in God's protective hand and that He would grant victory. That day he received the award of the Swords to the Oakleaves of the Knight's Cross; and three days later was promoted to the rank of Colonel-General.

Rommel's counter-attack, which succeeded beyond expectations – and well beyond any orders he received – saw him once again restored in heart and, his staff noted, at his best. He led the Marcks group towards Agedabia and entered the place at eleven o'clock next morning, 22 January, while his right wing, the Afrika Korps, smashed through the 1st Armoured Division Support Group on the desert flank, and converged on Agedabia from the south-east. Rommel's plan had been to send the left wing northward towards Benghazi; now he ordered it eastward to try to encircle the British in the Jebel east of Agedabia, while the Afrika Korps established blocking positions on the general line of Agedabia–Antelat–Saunu.

The encirclement was only partially achieved, but by now Rommel already sensed disorder and disorganization among his enemies and reckoned he was winning. Late on 24 January he made his plan for the next day. He would, once again, lead the Afrika Korps north-eastward to Msus. He had been inhibited from attempting a deep thrust across the arc of Cyrenaica by his fuel situation, but he believed from what he had seen, from his *Fingerspitzengefuhl*, that a short, violent jab towards Msus might convince the British that they had lost a decisive battle.

And so it was. As Rommel's two German divisions advanced towards Msus they overtook much of 1st Armoured Division, tanks and soft-skinned vehicles alike, and did enormous execution. Rommel had the satisfaction of seeing British vehicles racing in confusion across the desert in all directions, trying to get away. It was a tactical victory – no very subtle operational manoeuvre but a hectic race here and there all over the Jebel with friend and foe often commingled and unidentified. Were they hunters or hunted? a member of Rommel's command group wondered, seeing what he thought were British vehicles all around. Rommel had few doubts. 'There's the enemy!' he yelled, 'Take him prisoner!' The race went on and the prisoners poured in.[5] Rommel had simply sensed the situation, adjusted his plan, exploited the tactical superiority of the Afrika Korps and now had Cyrenaica at his feet. At eleven o'clock on the morning of 25 January he entered Msus, capturing ninety-six British tanks. Cavallero had arrived from Rome on a visit to North Africa and forbade further advance: but Rommel knew that he could take Benghazi at least and was determined to do so, to chance his fortune, seek German support, win time and be justified by events.

Cavallero was, like Bastico, particularly incensed at Rommel's performance since Rommel, who had concealed his intentions from Bastico, had posted his attack order at every German supply depot in Tripolitania on the day he went forward, knowing that Bastico would learn thus about his nominal subordinate's offensive. 'Cavallero implored me not to go on,' Rommel wrote afterwards. 'I told him that nobody but the Führer could change my decision.' There was no occasion on which Rommel more flagrantly ran against Italian directions; he succeeded, so was forgiven. Indeed, a telephone call from Mussolini himself on 26 January was described by Rommel's interpreter as *Scheibenhonig* – 'honey in the comb'.[6]

Rommel now made a feint towards Mechili – his fuel allowed for little more – and then moved on Benghazi from the east. His feint was soon rewarded by anxious signals from Eighth Army asking XIII Corps for information on Rommel's columns marching towards Mechili. This feint and the fighting at Agedabia and near Msus were enough, and the British withdrew completely from Cyrenaica. By 6 February they were back on the Gazala position.

This was less cataclysmic than appeared. Auchinleck, in a letter to Ritchie on 19 January, had actually foreshadowed as conceivable a British withdrawal as far as the Egyptian frontier if things turned nasty. Now they had turned very nasty indeed. Rommel had reconquered the territory he had taken in the previous March, and reconquered it in eight days – all this in the face of a certain enemy air superiority. He had, as in March, largely destroyed the enemy facing him. He had certainly delayed for a while any possible British resumption of the offensive. He had recovered the initiative, although it would be hard work to retain it. He had not sufficiently avenged Crusader, and he was still many miles from Tobruk; but he was to a considerable extent restored. On 16 February he flew to Rome, continued on to Germany and received from Hitler the Swords at the Führer's headquarters. He then took some much-needed leave at home and did not return to North Africa until 19 March.

Rommel had, for a few days, again been in his element.

'The merit and value of the desert soldier,' wrote Fritz Bayerlein, still Chief of Staff of the Afrika Korps,

> can be measured by his physical capacity, intelligence, mobility, nerve, pugnacity, daring and stoicism. A commander of men requires these qualities in even greater measure and in addition must be outstanding in his toughness, devotion to his men, instinctive judgement of terrain and enemy, speed of reaction and spirit. In General Rommel these qualities were embodied in rare degree and I have known no other officer in whom they were so combined.[7]

From one very close to Rommel, and a fine commander as well as staff officer on his own account, this was a remarkable encomium.

Occasionally, perhaps, the stoicism slipped, but they were not occasions of which Rommel's command were particularly aware. Occasionally his aggressive optimism betrayed his better judgement. More than occasionally, Rommel's determination to take control at the forefront of the fight was inappropriate to an army commander and led to errors, mismanagement, professional lapses for which he may, without pedantry, be criticized. But one cannot be prim about Rommel. He was what he was, and the courage and energy which led him sometimes towards unattainable targets and into avoidable trouble were the same qualities which gave him victory. After the reconquest of Cyrenaica congratulations poured in. 'When you were our commanding officer,' wrote a former Goslarer Jäger, Forstmeister Schluter, who had spent time working in forestry in equatorial Africa, 'you said, "Always come to me openly with your requests!"' and now the writer, a reserve Lieutenant, had a request. He wanted to join Rommel once again, to join Panzerarmee Afrika.[8]

Rommel knew that British strength was growing and that the British overall position in the Mediterranean would be increasingly superior to his own unless the sea-supply situation were radically changed in his favour. There was one drastic way to achieve this – to eliminate Malta as a base for British sea and air action against the Axis convoys running between Italy and North Africa; the importance of Malta was paramount, and was generally accepted as such. Rommel was only getting eighteen thousand of a Panzerarmee monthly requirement of sixty thousand tons of supplies, and he thought the German authorities were being too weak in pressing the Italians to increase their maritime efforts in the Mediterranean.

He also privately suspected that the Italians were being treacherous – that their hearts were not in the job. It was, of course, undoubtedly true that many Italian officers disliked both the Fascist regime and the war (and their German allies!): but Rommel was unaware of the benefits ULTRA was bringing to the British in the maritime contest. The German and Italian naval codes had been cracked – the 'Hagelin' machine used for maritime encoding had been 'broken' for some time, and the British had appreciated that one of the most fruitful methods of waging war against Rommel was to attack the convoys to Africa. A special intelligence cell had been created in Middle East GHQ to target Rommel's logistics, and one outcome was the excellent use to which the British were able to put their own stretched naval and air resources. Neither Rommel nor Hitler knew that they were confronting ULTRA. It was the joker in the pack.

During his March visit to Rastenburg Rommel discussed his situation and possibilities with Hitler, and found the Führer personally sympathetic to his representations. Hitler's geopolitical dreams were consonant with the sort of strategic advance in and from Africa which Rommel believed – and believed until the end of his life – was both possible and attainable, given more backing. Rommel was convinced that with a small reinforcement of German Panzer forces, and a strategic decision to take Malta, he could conquer Egypt, maintain the Panzerarmee there, and ultimately move north-eastward to threaten the Russian position in the Caucasus. At the time of his visit, Hitler was preparing his next War Directive, which, among other objectives, specifically ordered the German armies of the south to break through into the Caucasus, and allotted this task priority.[9]

Hitler's plans were not, therefore, remote from Rommel's visions. There was indeed a 'Great Plan', 'Plan Orient', to eliminate the British from the Middle East and conquer their main sources of oil by a vast convergent movement of which Panzerarmee Afrika would constitute the southern wing; and the same visions, indeed nightmares, periodically haunted the minds of the British Chiefs of Staff, as the German campaign of 1942 carried the Wehrmacht triumphantly into the Caucasus.

Rommel persuaded himself that with more support he could harry the British from Egypt – which was possible: but he also believed in the Great Plan, believed that with the consequent collapse (as he saw it) of the British position in the eastern Mediterranean nothing could seriously inhibit a victorious Panzerarmee from driving across the Syrian desert, to and across the Tigris and Euphrates, thereafter driving through the Persian mountains to join hands with the German armies advancing south through the Caucasus in an attack on the Russian oilfields.[10]

It must be doubtful, at the very least, whether such a strategic advance could ever have been practicable. Attractive though the idea was of closing the Middle East to the British and securing the whole southern periphery of Europe against amphibious intervention, to say nothing of the effect on British communications with Asia, there were too many preconditions, all of which needed to be met. The Germans and Italians would have needed total supremacy in the Mediterranean – not only by the seizure of Malta, but by building up such maritime power that any Anglo-American attempt to dispute that supremacy could have been defeated: the subsequent course of 1942 did not run in a way to make that very credible. They would probably have needed Spanish co-belligerency. They would have needed not only to eliminate the British, and to control an eastern Mediterranean port – whether Alexandria or in Palestine – but to control a secure line of communication forward from that port over a distance of more than a thousand miles of desert and mountain. The logistical effort, combined with the maritime effort, would have probably been unsustainable. And all this, of course, assumed a German victory in 1942 in southern Russia, where the Wehrmacht, after early successes, was in fact marching towards its most resounding defeat.

Nevertheless the timing of Rommel's visit to Hitler in March 1942 was opportune. He received a general, albeit limited, blessing on his efforts. Hitler was genial and Rommel sat next to him at a small dinner attended also by Keitel, Jodl, Schmundt and Westphal, who had accompanied his commander. Heinrich Himmler was also present, and raised the matter of a possible Hindu Legion, formed from Indians disaffected from the British Empire, as part of the SS. Discussion turned to Churchill, and Hitler abused him as a drunkard. The general tenor of conversation Rommel found depressingly unrealistic, and he was unsurprised though disturbed to hear the senior Adjutant, Engel, murmur to take no notice, that it was the same at every meal.[11]

OKH and Halder were predictably discouraging, however, concerned as they were with the huge demands of the Russian front. 'Rommel,' Halder observed to him, 'you're fighting a lost battle!' Nevertheless Rommel left Rastenburg with the promise that an airborne attack on Malta – 'Operation Hercules' – would take place by June if nothing unexpected interfered with it; and that meanwhile there would be an intensive bombing campaign against the island, to inhibit Allied action against the convoys. This took place. The 2nd Air Fleet had been moved from Russia to Sicily, bringing about a considerable improvement in the air situation, so that for the forthcoming battles Rommel could hope for something like air parity in the combat zone, as well as more regular supplies. Indeed, for the next three months Rommel's logistic position improved dramatically.

Rommel returned to the desert, not wholly dissatisfied with his visit although disappointed that the General Staff still refused to share his (and, to some extent, Hitler's) view of the importance of Africa: their belief in the 'Great Plan' was distinctly qualified, since it depended upon victories in Russia which were as yet unwon. It was, as ever, a question of priorities; and of keeping in balance the strategically desirable with the operationally and logistically attainable. Nevertheless, on 29 March, soon after returning from leave, Rommel gave his officers his view of the situation and of what lay ahead. The British planned to attack soon, but instead the Panzerarmee would itself undertake an offensive in about two months. Its objects: to prevent the reinforcement of Tobruk; and to take Tobruk.

A few days later an officer joining the Afrika Korps who knew

Rommel well from earlier days noted his air of health, animation and optimism. He had come just in time (from Russia), Rommel told him, just at the right moment. Rommel had a new offensive in preparation, to forestall the British.[12] At the end of April Hitler received Mussolini, Cavallero and Kesselring at the Obersalzberg and formally agreed that Rommel could attack in Cyrenaica in May, even if Malta had still not been assaulted. After taking Tobruk, however, he was to pass to the defensive.

The task was the most formidable Rommel had yet faced. The British had fortified a strong position running into the desert for forty miles from Gazala south to Bir Hacheim. Its extent was known to the Panzerarmee, although there continued to be uncertainties about some details of British deployment, despite the helpfulness of *die gute Quelle* at this juncture. In particular the Germans thought the southern part of the British position more strongly held than was the case – and underestimated the strength of the southernmost fortified point, Bir Hacheim.

The German estimate of comparative tank numbers and types – always the critical factor in Rommel's eyes – was somewhat optimistic. But Rommel knew that he was facing a strongly entrenched and strongly mined defensive zone – his staff estimated 500,000 mines, and although assessments of British armoured strength fluctuated he knew that he would, in the next battles, be outnumbered in tanks. He knew that the British were under pressure to attack themselves and retake Cyrenaica, and there were frequent reports that such an attack was imminent. On one such occasion Rommel was told that the British were certain to attack on Easter Monday, 6 April. He refused to believe it, and taking one tank as escort drove personally into the desert towards the distant British. There were no signs of British preparations or advanced deployment. Eventually Rommel stopped and heard the sound of distant shell fire. Seconds later there were shell bursts round his vehicle, the windscreen was shattered and a shell-splinter penetrated Rommel's tunic and bruised him severely.

He drove back, covering twelve miles in record time. He had seen and sensed enough to be sure that no serious British attack was about to be launched. 'I only wanted you to know,' he explained to his accompanying officer, who had just returned from leave in Germany, 'that the English aren't preparing an attack. Two advanced batteries, that was all. Once again it's all bluff!'[13] In his Easter letter to Lucy he wrote, mendaciously, that a shell-splinter had come through the window and landed in his stomach after going through his overcoat and jacket. Leaving a multi-coloured bruise the size of a plate it was finally stopped by his trousers![14] It is doubtful whether Lucy was deceived. Rommel's personal forays continued, and on 5 May Goebbels wrote in his diary of an English report that they had almost captured Rommel – unfortunately true, wrote the Propaganda Minister, deploring Rommel's carelessness of his life and security. Rommel was ebullient, but he had a big job on his hands.

A few – a very few – great soldiers have been modest men. Turenne is said to have mocked generals who were unable to admit when defeats derived from their own failures. In the resigned, philosophic calm of Marlborough in adversity there is, perhaps, a grain of modesty. In more recent times Eisenhower conveyed an impression of selflessness and unostentation. The wry self-depreciation with which Slim wrote of his own achievements contains humility as well as consummate art, and is as rare as it is impressive. But, on the whole, successful commanders have enjoyed the limelight and sung their own exploits very audibly. Even Wellington was said to listen with remarkable composure to exaggerated praise; while the vanity of some has crossed the frontier of the absurd.

Rommel, while certainly never absurd, relished the publicity accorded his exploits. Although unpretentious, he had a strong streak of vanity. From an early point in his career Josef Goebbels, inspired publicist of the Third Reich, had befriended him, and Rommel had reciprocated – unsurprisingly, for Goebbels, characterized by enemy propaganda as the 'father of lies', was also a man of considerable charm and intelligence as well as an orator of brilliance. Goebbels had undoubtedly marked Rommel as possessing, from the viewpoint of publicity, star quality. Here was a commander far removed in personality from the coldly remote (and, despite formal subordination, often coldly contemptuous) senior officers of the old school with their manifest inner rejection of the 'German Revolution' which National Socialism had brought. Here was a rough-tongued, warm-hearted, obviously sincere and direct soldier of genius; a man who was renowned for his ability to talk to soldiers as one of themselves, who clearly disliked snobbery and pretension, who never dissimulated but spoke from the heart; a man who, equally clearly, felt personal devotion to Adolf Hitler; a man, although obviously without interest in politics or political ideology, who was suitable for exaltation as a hero of the new Germany. Goebbels, therefore, put his heart into the adulation of Rommel for his real achievements, which the German press and media fostered. Rommel, despite the remoteness of North Africa from instinctive German concerns, despite the comparatively tiny number of German divisions and German soldiers under his command, became an important and popular public figure.

This was congenial. Rommel began to receive fan mail, which amused him and tickled his vanity. His appearance had always been photogenic – the compact figure, the soldierly smartness of bearing, the regular features, the expression conveying frankness, directness and more often than not humour. (Rommel had always himself enjoyed photography, and Goebbels had given him a camera as a present early in the war.) Rommel, too, understood the value of publicity, even within his own command. German soldiers were as readily sceptical of personal vanities as the soldiers of most other armies, but there was no doubt that to be led by one who was becoming a world-famous figure was good for Panzerarmee Afrika. His family were naturally and ecstatically proud to hear him referred to in speeches on German radio as 'our popular hero, Colonel-General Rommel'. 'It's all like some kind of dream,' Lucy wrote to 'my dearest Erwin'; 'All my prayers go only to the Lord to be with you and keep helping you towards your goal for *Führer, Volk und Vaterland.*'

Nor was Rommel's fame confined to the Axis. His opponent, Auchinleck, found it desirable to send a letter to all commanders and chiefs of staff in the British Middle East Command, in which he referred to the 'real danger that our friend Rommel is becoming a kind of magician or bogeyman to our troops, who are talking far too much about him . . . I wish you to dispel by all possible means,' the letter ran, 'the idea that Rommel represents something more than an ordinary German general.' But it was beyond the power even of Auchinleck, himself a fine, modest and soldierly personality, to convince British soldiers that Rommel was not 'something more than an ordinary German general'. Rommel had imprinted his image on the enemy, who were certainly not susceptible to German propaganda or media promotion. He had done so by his successes, by the impression he conveyed of an extraordinary resourcefulness and energy, by his (to them) painfully obvious superiority in speed of reaction and flexibility of mind. There was something else, too, something unprovable, something experienced only by the unlucky few who were captured, and then incommunicable, something which became part of the Rommel legend and justly so: his chivalry, his essential decency, what the British thought of as his fairness.

It may be that in Germany, still something of an atavistic warrior society, hatred of an enemy in war was rarer than among their democratic adversaries, to whom war itself was so disagreeable a phenomenon that the foe had *ipso facto* to be demonized. Rommel certainly found hatred of the enemy wholly alien. They were doing their duty, as he was. And something like this was the impression he conveyed.

How such an impression leaps across the gulfs of war, transcending propaganda, transcending the vengefulness which war so easily engenders, is to some extent a mystery; but in Rommel's case it happened. When, during the summer battles of 1942, Rommel's staff showed him a captured British brigade order on the treatment of prisoners, laying down that no food or water should be issued to them before interrogation and that they should not be allowed to sleep, he reacted immediately and at the highest level. OKW ordered similar treatment for British prisoners of war and stated why; then British radio gave out that no such order had, from the British side, been given; OKW asked Panzerarmee Afrika for facsimile of the British order, received it and published it: and the British order was expressly revoked. But Rommel had acted before these difficult and indirect exchanges achieved their object, by authorizing his staff to relay on German wireless nets, in clear, details of the captured order with its official reference number, together with an instruction that the same would happen to British prisoners unless the order were withdrawn; and shortly afterwards he had the satisfaction of hearing through his interception service a British order – also in clear – withdrawing it.[15] Rommel was prepared to threaten reciprocal action in a case like this, but his every instinct was to reduce the barbarities of war, to promote correctness and humanity towards an enemy. Reputation in this case did justice to reality. In England, British soldiers, reading newspaper reports of their comrades' doings in the desert, would say to each other, 'That Rommel seems to be a decent bloke, in spite of everything.'[16] Rommel was a tough professional, and he played to win: but he played fair.

But of his understanding and enjoyment of publicity there was no doubt, and he took trouble to give his public relations staff what they needed, in terms of favourable photographic opportunities, privileged access to himself and punctual information. Lieutenant Berndt, an official of the Propaganda Ministry serving on Rommel's staff and responsible for publicity, did his duty well. Berndt wrote frequently to Goebbels, and there were occasions when Rommel used Berndt's status with Goebbels to employ him as a personal liaison officer: a burly and convinced National Socialist, Berndt saw that the exploits of Panzerarmee Afrika and its commander were vividly recorded, and he was loyal to Rommel.

Some people were so struck (often resentfully) by this aspect of Rommel's personality that they derided him as vain and publicity-inflated beyond any realistic assessment of achievement. Denigration had begun in the wake of the campaign in France, where Rommel's undoubted exaltation of his division's exploits irritated some contemporaries. Now, in Africa, the flattering coverage of his campaigns gave comparable offence in certain quarters. It was muttered that Rommel was a useful vehicle for the Goebbels propaganda machine, little more. It was pointed out – and would continue to be pointed out – that he was commanding small forces in a very secondary theatre; that he was untried by the great and all-important tests of the Russian front. It was said – particularly in retrospect, and not without a certain basis in truth – that Rommel's prowess was exaggerated by his British enemies because at the time of his desert victories they were themselves fighting nowhere else on land, and needed to sustain their own self-importance by increasing the stature of their enemy.[17]

That Rommel was somewhat vain is certain, but he was a realist. He had no absurdly inflated picture of himself. That he disliked admitting error and tended to ascribe setbacks to the failings of others is also certain, but is not a weakness confined to the great captains of history. That he operated in a more limited sphere and with smaller forces than his comrades on the Eastern Front is also true; and that the Eastern Front was the vital one for the German nation is self-evident. But Rommel's military renown rested on a sure foundation; his exploits, at every level, already spoke for themselves. The men who served him – many of whom had also experienced war in Russia or would in the future – recorded their views of Rommel as one almost without rival, a master of manoeuvre, a master of war. His reputation was no bubble, and his greatest achievement was imminent.

Rommel's spirits were now high. He had received assent for a great attack, he believed in it and his subordinates believed in it. His letters home were relaxed and affectionate: 'Your spring-cleaning will finally be over,' he wrote on 2 May, 'and you can breathe freely again.' He wrote frequently about Manfred's educational progress, like most fathers finding it uneven, rejoicing at the successes and fussing at the criticisms, albeit always with understanding. Rommel's link with home – letters took about ten days between Germany and North Africa – meant everything to him. He wrote of daily trivia, visits, new faces, arrivals and departures from the Staff. He wrote often of the weather, which can so dominate desert life ('Hardly a day without a sandstorm,' he told his wife on 5 May). He consoled Lucy about his health – more than was justified. He chatted about the wider war: 'What do you think of the Japanese successes in Burma?' he asked on 4 May, 'India will soon be free of England and America [sic].' He referred often to the Russian front, noting with satisfaction at this time how things had gone for the Germans on the Kertsch peninsula, on the Black Sea ('The news from Russia is marvellous'). Often he simply wrote, 'No news here.'[18]

His excellent, placid and affectionate soldier-servant, Herbert Gunther, also wrote to Lucy from time to time. Gunther always spoke of Rommel as the best of masters, kind, never losing his temper, appreciative; and Gunther was devoted to the Rommel family.[19]

Rommel's plan for the operation, agreed in April between Germans and Italians at the highest level, was simple. It was also bold and remarkably optimistic. The optimism almost led him into disaster – he overreached himself. Thereafter, as so often, his audacity and sheer quickness on his feet, his tactical judgement, his energy in crisis, saved him from disaster and brought him to glory.

Rommel outlined his ideas to his commanders on 15 April in a talk which lasted just over an hour. He intended to make a strong demonstration frontally against the British positions south of Gazala. He hoped to deceive the British Command into thinking that he planned a main effort through the minefields and then an advance by the shortest route to Tobruk, the obvious strategic object of any attack. Having, he hoped, concentrated British eyes on the centre and north, he proposed to lead the mass of his mobile forces on a great drive round the southern flank, bypassing the southern end of the British line at Bir Hacheim and driving north-east to Acroma and El Adem. He then planned to turn and to attack the main British forces holding the Gazala line from the east, having routed the British mobile forces and armour in open desert where he assumed that his superior skill in mobile operations would give him victory. After the defeat, by these means, of the British field army Rommel would move to the assault on Tobruk, having (he hoped) cut off any British formations seeking to withdraw into it – or past it, to Egypt. '*Die Englischer Feldarmee*,' he said, '*muss vernichtet werden, und Tobruk* *muss fallen!*'*[20] He spoke again at a training and study session on 12 May, with most seniors present including all divisional commanders and the *Fliegerführer Afrika*, General von Waldau. The first, the main, task was to beat the British army somewhere west of Tobruk. The second main task – Tobruk itself – depended on achievement of the first. The enemy, Rommel said, was not very mobile by instinct but possessed formidable material strength.[21]

For this operation Rommel divided the Panzerarmee into two wings. In the north, to make the initial demonstration frontally against the British line, he gave Cruewell† command of a predominantly infantry left wing, consisting of two Italian corps, X (General Gioda) and XXI (Navarrini), comprising four infantry divisions (Brescia, Pavia, Sabratha and Trento, commanded respectively by Generals Lombardi, Torriano, Soldarelli and Scotti), and two German regimental rifle groups drawn from 90th Light Division, operating together as 15th Rifle Brigade. In the south the right wing, commanded by Rommel himself, would consist of the Afrika Korps (General Nehring) – 15th Panzer (General von Vaerst), 21st Panzer (General von Bismarck‡) and 90th Light (General Kleemann) less the rifle regiments with Cruewell – together with two Italian divisions, Ariete (armoured, General de Stefanis) and Trieste (motorized and under General La Ferla).

The plan presupposed victory by German mobile forces over British mobile forces in a battle of manoeuvre somewhere behind the British fortified front. Rommel, like his adversaries, regarded the comparative tank states of both sides as the most relevant statistic in desert warfare. It was certainly significant, although a comparison of other arms – notably artillery and very notably anti-tank guns – was also important and did not always lead to the same conclusion. But in tank numbers the odds were against Rommel. They had been against him before.§ He believed that in Eighth Army (General

* 'The English field army must be totally destroyed, and Tobruk must fall!'

† Cruewell had just lost his wife, who died from scarlet fever leaving four children, as Rommel wrote to Lucy, grieving.

‡ Neumann-Silkow had been wounded, and von Ravenstein taken prisoner, during Crusader. Summermann had been killed in an air attack on 10 December.

§ Rommel's Ic, von Mellenthin, described after the war his underestimate of the British order of battle, and it has been exaggerated since. Rommel knew how many armoured brigades he had against him, but believed there was only one, rather than two, army tank brigades supporting the infantry of XIII Corps; and his staff 'lost' one Indian brigade from the forward area and deployed another mistakenly. All in all his issued enemy order of battle was not far out, and it is hard to agree with von Mellenthin that had Rommel known all he might not have attacked. For the sort of

Ritchie) he would meet about seven hundred tanks, while he himself had a total of 560. In fact Ritchie had nearly 850 tanks of all kinds, so that the odds were rather more against Rommel than he thought. Rommel, however, had what both sides regarded as potential match-winners; his forty-eight 88-millimetre anti-tank guns. He also brought captured British weapons – and large numbers of vehicles – into his service, and the war diaries of his divisions abound with safety instructions for specified areas of desert when these unfamiliar arms were being tested and the troops trained. Some Russian guns, captured on the Eastern Front and initially sent to Germany for scrutiny, were also received: 76.5-millimetre and 50-millimetre calibre, they did useful work.

In tank quality there were considerable variations on both sides. The British had two army tank brigades, established for cooperation with infantry divisions and equipped with a mix of Valentine and Matilda tanks (276 in all), types well-known to the Germans. In their two armoured divisions – 1st and 7th, totalling three armoured brigades – the British had 573 tanks, of which 167 were the American Grants, new to the desert. The Panzerarmee intelligence staff had learned of the arrival of the Grants a few days before Rommel opened his attack; the Grant had strong frontal armour and a 75-millimetre gun which, at that time, hit harder than any other tank-mounted gun in the desert, and caused a disagreeable surprise to Rommel's troops when they met it. The Grant's gun, however, was mounted in a sponson, rather than in a turret with all-round traverse; a disadvantage quickly perceived by its users but less apparent to those on the receiving end of its fire.

Rommel had 228 Italian tanks of a light cruiser pattern, generally outgunned by enemy types. He had, in the Afrika Korps, 242 Panzer IIIs, of which only nineteen were of the latest pattern, with a long-barrelled 50-millimetre gun; forty old Mark IVs; and fifty light tanks. In reserve he had less than eighty replacement tanks, and he knew that the British capacity for replacement was far greater than his own. He also knew that the British had extended a railway from Mersa Matruh in Egypt to Belhamed, just east of Tobruk. Overall he knew that he was facing greater tank numbers, like his own of variable quality, but he trusted his anti-tank guns, he trusted the combined arms tactical training of the Panzerarmee, and he

---

battle he proposed to fight the discrepancies were hardly critical, although the further 150 Matildas or Valentines might have caused him to suck his teeth a little.[22]

trusted his own powers. In the air the Luftwaffe had a numerical superiority in serviceable aircraft, and a qualitative advantage in the Messerschmidt 109F.

Rommel planned personally to lead the right wing, the mobile force, first towards the centre of the British front, eastward from an assembly area near Rotonda Segnali, soon after Cruewell's demonstration further north. This would be a feint. Not until dark, he decided, would the armoured strength of the Panzerarmee change direction and march south. The distances were such that there would need to be a refuelling halt after rounding the bend, somewhere south-east of Bir Hacheim.

After refuelling, the right wing, with its several thousand vehicles, would swing north-east; the main body would march towards Acroma, twenty miles west of Tobruk, while 90th Light Division would move on a divergent course towards El Adem, about the same distance south of Tobruk and twenty miles south-east of Acroma. Rommel's forces would, therefore, dilute their initial concentration after rounding Bir Hacheim. Bir Hacheim itself would be overrun by the Ariete Division, marching on the inner flank of the Afrika Korps.*

The essential tactical object of the right wing, before attacking the Gazala line from the east, was, of course, to destroy the counter-offensive strength of the British armour. Where exactly this might take place depended on British deployment and British reactions. In all the fighting so far Rommel had gained a poor impression of the tactical capacity of the British Command, and in particular of their understanding of the importance of concentration and of their ability to respond with timely energy to new situations. He had also gained a poor impression of British tactics in mobile warfare, especially of their ability to combine armour, anti-tank guns and artillery with appropriate effect. Rommel respected British fighting capacity at the individual or unit level, and he seldom underestimated British resolve or courage, especially in a static or semi-static situation, a slogging match; but he reckoned that in manoeuvre in open desert he would always have the edge.

His appreciation – issued with the Panzerarmee's operation order – was that the British armoured forces would employ elastic defensive tactics and would probably try to concentrate for counter-attack

* Bir Hacheim's defensive strength was underestimated, and Rommel reckoned that its suppression would only take an hour. It took two weeks.

somewhere north-east of Bir Hacheim. He referred to the possibility of a British eastward withdrawal to the area of Bir El Gubi, followed by a concentrated attack against the German right flank, but dismissed it as requiring more flexibility than the British had yet shown. They would, he wrote, be more likely to fight it out 'behind the line Bir Hacheim–Bir El Harmat'.[23]

Most of Rommel's assessments were justified by events, but he was mistaken – both at the time and in retrospect – in supposing that the enemy commanders were insufficiently alive to the virtues of concentration in war. British plans for the German attack which they knew to be imminent (forestalling their own, scheduled for June) were largely dominated by arguments about where the three armoured brigades of the two armoured divisions should be deployed in the desert, so that they could be concentrated against the German *Schwerpunkt* when it became clear where that *Schwerpunkt* lay. The fault was not in British neglect of military theory, as Rommel supposed, but in the lack of authority in the British Command which permitted hesitation and delay in the execution of orders, together with that lack of tactical coordination at a lower level which he rightly assumed.

Rommel, however, was assisted by a curious circumstance. His own staff were sceptical as to whether the enemy could possibly be deceived by a demonstration against the British north and centre, such as that planned at the beginning for Cruewell and reinforced by Rommel's own intended feint eastward before he turned south. Rommel, in his various briefings, had placed much importance on his deception plan but his staff, or some of them, believed that a frontal attack through the British minefields must seem so obviously protracted, laborious and expensive a business that no commander, especially so famous a master of manoeuvre as Rommel, could possibly be imagined by his enemies to choose it. They therefore believed that British plans would inevitably assume a great wheel round the south, as was actually intended. The prospects for deception seemed to them dim.[24]

They were partially mistaken. On 20 May Auchinleck had written to Ritchie, on balance directing his eyes towards the centre, towards the possibility of a breaching of the minefields and a subsequent thrust down the line of the Trigh Capuzzo. Such a move by Rommel would have given Ritchie plenty of time to concentrate his armour centrally, for the breaching of the minefields in that sector necessitated an attack on the British 50th Division, and Rommel's subsequent advance would then have encountered a 'box' held by the 201st Guards Brigade at 'Knightsbridge', the area where the Trigh Capuzzo is crossed by the Trigh Bir Hacheim (a 'box', incidentally, not known to Rommel). Rommel, in fact, would have been attacking a sector heavily mined and held strongly in some depth. Nevertheless, although the ULTRA secret was kept close below Auchinleck's own level, the Commander-in-Chief was widely believed to have access to intelligence of a special kind and his advice was respected accordingly. Ritchie's dispositions, therefore, covered this possibility and led to a centre of gravity further north than his own instinct suggested.[25]

In fact ULTRA, although making clear to Ritchie that an attack by Rommel was imminent, gave no indication of where. ULTRA, indeed, had yielded no particular information relevant to this after the end of April, and as late as 18 May was interpreted as supporting the idea of a German concentration in the centre. This contradicted certain other indications from the British battlefield signals interception service and from prisoners which pointed to the possibility of a move south, at least by 15th and 21st Panzer Divisions, and a subsequent swing round Bir Hacheim. The latter was, in effect, overruled (although all contingencies continued to be thought possible). On 26 May British intelligence reckoned the weight of German armour still in the northern sector.[26]

British appreciations, therefore, were defective, and Rommel's feint might for a few hours have been taken to support Auchinleck's advice – advice (after sensible demurral by Ritchie) followed up by a further Auchinleck letter in the same sense on 26 May, the day Rommel had set for the start of his offensive.

Little of this need have mattered. In the case of a sweep by Rommel round the south Ritchie's armoured strength, tank for tank and taking quality into account, was such that the southernmost British armoured brigade (4th), deployed some fifteen miles east of Bir Hacheim, was capable of dealing with the whole of the Afrika Korps, if either of the other two joined it; and the nearest (22nd) was only about ten miles away to the north. As to a frontal attack, given the length of time a German assault through the minefields must take, it would have been perfectly possible to concentrate the two southern brigades with the northernmost (2nd) astride the Trigh Capuzzo in plenty of time to achieve crushing superiority against a Rommel emerging from the defended belt.

Against a Rommel swinging round Bir Hacheim Ritchie's contingency plan ordered engagement by 7th Armoured Division (4th Armoured Brigade), on the general line of the Bir Hacheim–Bir Gubi track, that is, well to the south; reinforced as soon as possible by 1st Armoured Division (2nd and 22nd Armoured Brigades) twelve and ten miles away respectively. It was, of course, unfortunate for Ritchie, albeit a self-imposed misfortune, that to concentrate two or more brigades in the south he might have to pass command of a reinforcing brigade from one armoured division (1st) to another (7th); in the British army an untidy exercise if avoidable, and particularly tedious if strong-minded Divisional Commanders, anxious to retain their own commands intact, were to press their own views of the impending battle on insufficiently authoritative superiors.* But the dispositions and distances from each other of the British armoured brigades (and the contingency plans made for them) are inconsistent with a criticism that the virtues of concentration were not understood. The fault, when it came, lay elsewhere.

Whatever their dispositions, therefore, the British had taken into account in their plans the contingency of Rommel seeking to drive round the southern flank, but had inclined to the belief that this would be a diversionary or tentative move rather than the *Schwerpunkt* of Panzerarmee Afrika; and certainly reckoned that the time taken to cover the distance (including refuelling) would allow reaction. Rommel's concentration at Rotonda Segnali was noted and reported, and by nightfall on 26 May British armoured cars† reported a large-scale movement of German vehicles eastward from Segnali towards the Gazala line.

Neither this nor Cruewell's diversionary attack further north had the desired effect of making the enemy believe that the real *Schwerpunkt* was now disclosed as coming in the centre-north. Neither was taken as convincingly indicating confirmation for Auchinleck's thesis. And after dark the same armoured cars were soon reporting what sounded like a major move in a south-eastward direction (Rommel set the right wing on the thrust line towards Bir Hacheim at nine o'clock in the evening), although it was unclear to the British for some hours how large or how menacing this move was. It was at

* And in this instance the commander of the British 1st Armoured Division (General Lumsden) mistrusted the ability of the commander of the 7th Armoured Division (General Messervy).

† In fact South African cars: but the British, South African and Indian forces were fighting in a British army, and the term 'British' will be used throughout, except in the designation of formations.

about five o'clock in the morning of 27 May that Rommel's interception cells first heard British armoured cars calling – 'Enemy tank columns moving towards us. It looks as if it's the whole damned Afrika Korps!'[27] Some time later British intercept services heard a German enquiry about where a Panzer division had reached, to receive a coded reply accompanied by the electrifying words '*Rommel an der Spitze!*' – 'Rommel leading!'[28] So it was already at dawn on 27 May. This was *Fall Venezia*.* The armoured and motorized forces of Panzerarmee Afrika, covering forty miles of desert and moving to prearranged refuelling points with routes and timings very exactly worked out, had been streaming towards and round Bir Hacheim in the moonlight; and Rommel had placed himself at their head.

* 'Venezia' has been mistakenly supposed to have been the whole plan for the Gazala battle. It was in fact the codeword for only part of it – the deep penetration, swinging round Bir Hacheim.

CHAPTER FIFTEEN

'Heia Safari!'

'IT WILL be hard,' Rommel wrote to Lucy on 26 May, 'but I have complete faith that my army, above all my German soldiers, will win. Every one of them knows the significance of this battle.' His heart, as always, was with her as he exhorted her to look after herself, 'for *both* of us, your menfolk! and face your destiny as a soldier's wife as bravely as you always have!'[1]

What would be known as the battle of Gazala was the high point of Rommel's military achievement, an adventure in which he displayed all his best as well as some of his worst characteristics and one which culminated in the defeat of an enemy superior in every material sense, an enemy who had been expecting his attack and had had ample time and resources to prepare for it. The battle of Gazala culminated, too, in the capture of a strategic object – Tobruk – on which Rommel's eyes had long been fixed and which was thought to have the capacity to transform German chances in the North African campaign. And it culminated in Rommel's advancement to the rank of Field Marshal.

As in almost all fighting in the North African desert (and elsewhere), the imposition of orderly narrative and chronological sequence to some extent falsifies the confused obscurity which, for much of the time, covered the faces of battle. To read contemporary signals and study the reactions on the spot of the main opponents is to have a sense of reading about completely different actions, so unrelated to the actual situation and the objects of the enemy were, often, the perceptions of each side. But the battle of Gazala may be divided to some extent into phases: and the first phase lasted from the start of Rommel's great advance in darkness on 26 May until the end of the month: five days.

* * *

During those first five days Rommel swept round Bir Hacheim; had his initial encounter with British armour, an encounter from which both sides lost heavily; pressed his depleted Panzer divisions northward, on their original thrust line towards Acroma; found that they were difficult to resupply and managed to lead a replenishment column northward to them, but also found them to a large extent surrounded and cut off, hemmed by the British minefields to the west and the undefeated (albeit dispersed and similarly depleted) British armoured brigades to their east and south-east. He recalled 90th Light Division, which had been set on a divergent thrust line towards El Adem, in order to concentrate the mobile forces of the Panzerarmee; and resolved to stand for a short while on the tactical, east-facing, defensive while he destroyed a British infantry brigade on the main British position and cleared a shorter route for resupply through the minefields. Throughout this first phase – and thereafter – the southern point of the British position, the strongpoint of Bir Hacheim, resisted Rommel's assault, restricting his manoeuvres and complicating his resupply from the west: and throughout this phase the attacks by Cruewell against the north of the Gazala line – for his initial 'demonstrations' were succeeded by attacks in earnest – were held, and no ground won.

Rommel's initial advance in the first hours of daylight on 27 May went well, although the British were aware of his movements and had every reason to be prepared for his onslaught. Whatever Rommel achieved on the first day, it was not surprise. Armoured cars had been warning of his movement, and from before dawn had made clear that the movement was a massive one. The British had reckoned that they needed two hours' notice of the direction of Rommel's main thrust in order to have time to concentrate most if not all armoured strength in battle positions. They got it.

The British got sufficient notice, but did not act on it with sufficient urgency. They had stationed near what transpired as Rommel's thrust line two motor brigades in 'boxes' – 'pivots of manoeuvre' to assist the armoured battle and canalize an enemy advance. The first of these, 3rd (Indian) Motor Brigade, deployed a few miles south-east of Bir Hacheim, was smashed by the concentrated power of the Afrika Korps in whose path it lay. The second – 7th Motor Brigade – was in position at Retma, twenty miles east of Bir Hacheim and near the route taken by 90th Light Division, towards El Adem. 7th Motor Brigade retreated hastily eastward towards Bir El Gubi and 90th Light Division raced on to reach the area of El Adem at about eleven o'clock, overrunning the headquarters of the southernmost British armoured division (7th) on the way. Even this advance, spectacular though it was, ran at least three hours behind schedule.

Meanwhile Rommel, with the Afrika Korps of over five hundred tanks, was on his principal thrust line northward towards Acroma. The first British armour to challenge him was 4th Armoured Brigade, the armoured brigade of 7th Armoured Division whose headquarters was in process of being overrun far to the east. The two brigades, 2nd and 22nd, of 1st Armoured Division, some miles to the north, had been ordered to move – or, to be exact, 1st Armoured Division headquarters had been ordered to move them – at seven o'clock, when it was clear beyond argument that Rommel's main thrust was in the south. These two brigades had been intended to move to join 4th Armoured Brigade. They had still not moved two hours later, by which time the Afrika Korps had already driven 4th Armoured Brigade, badly mauled, eastward towards El Adem, and was sweeping down on 1st Armoured Division itself.

Rommel, in engaging 4th Armoured Brigade with his 15th Panzer Division, had first met the Grant tanks. He had lost heavily and suffered a disagreeable surprise. It was a diminished Afrika Korps which pressed on now towards Acroma and towards the next armoured clash, with 22nd Armoured Brigade. And here again – although he forced 22nd Brigade to withdraw northward, to the area round 'Knightsbridge' on the Trigh Capuzzo – Rommel suffered severe tank losses. His planned timing here, too, was several hours adrift.

The third British armoured brigade, 2nd, now came up on the left or southern flank of 22nd, facing west, and attacked the Afrika Korps from the area east of Knightsbridge as it crossed the Trigh Capuzzo. At the end of the day Rommel had the Afrika Korps Panzer divisions united north and west of the Knightsbridge box, itself held by the British 201st Guards Brigade. To his immediate west was the main British west-facing position, and in particular, at Sidi Muftah, the 150th or southernmost brigade of the British 50th Division, deployed behind minefields mainly for west- and south-west-facing defence. To Rommel's east, north and south of Knightsbridge, were two British armoured brigades, depleted no doubt but almost certainly still numerically superior to his own forces; and to the south-east 90th Light Division (which 4th Armoured Brigade, having withdrawn eastward after its original encounter with Rommel and having lost most of its Grant tanks, had been ordered to attack), had formed a hedgehog two and a half miles south of El Adem. The Ariete Division was at Bir El Harmat, south of Knightsbridge, and found itself attacked not only by 2nd Armoured Brigade from the east but by 1st Army Tank Brigade (deployed in support of 50th Division in the line) from the west.

Rommel had lost a great many tanks – approximately one in every three of the Afrika Korps was a casualty. Bir Hacheim was still held by the enemy – the stout-hearted 1st Free French Brigade – and from it the Panzerarmee's supply columns were being harassed by periodic raiding parties. Rommel's forces were hemmed in to both east and west between the boxes at Knightsbridge and Sidi Muftah: the main British Gazala line was intact; and the British mobile forces, although he knew that he had inflicted heavy casualties, were still in the field and still active in harassing soft-skinned convoys of replenishment vehicles. 21st Panzer Division had been replenished but 15th Panzer Division had no fuel, and attempts to get supply columns northward towards the Afrika Korps were failing. Nothing was getting across the Trigh Capuzzo.

Rommel's position was precarious. His northward movement, consistent with his original plan, still seemed possible; but the enemy 'fortresses' at Sidi Muftah (150th Brigade), Knightsbridge (201st Guards Brigade) and Bir Hacheim (1st Free French Brigade) were still holding, his extended supply line from the south was vulnerable, his mobile divisions were far from concentrated because of the divergent thrust line given to 90th Division, and half the Afrika Korps was out of fuel, immobilized. Rommel had every reason to suppose that a British operation with concentrated armoured forces might on the next day, 28 May, surround and destroy the mobile wing of Panzerarmee Afrika. His staff expected it.

At first Rommel refused to believe that the initiative could have passed from him. He ordered 21st Panzer Division, the only Panzer division with fuel, to continue northward towards Acroma. Rommel had overreached himself but would not accept that fact until it was forced upon him. Characteristically, he had set over-ambitious objectives, believing that the menace posed to the enemy by the depth and impetus of his thrust would, somewhere, somehow, force a reaction whose exact nature none could foresee: and that in the consequent unpredictable fighting his own swiftness of action and the training of his troops would bring victory. 'No plan survives contact' – the great Moltke's aphorism was wholly congenial to Rommel.

Nevertheless by dawn on 29 May Rommel was in difficulties, although he had just solved one of them. 21st Panzer had advanced north as ordered on the previous day, had scattered an enemy regiment, and had reached a point ten miles from the sea, on high ground overlooking the Via Balbia; but 15th Panzer was still grounded for lack of fuel. Rommel had ordered Cruewell to deliver a major frontal attack against the 1st South African Division near Gazala, to distract the enemy and if possible force a way through the minefields and open a direct supply route. The attack, by the Sabratha Division, took place at dawn: and failed. And now at last the British 2nd Armoured Brigade was moving west from Knightsbridge, into the area south of 15th Panzer on the Rigel ridge.

Supply was the crux. Rommel, during 28 May, had been ceaselessly attempting to find a usable route from the south to the Afrika Korps, a route not instantly exposed to British fire from west or east. He believed that he had found one, and at four o'clock in the morning of 29 May he personally led a supply column by this route and led it to 15th Panzer which was thus replenished.[2] Fresh supplies also reached 21st Panzer. The immediate logistic crisis was over.

Rommel has been criticized for this departure from the usual duties of an army commander, but it was wholly in character. He always believed in personal intervention and inspiration at the critical point in battle. The critical point may not always be where shot is flying; on 29 May the critical point for the Afrika Korps was where a supply column was attempting to reach 15th Panzer, and Rommel placed himself at that critical point and ensured success. Quickly the tanks and vehicles of 15th Panzer were fuelled and faced east to deal with 2nd Armoured Brigade. During the ensuing fighting the enemy brought up 22nd Armoured Brigade, in line with 2nd; and Rommel recalled 21st Panzer from the north and ordered 90th Light from the east and Ariete from the south into the same area as 15th Panzer. His mobile forces – what was left of them – would soon be almost concentrated in the area west of Knightsbridge. On the same day he heard that Cruewell had been shot down in a light aircraft and captured.

Rommel now realized that there could be no further advance northward and certainly no early attempt to attack the enemy-held Gazala positions from the east. His original design had failed. He had on his hands a new situation. Never a slave to preconceived pictures of how battle would go, he now concentrated his energies on withdrawing into a defensible position and opening a direct line for resupply which would not necessitate the journey round Bir Hacheim. In the latter object he was fortunate; the minefield belt between Sidi Muftah and Bir Hacheim was largely undefended, and the Trieste Division established a route through. Rommel now had the Afrika Korps complete, together with Ariete, in the area between the minefields and Knightsbridge.

This, however, was still extremely vulnerable. The supply route was threatened by the Sidi Muftah 'box', held by the British, just as the route round Bir Hacheim was threatened from that quarter. Before he could regard the Panzerarmee, gravely weakened in numbers as it was, as balanced and ready for further operations, Rommel knew that he had to eliminate the British at Sidi Muftah. On 30 and 31 May he attacked with 90th Light Division, the Trieste Division and the Panzer grenadiers of 21st Panzer.

On the first day of the attack on Sidi Muftah Rommel had driven west through the minefield and gone to the headquarters of the now captured Cruewell, where he found none other than Field Marshal Kesselring, the German 'Commander South-West' on whose deftness of touch with the Italians and others much of Rommel's success depended. It fell often to Kesselring to reconcile the orders Rommel was receiving from the Italian High Command in North Africa, his nominal superiors, with the directions he was also receiving (or soliciting) from Berlin, and in discharging this duty Kesselring – an intelligent, genial man known as 'smiling Albert' – was generally wise, firm and morally courageous as well as tactful. He had been visiting the front and suddenly found himself, as the only officer of seniority, requested to take over command of the left wing of Panzerarmee Afrika. He was amused, and for the while consented to take General Rommel's instructions. He and Rommel discussed and agreed the next phase of battle.

The attack on the Sidi Muftah box was the critical point. If Rommel could destroy the enemy at Sidi Muftah he held a broad salient into the British line and could supply and reinforce at will along the line of the Trigh Capuzzo. He was fortunate in being given time for this operation, time which greater and more urgent British activity from the east might have denied him. He personally led the foremost Panzer grenadier platoon, and by 1 June was master of Sidi Muftah, with three thousand prisoners and 124 guns. This victory, after exceptionally fierce fighting, marked the end of the first phase of the battle of Gazala; and Rommel wrote to Lucy that the great crisis of the battle was over.

During this phase Rommel's battle had been dominated by the problem of getting his replenishment columns to his mobile forces. His initial advance had been strongly controlled and powerfully driven, but had run out of steam well short of its objectives. He had, with a certain amount of justification, relied upon the tactical superiority of his forces to deal with the opposition in any battle of manoeuvre, and had realized that resupply would be the paramount consideration, but he had miscalculated, in particular, the disruptive effectiveness of Bir Hacheim and the vigour with which it was defended. He had miscalculated possible timings, and thus difficulties.

Rommel may not, however, have miscalculated the indifferent powers of coordination shown by the British Command and the British armoured formations; to the end of his life he was sure that the British had missed an opportunity to destroy him by well-orchestrated action on 29 May, and such missed opportunities provided exactly the chance he always assumed the British would give him. But at another level he had miscalculated and underestimated the fighting strength of the enemy armour, a miscalculation which led to his dangerous situation on the second day of battle, with one Panzer division pressing on towards its original objective, another grounded for lack of fuel some miles to the south, and Ariete and 90th Light Divisions in separate parts of the desert. These dispositions would have been unimportant had the British armour itself been scattered, mortally wounded; whereas in fact, although hard hit, all British armoured brigades were still in hand and at least capable of effective and concentrated manoeuvre.

Rommel has been criticized for the divergent axis of 90th Light Division. In fact, although he did not know it precisely, it was directed on a line acutely sensitive to Ritchie. Eighth Army's stockpiles for the forthcoming British offensive were in the area of Belhamed, and Ritchie's deployment and subsequent manoeuvres could not risk uncovering a direct run from Bir Hacheim to that place – a direct run which the move of 90th Light seemed to presage and to justify Ritchie's counter-measure with 4th Armoured Brigade.

Rommel had also miscalculated the toughness with which the box at Sidi Muftah – and, in the days ahead, Knightsbridge – fought. He had not, however, miscalculated the flexibility and resilience of the Afrika Korps. As a force generally capable of exercising tactical dominance over the enemy in open desert it had once again shown its superb form, in the first clash with 4th and then 22nd Armoured Brigades, the triumphant run northward – and then withdrawal – of 21st Panzer, and the skill and handiness with which it had regrouped between the Azlagh and Sidra ridges west of Knightsbridge while the infantry broke into the Sidi Muftah box and destroyed 150th Brigade. Rommel, furthermore, was being admirably supported by the Luftwaffe. *Fliegerführer Afrika*, General von Waldau, was reporting to Kesselring and his air reconnaissance was showing, with remarkable clarity, the development of the inevitably confused situation. Von Waldau, however – the complaint would recur and become angry – was already referring to the bewildering changes of plan and situation to which the ground forces were expecting his airmen to adapt, and he knew that enemy losses, although considerable, would be made good.[3]

Rommel had not miscalculated his own abilities. He had not triumphed in a planned, set-piece manoeuvre; nor had he so attempted. But he had set far-reaching aims, established his forces in the rear of the enemy and fought his way to a situation in which reinforcement and resupply could now flow safely eastward to feed his next attempt. It was not exactly the battle he had forecast, but immediate danger had passed. The British had not attacked him in a coordinated or successful way during his moments of greatest vulnerability and he could have ascribed this, to some extent and without vanity, to the demoralizing effect of his own presence and energy. 'No plan survives contact,' and Rommel's had certainly not done so. The important thing was not to pretend that a battle had run as predicted or seek to press events back into a previously conceived mould, but to accept present reality and move from there. Present reality meant that Rommel's mobile forces were concentrated in an area to be known as the *Hexenkessel*, the Cauldron, and must obviously, before further advance, be prepared to receive and defeat a British counter-attack.

Above all Rommel had to a very great extent set the pace and pattern of events. He had seldom been at a loss. Although often as ignorant as his opponents of exactly who was where or doing what, he nevertheless always gave the impression of being in control, of deciding with great rapidity, of drawing events in his wake, of recovering when he stumbled, of reasserting his will upon battle, endlessly resilient. To read his diaries or those of his staff officers during those Gazala days is to accompany a man seemingly in perpetual motion, observant and decisive to a remarkable degree even for Rommel, tireless.[4] A British officer who drove in error into the Sidi Muftah area, not knowing that 150th Brigade had been overrun, found himself suddenly a prisoner, standing under guard next to a car in which Rommel sat, directing the battle of the Cauldron, an armoured radio car on either side of his own, handing scraps of paper with his orders scribbled on them first to left then to right, reading the battle and very obviously directing it entirely personally, utterly calm, utterly in control, as it seemed utterly confident. The contrast with what the captive had seen of British command was sharp indeed.[5]

To Ritchie, and even more to Auchinleck in Cairo, the situation near the end of the first phase of the battle of Gazala looked very different. Rommel's attacks on Sidi Muftah were taken as the frenzied efforts of a captive who sees a prison door about to be slammed behind him and is straining to hold it open. His clearing of a route through the minefields was thought to be the desperate convulsion of one who had been trapped and was trying to escape westward. 'Well done!' Auchinleck signalled to Ritchie on the evening of 29 May. 'If he tries breaking out take any risk to prevent him. He must not get out!' And at one o'clock in the afternoon of 31 May 1st Armoured Division sent a signal to its superior corps headquarters that the Germans were 'streaming out of a gap in minefield to west'.

On the morning of 29 May, before a resupply line was cleared and established, Rommel was undoubtedly anxious. His offensive seemed to have failed and there were voices, angrily rebuffed, which reckoned he should somehow withdraw. But by 31 May nobody was streaming west. Nobody was trying to break out. Ritchie undoubtedly held the initiative and Rommel was standing on the defensive, but any idea that he was beaten was false. He showed Swabian tenacity. The British Gazala line would shortly be breached by the defeat of 150th Brigade. Between the southernmost intact British position and the redoubt of Bir Hacheim was a gap of fifteen miles; in the area west of Knightsbridge, between the Sidra and Azlagh ridges, the Cauldron, Rommel had the mass of his remaining mobile forces; while further north the Italian divisions and German rifle brigade of the left wing of Panzerarmee Afrika were still standing ready.

Rommel now did a curious thing. He moved personally to Bir Hacheim. He had agreed with Kesselring that the next phase of battle should be a clearance of the southern half of the battlefield, so that there would be left to the enemy only the minebelts immediately south of Gazala, the Knightsbridge box, such mobile forces as survived after the British counterattack (judged imminent) and the fortress of Tobruk: and this agreement involved the elimination of the enemy at Bir Hacheim, whither Rommel now deployed for the attack 90th Light Division and the Trieste. He described Bir Hacheim as one of the hardest fights of his career. The French fought with considerable *élan*. The place was garrisoned by four battalions with twenty-four field guns and eighteen anti-aircraft guns. It was well stocked with supplies, mined, wired and well-prepared; and it would not be until the night of the 10 June that the garrison ultimately broke out and withdrew eastward. By then Rommel had reinforced the attacking troops from 15th Panzer Division and taken personal direction of the battle, dividing the assault forces into three groups and leading one of them, a force of three battalions, himself.

Yet Rommel's personal move was strange. By 1 June he had a secure route for supply from the west, by the Trigh Capuzzo, so that the elimination of Bir Hacheim would not particularly ease his replenishment problems. Clearly Bir Hacheim, while the enemy held it, was a complication to manoeuvre, another '*Dorn im Fleische*' as Rommel's staff described it.[6] Bir Hacheim had originally anchored the south of the enemy line and forced the Panzerarmee to an extended march when seeking to move round the southern flank – somewhat as Hougoumont, anchoring Wellington's right, had complicated French outflanking manoeuvres at Waterloo, and similarly irksome to Napoleon as to Rommel; but by the time Rommel was devoting a week to subduing it Bir Hacheim had lost its earlier significance.

Rommel, nevertheless, was determined not to leave it in enemy hands: and no doubt German information that there were plentiful supplies and water therein contributed to the decision, for the Panzerarmee was extremely short of both. At one point on 2 June Rommel thought a British officer under white flag might convey a summons to General Koenig, commanding at Bir Hacheim: a summons to the garrison with a promise of honourable treatment as official combatants (not insignificant, for Germany, of course, was formally at peace with France, and these were 'irregulars').[7] That summons was not carried. Two further messages from Rommel, however, reached Koenig on 3 and 5 June, the latter carried by a German officer. Both contained invitations to surrender and both were rejected.

Rommel had greatly underestimated Bir Hacheim, and had to invoke maximum effort from the Luftwaffe on 7 June, effort in which both Germans and British lost heavily – fifty-eight German aircraft and seventy-six from the Royal Air Force. Von Waldau reported angrily. He understood that Rommel had criticized Luftwaffe support and he gave him details of the enormous number of sorties flown, the losses of aircraft, and the inadequate (as he described it) actions of the ground forces, actions which could not possibly lead to success and to support which the Luftwaffe was being asked to undertake, he said, useless and sacrificial operations. On 9 June he sent a curt signal to Rommel that his airmen had flown 1030 sorties (460 bomber and 570 fighter-bomber attacks) over Bir Hacheim. After it was all over, and adverting to what he called 'inaccurate statements', he proposed that his own conduct be examined by court martial, a proposal Kesselring presumably smothered with his generally deft hand. Von Waldau was as sharp to deal with as Rommel himself.[8]

The enemy had held Bir Hacheim in the latter days at Auchinleck's insistence – Ritchie had wished to withdraw the garrison several days earlier. In the event Rommel took a thousand prisoners when he finally entered the place on 10 June, but 2700 men got away.

While Rommel, wisely or not, was devoting considerable resources to the fighting at Bir Hacheim, the British had assembled forces for the expected attacks on his position in the Cauldron.

These attacks were made on 5 June, and achieved little despite some remarkably hard and gallant fighting by both sides in certain parts of the field. To the Panzerarmee the enemy operations appeared unco-ordinated and tactically inept, and were defeated without especial difficulty. One British attack came from the east, and another – commanded by a different corps – from the north, but in both cases Rommel's tried tactic of facing advancing enemy armour with a line of anti-tank guns, keeping his own armour in reserve for a counter move, paid a rich dividend.

Once again the men of the Panzerarmee noted a lack of tactical cooperation between British arms and weapons. The German anti-tank guns were vulnerable to British artillery, or to infantry by night. The German tanks were vulnerable to British anti-tank guns, and tank guns, for British armour was not qualitatively inferior to the Panzerarmee's overall. Infantry, in the desert, were only of use at night or in places, particularly in defence, where terrain or fortifications (like minefields) actually canalized movement; such places were rare once fighting had become fluid, and infantry were vulnerable when moving, whether on foot or in soft-skinned vehicles, and were thus hostages to fortune, whose protection or redeployment posed problems. These factors had led to infantry deployment in the 'boxes' which characterized the Gazala battles – well-stocked strongpoints, with infantry and artillery sited for all-round defence and thus creating those 'pivots of manoeuvre' which looked so formidable on the staff maps of higher commands but whose influence on the course of the campaign was inevitably limited by numbers, vision and the range of weapons. All these realities, common to both sides, meant that success in the tactical battle, and especially in the tactics of the attack, necessitated first-class, well-thought-out and intelligently practised cooperation between all arms.

This necessity was not met in the British attacks on the Cauldron, named by them Operation 'Aberdeen'. A large number of British tanks were destroyed, bringing the opposing sides to something like equality overall. At midday on 5 June Rommel, judging that his defence had succeeded in essentials and that the moment was ripe for a countermove, took time off from Bir Hacheim and led 15th Panzer Division first south-east, then south, out of the Cauldron area to south of Bir Harmat; and then wheeled north-east against the left flank and rear of the British attacking the Ariete Division on the Azlagh ridge; simultaneously 21st Panzer Division attacked from the north, from the Sidra ridge, against the British right flank. By the end of the day Rommel had taken over three thousand prisoners, had overrun several British formation headquarters, and could regard the British counter-attack against the Panzerarmee's salient as decisively defeated.

Rommel, however, had also lost heavily in the fighting. He had lost some key officers – Gause, Westphal, and two other staff officers of the Panzerarmee headquarters had been wounded on 1 June, the latter two fatally; Gause was succeeded by Bayerlein from the Afrika Korps, and Westphal (Ia) by Mellenthin (Ic), but the strain was telling. Rommel was very short of German infantry. He had 160 German tanks and seventy Italian but he reckoned the enemy still numbered more overall, and he reckoned right – just. Nevertheless by 11 June Rommel had eliminated Bir Hacheim and was ready for the next, the third, phase of the battle of Gazala.

During the second phase, the fighting at the Cauldron and Bir Hacheim, Rommel had triumphed over numerically superior forces. At the Cauldron he had triumphed by the superior tactical skill of his troops and by the immediate grip he had exercised over the battle, leading, very personally, one expert counter-move on the afternoon of 5 June. At Bir Hacheim he had triumphed despite miscalculation. He had been forced to deploy wholly disproportionate strength to subdue a place whose importance had largely been overtaken by events. The second phase of Gazala, however, had shown once again that Rommel's power lay in his understanding of the tactical battle and in his skill and energy in conducting it. In writing afterwards he criticized the British for their slowness of reaction, their failure to take advantage of his own moment of greatest weakness at the end of May; and for their frittering of forces, their apparent incomprehension of the importance of concentration of armoured strength. Some of this was justified; but far more significant was something Rommel probably did not know. His enemies were plagued by feebleness, by lack of instant authority in the high command. Intentions were too often obscure. On the British side orders at army, corps or divisional level were too often treated as bases for discussion, matters for visit, argument, expostulation even. The result was a system of command too conversational and chatty, rather than instant and incisive.

Examples abound. A divisional commander would be ordered to transfer a brigade to another's command. He might reckon the order unwise, or premature, or unfair to the troops because admitting too little time for orderly execution, and would – no other word suffices – prevaricate, in perfectly good faith, while sure that such prevarication was in the true interest of the troops and the battle. By the time the order was executed (with compounding delays further down, for such attitudes are infectious), Rommel would have acted and acted decisively. Such indiscipline at the top of the hierarchy of the British Eighth Army gave Rommel a peculiar advantage, an advantage increased by a general suspicion on the British side that Ritchie was not wholly his own man but was very strongly overseen by Auchinleck; and it was indeed true that Auchinleck interfered often with Ritchie's conduct of operations, often unwisely.

What Rommel did know, however, was that in the tactical battle his troops again and again showed themselves superior. Their coordination, training, initiative and discipline were superior. And this superiority was sensed by their opponents and had inevitable effect on the minds of their opponents' commanders. The Panzerarmee had achieved formidable stature in the awareness of its enemies. It seemed a natural winner, difficult to challenge. Local British victories seemed insubstantial and temporary. Local German victories seemed inevitable and alarming well beyond their actual scale. Behind them all there seemed to loom the figure of Rommel, ubiquitous, menacing and indestructible.

The third phase of the battle of Gazala began on 11 June. Rommel had visited 15th Panzer on the morning of 9 June and had given orders for exactly what he wanted to do the instant Bir Hacheim fell. Every man must be brought forward from reinforcement or reserve units – Rommel was by now very short of German soldiers. His plan was to lead a direct advance from the south-west towards El Adem. He had under immediate control 15th Panzer, Trieste and 90th Light Divisions, and he saw this as the southern pincer of a movement supported by 21st Panzer advancing from the Cauldron area in the west.

By noon on 12 June the Panzerarmee was threatening part of the enemy armour (4th Armoured Brigade, south-east of Knightsbridge) with encirclement. 21st Panzer had moved eastward, south of Knightsbridge, without interference and was attacking the right rear of 4th Armoured Brigade while 15th Panzer was threatening its left flank. By dusk both 4th and 2nd Armoured Brigades had withdrawn northward, getting north of the Raml escarpment as quickly as they could to escape the pincers of the Afrika Korps. The British, pursued northward by 21st Panzer, lost 120 tanks in the running fight. The remaining British armour, somewhere east of Knightsbridge, had little thought or capacity left for offensive action.

It was a day in which Rommel had immensely profited from that weakness of authority in the enemy command which had so often been his greatest unknown asset. Rommel's mobile troops had not been concentrated. The enemy, whether fortuitously or by design, had been largely deployed in neighbouring parts of the desert and could certainly have achieved concentration of superior force against either arm of Rommel's movement. Ritchie and the commander of the British XXX Corps, Norrie, recognized the opportunity. The failure to take that opportunity lay in the leisurely and questioning way in which orders were treated. As an example Lumsden, the commander of the 1st Armoured Division, was asked at midday by Norrie when he could take command of the remaining armour, including brigades from the other (7th) Armoured Division, whose commander was missing. Forty minutes later XXX Corps staff were told that Lumsden was thinking it over. Twenty-five minutes later again it was agreed by Lumsden's staff (having consulted the staff of 7th Armoured Division) that command of the two 7th Division brigades could be assumed at two o'clock that afternoon. In fact it was nearly half-past three before command had been assumed – and neither brigade had yet moved. Shortly afterwards Lumsden, attributing this to tank casualties, made clear that he anyway could not operate offensively against the Afrika Korps (Rommel's right-hand pincer) as ordered, and did not propose to.

Such reactions, or lack of them, may be compared to the instant vigour with which Rommel saw, decided and acted, and the equal vigour with which his own mobile divisions complied with his will. It should not be thought that he did not discuss with his subordinates – he visited his divisional commanders ceaselessly during the Gazala fighting, always explaining his intentions, what he hoped to do and why. They were in his mind. But when Rommel gave an order it was an order.

He now again, and without question, held the initiative. On the next day, 13 June, he directed 15th and 21st Panzer Divisions towards each other along the Rigel ridge, north of Knightsbridge, defended by part of the British 201st Guards Brigade; by evening that ridge was taken, Knightsbridge was completely isolated and its garrison ordered to break out. Rommel could now roam the desert virtually at will between the Gazala line and Tobruk. His intelligence staff told him their estimates of enemy tank losses and their assessment that the British armour was no longer a serious factor to be taken into account. Nevertheless even on so apparently victorious a day matters, as ever, looked somewhat different from different viewpoints. Rommel signalled 90th Light Division at 12.30 to move west at once, as the position was very favourable. The Divisional War Diary recorded gloomily that the division was not in a position to carry out the order, being short of everything – ammunition, water, rations; and that although the general situation might be favourable, the division's situation seemed to it to be rather precarious.[9]

Rommel, however, was sure that he had won, and he was right. Next day, 14 June, he attempted to drive the Afrika Korps once again northward, to cut off what he believed must soon be the retreat of the British divisions from the Gazala line; and on that day Ritchie, recognizing that the battle was lost and that his mobile forces were no longer capable of recovering any sort of initiative, ordered the abandonment of that line. 'Panzerarmee headquarters,' ran Rommel's personal diary, 'reports to the commander-in-chief at 1559 hours rearward movement from the Gazala position.'[10] His letter to Lucy on the following day stated flatly and triumphantly, 'The battle is won. The enemy is breaking up.'

Rommel's troops were utterly exhausted. The fighting of the last eighteen days had been ceaseless and expensive in men and materiel, and the year's heat was approaching its peak. Rommel had concentrated the Afrika Korps west of the Trigh Bir Hacheim for a final drive towards Acroma and the Via Balbia, the coast road. It was his original planned thrust line of eighteen days before, and he rode with the leading tanks himself. On 16 June his diary recorded his criticism about the weakness of troops deployed in the coastal area, despite his orders for a main effort to trap,[11] but even Rommel was unable to galvanize the German divisions that day, or make the sort of speed and distance which would have cut off XIII Corps' retreat from Gazala; most of the British got away, their 50th Division moving westward, clear of the minefields, and then swinging away south and south-east past Bir Hacheim towards Egypt and safety. By 15 June Eighth Army was in full retreat. In the following two days Rommel harried his worn divisions, personally rating 21st Panzer and placing himself at its head for a final effort, but it was to little avail. The trap was evaded.[12]

Rommel, in this third phase of Gazala, had once again been in his element, handling and directing and leading his remaining armour in a series of encounter battles which soon turned into pursuits. Throughout – and indeed throughout the battle – Rommel showed his uncanny sense of where to drive in the desert, how the enemy would react, how the game would go. Amidst confusion he saw clearly and acted decisively, setting the pace of events. He failed to destroy the British in the Gazala line – perhaps, as some of his staff believed, because he had at times pursued too diverse objectives, forcing upon himself that loss of concentration for which he readily criticized his opponents. Perhaps – it was a recurrent characteristic – he was unduly ambitious in his plans and optimistic in the timing of them so that his calculations were too often upset by the run of events. But Rommel had won the battle. The enemy were streaming eastward in a good deal of disorder. Tobruk lay before him. Crusader was avenged.

Although Rommel did not know it, there was confusion within the enemy's councils on the matter of Tobruk. When the battle of Gazala started it took the form of an attempted pre-emption by Rommel of a British offensive, and it was some time before the British recognized that they were being forced not only on to the operational but the strategic defensive. It was a profoundly unpalatable fact, both to Auchinleck in Cairo and to Churchill in London; and it brought into sharp focus policy concerning Tobruk.

The German siege of Tobruk, lifted by the British Crusader offensive, had been a demanding experience for both sides, and in the new and menacing circumstances Auchinleck ordered that Tobruk was not again to be 'invested'. For this to be effective Ritchie needed to hold a position covering the approaches to it – a project which the fighting of the last few days made highly improbable. Ritchie's armoured formations were anxious, with reason, not to be caught in open desert by the now-superior forces of the Afrika Korps, while his infantry had evacuated the Gazala line and were moving east as fast as possible.

Policy was discussed in a long exchange of letters between Auchinleck and Ritchie, with a good deal of interference from the former in the latter's business. Auchinleck had originally intended that, if the Gazala line could not be held, the army should next stand on the Egyptian frontier, but he was under considerable pressure from London, where Churchill was somewhat weak in Parliament at that time, a weakness exacerbated by the British defeat: hence the ambivalence about Tobruk.

The prohibition on investment turned into an essay on the contingency of 'temporary isolation' of Tobruk; consistent with this, Ritchie hoped to assemble a sufficiently powerful mobile force in the area of El Adem to threaten Rommel's communications as Rommel threatened Tobruk. Ritchie's earlier instructions, to hold a general line from Acroma to El Adem, were impossible of execution almost as soon as they were issued. Churchill, meanwhile, had reacted with predictable indignation to the news that Tobruk might be sacrificed and had been reassured that it was not to be invested and that Rommel was to be held or fought west of a general line from Acroma to Bir El Gubi – a most optimistic expectation.

There was, in fact, no way by now in which Rommel could be prevented from investing Tobruk unless his forces could be beaten in open desert; and he and Ritchie were united in regarding this as most improbable. Rommel knew that the enemy were withdrawing and withdrawing fast. He knew that they were unlikely to be able to stand or recover west of the Egyptian frontier. And he reckoned that the best hope of taking Tobruk was to move on it immediately, while the general disorganization of Eighth Army could be exploited. He did not know that on 15 June the fortress commander of Tobruk, General Klopper, commander of the 2nd South African Division, had told subordinates that a siege of three months must be anticipated; but he knew that his own task was likely to become more rather than less difficult with each passing day. By the evening of 18 June his forces had completely surrounded Tobruk.

Rommel's plan was simple. He intended to use all the power which the Luftwaffe could assemble in a short but very intensive bombardment; then to attack in the south-east sector of the perimeter, to work forward with infantry and combat engineers and prepare a crossing of the anti-tank ditch which was a key to the defences. Thereafter Rommel's remaining tanks in the Afrika Korps would advance to the high ground dominating the harbour and, as soon as possible, smash into the town itself.

This happened as planned. The first Stuka attacks went in at 5.50 a.m. on 20 June, watched by Rommel. Although he did not know this (but may have suspected it), Tobruk was now a much easier nut to crack than formerly, as mines had been removed to strengthen the Gazala position.

In the event, no serious counterattack by the defenders was devised or executed, although several strongpoints put up savage resistance. Klopper's headquarters, nearly overrun, was ordered to disperse and effective control of the defence broke down. The leading German infantry had started working forward at seven o'clock in the morning of 20 June. Rommel drove in their wake, and at two o'clock in the afternoon sent Berndt back with an order to bring the motorized and armoured elements of the Ariete and Trieste Divisions as fast as possible through the minefields and lead them west in the tracks of 15th Panzer to gain the El Adem road.[13] By six in the evening 21st Panzer Division was in the town. That evening an onlooker described the scene and Rommel's message to the Panzerarmee:

> Rommel sits with Colonel Bayerlein by the light of a flickering candle at a hasty snack from captured English rations. Only his eyes gleam with deep and unalterable happiness. 'It's not simply command,' Rommel declares in the hour of his greatest triumph, 'which makes such victories possible! One can only achieve with troops on whom one can lay every burden, of deprivation, battle, need, ultimately death. My soldiers, I thank you all!'[14]

At six o'clock next morning Klopper sent emissaries to negotiate surrender. The fortress of Tobruk, with its port, its stocks, huge numbers of vehicles, on which Rommel would be increasingly dependent, and 32,000 prisoners, had fallen in twenty-four hours. At 9.45 on that morning of 21 June Rommel signalled to the whole of the Panzerarmee Afrika: 'Fortress Tobruk has capitulated. All units will reassemble and prepare for further advance.' He had quickly settled the first details of the surrender. The South Africans had asked that the considerable number of black prisoners should be segregated from the whites, a request Rommel turned down flatly, saying that the blacks were South African soldiers, had fought alongside whites, worn the same uniform and were all captives together.[15]

It was, he wrote to Lucy on 21 June, a 'wonderful battle'. But Rommel, even in the exhausted aftermath of the battle of Gazala, would allow neither the Panzerarmee nor its enemies respite. 21st Panzer was rolling eastward that evening and Rommel's signal to the army ended: 'In these next days I shall once again ask of you a great performance, so that we achieve our goal.'[16]

That evening Rommel learned of his promotion. He had released one of his public relations officers at Goebbels's behest to fly in a Heinkel to Berlin and to brief the press at first hand. Hitler himself received this emissary with the news that he had just named Rommel for advancement to the dignity of Field Marshal. The Führer was impressive on this occasion, with his understanding, his knowledge of British improvements in anti-tank gunnery, his absorption in the details of the fighting. He listened, raptly, to the relation of French heroism at Bir Hacheim, calling out that he had always declared the French to be, after the Germans, the best soldiers in Europe.[17] The name of Tobruk made rousing headlines. The sun was shining. Rommel heard the news of his elevation on German radio at 9.50 p.m. on 22 June.

Congratulations poured in over the next few days. From Hitler, acknowledged with the customary signal of thanks ('*Vorwarts zum Sieg, fur Führer, Volk und Reich!*'); from Goering, expressing belief that the victory of Tobruk would for ever be associated with the name of Rommel in the 'war for freedom being waged', and also expressing happiness that the Luftwaffe had played its part, a happiness certainly justified but precarious, for in fact the RAF was now firmly getting the edge over its opponents; from Goebbels; from Mussolini and the Italian commanders; and from very many others including long-retired generals such as Ritter von Epp, the one-time liberator of Munich from the Communists, who congratulated Rommel on restoring the tradition of German arms in Africa.

It was easy for foreigners to forget that Africa had exercised a powerful, a magical, a romantic influence on a generation of German soldiers, explorers and administrators, comparable to that felt by generations of British and French. In the once-German colonial world Rommel's victories stirred proud echoes; and to the men of the Panzerarmee itself the African adventure did not lose from being in some degree heir of a tradition, the tradition of German east and German south-west, of Kamerun, of the legendary von Lettow-Vorbeck.[18] '*Heia Safari!*' – 'Good hunting!' – soldiers of the Afrika Korps would say, as the great game rolled east and west and now east again. This seemed the opportunity of a lifetime. Rommel, his interpreter noted, was overjoyed, like a little child, '*wie ein kleines Kind*': and added, '*Vielleicht kommen wir bis Kairo*.'*[19]

* 'Perhaps we'll reach Cairo.'

CHAPTER SIXTEEN

The End of the Line

ALTHOUGH ROMMEL was operating within an already flawed alliance command system, he has been criticized for his tactlessness to and about the Italians, tactlessness which made a bad system worse.

It was true that in the early days he had found the Italians unreliable and said so, making it clear on more than one occasion that he could rely only on Germans for important tasks; this was unwise as well as partly unjust. It was true that he treated his Italian higher commanders – first Gariboldi then Bastico – in a somewhat cavalier fashion, especially in respect of his future intentions: he had little faith in their discretion or security, and often wrote of his distrust of '*den Brudern über den Weg*', the brethren across the way. It was true that he sometimes spoke to his subordinate Italian generals with painful frankness, signalling to the commander of X Corps on 27 June 1942, for instance, that he urgently requested corps command posts to be established not more than ten kilometres from the front line – and to XX Corps three days later (when told the advance couldn't be carried further) that he hoped they would soon dispose of what sounded laughable opposition.[1] It was true that he gave absolute primacy to the operational needs of the campaign and was disinclined to take much account of Italian colonial susceptibilities.

It was certainly true that Rommel fulminated over what he saw as Italian lack of overall effort in the Mediterranean, and thus inadequacy in the matter of supply, an inadequacy he sometimes ascribed to treachery. He thought – and he was not alone – that the war was not taken as seriously in Rome as it should be. And it was true that occasional Italian excesses towards Arab women enraged him, especially for the alienation it could create among Arabs from the

Axis cause: Rommel fostered throughout the German troops in Africa good relations with the Arabs, who often showed considerable friendship and preference for the Germans over other Europeans – not least, however perversely, because Hitler's antipathy to Jews was known even to the Bedouin, to whom it was sympathetic. The British were confident that the Arabs were well-disposed towards their side, but the confidence was sometimes misplaced. The Arabs possessed a genial capacity for appearing to share the sympathies of any companion, although they were temperamentally hostile to the Italians.

But Rommel felt for the Italian rank and file. He was not so single-minded a soldier as to exclude the less martial qualities from admiration: 'There are other virtues than military ones,' he observed to his staff *à propos* his Italians.[2] He knew more Italian than he admitted, although he always had his remarks translated. When the Italians did well he was generous in publishing messages and orders telling them so. He was sometimes brusque with their superiors, but to the soldiers themselves he showed humour and kindness. He sympathized with them and they reciprocated. Hitler had written to Mussolini, soon after Rommel's arrival in Africa, that General Rommel would surely earn the loyalty and affection of Italian soldiers; and Hitler had been right. But Rommel thought they were too often let down by the indifference and selfishness of their seniors. He detested the way Italian officers often enjoyed different rations and a far better standard of living than their men on campaign. The Italians were skilled at contriving creature comforts, even in the desert, and Rommel despised such comforts. His own regime was Spartan – the same rations as the ordinary German *Landser*, one glass of wine in the evening if in his tiny mess, as little sleep as any man in the Panzerarmee; and he never smoked. Austere by nature, the sybaritic as well as the half-hearted facets of the Italian military character irritated him.

Yet the Italians had played an active and often a distinguished part in the battle of Gazala, both in the actions against the northern minefield belt and in the mobile operations. Trento Division had first breached the minefields and forced a supply route eastward north of Bir Hacheim. Ariete had fought vigorously in the Cauldron, at Bir Hacheim, throughout. And Rommel cared for the safety, the security, of his Italian divisions as conscientiously as for his Germans, and was undoubtedly plagued by concern for their vulnerability. For the majority of Italian divisions were still unmotorized, and had to be carried by scarce transport from one deployment to another.

Their most practicable function was in positional warfare, that rare and partial phenomenon in the Desert War. And now, as the Italians, following up the Afrika Korps, moved eastward towards Egypt, they were moving towards what would become the greatest act of positional warfare, as opposed to manoeuvre, in the whole drama of the North African campaign.

It did not appear so initially. Rommel said buoyantly to Kesselring, Cavallero and the Italian senior commanders on 26 June that a break-through on the frontier should see the Panzerarmee in Cairo and Alexandria by the end of the month.

Rommel had always wished to invade Egypt. He was still convinced that the 'Great Plan', Plan Orient, was practicable, whereby a German army, masters of the Middle East, could control the area's oil and act as the southern pincer of a huge operation in conjunction with the Wehrmacht's advance into and south from the Caucasus. In June and July 1942 it was still possible to believe in such dreams and Hitler himself, in a letter to Mussolini written on the day following the capture by Rommel of Tobruk, referred to them. Nor were these visions without rationale. Hitler (and Mussolini) reckoned that it was essential to keep the enemy pinned on an outer perimeter – in Russia, in the Atlantic, in Africa – within which the Axis powers could organize an economically viable and defensible region before the ultimate menace to Continental Europe, the intervention of the United States, evolved. This implied a forward policy in the Middle East and the Caucasus, producing both oil and a distant glacis for the security of Europe itself.

Hitler had already issued the relevant directions, and on 28 June the opening phase of the German southern offensive in Russia began, directed on Voronezh and aiming to encircle and destroy all Soviet forces west of the Don. The offensive was led by General Hoth, commander of 4th Panzer Army and Rommel's old corps commander in France. Less than two months later, in the extreme south, the Germans had forced some of the Caucasian mountain passes and were standing on the east coast of the Black Sea; by which time Rommel and the Panzerarmee were within sixty miles of Alexandria.

Rommel believed that the Panzerarmee might be assisted by Egyptian anti-British sentiment, particularly in the Egyptian army, and he hoped for something from two operations of the *Abwehr*, the German espionage and counter-espionage service, which were undertaken at this time. He was disappointed. In the first of these operations a former Chief of Staff of the Egyptian army, Masri Pasha, who had expressed a desire to work with the Germans, was to be 'lifted' from the desert near Cairo by an intruding Heinkel aircraft; but he was arrested before this could happen. In the second operation two German agents were to be introduced into Egypt overland, by a cross-desert expedition covering nearly two thousand miles from Tripoli, via the Jalo and Kufra oases to Asyut in upper Egypt. The expedition was conducted by Count Almasy, a Hungarian air ace of the First World War, who had extensive knowledge of the North African desert from pre-war expeditions. This attempt also failed, although Almasy completed his journey and the agents reached Cairo. Once there they were quickly detected by the British security services, arrested, and sufficiently suborned to make sure that such information as they passed was of no use whatsoever to Rommel.[3]

But whether or not some sort of collaboration might assist him (and it could never be more than a slender hope, in view of the total control the British were exerting over Egypt), Rommel was now, once again, on the Egyptian frontier. When it became clear that the enemy had no intention of contesting the frontier area he had to decide whether to stand there or go forward. To go forward required, even for Rommel, the approval of the Italians, for all the time he was under Italian authority in major matters of policy, and in spite of Italy's own foray into Egypt in 1940, for German forces to drive deep into the heart, perhaps the capital, of another country (nominally neutral) from Italian colonial territory was undeniably a major matter of policy. Rommel received the necessary authority from Mussolini himself, and advanced. The Panzerarmee had only forty-four running tanks left.

Rommel's heart was in the enterprise. Opinion within the Axis hierarchies was again divided about the practicability of strategic advance without undertaking the assault on Malta: but the latter was always – and inevitably – assessed as a costly and uncertain operation and Rommel's strength had been boosted by the pickings of Tobruk. He pointed to the possible benefits of exploiting victory, of acting rather than waiting. His arguments were congenial to both Führer and Duce. He believed that it was essential, once again, to give the enemy no respite; that to adopt the defensive, albeit temporarily, on the frontier would be to surrender the initiative; and that the frontier position could anyway easily be outflanked by enemy mobile forces. For despite Rommel's confidence in the superior tactical effectiveness of the Panzerarmee he knew that the enemy would always have more fuel, more replacement vehicles and thus more potential mobility; and an open flank, he felt, would be to the enemy's and not his own advantage. He therefore concluded that before undertaking the next and, as he hoped, final act in the drama of North Africa he should establish the German–Italian forces on a position with flanks secured by terrain, a firm launch-pad for the conquest of Egypt. This meant going forward.

Having surrounded the defended place of Mersa Matruh by 28 June (but having failed to prevent many of the garrison from escaping eastward), Rommel now held every port in North Africa west of the Egyptian delta, had overrun enormous quantities of stores and felt increasingly confident that his instinct had been right. The British troops deployed round and near Mersa Matruh were racing eastward, sometimes only narrowly ahead of Rommel's columns, sometimes actually behind them, urgent to get away. There had been British confusion at Mersa Matruh, divided counsels, equivocal intentions, from which Rommel had profited often. On 1 July he mounted his first attack on the British in position south of El Alamein – a railway station on the line from Alexandria to Mersa Matruh, and only sixty miles west of the former.

Rommel knew that within Eighth Army Auchinleck had relieved Ritchie and taken command himself. Despite his confidence that the advance had been the right choice, he knew that time was running against him. The enemy's strength would inevitably increase at a greater pace than his own, backed as it now was by the immense potential of the United States as well as the British Empire. Meanwhile there might, just, be time to win. One of *die gute Quelle*'s last reports had been of considerable British panic in Cairo.

The Alamein area was one of those few positions in the Desert War where the front could rest on topographical features which prevented outflanking: in the north the Mediterranean, and in the south the huge sand sea of the Qattara Depression, seven hundred sheer feet below the adjoining desert; a front of about thirty-five miles. A number of low ridges, mostly running east and west, broke the landscape and afforded relatively commanding observation: these ridges were features of inevitable significance. But for the most part, north of the Qattara Depression there were few obstacles to movement, although as everywhere in the desert some terrain yielded better and some worse going for vehicles, and in the central and southern part of the area were a number of sand cliffs and escarpments. It was in this bleak landscape, in July 1942, that the British turned, stood and fought.

The fighting was scrappy and bitter. The strength of both sides had been drained by the huge exertions of the preceding weeks. The restricted front and the defensible features restored to infantry a more important part, and Rommel was particularly short of infantry. His first attack on 1 July failed completely. He found the enemy ready for him and supported by powerful air strength; he lost a significant number of his remaining tanks, and 90th Light Division suffered so heavily from artillery fire that they were not far from panic, and were held by their commanders only with difficulty. A renewed effort on the following day achieved little, the attacks repeatedly broken up by British guns. Although reluctant to admit it to himself Rommel was now losing the initiative which he had, with few breaks, held since 26 May.

Rommel planned his next attack for 10 July, giving out orders at a personal visit to Afrika Korps headquarters in his usual vigorous individual style – a battle group of tanks, guns and anti-tank guns would be formed immediately, to operate directly under army command, to move on a particular hill; battle group instantly detailed, from 15th Panzer Division; battle group commander received precise orders from Rommel himself.[4] The attack had to be cancelled in face of a strong British offensive in the coastal sector against two Italian divisions, Trieste and Sabratha. 'No day goes by,' Rommel wrote to Lucy on 11 July, 'without the most appalling crisis. It makes one howl!'[5] He again attempted to advance with 21st Panzer on 13 July, and for once German coordination of tanks and infantry was faulty and the infantry, isolated, were broken up by enemy artillery fire with nothing achieved – an experience repeated on the following day. On 16 and 17 July Rommel just managed to check an Australian attack, also in the northern sector towards the Miteriya ridge, by moving German units northward from the centre. On 15 July he had narrowly held an attack by the New Zealand Division in the area of the Ruweisat ridge, near the centre, and had counterattacked, but only regained part of the ground lost. This was becoming a battle of small-scale, often expensive, pushes and jabs by one side or the other and little opportunity for manoeuvre. On 17 July, in conference with Kesselring, Cavallero and Bastico, Rommel painted a black picture.[6]

His letters home inevitably reflected his deep unease, although he found relief in turning his mind, however dark the day, to domestic matters, enquiring on 21 July, 'What's Manfred doing with his holidays?' and giving Lucy details of his latest royalty statement from his publisher, Voggenreiter.[7]

Another New Zealand attack, on 21 and 22 July, was preceded by a very heavy bombardment and matched by yet another Australian move in the north, but was not accompanied by armour in the sort of coordinated tactical handling at which the Afrika Korps still generally excelled, and was defeated by a 21st Panzer counterattack inflicting heavy losses. For Rommel this was a critical moment in the July fighting, in what became known as the first battle of Alamein. He wrote on 25 July, 'The difficulties of the situation in these last days can hardly be described.'[8] Although on 26 July he could write that the worst of his troubles were disappearing, the July losses had been approximately even on both sides, and Rommel was able to afford them a great deal less than Auchinleck. He wrote on 2 August that these weeks had been the severest fighting yet seen in Africa, and that even he was feeling very 'tired and limp'. The feeling was justified, and by more than physical condition. Rommel's bid to exploit the fruits of victory at Gazala had failed. The tide was turning and he recognized it.

There was a lull, and both armies needed it. This had been exhausted swordplay, wrists tired on both sides. Some fresh players were arriving, but some expert players were casualties – Rommel, on 10 July, had lost his radio interception unit and its brilliant leader, Captain Seebohm. Seebohm had been able, on occasion, to place his vehicle next to Rommel's and to pass to him translated versions of British radio messages even before these were formally acknowledged. This was indeed a loss to Rommel's intelligence. He had also, since 29 June, been without the helpful transmissions from Cairo of Colonel Fellers, *die gute Quelle*.

Rommel had received reinforcements by air: the 164th Light Division (General Lungershausen) from Crete, quickly blooded by being thrown into the fighting of 11 July; the Ramcke Parachute Brigade, originally formed for Operation 'Hercules', the invasion of Malta – especially tough soldiers and including many veterans of the immensely expensive airborne invasion of Crete in 1941. He had also been reinforced by the Italian Littorio Armoured Division (General Bitossi) which arrived on 28 June. His tank strength had increased again, including a number of Panzer IVs, strongly armoured and carrying the excellent long 75-millimetre gun.

Nevertheless, of his German soldiers, seventeen thousand had been in Africa for sixteen months. The sickness rate was high, and they needed relief by fresh men. The fresh men started to arrive, but the fatigue and the debilitating effect of heat, poor diet, flies in sickening quantity, frequent stomach disorders and fatigue was taking its toll of Rommel's commanders; and of Rommel himself. His blood pressure was troubling. His digestion was again uncomfortable. He often felt exhaustion now, and knew that Africa had weakened him severely. He realized that he needed a decent rest in European conditions, hard though that was to envisage. 'You can well imagine,' he wrote to Lucy on 28 July, 'how extraordinarily sad I am not to be able to make leave plans.' On 7 August he wrote: 'How much too happily I would jump across to Germany! But too much depends on these next weeks.'[9]

He planned to resume the offensive in the desert as early as possible: in that, he believed, lay hope. Minelaying was being undertaken by both sides with great intensity, and the longer an offensive was delayed the more difficult the opening 'break-in' phase would be. For Rommel knew that British strength was increasing all the time as he had always appreciated it would. His staff estimated that by 20 August the enemy would have nine hundred tanks, 850 anti-tank guns and 550 guns. To counter this Rommel would have fewer than five hundred tanks, of which only two hundred were German. He expected American units to appear in Egypt; that had, indeed, been mooted in Allied councils, but in the event American tanks arrived – in considerable quantities – rather than American troops.

The air war, too, was going against Rommel, and had been since the end of the Gazala battles. The Royal Air Force were active and aggressive. The front-line troops – and Rommel's own headquarters – were heavily attacked at the beginning of July, and on 8 August the port of Tobruk was bombed. Although Tobruk's capacity was limited, it had the enormous merit of being comparatively near, and now the cargoes it could handle were reduced by 20 per cent. Rommel's supply requirement was further exacerbated by the fact that, although he had captured huge numbers of enemy vehicles in Tobruk (without which he could not have continued), his dependence on them – they constituted over 80 per cent of the Panzerarmee's vehicle train at this time – meant that he was chronically short of spare parts. Replacement vehicles from German and Italian sources were only arriving slowly.

* * *

Thus, once again, Rommel's logistic situation was extremely serious. His troops were actually short of food and the Panzerarmee's bread ration was halved. In mid-August they held between eight and ten days of ammunition expenditure at combat rates, and enough fuel to take the Panzer divisions sixty miles for three successive days; but the prospects were grim.

Replenishment prospects had always swung between grim and uncertain, and had never been far from Rommel's mind. On the one hand he has been excoriated as neglecting logistic factors, and on the other he himself devoted a good deal of time and ink to complaining about the incompetence or worse of the authorities, both German and Italian, who failed to keep his army better provisioned, so that he has been criticized for unconstructive nagging. But the factors controlling supply of an army fighting in North Africa (where virtually every commodity had to be brought in by sea), although easy to comprehend, were variables. They could be affected, and win a different significance in the equation, as a result of action taken by one side or the other.

Of these factors the first was the length of sea passage and (almost in inverse proportion) its vulnerability. The shortest sea passage from Italy to Libya was to the port of Tripoli. The most serious threat came, as elsewhere, from the submarine and the aeroplane, and in this case from British possession of the island of Malta and its use as a base for sea and air operations against the convoys; but for most of the period, even with Malta untaken, losses at sea were narrowly manageable so long as convoys were routed to Tripoli. Possession of the Cyrenaica airfields was also, of course, a key factor in this particular equation; and there were bad periods, with Tripoli virtually blockaded. Success (comparative) or failure (also comparative) naturally fluctuated in proportion to the air attacks which the Axis air forces were able to mount against Malta and to the extent that the island's stocks could be reduced by blockade. Malta always loomed large, and the arguments about Operation 'Hercules', the planned airborne assault on it, were inevitably sharp.

If, however, Axis convoys were routed further east, they became more vulnerable as they came within range of enemy airfields in Egypt and as they spent longer on sea passage. Escorts were used, and in August 1941 OKW had authorized German submarines and destroyers to be sent to the Mediterranean, to reinforce the Italian Navy; but in terms of the proportion of fighting ships to cargo escorted they were regarded as uneconomical – an understandable judgement whose merits essentially depended on the strategic importance ascribed to the North African theatre.

The second factor was port capacity. The vulnerable ports of Benghazi and Tobruk – vulnerable to Egypt-based air attack and in the sense of time on passage – also had very limited capacity compared to Tripoli, which could handle five cargo ships simultaneously. Benghazi (2700 tons) and Tobruk (1500 tons) were of much less significance.

The most secure port, with the greatest capacity and the greatest chance of being reached safely, was, therefore, Tripoli. But the German–Italian army was attempting at all times to defend as far to the east as possible, and then to advance – in Rommel's ambitions, also as far to the east as possible. This introduced the third and arguably greatest factor in logistic support – the length of the land line of communication forward from port, and the consequential demand for road transport. Air freighting was used to the limit practicable, but the great mass of supply inevitably moved by sea and road.

Every mile that Rommel advanced increased this demand; and the further the front from the port of commodity entry into North Africa the greater the demand would be. Not only, of course, did the demand for road transport thus vary, but the fuel consumed by that road transport (and the spares required for a transport fleet travelling over huge distances on indifferent surfaces) also varied.

Rommel's problem therefore, and the problem of those charged with supporting him, was how to supply a fighting force sufficiently large to achieve the strategic objects of the campaign, over a line of communication from Europe, vulnerable (at sea) from sea and air attack – but diminishingly so the further west the port of disembarkation; and vulnerable (on land) from air attack as well as from transport inadequacy, vehicle fatigue and fuel shortage – but diminishingly so the further east the port of disembarkation. The problem, according to von Rintelen, German Military Attaché in Rome and OKW Liaison Officer with Commando Supremo,[10] was insoluble. It was not insoluble, but to solve it at any particular moment required favourable alteration in one of the variables – more wheeled transport (and the German motor industry was not comparable to the sort of phenomenon represented by the United States); the capture of a port nearer the front (but with the possible consequential of increased vulnerability of convoy and disappointing port capacity); the importation and stockpiling forward of much larger quantities of fuel and other commodities; or withdrawal of the front to nearer base.

Rommel was frequently exhorted to the latter expedient – or, a variant of it, exhorted not to advance and thus avoid exacerbating his supply situation. He had been exhorted by the Italian Command in February and March 1941, before his first offensive which swept the British from Cyrenaica. He had been exhorted after the retreat to Mersa El Brega following Crusader, before his second great offensive which took him to Gazala and Tobruk. He had been exhorted again, as he stood at the edge of the Nile delta. On each occasion those who urged him to stand still, to remain within more comfortable reach of his supplies, had much reason to prove their point. Yet, Rommel could demand, what was the point of his presence and the presence of the Panzerarmee in Africa at all? If, by taking great risks, he could exploit the enemy's tactical defeat, might not yet a little more effort, a little more ingenuity, actually send the British packing and achieve a huge strategic object? With the doubters, the sceptics, he could point to the actual record. They would have held him near Tripoli in February 1941. Instead, within two months he was on the frontiers of Egypt. They would have held him there again in January 1942. Instead, by the middle of June, he was master of Tobruk and within sixty miles of Alexandria. He had ignored warnings and he had been justified.

Rommel, furthermore, believed that the variables in the equation could be favourably changed, given more effort. He wrote bitterly of this afterwards, complaining of lack of Italian vigour and will in convoy provision, of the lack of coastal vessels to bring supplies forward by sea from Tripoli and Benghazi, of the unenterprising approach to port capacity. These became constant reproaches; like many men, Rommel's sense of priorities and his estimate of difficulties were highly coloured by where he happened to be.

Nevertheless logistic factors have a disagreeable habit of refusing to disappear. The monthly 100,000 tonnage required by the Panzerarmee, the fuel situation in Africa overall, the strain on Rommel's vehicle train however organized,* and the disappointing palliative represented by Tobruk (the voyage to which was so hazardous that it was described as the cemetery of the Italian Navy), had to be balanced against Rommel's certain knowledge that Auchinleck was receiving powerful reinforcements of men and equipment, and against the obvious fact that Auchinleck's army was standing within

* The journey between Benghazi and the front was eight hundred miles, from Tripoli 1400 miles.

a very short distance of its base and could have no resupply problems whatsoever. When the balance was struck it was near impossible to see how Rommel could sustain another protracted campaign, even perhaps another battle, unless there were a significant change for the better in one of the logistic variables. Only one of these was under his direct control – the position of the front, the length of his land line of supply. He could go back – a long way – could delay any further ambitions until there had been the so-often-deferred assault on Malta and consequential improvement in the long-term supply situation, until major supplies had been built up in North Africa, including supplies of replacement vehicles. Or he could go forward, consummate the campaign quickly and supply the Panzerarmee through the port of Alexandria and the airfields of Egypt.

As the July weeks passed Rommel knew that he was too weak for the second alternative to have a particularly good chance. He had captured two thousand vehicles and 1400 tons of fuel in Tobruk, but his problem, although lessened, was not solved. In mid-August the British forced, at considerable cost, a convoy through the western Mediterranean to restock Malta and thus to increase, once again, the capacity of the island fortress to threaten and reduce the Axis convoys to Africa. For all the reasons given, as well as severe shortages of naval fuel, cargoes were again mostly being directed to Tripoli. Rommel would be invading the Egyptian delta at the end of a line of communication stretching halfway along the length of the North African littoral.

He believed, nevertheless, that to wait or go back would be to surrender the initiative to an enemy soon to become irresistibly strong, while to attack was at least to make a grasp at strategic victory. Kesselring, a shrewd calculator of military factors albeit by temperament an optimist, now concurred. Rommel's late-summer campaign in Egypt was supported by the hopes and directions of his superiors.

It was decided to launch a major attack at the end of August. As at Gazala it would be an attack on a mined and fortified position. As at Gazala, Rommel decided that it would involve a demonstration in the north and a *Schwerpunkt* in the south, an enveloping move round the enemy's left flank, followed by a northward drive to the sea. The morale of the Panzerarmee was still high. The day fixed for the attack was 30 August and, perhaps more comfortingly than accurately, Rommel wrote to Lucy early that day: 'as for my health

I'm feeling quite on top of my form. There are such big things at stake. If our blow succeeds it might go some way towards deciding the whole course of the war.' In Russia the Germans had reached the Volga and were already in the outskirts of Stalingrad. And here in Africa, if the Eighth Army could be dealt with as at Gazala, Rommel could be on the Suez Canal, master of Egypt. He was, however, feeling uneasy, and his companions noted that uneasiness. He suspected that his last chance was imminent.[11]

Before the advance began at about half past eight in the evening of 30 August Rommel joined the headquarters of the Afrika Korps, still led by General Nehring. He knew that since 15 August the British Eighth Army had been under a new commander: General Montgomery.

The military talents of Bernard Montgomery differed in almost every particular from those of Erwin Rommel. Both had taken part at impressionable ages in the First World War, but had emerged from it with differing military philosophies. Montgomery – after initial platoon command experience at First Ypres, a gifted staff officer on the Western Front throughout – had carried into life a preoccupation with efficient organization, careful preparation of battle, husbanding of material and human resources, application of certain simple and cogent principles of war. Rommel, no less thoughtful but a regimental officer or commander almost throughout the war, had carried forward essentially tactical lessons – a high belief in the efficacy of surprise, in the primacy of attack, in personal leadership at the critical point of battle, in unrelenting physical energy, in command from the front. Where Montgomery was deliberate, a cautious calculator of hazard and advantage, prone to overinsurance, Rommel was an optimist who reckoned that in war nothing is certain and that the uncertainty factor could turn as easily to his advantage as not, especially if he 'took time by the forelock' and set the pace. Montgomery thus refused battle, where possible, unless the odds were highly favourable, whereas Rommel, although he strongly denied being a gambler, took undoubted and sometimes breathtaking risks if he reckoned the stakes were high enough and that he had a way out if things went wrong. Montgomery excelled in the carefully planned, set-piece battle, was usually a master of organization and logistic provision, aimed to pick trusted subordinates and give the detailed battle to them to conduct. Rommel, quicker in mind and action, commanded from the saddle, interfering often in that tactical battle at which he was adept, and aimed to be present himself, as often as possible, at the point of decision.

Rommel was a master of battlefield manoeuvre, of improvisation, a thruster. Montgomery never gave himself that challenge, and would probably not have excelled at it. Instead, he fought precisely the sort of actions at which he did excel; and conquered thereby, for he had the resources to do exactly as he chose. Montgomery, unlike Rommel, made good his victories, prepared his consolidation of them, never went back. Rommel's triumphs were by comparison ephemeral: but after 1940 Rommel won, when he did win, with numerical and material inferiority, offset by an energy, a speed and a dynamic skill uniquely his own; characteristics different in kind rather than degree from those Montgomery displayed.

Yet if the military aptitudes of these two formidable soldiers were wholly different, their human qualities and characteristics were strikingly similar. Both placed enormous importance on the confidence engendered in their troops. Montgomery took endless trouble to communicate, to show himself, to talk to soldiers, to reassure: Rommel, more a 'soldiers' soldier', was equally and robustly communicative, joking and shouting like a sergeant-major, unsparing of himself, renowned for his Spartan lifestyle, his sharing of the army's discomforts, his reckless personal courage. Attaining the object in different ways, both men rated very highly the human factor in war, and their armies knew it. Both men were fanatics for physical fitness, in themselves and their subordinates; and both were austere in their personal habits. Both were devoted to their families, each with an only son. Neither had any very great interest outside the military profession, although both were skilful photographers, and keen on winter sports.

Both men were very conscious of the merits of publicity, of public relations, of self-advertisement, of the creation of a recognizable 'persona' within both the army and the nation which sustained that army. Rommel's image was to an extent nurtured by Goebbels, but Goebbels recognized that he was handling a general with star quality from the publicity angle. Montgomery's acquisition of public relations skills came rather less naturally to him and was, perhaps, a more contrived, a more deliberate policy – and none the worse for that. Rommel's flair for publicity was more a consequence of personal penchant, shrewdly exploited by the German media thereafter.

For Rommel was vain. And Montgomery was also – or became – vain. Both men shared a relish for popular acclaim. Both men increasingly valued their own achievements and came to believe that those who disagreed with them or criticized them were fatally flawed. Both were egocentric, competitive, intolerant of opposition and, on occasion, susceptible to flattery. In consequence, both were widely criticized within their own armies, for exhibitionism, self-assertiveness, showing off. This trait did not spring, instantly flowering, when they became generals. It had long been recorded.

Both men were harsh with failure and sometimes hasty in presuming it: it was a facet of the high standards which both demanded of themselves and of subordinates. But both were deeply admired by their own immediate staffs, those best able to judge performance under pressure, performance in crisis. Both tended to be uncharitable about the qualities of military contemporaries – Montgomery more so. Both were ready to bypass the channels of command communication, to the very borders of loyalty – Rommel more so. And both made up their own minds and took their own line, after listening to the best advice they could, and thereafter held to it with a maximum of independence and moral courage, and a minimum of subordination.

Both were on occasion discourteous and impatient with allies – Rommel with the Italians, Montgomery with the Americans – to a point potentially harmful to the causes they served. And both needed, on such occasions, a more emollient superior to repair the damage; Brooke[12] in Montgomery's case, Kesselring in Rommel's.

Both men identified wholly with their commands, would defend fiercely the interest and the reputation of the men they led – Eighth Army in Montgomery's case, in Rommel's his 'old Afrikaners' of the Panzerarmee. Both men were exceedingly ambitious. Both were essentially principled and decent. Both had exceptionally strong wills. They were now about to meet.

Montgomery's positions were manned by four infantry divisions on a frontage of about twenty miles – from the north: 9th Australian (General Morshead), 1st South African (General Pienaar), 5th Indian (General Briggs) and 2nd New Zealand (General Freyberg) – deployed behind minefields; and a fifth, 44th British (General Hughes), deployed *en potence* on the Alam Halfa ridge which ran east and west some fifteen miles behind the New Zealand Division. There was left a deliberate gap south of the New Zealanders, a door through which Rommel could pass (and which, from ULTRA, it was known that he would indeed pass). In this gap were stationed a motor brigade and a light armoured brigade, with orders to cover the minefields but thereafter give ground when necessary. The main body of the British armour – three more armoured brigades – were in position west and south of the Alam Halfa ridge, ready to engage from defensive positions (adjusted as necessary when Rommel's moves became clear) and to take the Panzerarmee in front and flank when it struck and turned at that ridge. The British armour would be employed in an almost entirely anti-tank role. Rommel was to be enticed into a killing ground.

Rommel's intelligence staff gave him a reasonably accurate picture of British deployment, although they substituted their old opponents, 50th British Division for 1st South African, and they did not know of 44th Division on the Alam Halfa ridge. Rommel was attacking with an armoured strength of 470 tanks, excluding the Italian light tanks; and of these two hundred were the German Panzers of the Afrika Korps. He faced just under seven hundred tanks. Rommel thought British tank strength higher than was in fact the case, but he knew that there would anyway be considerable and readily available reserves with which the enemy could make good casualties, and in this he was right.*

Rommel was, therefore, facing superior numbers, as he so frequently and successfully had done before. His technique – once again, as before – would be to drive deep, to threaten the enemy's vitals, his rear areas, his line of communication to the Nile delta, and thus force him to react, to expose his mobile formations to the superior power of manoeuvre, the superior, more dynamic and skilful leadership and tactical handling of the Panzerarmee.

Rommel's operational plan was very similar to that at Gazala three months earlier. There would be a frontal demonstration in the north, where the Italian infantry divisions were facing the British minefields behind minefields of their own. Reinforced by the Ramcke Brigade and the newly arrived German 164th Division these divisions would aim to hold the British defenders from reinforcing southward. At the same time the armoured strength of the Panzararmee – the three divisions of the Afrika Korps, the three armoured or mechanized divisions of General De Stefanis's XX Corps (Trieste, Arite† and

* British strength was about to be augmented by the arrival of American Shermans – robust workhorses, armed with the same 75-millimetre he had met in the Grant at Gazala. It was not an anti-tank gun comparable to his own long 75-millimetre, and the Sherman was easily inflammable, but it was mechanically first-class.

† Now commanded by General Arena.

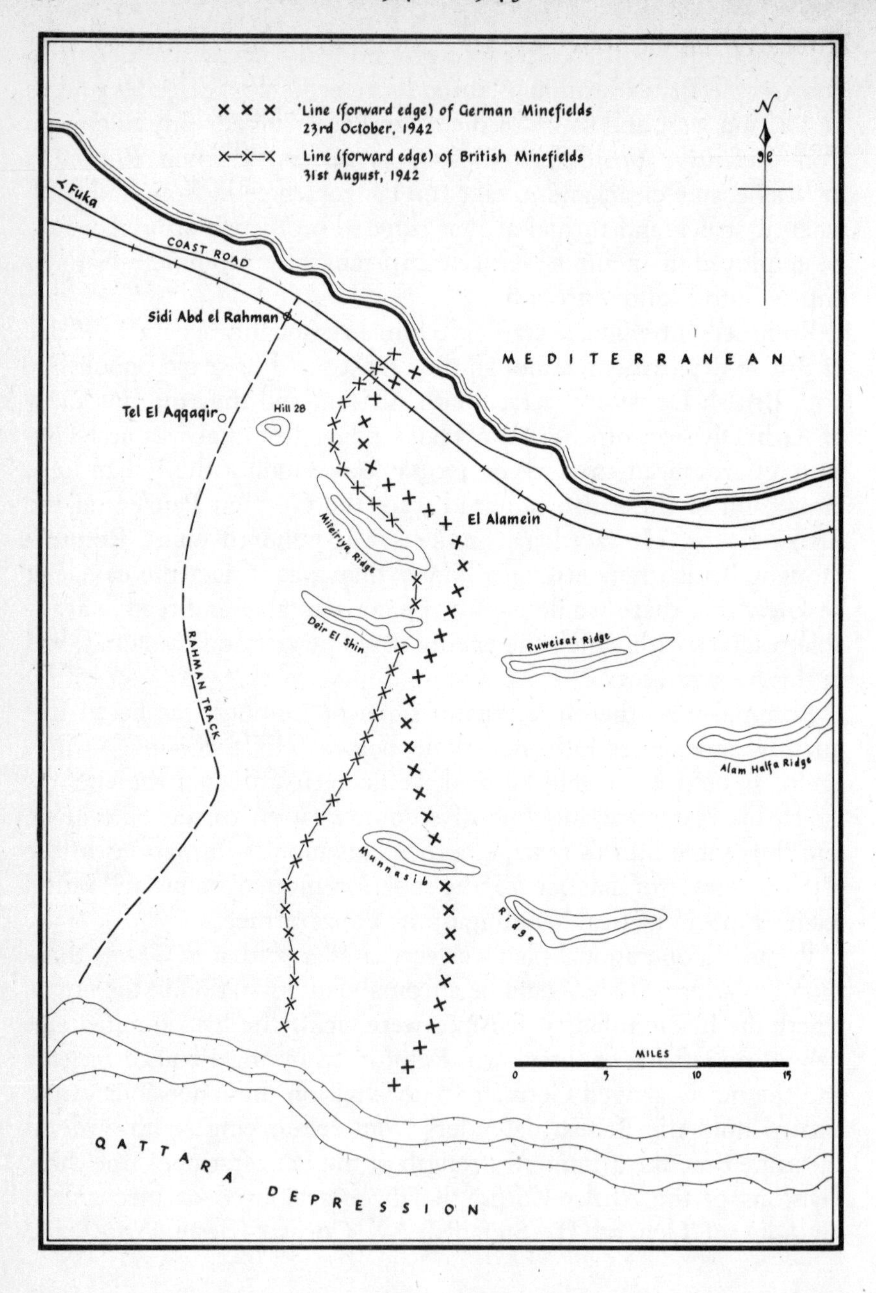

XVI Area of operations, Alam Halfa (August 1942) and El Alamein (October 1942)

Littorio) – would breach the British minefields in the south of the front, advance deep (some twenty miles) eastward, and then swing north behind the main British positions, dealing with the British armour, which would surely be bound to react, with the same unrehearsed and improvised vigour which had so often brought success in the past. Rommel would cut off the British withdrawal this time, would reach the sea and would then turn east and set his mobile divisions racing into the delta, to Alexandria and Cairo, to the limitless reserves of fuel and vehicles known to be based there. The 'Great Plan', Plan Orient, would have been launched in its southern arm. This mobile operation – the *Schwerpunkt* – would begin as soon as the minefields were breached, once again in the moonlight of the night of 30 August. It would get under way with the Italians on the left (and with 90th Light Division, detached, on their left again) and with 15th and 21st Panzer Divisions of the Afrika Korps constituting the right wing, to make the deeper wider thrust and swing.

Within three days it was clear that Rommel's operation was a total failure. No greater difference in course and eventual outcome from the battle of Gazala could be imagined.

Firstly, the Panzerarmee took a considerable time to clear a passage through the enemy minefields. These had been underestimated, and from the start Rommel's timetable was thrown into disarray. Second, the enemy's command of the air was used with devastating effect against Rommel's concentration areas and supply columns. The air war had once been more or less evenly balanced; there had been periods of Luftwaffe superiority, there had been periods of Royal Air Force superiority. Rommel had, in the previous year, suffered heavily from bombing attacks by British Blenheims during his first investment of Tobruk; but the Messerschmidt 109F2, introduced to North Africa in August 1941, had been remarkably effective, and during Crusader honours and losses had been approximately even. During the Gazala fighting, however, and in the subsequent pursuit and initial battles on the Alamein position the Luftwaffe and the Italian air squadrons, although numerically superior to their enemies, had lost heavily and there had been enormous and unremitting strain on aircrew. By the end of the Gazala battles the British had achieved the initiative; from mid-July '*Fliegerführer* Afrika' General von Waldau's reports had been increasingly pessimistic; and the British had few concerns about aviation fuel. The consequence was that Rommel now received comparatively little air support, while the Panzerarmee was subjected to ceaseless attack night and day from squadron after squadron of bombers – Wellingtons, Albacores, Bostons, Baltimores and Mitchells; supported by twenty-two squadrons of fighters.

The right wing of the Panzerarmee, through both of these (and perhaps other) causes, began limping rather than racing forward, taking over three hours to cover twenty miles. This was not the Afrika Korps which had swept round Bir Hacheim – *Rommel an der Spitze*. Nehring, their commander, was wounded, and authority was temporarily exercised thereafter by his chief of staff, the admirable Bayerlein. A few minutes earlier von Bismarck, the gifted and experienced commander of 21st Panzer Division, had been killed by a mortar bomb. Already at eight o'clock in the morning of the first day, 31 August, Rommel, undoubtedly shaken by the effectiveness of the enemy air attack, considered stopping the operation and withdrawing.[13] His left wing, the Italian XX Corps with 90th Light Division, were making slow progress, a fact which should not have upset and might even have assisted the battle had it turned, as Rommel hoped, into a true battle of envelopment; but on this occasion it disturbed him. His famous *Fingerspitzengefuhl* told him without question that things were going wrong.

The Afrika Korps halted to refuel in the desert south-west of the Alam Halfa ridge. Bayerlein tried to persuade Rommel that they should carry on, and Rommel, uneasy, decided to continue. He had intended to move some way further eastward and then turn north towards and past the east extremity of Alam Halfa; now, at 8.15 a.m. on 31 August, he gave orders for the Afrika Korps to swing north earlier than originally planned and march towards Alam Halfa's west end.[14] Unbeknown to Rommel, the Alam Halfa ridge was itself defended by the British south-facing 44th division: while west and south of it the British armoured brigades were deployed, facing south and west respectively, ready to take the advancing Panzers with tanks essentially fought as anti-tank guns.

The advance continued slowly to a culmination designed not by Rommel but by Montgomery. The British armour remained in position, made no sortie, no attempts at manoeuvre. They gave to Rommel no opportunities whatever. They simply stood firm.

On 1 September, the second day of the operation, Rommel, amidst incessant air attacks (several were killed in his own *Kampfstaffel* and he himself narrowly escaped death from a bomb splinter), visited the Afrika Korps in the morning and in the afternoon again considered breaking off a battle which was failing. His Panzers had reached the British armoured brigades and were butting their heads against numerically superior and well-deployed forces on defensive ground of the defenders' choosing, with fire engagement ranges selected and contingencies planned. His fuel situation was tenuous, which seemed to make impossible any far-reaching manoeuvres even if such had been operationally practicable – already at seven in the morning General von Vaerst, now commanding the Afrika Korps,* told him that he foresaw the attack soon foundering for lack of fuel. His troops were suffering severely from air attack, wherever in the desert they were. He made one more effort against the Alam Halfa ridge, the key to the position, that day, – with 15th Panzer alone, a most improbable venture and one which gives the impression, justly or not, of a Rommel resigned to failure, limiting his bets but deciding to make one final probe in case the enemy (as often in the past) should at last form a pessimistic impression of the situation and start to withdraw. It didn't happen. It was never likely to happen. The 15th Panzer attack made no significant headway and next morning, at 8.25 on 2 September, Rommel's diary recorded bleakly the decision to break off the battle – '*Entschluss zum Abbrich der Schlacht gefasst.*'[15]

Withdrawal began. During the night of 1 September Rommel's faithful interpreter had recorded a bombardment such as never before experienced, '*so wie heute Nacht sind wir noch nie bombardiert worden!*'[16] Rommel may have exaggerated in his later writing the cataclysmic effect on his troops of the enemy's dominance of the sky, but it was very great. The Luftwaffe had, in fact, scored some significant successes on 1 September, so the air battle had not gone all one way, but it had gone enough to have a decisive effect on the battle. News came through during 2 September of the loss of a tanker with eight thousand tons off Tobruk, while a huge consignment of fuel, amounting to ninety thousand daily gallons flown into the theatre by the Luftwaffe, had been to a large extent frittered away through inadequate issue control on the line of communication.[17] But although the loss of the tanker was a bitter blow (one of many such), Rommel's decision to break off the fight had already been

* Transferred from 15th Panzer Division, temporarily handed over to General von Randow. Von Vaerst resumed divisional command on 17 September when General von Thoma arrived to take command of the Afrika Korps and von Randow took over 21st Panzer command.

taken. It has been said that his famous instinct for battle was not in evidence at Alam Halfa. Arguably, it was never more trustworthy. He had been unhappy before the battle, and he knew when it was lost. Surprise and shock, his two battle-winning weapons, had been totally denied him. His enemy, forewarned in detail by ULTRA and by an effective battlefield signals interception service, was ready. All Rommel's antennae warned him of a weighted situation where he could not, as in the past, overcome inferiority of numbers by speed and ingenuity; airpower, ULTRA and fuel were against him. And Montgomery.

Montgomery decided to abandon earlier schemes to exploit his defensive victory, and instead to wait, to consolidate, to husband his army's strength for the major offensive battle which he knew he must next fight. Rommel withdrew westward, an efficient withdrawal, comparatively unharassed. He still had 160 tanks in the Afrika Korps and the Italian divisions had 270. Losses of equipment had been comparatively small, although he had lost nearly three thousand men, including 570 prisoners. But Rommel had lost the battle of Alam Halfa, and with it the last hope of initiative in Africa. Plan Orient, in its southern arm, was dead.

Rommel largely blamed his fuel supplies for the frustration of his plans. Kesselring visited him at 5.30 p.m. on the day he ordered withdrawal, and argued for optimism. With at least half a point in his favour he observed that Rommel had enough fuel for the Afrika Korps' and XX Corps' withdrawal, and that that same quantity might have carried him to Alexandria and its near-infinite resources if he had won instead of losing the battle; and certainly, although Rommel's fuel situation was constraining if not desperate, he had appreciated it before starting. But Rommel also, and more convincingly, ascribed the failure to enemy air power – air power used in manifestly closer cooperation with ground forces than heretofore. He wrote afterwards of the futility of troops seeking to operate, to manoeuvre, in a hostile air environment. The savagery and efficacy of the attacks by the Royal Air Force on the Panzerarmee probably inflicted fewer casualties than at first appeared, but they enormously affected its commander's mind; and this indelible impression led Rommel to conclusions which some would call defeatist and some would regard as exaggerated, but which he would carry with him to the end. He wrote that the Panzerarmee had no alternative but to plan the forthcoming defensive battle from the strongest possible fortified positions. In the face of an enemy whose air superiority would increase with every convoy reaching the delta manoeuvre would be impossible, as would all manoeuvre in similar circumstances. The first exchanges of the battle of Normandy undoubtedly took place before the ridge of Alam Halfa.

And Rommel was confronting a new enemy. Montgomery kept a close control of the battle. He knew what to expect, his plans were thorough and sound, he attempted no manoeuvres beyond a predicted reinforcement of the southern battlefield on the first afternoon, and he resisted the temptation to make victory more decisive by ambitious forays on Rommel's withdrawal. But a certain indefinable electricity ripples across a battlefield, a current flowing from personality, intangible, unproveable: and Rommel, ascribing it to whatever cause he chose, found that the enemy forces were unflustered, in position, not overreacting to his own moves, confident. This confidence, in such measure, was new. Rommel had admired much of Auchinleck's handling of Eighth Army, especially in the previous month when first frustrated at the Alamein position. He had admired Auchinleck's coolness and skill, and sensed his essential strength. He would criticize what seemed to him Montgomery's slowness to grasp opportunity and exploit success when he, Rommel, had lost a battle. 'If I were Montgomery,' Rommel remarked bitterly to his staff in September, '*we* wouldn't still be here!'[18] It was the comment of a man who felt he had been beaten by material superiority rather than enterprise and dexterity. Rommel did not – he could not – directly connect the enemy's performance at Alam Halfa with a new spirit which had infected them from the hour of Montgomery's first arrival, but it was so, and after Alam Halfa the infection grew powerfully.

It was strengthened by the fact – far from new but able to be put to increasingly good account – that British intelligence now hugely surpassed Rommel's. ULTRA was performing admirably. It had greatly contributed to the sinkings which had helped strangle Rommel's supplies, and it was giving invaluable insights into the morale, strength and intentions of the Panzerarmee: and with the silencing of *die gute Quelle* Rommel had no comparable insights. Henceforth his operational intelligence about the enemy troops facing him was to be largely limited to gleanings from the small number of interrogated prisoners of war who yielded something worthwhile. He was comparatively blind, while Montgomery was like a man playing a hand of cards with a mirror behind his opponent's back. Every move of Rommel's at Alam Halfa had been predicted.

But there was something else. The performance of the Panzerarmee lacked that *élan vital* with which Rommel had so imbued it, and the reason lay not only in its tiredness – and it was very tired – but in its commander. Rommel, habitually so resilient, was uncertain before Alam Halfa. He was of course troubled about logistics – a rational unease, but one he had often faced before and successfully. He alleged, in recounting the story of the battle, that his supplies were being harassed by the southernmost British armoured brigade, which was not the case and which betrayed a most uncharacteristic nervousness, a readiness to look on the dark side of the possibilities of the situation. He certainly found good cause for his fears in the air superiority which the enemy used so effectively. But his dejection – for, by his own words, it approached dejection – ran deeper. There was little opportunity at Alam Halfa for Rommel to find relief in personal leadership, in some inspired initiative as often before – on the contrary he largely (and no doubt wisely) remained with his own headquarters command group except when visiting. Certainly his health was poor, which no doubt affected his whole being. But his famous sixth sense for battle was perhaps working in a novel and disagreeable way. Rommel, for the first time, did not really believe in what he was doing. Two months earlier, after Tobruk, the sun had been high in the heavens. Now the future – disturbingly for so normally vigorous an optimist – felt as if it might be dark.

Rommel wrote yet another desperate report for OKH on the supply situation in North Africa; and resolved at last to mend his health and to apply for sick leave and treatment in Germany. This involved handing over command of the Panzerarmee for a significant period – it was clear he needed a complete change and a complete rest if his stomach troubles, his blood pressure and derived ailments were to be dealt with. Six weeks had been mentioned. Gause had returned to duty on 5 September, relieving Westphal. It was ordered that General Stumme, a veteran of the Eastern Front who had been Rommel's predecessor in command of 7th Panzer Division in 1940, should be flown to Africa and appointed Rommel's deputy, to take command during the Field Marshal's absence. Rommel longed for his arrival, yet dreaded it. 'With one part of me,' he wrote to Lucy on 9 September, 'I rejoice at the prospect of getting out of here and seeing both of you again. With another part of me I dread the difficulties of this campaign if I can't be here myself.' Two days later he wrote with evident happiness that he should arrive before the letter; and he tracked Stumme's progress – leaving Berlin 15 September – arriving Rome 16 September – due with Rommel 19 September – in daily letters, regardless of his own impending arrival at home, like an impatient schoolboy.[19]

Stumme duly arrived on 19 September, and Rommel inducted him into the problems of the Panzerarmee at a conference the same afternoon with Cavallero (also visiting) and the Panzerarmee quartermaster staff, a conference at which harsh truths were spoken about the supply situation.

By then Rommel had given orders for the preparation of a greatly strengthened defensive position at Alamein, a position incorporating nearly half a million mines, laid in great depth. He planned this in considerable detail and personally – subordinates who had hitherto largely known him as the master of manoeuvre were deeply impressed by his skill and wariness in the siting of positions, with '*die Technik des Stellungskrieges*' (positional warfare) and with his eagle eye. This was the enormously experienced infantry officer who said that 'Well-chosen positions save blood and alone give soldiers confidence.'[20] Rommel knew well that if the Panzerarmee were to remain in Egypt the next battle would be against a massive British attack. He presumed that any frontal attack would be accompanied by landings from the sea, and gave orders about early counterattack and the deployment of troops with heavy weapons to engage the beaches instantly – orders which, again, foreshadowed the battles of Normandy.[21] His intelligence staff provided detailed and timely information on the build-up of British equipment and supplies through both Suez and Alexandria.

But now Rommel's own earlier faith in the possibilities of the situation had infected those above him, while he himself was losing it. The march of the Panzerarmee along the Mediterranean had been something of an intoxicant. Everywhere the Wehrmacht was going forward, was on the brink of tremendous possibilities. Now there was firm belief in OKW that, having got so far, the Panzerarmee must certainly stand at Alamein, that the great opportunity would recur. A visitor from Berlin during the previous month – General Warlimont, Deputy Chief of Staff – had spoken optimistically of Plan Orient. General von Kleist, with Army Group A, would soon be advancing south from the Caucasus. The enemy position in the Middle East would soon be threatened from both west and north. There could be no question of the Panzerarmee going back. But that had been in August, before Alam Halfa.

Rommel paid a few visits before his intended temporary departure. He flew to the Siwa oasis and was enthusiastically entertained by the local chiefs. He flew to Tobruk and congratulated the garrison on defeating a British raid from the sea. On 22 September he handed command to Stumme, telling him (somewhat to Stumme's discomfort) that in the case of a British offensive he, Rommel, whatever the state of his health, proposed to return to Africa. Then on 23 September he flew to Rome and had an animated discussion with Mussolini about the problems and importance of improving supplies to the North African theatre. Mussolini believed that the following year, 1943, would see American attempts to land in North Africa. It was essential to reach the Nile delta and improve the overall position in the Mediterranean before the Americans arrived.[22] He gave such assurances as Rommel had often heard about fuel supplies, and the conference minutes recorded, no doubt with Cavallero's resigned acquiescence, that '*Marschall Cavallero will tun was er kann*',[23] Cavallero would do what he could. It was the German way, Cavallero noted somewhat disagreeably in his diary, to blame others for their own difficulties.[24] From Rome Rommel flew to Berlin, and several days later reported to Hitler.

Hitler believed that Rommel need not have called off the battle at Alam Halfa, but nevertheless received him with great kindness, and was solicitous and congratulatory on Rommel's extraordinary achievements. Alam Halfa may have been – indeed was – the crucial turning point in the North Africa campaign, but to the German Führer and people it had been a minor and temporary check. Rommel was triumphant hero, master of virtually the entire extent of the North African littoral, the Marshal who, with small forces, had apparently come near destroying the prime effort of the British Empire. He was accorded a hero's welcome, staying as a guest in the Goebbels's house. A great reception was held for him on 30 September in the Sportspalast in the presence of the highest dignitaries of the Third Reich; he was handed his ceremonial baton; and, which gratified him more, he was assured that matters really were in hand to improve the flow of materiel to the Panzerarmee.

Rommel spoke to the press, uttered brave words. The German soldier, he declared, had now reached Egypt, and where the German soldier took possession of territory he would never be driven out. This was the sort of thing Germany wanted to hear. On the day Rommel reached Germany Halder, Chief of the General Staff, was dismissed, largely in reflection of Hitler's disappointment at the course of the campaign in southern Russia. Halder had always viewed both Rommel's achievements and the North African campaign with what Rommel regarded as infuriating scepticism. Now he was gone, his place taken by General Zeitzler.

Rommel then travelled to Semmering, in the Austrian Alps not far from Wiener Neustadt. Here he was intended to find rest, restoration and peace.

Inner peace eluded him. Rommel's anxious mind was with the Panzerarmee. He had been told by Hitler that he should have a complete change, that he had endured enough of Africa. Command of an Army Group in southern Russia was mooted. Rommel, however, was unhappily aware of having left his 'Afrikaners' in a dangerous situation and at a menacing moment. He wrote to Stumme a full account of Hitler's assurances about supply.

Nevertheless, and in spite of grave misgivings within OKW about the Stalingrad advance (where General Paulus had already reported his concern about the length of the salient at whose tip was his Sixth Army), for most people in the German Reich at that moment the war was going well. Several weeks were to elapse before the beginning of that great counter-movement which would see the British and Americans take control of the North African shore while the Russian tide lapped round Stalingrad and isolated Paulus's unfortunate troops. For most people in the Reich the sun was still out. For most people, too, there was little knowledge of what was now being done in their name.

This last was appalling. Deportations of various groups – largely the Jewish populations – from Germany, Austria and German-occupied Poland as well as from Bohemia and Slovakia, had been going on for some time. These deportations were presumed by the ordinary citizens of the Reich to be matters simply of resettlement, of a separation of races consistent with National Socialist theory, and as the confinement of certain essentially anti-patriotic sections of society in the interests of security. Few asked too much about the subject, or would have received much enlightenment had they done so. The deportations were protected by a wall of secrecy and fear, and in the Third Reich too many people disappeared, too often, under suspicion of infringing state security for the practice of probing enquiry into such matters to be widespread. On the day Rommel flew back to Germany an instruction emanated from the German Foreign Ministry to press on with negotiations for the evacuation of

Jews from certain neutral countries in the German sphere of influence – negotiations which had mixed success. But the deportations from within the greater German Reich had been going on now for several months. The critical Wannsee Conference which provided the framework for genocide had been held in January.

'Deportations', 'resettlements': euphemism, as ever, was the most shameless of liars. On the same day that Rommel left Africa – 23 September – two thousand Jews were deported eastward to a place called Maly Trostenets, near Minsk. All were murdered. Three days later a railway planning conference took place in Berlin which ruled that three trains daily, each train carrying two thousand people, would go from various districts to Treblinka and Belzec.* These quotas were not filled, but they implied capacity targets; and in the following month, October 1942, eight thousand reached Treblinka, where gas chambers for mass extermination had already been established. These frightful undertakings, still almost beyond comprehension, so sickening are they to the mind, took place while Rommel was seeking peace and restoration in the mountains of Semmering.

As yet Rommel knew nothing of such things. Like the great majority of his fellow-countrymen he supposed that the German authorities, preoccupied as they must be by the logistic and other problems of wartime, needed considerable powers and were using them, by and large, in the interests of the security of the state. Germany was fighting a life-and-death struggle on the Eastern Front. Europe was in a fair amount of turmoil, and that there were a good many enemies and ill-wishers to Germany none could doubt. His own mind was in Africa.

On the home front, Rommel had never known or liked many of the more influential Nazis, except Goebbels; but he presumed that the Führer was keeping them under control. And as for the Führer himself, Rommel had once again felt a rapport established. He had tried to impress Hitler with the seriousness of the situation he knew best, and Hitler, the great encourager, the great consoler, had made him feel that his problems were understood, that they would be addressed with more urgency now. Rommel had found, he wrote later, a certain unjustified optimism in OKH, but Hitler himself, as ever, had restored his confidence, shown him trust and affection, made him feel better.

* * *

* The total murdered at Treblinka, in Poland, was said to be 840,000: at Belzec, also in Poland, 600,000.[25]

At Semmering Rommel received a long letter from Africa, written on behalf of Stumme on 13 October. An enemy offensive was expected in the near future. A lot of detail was included: in the primarily Italian sector of the front difficulties had arisen because the defensive minebelts were planned to incorporate many British mines, some of which proved to be dummies. The supply situation was still poor, but should suffice for purely defensive battle – the Field Marshal should rest assured that in spite of difficulties the Panzerarmee was ready to receive a British attack. Stumme had written ten days earlier to Cavallero telling him that when certain deployments ordered by Rommel had been completed (due on 20 October) the army should be capable of meeting the expected frontal assault.[26] Thereafter an Axis offensive might again become possible. Meanwhile, Stumme's letter continued, a quarter of a million anti-tank mines had been laid – a grand total, including British and anti-personnel mines, of 450,000 on the Alamein position.

On the afternoon of 24 October, Rommel was telephoned by Field Marshal Keitel, Chief of OKW. Would Rommel, Keitel enquired, be fit to return to Africa immediately? The British had begun what seemed to be a major offensive on the previous evening, at Alamein. Stumme was missing.

Rommel answered that he was ready. That evening Hitler telephoned personally, and telephoned again at midnight. He first expressed anxiety that Rommel should not interrupt his cure unless the situation were really serious: but in the second call he said that it was indeed serious. He asked Rommel to fly as soon as possible, to resume his command. Next morning, 25 October, Rommel took off.

CHAPTER SEVENTEEN

Watershed

AT 8.40 P.M. (German time*) on the evening of 23 October the men of the Panzerarmee saw the eastern sky suddenly illumined, and seconds later their ears were assailed by the sound of a barrage fired by 456 pieces of British and Imperial artillery. The battle of Alamein had begun.

It had been long expected, although date and detail were certainly unknown and the enemy had taken considerable pains to conceal them. The most recent OKH intelligence estimate – a representative of the General Staff's *Fremde Heere West* (Foreign Armies West) had visited the Panzerarmee's headquarters two days before the battle – had been that the British attack would be launched at about the beginning of November. On the other hand it had been noted that at least one earlier British offensive had been preceded, eight days before, by a specific radio order relating to reserves of stretcher-bearers. An exactly similar transmission was monitored on 15 October, and the Panzerarmee's Ic, with commendable attentiveness and imagination, deduced therefrom an attack on the twenty-third, and advised the operations staff accordingly.[1] This was far-fetched, albeit accurate; but radio silence on all enemy nets from nine o'clock on the morning of the twenty-third seemed to give it credibility; and at 8.40 p.m. the guns spoke.

Rommel, although he had been away for a month, had laid down every detail of the German–Italian defence. He was under no illusions: if he fought on the Alamein line (and the strategic importance

* One hour earlier than British time. For the British the Alamein opening barrage was timed for 9.40 p.m. Somewhat inconveniently, German time reverted (by an hour) during the battle, at the beginning of November.

of so doing had been emphasized by OKW) he would be fighting a static battle of attrition, a *Materialschlacht*. The prospect was ominous. First, Rommel (although it was uncertain that he would be personally conducting operations, he had every intention, somehow, of so doing) disliked that sort of battle. Although he never ceased to be a highly skilled and thorough infantry commander, whether in siting a defensive position or leading an attack, his genius was for manoeuvre, for movement, for the unexpected: at Alamein there was little chance of such. Second, the winner of a *Materialschlacht* tends to be the one with the most materiel and the more certain methods of replacing it. This would not be Rommel.

Yet no other sort of battle was likely to be possible. Flanks rested on sea and Qattara Depression, and the forces available to both sides filled the available terrain; attack, at least initially, would be frontal. Rommel's Italian infantry divisions had no means of mobility except their legs, and legs tire more quickly in desert going than in most. Worse, Rommel's perennial shortage of fuel would inhibit if not altogether prevent any but the most limited and local movements and manoeuvres. Worse still – and indelibly marked on Rommel's mind – the enemy air forces now had absolute mastery of the air. Movement as well as resupply would be dominated by that fact. It meant that the troops, to have hope of victory, must be deployed in particularly well-prepared and well-protected positions: or have short, well-reconnoitred and rapidly accomplishable attack or counterattack roles.

Rommel examined his conscience afterwards as to whether he might have prepared a strongly mined position for his non-motorized infantry far to the rear – a line running south from Fuka, sixty miles west of Alamein. He might then, he worried, have manned the existing Alamein line (mined as it was) with motorized formations – presumably 90th Light, Trieste, and other divisions for whom he could have assembled transport – so that they could, if and when it became inevitable, 'break clean' and withdraw to fight another day in good order, with a firm fall-back line behind them; while between the two positions he could have held the concentrated armoured strength of the Panzerarmee – two German and two Italian divisions – to challenge the enemy in open desert. The merits of such alternatives can be nothing but speculative. It is unlikely that the mines and effort available would have sufficed for two strong positions, so that one or other would have been an easier task for an enemy than was Alamein. The 'motorized formations' held forward would have been a good deal thinner on the ground than was the German–Italian army on 23 October, and their overwhelming by the enemy presumably the more rapid. Above all, however (and it is surely the only comment which can be regarded as certain rather than theoretic), Montgomery's plan would have been wholly different; and it is hard to believe it would not ultimately have been as efficacious. Rommel's *post facto* alternative concepts might have led to a more interesting battle, but would probably have done little except postpone defeat; if that.

The Panzerarmee, therefore, was deployed with infantry formations in line – two of them German, 164th Light Division (General Lungershansen) and the formidable General Ramcke's 288th Parachute Brigade; and five of them Italian infantry divisions – from the north, Trento (General Masina), Bologna (General Gloria), Brescia (General Brunetti), Folgore (General Frattini) and Pavia (General Scattaglia). These Italian divisions were interleaved with the Germans, in some cases at regimental rather than divisional level, so that there were three German–Italian 'pairs'; from north to south 164th/Trento, part of Ramcke's brigade with Bologna, and the other part with Brescia. In the south Folgore and Pavia were untwinned. Rommel was short of German infantry. The northern half of the front, from the sea to south of Ruweisat ridge, was under the command of General Navarrini's XXI Corps; the southern half of General Nebba's X Corps.

The armour was also to some extent twinned. Behind the northern part of the front, around ten miles south of the coast, were 15th Panzer (General von Vaerst) and the Littorio (General Bitossi) Divisions; while between fifteen and twenty miles further south were 21st Panzer (General von Randow) and Ariete (General Arena) – the latter regarded by the Germans as the most reliable of the Italian divisions, the bravest and best of comrades. Behind the left wing and near the coast were 90th Light (General Graf von Sponeck) and the Trieste motorized division (General La Ferla), in reserve. A total of five hundred (two hundred German) tanks confronted an enemy estimated (correctly) to have around one thousand.

The minefield area was several miles deep, and to break through it was always, Rommel reckoned, going to take the enemy a lot of time – time during which his own armour could concentrate opposite any threatened breach. The minefields were covered by forward positions, outposts, while the main defensive line was sited between one and two thousand yards west of the minefields and was itself laid out in considerable depth – between two and three thousand yards. Every position was prepared for all-round defence – Rommel knew that there would, at the least, be enemy penetrations, probably in considerable strength. Reconnaissance patrols and dogs were deployed to give warning of approach to the minefields.

The Panzerarmee's intelligence of the enemy's strength was accurate overall, in terms of numbers. Identification (and placing) of enemy units and formations was also not far from accuracy, although there were defects of no great significance. British radio security had been good and revealed little. German air reconnaissance had been scanty. A few prisoners had been taken on patrol during October and had marginally filled out the enemy order of battle; there were few surprises to come on that score. Nor was there surprise, although there was consternation, at the intensity of the bombardment when it started.

The bombardment had been preceded by several days of increasing air attacks, and now it had, in many cases, its planned effects of neutralizing the German artillery response by counter-battery fire, and of destroying communications. To the Germans it was as the British experience of the opening of the great German offensive on 21 March 1918 – a pulverizing bombardment, aimed at the nerve-centres and response capacity of a defending army.

At 9 p.m., German time, the assaulting divisions advanced, their mine-clearing parties well organized, their infantry designed to assault the German forward positions and assist the path of the British armour towards what must be their objective – open desert. As the hours passed it became clear that the main enemy thrust, at least initially, was in the area held by 164th and Trento Divisions, the area north of the Miteiriya ridge. The attack seemed to be coming in on a comparatively narrow front, in a darkness choked with sand and smoke. As the infantry of 51st Highland Division* (General Wimberly) advanced their pipers played them forward; just as von Bismarck had paraded his divisional band to play 21st Panzer towards the start line at Alam Halfa two months previously, with the stirring marches of German history. The conduct of war is eased, even inspired, by the trappings of war.

At daybreak on 24 October the enemy were still well within the defended area; there had been no breakthrough, no break-out. But

* Reconstituted since meeting Rommel at St Valéry in 1940.

a number of Italian units were reported to be in a state of panic, two infantry battalions of the German 164th Division in the north had been, it appeared, almost destroyed, and the enemy's deep penetration was on a frontage of about six miles. In the far south, too, a major attack was reported and partial penetration of the minefield. Montgomery had carried out deception measures of great thoroughness to mislead the Germans as to his main thrust's direction, and these were at least partially successful. With the general situation obscure the Panzerarmee Commander, Stumme, had driven forward towards 90th Light Division headquarters to try for a better picture of what was happening. Stumme, corpulent, experienced, was not in the best of health. He was next day discovered dead – apparently from a heart attack during a British air strike. In and behind the minefields, in the sector of attack, there was a turmoil of fire and confusion but the Panzerarmee's line appeared to be holding. Just.

It was after dark on 25 October, the second complete day of battle, when Rommel arrived at the headquarters of the Panzerarmee. He had flown to Rome that morning and been greeted by the German liaison officer, von Rintelen, at eleven o'clock. Von Rintelen had told him that the Panzerarmee's fuel situation was again critical – another tanker had been sunk on 20 October with 1650 tons. It appeared that there were only three days' fuel with the troops, at anything like combat rates of calculation.* The situation was significantly worse than Rommel's predictions when he left, but beyond the customary fulminations to Rintelen (who had himself been away and was in poor health) there was little he could do. Rommel had flown on to Crete and been greeted at three o'clock in the afternoon by General von Waldau, *Fliegerführer Afrika*, who gave him the latest news. Stumme's body had now been found. General von Thoma, the newly arrived commander of the Afrika Korps,† was temporarily in overall charge.

Rommel had flown on to the Qasada airfield in North Africa, had climbed into his Storch and flown eastward until dusk, then landed

* This statistic varies with different accounts, including (retrospectively) Rommel's. 'Three days' fuel' is certainly an exaggeration of the seriousness of the position. The Panzerarmee daily report on 19 October, since when there had been only one intensive day of battle (and little manoeuvre), described the fuel situation as likely to last eleven days. A signal to von Rintelen on the same day, however, cited three days. Von Rintelen was not unused to exaggerated cries of pain.[2]

† He had joined a few days before Rommel's departure to Europe.

and transferred to his car. He reached Panzerarmee headquarters and at half past eleven that evening sent a signal to all: 'I have taken command of the army again. Rommel.' It was a gesture clearly intended to give confidence, and it very probably did. Rommel was undoubtedly, however, still a sick man.

Rommel was briefed by Westphal at eight o'clock in the evening. He reckoned from the Panzerarmee situation map that the enemy must be on the point of break-out, and that the most threatened sector was around a feature marked as Hill 28,* two miles east of Tel El Aqqaqir; at an equivalent distance west of Tel El Aqqaqir the two northern armoured divisions, Littorio and 15th Panzer, had been deployed. A few miles south-east of Hill 28 was the northern extremity of the Miteiriya ridge, a ten-mile-long feature running north-west to south-east which was also, it appeared, in enemy possession. During the day (25 October) there had been repeated attacks by 15th Panzer on the enemy lodgement at Hill 28 and at enemy tanks seeking to advance west of Miteiriya – both positions being inside the main defensive area although not far from its western crust. In these attacks 15th Panzer had lost heavily; Rommel learned that it had only thirty-one tanks left serviceable from an original total of 119. This was indeed a *Materialschlacht*. At midnight Hill 28 was reported in British hands.

Rommel's immediate concern was to bring every armoured unit he could against the British in the north-central region – Hill 28 and Miteiriya. Part of the orders for the conduct of the defence had been concerned with counterattacks, planned on a contingency basis, particularly in the coastal sector and either side of Ruweisat as well as at Deir El Munassib in the south.[3] The critical sector was now at least ten miles north of Ruweisat; Montgomery's deception measures and British security had worked well in this regard. Rommel knew very well that assembly and attack would be difficult and costly – the enemy air forces were bombing round the clock, their attacks being assisted in darkness by pathfinder aircraft dropping flares. The enemy, however, had established a salient in the north-centre of the front and to deal with it needed every counter-punch he could produce. On his first day back in command, 26 October, Rommel personally directed every weapon system he could assemble from 15th Panzer and Littorio to the threatened sector; and ordered 90th Light Division to march south-east to the same area.

* The British knew it as Kidney Ridge.

To Rommel the day was one of unremitting battle as fire was exchanged around Hill 28 and west of Miteiriya, and as the enemy air forces took their toll and the enemy artillery pounded away. To Montgomery it was a moment for reflection. His main thrust was being held – an unpalatable fact to which his own command arrangements, involving two corps in overlapping responsibility, had significantly contributed. He accepted, now, that British armour would find it intolerably expensive to try to push much west of the Miteiriya ridge. He determined, therefore, to hold it defensively; to strengthen and enlarge his position at Hill 28, both westward and north-westward; and to prepare a new attack northward, from the same general area, for the night of 28 October.

Rommel, as yet unaware of a change of enemy thrust line, still had his eyes on the centre of what had become the main battlefield – the northern half of the front. He reckoned that the enemy's attacks westward had been halted but would certainly be renewed, and his own reserves for counterattack were weak indeed. Aware that he was denuding the south of mobile troops, and that if an enemy *Schwerpunkt* were now ordered in the south he would probably have insufficient fuel to move them back again, he ordered 21st Panzer Division to march north on 26 October.

Very early next morning, 27 October, enemy armour began attacking south-west from Hill 28. This was the sector Rommel felt most threatened, and he reacted with local counterattacks – comparatively short jabs – by elements of all his armoured divisions, he himself orchestrating the battle. These enemy attacks were in fact part of Montgomery's attempt to enlarge his position round Hill 28, operations carried out by the armour of his X Corps (General Lumsden), pushing west and north-west to the next ridge beyond Hill 28, aiming to establish an anti-tank screen (in darkness) through which further British armour had been ordered to advance just before dawn. They were not, however, the principal operation with which Montgomery was now concerned – that would be his northward attack from Hill 28 towards the coast the following night, 28 October. The attacks to which Rommel responded with such desperate energy – and with some success, although his armour, halted by enemy anti-tank guns round Hill 28 during the afternoon, could not advance far to the east – were to an extent a probe. They might have led to break-out, but Montgomery doubted it. For that to come there had to be more fighting in the Panzerarmee's main defensive position itself, more of what Montgomery had called 'crumbling'. By nightfall on the twenty-seventh he had given up the idea of further westward advance from Hill 28. Rommel, for a brief while, had stopped him.

But Rommel was now having to commit elements of his armoured divisions to defensive positions, where there were clear gaps in the line through which an enemy might march unopposed. No one could conceive, he wrote to Lucy that day, the burden that lay on him. And the next day his letter ran: 'Should I never leave this place I want to thank you and our son for all the love and happiness of life.'[4] He was convinced that the scale of the crisis was not yet appreciated by OKW – he knew it was not by Commando Supremo – and he considered sending a staff officer, Berndt, a Party man with the appropriate entrée, to report to the Führer personally. Rommel still cherished the idea that if only Hitler could hear from him directly he would understand.[5]

Yet another tanker, the *Proserpina*, with three thousand tons of fuel, had just been sunk. In Rome Mussolini told Cavallero that the problem of fuel for the Panzerarmee 'gnawed at his liver, day and night',[6] and there is no reason to doubt him. The mood in some quarters in Rome, however, was sceptical of Rommel's difficulties. Kesselring and Goering were both there, the latter speaking of Rommel as over-influenced by passing events.[7]

At nine o'clock on the evening of 28 October the battle had been going on for five furious days, and the Panzerarmee heard what sounded like another concentrated artillery barrage west and north of Hill 28. This was the precursor of Montgomery's next main move – an attack northward by 9th Australian Division (General Morshead), aiming to smash the northern part of the Panzerarmee's line. It was to be followed by a westward attack, probably two nights later, along the coast. These two consecutive operations would, it was reckoned, open a major breach in the extreme north of the front.

Rommel assessed that the battle's centre of gravity, obviously in the north, was likely to remain there. He now had virtually all of his own remaining offensive strength also in the north.* The effect of enemy air superiority was as awe-inspiring as he had predicted, and the Luftwaffe were barely able to fly. Throughout the first day of the battle the Royal Air Force had, without interference, actually maintained fighter patrols over every German forward landing ground. And now, following the barrage, Rommel learned of yet

* Except Ariete.

another formidable enemy night attack – he had always admired the enemy's use of infantry at night, and these were Australian infantry, for whom he had conceived an especial regard. He soon heard that the enemy's attack appeared to be going well. A German battalion of 164th Division and an Italian Bersaglieri battalion seemed to have been overrun. During that day he had seen an order – notorious as the 'Commando Order', the *Kommando befehl* – which stated that members of enemy raiding forces captured behind the lines were not to be accorded prisoner-of-war status: they had used the methods of gangsters, and were to be summarily executed. Rommel burned the order in the presence of his Ia.[8] The time would come when he would draw the correct conclusions from such orders; meanwhile they were simply to be ignored and destroyed.

A few hours later, early in the morning of 29 October, Rommel was pacing up and down his headquarters dugout in an agony of mind. He knew that very soon now the whole Alamein position would be untenable. He signalled Cavallero that if he had an urgent delivery of supplies and reinforcements of some six thousand trained and equipped men he had a chance of holding, but he knew that this, modest though it was, was whistling in the wind.[9]

For although Montgomery had nowhere broken clean through, had nowhere been able to launch his tanks into open desert where fuel and numbers must give them the victory whatever the balance of tactical skill, the defences were being inexorably ground to pieces, and this latest Australian attack was but the most recent example. The issue could only be a matter of time, and probably very little time. The breach might come, as now seemed likely, in the far north whither the Australians were fighting their way. Or it might come at the earlier point of maximum pressure, west from Hill 28 and Miteiriya; but come it must. There was insufficient strength to halt it. This was a *Materialschlacht*, and the enemy's materiel and numbers were by a good margin the stronger. And even if fuel and tank numbers and terrain were to admit of some operational counter-stroke behind the minefields, the air situation did not. This was a battle of attrition, of destruction, of wearing down, exactly comparable to the battles of the Western Front in the war of 1914–18. Montgomery, like Haig on the Somme or at Third Ypres, had assembled a huge quantity of men and guns and was pounding and stabbing at his enemy until part of the front collapsed and he could hope to move through to operational victory. Like Haig he was persevering, in face of what were clearly heavy losses. Unlike Haig, it looked as if that perseverance would actually result in a breach, so great was the disparity and ability to replace human and equipment losses between the two sides. And, unlike Haig, Montgomery possessed the means for rapid and sustained exploitation and pursuit; for decisive victory, if he had the energy and boldness to grasp it.

Rommel knew that he must, somehow, prepare for withdrawal. Withdrawal must be carried out when pressure became so great that enemy breakthrough was imminent, but before it had actually occurred. It would be difficult and in some cases impossible – he had insufficient transport to lift his Italian infantry divisions on wheels, and he hated the thought of abandoning them. He felt great loyalty to his Italian soldiers, and he knew how bravely they had often fought, against considerable odds, in circumstances wherein few soldiers could accomplish much, and in a cause for which, as was generally known, not all had enthusiasm. Now many would become prisoners of war and the responsibility was his.

Withdrawal could not be allowed to lead to mobile warfare – that, Rommel's natural medium, was made impossible by the constraints upon him and by the British command of the air. Instead, it must be rapid and direct to a new and reconnoitred position. He decided to have such a position inspected at Fuka, sixty miles behind the Alamein line. At eleven o'clock on 29 October Rommel heard that yet another tanker, the *Louisiana*, sent out as replacement for the destroyed *Proserpina*, had been sunk. The lifeblood of the Panzerarmee was being cut off.

The Australian northward night attack of 28 October had gone slowly, with success in some areas, confusion, loss of direction and casualties in others. For Montgomery this was disappointing. Australian success was a necessary prerequisite to his next main move, a westward attack parallel to the coast by General Freyberg's New Zealand Division. Nevertheless Montgomery decided to persevere with the Australian attack during 29 October. It was a part of those 'crumbling' operations which he still saw as necessary if a real and unsealable breach were to be made in the German line. In a *Materialschlacht* the will of the commander to maintain a policy, regardless of the misgivings of subordinates, often in spite of casualties, is very important, and Montgomery knew it. His will was strong.

During the evening the staff of the Panzerarmee suffered a mild distraction, one of those inexplicable pieces of false information which cause a large and temporary stir at certain moments of most wars and which are more entertaining in retrospect than at the time. A message arrived from the Italian Commando Supremo at six o'clock. Two British divisions had been identified by radio interception, moving westward through the Qattara Depression, far to the south! They had reached a point one hundred kilometres south of Mersa Matruh – many miles behind the right shoulder of the Panzerarmee. Rommel's Ic observed that this seemed improbable, and – flying one of their too-occasional and inhibited sorties – the Luftwaffe were soon able to report that it was demonstrable nonsense.[10]

Rommel had already arranged for the Fuka position to be reconnoitred. He now ordered 21st Panzer Division, which had been plugging holes in the front, to be replaced by the Trieste motorized Division during the night of 30 October so that he had 21st Panzer manoeuvrable and in hand. His eyes were still on the northern sector, sensitive to a massive attack towards and then along the coast, an operation which Montgomery had intended but, unbeknown to Rommel, had now switched. Montgomery had made a new plan. He had, once again and skilfully, decided to vary his main point of attack.

When Rommel heard, on the morning of 31 October, that British tanks had actually reached the coast road he drove there himself and, in the old style, personally organized a counterattack at midday by elements of 21st Panzer and 90th Light Divisions. This met stiff opposition but a little later managed to drive the enemy south of the railway line which runs parallel to coast and coast road. Rommel now had 230 tanks left, ninety German and 140 Italian. He estimated – with some accuracy – British residual strength at eight hundred: the odds were worsening. His position was tenuous in the extreme, the Alamein line was now nothing but a thin and uneven crust and the *Materialschlacht* balance sheet was moving inexorably against him. The Fuka reconnaissance had been satisfactorily completed.

The British named the operation which was to be the culmination of the battle of Alamein Operation 'Supercharge'. This was Montgomery's new plan. The attack had been planned to take place during the night of 31 October but had been delayed for twenty-four hours. It came in at one o'clock in the morning of 2 November – another night attack, preceded and accompanied by another formidable artillery programme, delivered on a four-thousand-yard front by the New Zealand Division with an armoured brigade and reinforced by two brigades from 50th and 51st Divisions. The attack was intended to secure infantry objectives, clear, at last, of the German main defensive area; and these objectives were to be reached in less than three hours. Thereafter the British armour, having formed up behind an anti-tank screen deployed by the successful infantry battalions, would pass through towards the Rahman track and Tel El Aqqaqir; to open desert. The westward thrust line of Supercharge was immediately north of Hill 28, scene of the opening phases of battle. It would strike remaining elements of 164th and parts of the Trieste Division. Montgomery had responded to new and accurate intelligence and had shifted his weight southward from his earlier intended line.

Rommel had spent the preceding day, 1 November, in the northern sector, stiffening the resistance and leading counterattacks against the Australian breakthrough northward towards the coast. During the night he only received confused and uncertain reports, but by dawn he had a tolerably clear picture of where the main threat was now coming: once again from the area of Hill 28. He determined on a counterattack as early as possible against what was now a west-pointing British salient created by Freyberg's infantry advance; it should be a concentric attack by the Afrika Korps – 21st Panzer from the north and 15th Panzer, supported by what tanks could be assembled from Littorio and Trieste, from west and south-west. Rommel also gave orders for the Ariete Division to move north. The southern sector of his front was now entirely denuded of armour. On that day he was angered by a report that an advanced dressing station had been bombed despite prominently displayed red crosses, and ordered that captured enemy officers would be regarded as hostages against such behaviour, the fact to be relayed to the British;[11] such exchanges of barbarity were uncommon in the Desert War, and indicate the strain on the generally and outstandingly chivalrous Rommel.

Rommel's anti-tank line west of the British salient had managed to inflict a good many casualties on tanks already seeking to advance westward – armour ordered to 'pass through' the positions secured by the Supercharge infantry advance. His counterattack against the salient was mounted at eleven o'clock in the morning of 2 November – ULTRA had kept Montgomery informed and he believed, and passed the information on to Freyberg, that it would be mounted at half past nine. Rommel's attention had been momentarily distracted by a signal from Commando Supremo with information of a major enemy landing from the sea 'somewhere behind the front', but his main preoccupation was with the British salient and his efforts to smash it.

Rommel's tank strength was now seriously reduced, and by evening would amount to only thirty-five in the Afrika Korps; by then he knew that although he had inflicted considerable casualties on the British there was no question of having eliminated their salient or seriously frustrated Montgomery's will. In seeking even a local decision the counterattacks had failed. The British would replace their tank casualties. The line of the Panzerarmee's defensive positions, which had been, somehow, holding since 23 October, was finally breached with the loss of the defended ground north-west of Hill 28, now in Freyberg's hands. The Afrika Korps reported that the Littorio Division was panicking and no longer under control; and similar reports were coming in about the Trieste.[12] Rommel's thin line of anti-tank defence along the Rahman track could be broken whenever Montgomery chose to make a further coordinated westward move. It was, Rommel knew, the end.

Rommel had been thinning out rear elements of the Panzerarmee during the previous forty-eight hours. He knew, now, that the moment had arrived which he had for several days recognized as inevitable. There could be no sustained defence for more than a few hours. His duty was to save what he could. He was pessimistic about how much this would amount to, and that evening, 2 November, he signalled Commando Supremo that only part of the mechanized formations of the Panzerarmee would be able to disengage and that it was improbable much of the Italian infantry, for lack of transport, would avoid destruction. He knew that there were likely to be exaggeratedly optimistic estimates of the situation in Rome – and perhaps in Rastenburg, where Hitler was in his War Headquarters and would learn of the signal; but his way was plain. During the night of 2 November German and Italian infantry started marching west.

Next morning, Rommel took stock. He had in the Afrika Korps about thirty tanks. He hoped that if his armour withdrew as slowly as possible, contesting the ground against enemy frontal advance, avoiding attempts at encirclement, he could sufficiently delay matters for at least some of his infantry divisions from the line to be got clear away. Had he been in Montgomery's shoes, with Montgomery's reserves and knowledge, Rommel would probably have already shifted his entire armoured weight far southward on learning of the enemy's denuding of the southern front; and would have then attempted to force another, final, breach in the defenders' line in the south and lead a great concentrated march north-west towards the coast, a battle of encirclement, a Cannae. Montgomery had originally envisaged such a dénouement, but now he had no mind for such manoeuvres. He was fighting the sort of battle in the north he had planned, albeit not exactly as he had planned it. The balance of forces was moving ever further in his favour. He was grinding his enemy to pieces, and there would be no battle of encirclement.

But for Rommel the battle was nevertheless lost. At half past eleven he signalled to OKW direct. During the preceding twenty-four hours, he now said, the enemy had attacked with four or five hundred tanks and had broken into the Panzerarmee's positions to a depth of fifteen kilometres, on a ten-kilometre-wide front. He said that his losses were such that to hold a coherent defensive front was no longer possible. The Italian troops were no longer battleworthy and their infantry were giving up their positions without orders.

Rommel signalled that he was perfectly aware of the strategic importance of holding the position at Alamein, but that the only hope of delaying and punishing the enemy, of making him pay dearly for ground won, would lie in mobile operations (he meant, of course, rapid withdrawal), leading to the adoption of a new position running south from Fuka. He requested approval of his intentions. He reckoned he just had enough fuel to get the army back to Fuka.

The signal Rommel received from Hitler at 1.30 in the afternoon of 3 November was not in reply to this request for approval of his intentions. It was the response to his earlier, and equally if not more pessimistic, signal to Commando Supremo – which had been read not only in Rastenburg but in London, thanks to ULTRA. But this signal of Hitler's on 3 November marked a turning point in Rommel's life and views.

Hitherto Rommel had apparently been trusted by Hitler. When he had advocated boldness, acceptance of risk for high stakes, Hitler had backed him. When he had appreciated that, temporarily, Cyrenaica must be given up in the preceding winter, despite Italian protest, Hitler had backed him. He had had, in the past, long and intimate conversations on military matters with Hitler and he knew that Hitler admired his judgement, was confident of his loyalty, shared much of his military philosophy. Now he had told Hitler that unless the Panzerarmee was withdrawn from a murderous *Materialschlacht* it would be destroyed and North Africa lost – and Hitler had replied that 'Not a step was to be yielded.' This was, as Rommel understood it, a flat rejection of his operational advice and a death warrant signed on his army. It was also, unmistakably, an order, a *Führerbefehl*. 'As to your troops,' Hitler's signal had ended, 'you can show them no other road than that to victory or death.'[13] The signal was read with interest by Montgomery.

Rommel spent twenty-four hours of indecision, hours with which he afterwards bitterly reproached himself, and with which Kesselring (who visited him early next day) in effect also reproached him. That evening of 3 November he walked up and down alone in the desert, wretched, until Westphal sensibly ordered a staff officer to go and join him, to keep the *Oberbefehlshaber* company. Rommel talked without restraint or discretion. He said that if the Panzerarmee remained where it was it would be totally destroyed in three days. Hitler, Rommel said frankly, was a lunatic, determined from sheer obstinacy on a course which would lead to the loss of the last German soldier and, one day, to the total destruction of Germany.[14]

Next morning Kesselring – and it was helpful to have a second opinion in such a dilemma, albeit belatedly – told Rommel that he ought not to regard Hitler's order as binding in detail. Kesselring knew that Rommel's stock stood high with Hitler, that Rommel exerted an almost hypnotic influence on the Führer.[15] Kesselring stiffened Rommel at that critical moment to an act of disobedience. Fortunately the enemy still appeared deliberate rather than dynamic in their advance. Something like a Panzerarmee line was still holding along the Rahman track – a line on which, Rommel was told, two hundred enemy tanks had been checked by twenty of the Afrika Korps at eight o'clock that morning.

Rommel was persuaded. He had already had Berndt flown to Germany with representations, a step he had been considering for several days. Berndt, he hoped, could through contacts and intermediaries somehow represent the truth of the situation more convincingly than Rommel's signals had apparently been able to do. Rommel later supposed that propaganda – the needs of public confidence and public presentation – had inspired the *Führerbefehl*, but however that may be he had handled it with wholly uncharacteristic equivocation. He had signalled to Hitler again, and had then passed the order – 'No withdrawal' – to the Afrika Korps and told von Thoma it must be obeyed. Von Thoma had said: '*Minor* withdrawals are presumably permitted?' (they had, of course, been going on for some time); and Rommel had agreed, while saying shortly afterwards that the Panzerarmee must 'stand fast'. But his staff heard him say aloud, 'The Führer must be a complete lunatic.'[16] And with a very untypical disingenuousness Rommel ordered that, while the rearward movement of transport must cease, only sufficient men to drive the vehicles should be retained at the front.[17]

But now Rommel was persuaded. To disobey or disregard a *Führerbefehl* was wholly unnatural to him. Despite his frequent backing of his own judgement, his straining to get his own way, his life had been based upon obedience – reasoned obedience, obedience after fully representing the facts, obedience in awareness that the authority giving the order realized as much as he of the situation, but obedience nevertheless. And the Führer was Supreme Commander of the Armed Forces of the Reich, armed forces facing perhaps the most dangerous moment in their history, on all fronts. At half past three on the afternoon of 4 November, nevertheless, Rommel ordered the withdrawal of the Panzerarmee and informed OKW. In Rome the authorities had no sort of comprehension of the situation on the ground, Mussolini sending a signal to Rommel on 2 November: 'Duce considers it imperative to hold present front at any cost,' and Cavallero remarking to Mussolini (on seeing Rommel's communication) on 4 November, 'If Rommel withdraws the army is lost.' [18]

Rommel had received an Afrika Korps report, two hours earlier, that enemy tanks had finally broken through in the centre of the contested front. He had heard a British radio signal earlier that day that von Thoma had been captured. There was little left of the Afrika Korps and the Ariete had been virtually destroyed. Soon the British would be reported as streaming westward through the breach. The battle of Alamein was over.

Later that evening Rommel received Hitler's approval of his decision. Kesselring – courageously and honourably – had himself telephoned Hitler's headquarters. The Panzerarmee was now withdrawing. It would clearly be in deadly danger of envelopment by a swift and resolute pursuit.

On a larger map Rommel's command was menaced by a new and enormous development within a few days of defeat at Alamein. In Rome, next day (where Mussolini, not unnaturally but with futility, was 'insisting' that every effort be made to take back the Italian infantry), the dominant subject of conversation was a huge Allied convoy detected at Gibraltar. The Italians derived at least a second opinion on the significance of this from their regular interception of conversations between Kesselring in Rome and Goering in Berlin, and reckoned the convoy might be heading for French North Africa (thought by Goering unlikely), for Italy or Corsica, or for Tripolitania. The convoy was shadowed, Hitler describing it as clearly the vital point of effort. Spanish sources suggested a division of the convoy between North Africa and an actual invasion of Italy. Tunisia was always, clearly, a possible destination and it was agreed between Italy and Germany that in that case reaction must be instant, but must initially depend on French attitudes.[19]

In the event, on 8 November British and American troops began landing in considerable strength on the coast of French North Africa, astride Algiers. It was clear to Rommel that this must mean the end of the army in Africa.

Rommel's withdrawal of his forces from the Alamein line and his subsequent march to Tunisia along the North African littoral was bitter for him, but it was extraordinarily successful. Inevitably a great many Italian infantrymen, without transport or hope, were taken prisoner, a matter of understandable remonstrance by Cavallero to Rintelen in Rome:[20] but the Afrika Korps, now under Bayerlein's command,* still retained the skeleton of its formations and its command structure, and by energetic driving of the remnants, day and night, Rommel managed to assemble a small force on the Fuka position. He had no intention of offering serious battle there. His plan was to constitute some sort of rearguard – for most of the march 90th Light Division – and to deter the enemy from seeking to overtake him directly. Against an enemy move round the desert flank to entrap him he could oppose little armoured or anti-tank strength. His tank state on 4 November when the great retreat began was about thirty German and some ten Italian. There could be no question of seriously undertaking mobile operations, however limited and defensive; and his fuel situation allowed little more than a direct move by the shortest route to the rear. Even that was frequently halted for uncomfortably long periods while fuel was awaited.

But Rommel got away. He evaded his pursuers, and he saved a remarkable amount of the German–Italian army from destruction. What started as a jumble of disordered columns, with units mixed, unguided, incapable of combat, with men crowded on to vehicles wherever space could be found, and with a large number of men from the southern sector of the Alamein line marching westward

* Temporarily. General Fehn took over command on 19 November.

across the desert – what started as dangerously near a panic-stricken rabble in the aftermath of defeat had had order restored to it by the time the Libyan frontier was reached on the afternoon of 6 November. There were a few welcome surprises. Ramcke's parachute brigade, although without allocated transport (and loudly indignant at the fact) had ambushed a British column, grabbed their vehicles and got away in them. Around Sollum Rommel took stock and found that he had about 7500 men, of whom five thousand were German, now only about twenty-one tanks, thirty-five anti-tank guns, sixty-five pieces of field artillery, and twenty-four anti-aircraft guns. He was facing the entire strength of the British Eighth Army and he knew it. He estimated the pursuit force as probably consisting of about two hundred tanks and as many armoured personnel carriers, while British armoured cars would be ranging the desert flank.

But he got away. He gripped the withdrawal as hard as he was used to grip every advance. When traffic chaos threatened at such bottlenecks as Halfaya he personally hurried a number of officers to establish traffic control points and impose ruthless authority. He shepherded fleeing columns of supply vehicles into areas where they could be organized and nerves calmed. He aimed to pause at Tobruk where there should be supplies, to hold the pursuit off there as long as possible in order to load up with those supplies, and thereafter to go back by the route he knew so well – Gazala, Benghazi, Agedabia, Mersa El Brega. He knew he could hold for a short while at the narrows of Mersa El Brega, perhaps receiving reinforcement from Tripoli. But he also knew that he would not for a very long time, if ever, recover the strength for any sort of counter-move. Alamein had near torn the heart out of the Panzerarmee.

Yet it got away. In spite of ceaseless air attacks, a chronic shortage of fuel, a large number of vehicles on tow, a pitiful strength compared to its pursuers, it escaped annihilation. The principal cause lay in the caution of its enemies. Montgomery, Rommel wrote, risked nothing: bold solutions were completely foreign to him.[21]

The comment on Montgomery, frequently repeated, has been challenged.[22] The timidity of Montgomery's subordinates has been blamed for the half-heartedness of the pursuit; the weather has also been accused – certainly both sides found progress hampered by heavy downpours from the night of 6 November. But, whatever the cause, to Rommel the lack of dynamism in the forces following him was very evident, and it became a factor in his planning. He recognized Montgomery's astuteness, careful preparation, determination not to court setback. He referred to him as 'the Fox'. He realized, with some bitterness, that Montgomery would never overreach himself, would never expose himself to a counter-stroke, was probably exactly aware of his opponents' weaknesses and constraints. But enveloping movements by the pursuers were invariably too limited and too tardy; they failed to trap; and the remnants of 90th Light Division fought off direct pursuit again and again. Rommel evacuated Tobruk on 12 November, and his leading elements were at Mersa El Brega the following day.

For Rommel the period of the withdrawal – ultimately to Tunisia – acted as a catalyst to his own reflections: about the North African campaign, about the strategic direction of Germany's war, and about the leadership of the Reich.

On 10 November German troops, airlanded, began what would soon be a major reinforcement of North Africa through Tunis. On 11 November Rommel asked Cavallero and Kesselring to visit him. Since the bruising defeat of a week earlier, and particularly since the Anglo–American landings in North Africa on 8 November, he had been thinking hard, and he believed that there was need for strategic guidance from the highest level. What was the Panzerarmee, insofar as it existed at all, now seeking to achieve? What was the object of the North African campaign, now that the whole might of American industrial and amphibious power could be deployed across the Atlantic sealanes, and must master whatever could be opposed to it unless the effort to defend North Africa were to be immense – and surely immensely out of proportion to its strategic importance? What, now, *was* its strategic importance? Cavallero and Kesselring declined yet to confer with him,* and Rommel, once again, sent the faithful Berndt flying back to Germany to seek an audience with the Führer, which Rommel felt was absolutely essential.

Berndt's mission was a total failure, and Rommel gloomily heard from him that Hitler had dismissed his anxieties with irritation. Hitler had been determinedly optimistic. Following the withdrawal closely, he had spoken of Mersa El Brega as being the 'springboard of a new offensive'. It was clear that Hitler, in the past so trusting of Rommel and his judgement, so sympathetic, so approachable, had come to believe that setback had turned his favourite general into a

* In fact Cavallero flew to Tripoli on 12 November and was in North Africa for three days, but did not see Rommel.

pessimist, and a pessimist worrying about the concerns of others. Hitler had told Berndt that Tunisia was nothing to do with Rommel, who should simply assume that the Tunis bridgehead 'would be held'. '*Der Führer bittet den FeldMarschall,*' Berndt reported Hitler as saying bitingly, '*Tunis ausser Betracht seiner Berechnungen zu lassen.*'*[23] There were messages, nonetheless, of Hitler's especial confidence in Rommel.

Rommel persisted in his angry scepticism. He knew that when Montgomery moved up supplies and advanced, moved them up again and advanced again, sooner or later the Panzerarmee would once more be overwhelmed. The reinforcements in men and materiel which he could hope for would never suffice to conduct a protracted defensive battle, another *Materialschlacht*, with Eighth Army. Still less were there likely to be the resources for mobile operations. The day might well be delayed, but sooner or later Tripolitania itself would have to be evacuated. There could be no question of a sustained defence when they reached Mersa El Brega; and ideas of a riposte, another offensive as in January, were fantasy. There were new players in the game and the rules had changed. Montgomery was a man who would attack when, and probably only when, he had the strength and the deployment which would enable that attack to be sustained and fed and fed again until it prevailed.

In Rommel's judgement the Panzerarmee should now withdraw to Gabes, west of the Tunisian border, where the terrain would constrict manoeuvre and make defence against superior numbers practicable. He believed he could for a while hold at Gabes. It would, furthermore, take Montgomery some time to move up the sort of supply train and strength that he would demand for an attack, so that if Rommel could take the Panzerarmee the whole way back to Gabes he would not only gain a significant amount of time but could unite with the German forces now building up in Tunisia (and receiving the lion's share of all supplies routed to Africa, Rommel noted). Once united, the German–Italian army could perhaps aim to deal a crippling blow at the Anglo–Americans in the west of Tunisia, and then turn in greater strength to face Eighth Army, having eliminated the threat to their western flank. All this suggested concentration earlier rather than later.

But even this, Rommel believed, should itself be regarded as no more than a holding operation. It should buy time, avert annihilation

* 'The Führer requests the Field Marshal to leave Tunis out of his calculations.'

– not now of one but of two Axis armies; and it should use that time for the conduct of a fighting withdrawal to Tunis and an evacuation to Europe. For what was now the point of the North African campaign? Plan Orient, the grand dream of conquering the Middle East from the Caucasus and from Egypt was utterly dead; had died at Alam Halfa and been buried when Allied troops landed in French North Africa, when the Americans had started to play a direct hand in the game. And if Plan Orient was dead what were they all doing in North Africa? Italian possessions were now past protecting. The Reich was threatened, if threatened it was, from the east; and one day, perhaps, from the west. It was dissipation of force to try additionally to hold a southern front in a piece of Africa.

Such strategic appreciations were not, Rommel recognized, a matter for him. Other considerations undoubtedly entered the equation – the stability of Italy as an Axis partner; the natural instinct, particularly alive in Mussolini, that a front in North Africa provided a glacis for the defence of southern Europe itself; the possibility that French possessions in North Africa might ultimately be allotted to the Italians if Tunisia could be held with the help of a more wholehearted Italian effort; more operationally and immediately, the favourable effect on air power, and thus on the campaign in Tunisia and its supply, if Eighth Army could be held as long as possible east of Tripoli. To Rommel, however, such matters would be academic unless he could save the Panzerarmee. If he were, anywhere, ordered to stand as at Alamein, without freedom of action then, once again and tragically, he would be sacrificing his non-motorized troops, and probably others as well. Once again he was commanding some non-motorized Italian infantry divisions. Elements of the Pistoia, Spezia and 'Young Fascist' Divisions* – without troop-carrying vehicles – were arriving at Mersa El Brega to come under his command. Rommel envisaged another Alamein tragedy. He had, however, also received the reinforcement of the Italian Centauro Armoured Division (whose commander, General Pizzolato, he discovered, had fought against him at Longarone, a congenial circumstance).

Rommel was now withdrawing through Cyrenaica, watching the Jebel flank near Msus which he had so often exploited; noting the confusion in Benghazi, which was about to change hands for the fifth time in the campaign; overseeing the booby traps and demo-

* Under, respectively, Generals Falugi, Scattini and Sozzani.

litions, what he called the 'neat little surprises' prepared by the Panzerarmee's engineers;[24] all the time planning how to persuade his superiors of the obvious (to him) truth that to take the German–Italian army to something like a defensible position at Gabes would require not only correct but timely operational decisions. Although he was a sick man, his verve and humour had not deserted him. When the men of his *Kampfstaffel* were watching with him a film in late November (mobile film shows were frequently managed, with films flown from Germany) the newsreel, the '*Wochenschau*', unfortunately showed Rommel at the great rally in Berlin when he had addressed the press and said that the German soldier would never be turned out of Egypt. Rommel joined in the roars of laughter with which this was greeted by his 'Afrikaners'.[25]

On 20 November Rommel received a signal from Hitler. The Mersa El Brega position was to be held 'at all costs'.

On the same day Rommel visited a reconnaissance Battalion of 21st Panzer Division, the first unit to land at Tripoli in February 1941, led by Major von Luck, one of his old commanders from 7th Panzer Division in France. He had seen von Luck on 8 November in the immediate aftermath of Alamein, when Rommel had been in despair at what he openly referred to as Hitler's 'crazy order', an order which had cost him twenty-four hours and many lives; and on that day von Luck had noted that 'we all felt that we had to stand by our Rommel'. Now Rommel was with them again, and speaking with equal vehemence and equal indiscretion. His tongue was doubtless loosened by gloom at what he could see coming, another fearful conflict between obedience and duty to his men; and all for an authority in which he had been losing faith by the hour. The war was lost, he said on this occasion; Germany must seek an armistice.[26] He even spoke of the necessity to force Hitler to abdicate, of the necessity for a complete change of the policies with which the Nazis were identified, the persecution of the Jews, the absence of concessions to the Churches. These sentiments, dangerous to hold let alone to express, were new in Rommel's mouth: but perhaps not in his mind. He had spent many weeks at home before Alamein, and he had doubtless heard and pondered more than any record shows.

Nevertheless Rommel had been hitherto apolitical, a faithful believer in the Führer's genius, quintessentially a patriotic and uncomplicated soldier. Something had happened; and the 'something' was what Rommel was coming to perceive as the total unreality with which Hitler was surrounding himself, and the indifference

Hitler was showing to serious professional advice and to the fate of German soldiers. Hitler believed that he alone knew best, that his will alone sufficed to turn the tide of military events. If that had ever been true it was so no longer. For Rommel the sky was dark. And together with this darkening of the sky, this growing realization of how great now were the strategic odds against Germany, there was also – and it would burgeon – a realization that the actions of the regime in some other fields were vicious and criminal.

Rommel's loyalty, personal devotion and gratitude to Hitler were not yet, however, permanently annulled by the disastrous turn of events. He undoubtedly continued to believe that Hitler was open to persuasion, that he was the victim of miserable and pusillanimous counsellors, and of bad men. He wrote to Lucy on 15 November: 'May Almighty God help me in the coming year, as in the past, to justify the faith in me of the Führer and the German people.'[27] He determined, once again, to report personally. He had always been able to communicate with Hitler. Despite what he heard, despite Berndt's discouraging experiences, he managed to suppose that, given really frank and objective advice based on fact and experience, the Führer could be brought to see reason – his instinct for military and strategic issues had, after all, always impressed Rommel in the old and friendly days. Rommel decided he would fly to face Hitler.

Meanwhile what was left of the Panzerarmee was back at Mersa El Brega, and there Rommel saw Bastico, still Italian Commander-in-Chief in North Africa and now a Marshal, on 22 November; and received what he regarded as the overdue visit of Cavallero and Kesselring on the twenty-fourth at the Arco dei Fileni, the great triumphal arch on the desert road at the border between Cyrenaica and Tripolitania. There he argued his view of the strategic realities.

Rommel gave his hearers an unsparing account of Alamein and the shortage of resources, both troops and supplies, both on land and in the air, which had, he said, made the outcome inevitable. He then assessed the present situation. In two or three weeks* the enemy could probably concentrate an army at the front including over four hundred tanks: the Afrika Korps had thirty-five. His Italian subordinate commanders agreed with him – Rommel had conferred two days previously and they were of one mind: the troops could not withstand a strong attack.[28]

* The estimate was creditable. On 13 December, two weeks and five days after Rommel's meeting, Montgomery reported to the Chief of the Imperial General Staff,

Rommel said that behind Mersa El Brega – which, he reminded them, he had direct orders to defend 'at all costs' – there was no effective defence possible east of Tripoli. He would, he said, naturally do his best at Mersa El Brega, but they should not deceive themselves. If the army were to be preserved he must be free to withdraw it when truly necessary.

Kesselring – who had just received categoric instructions from Hitler himself to be 'responsible' for all replenishment matters in the Mediterranean (responsibility with only limited power attached) – remarked that the defences of Tunisia would take time to prepare, and he was concerned at the implications for that preparation, as well as for supply of Tunisia from Europe, if the Tripolitanian airfields fell to the British. Rommel again and again adverted to the operational points he thought inescapable – if the Panzerarmee stood overlong at Mersa El Brega it would be destroyed; and west of that place there was no position offering much strength until Tunisia, and the chance of concentrating with the German forces now building up there, was reached. To think in terms of a counter-offensive was fantasy. The enemy might, indeed, offer a chance of some fleeting tactical stroke at the front, but behind him was the inexorable strength now massed in Egypt or moving forward from it; and Montgomery was taking no risks. And nothing made sense unless there was a strategic object. The only coherent strategy was, surely, to concentrate the German–Italian forces on the most defensible ground in Tunisia and plan a fighting retreat to a bridgehead: and, although he did not say it on the record, a retreat followed by an evacuation in order to husband the greatest possible strength on the mainland of Europe. That was where the war would be decided. Furthermore, Rommel argued, no other operational policy had any chance of military success. United States power was now deploying on this side of the Atlantic and would remorselessly increase. All this meant that Rommel must be free to begin the withdrawal from Mersa El Brega when he decided, or the Italian infantry would suffer the same fate as their comrades, and end up in British prisoner-of-war camps. Rommel knew that he might be able to stand at Mersa El Brega for a little longer than he chose to do – but for what object?

Rommel realized that to Cavallero, whom he stigmatized as an armchair general, sufficiently intelligent but weak of will, and to

---

Sir Alan Brooke, that he was now logistically in a position to 'set about' the enemy at the Agheila position (Mersa El Brega).

Kesselring, who, he wrote, thought almost exclusively in terms of air power, of the battle for the Mediterranean airfields, his views were defeatist. He was, they imagined and would say, downcast by one defeat, a pessimist, insufficiently resilient. Rommel knew of this spreading view and it made him angry. He knew that to the survivors of the Panzerarmee his conduct of the withdrawal had been masterly; he was still Rommel. He reckoned that whether in victory or defeat his policy and his advice had been realistic and rational – based on a sensible assessment of factors, not on wishful thinking. He had advocated boldness throughout the long advance to Alamein and he had not only been justified but had been encouraged. If others had thought it right to forbid his onward movement they had certainly not done so and had been glad to honour his victories. And now, too, he advocated realism, as he saw it. There was a new situation and to face it objectively was not defeatist but honest. He realized that Kesselring and Cavallero supposed he had already determined not to attempt much delaying action in Tripolitania, but to reach Tunisia as early as possible. Why not? He, not they, knew how to save and fight the Panzerarmee.

In fact, rather surprisingly, Cavallero had also come to the conclusion that the essential ground to hold was in Tunisia. If the Axis held Tunis they had at least the chance of dominating the Sicilian narrows, and of again operating eastward, if they chose, one day. If Tunis went, North Africa went. Cavallero's – and, indeed, Kesselring's – differences with Rommel were less on the strategic issue (although they did not yet share his conviction that North Africa itself was a lost cause) than on the operational question of the benefits a more protracted defence in Tripolitania could bring to the preparation of defence in Tunisia.[29]

The only outcome of these meetings was an order from Mussolini on 26 November in the same sense as Hitler's six days previously, to stand firm at Mersa El Brega. On the twenty-seventh – their wedding anniversary – Rommel wrote of his love and gratitude to Lucy, but also said, 'I fear the war is not turning to our advantage': justified words indeed. On 28 November he flew to see Hitler at Rastenburg, the Führer's headquarters in east Prussia, landing there at twenty past three in the afternoon.

For the first time, at first hand, Rommel experienced a different and alarming side of Adolf Hitler. When he left him he realized that Hitler was being governed by emotion and wishful thinking, and was surrounded by sycophants who would never dare challenge his fantasies. Hitler did not know the truth, and refused to accept it. This man was the supreme commander of millions of Germans and their allies, conducting a life-and-death struggle on the Eastern Front and threatened by a formidable conjunction of his enemies to the south and the west of the Continent. Yet he seemed beyond rationality.

Hitler gave Rommel an example of one of his notorious rages. Hitler's rantings have been, according to many witnesses, much exaggerated; in general he preferred to avoid confrontations, to defuse situations. On this occasion, however, he showed savagery. His first words were to ask how Rommel dared leave his theatre of command without Hitler's permission (in most circumstances a justified remonstrance, but now simply the preamble to a tirade). He stormed at Rommel, alternating between enraged shouting and terrifying silences.[30] He said that the Panzerarmee had thrown their weapons away, had not fought. He said that he, alone, had saved the Eastern Front in the first winter of the Russian war by his resolute orders to stand firm. It was simply such resolution that was required. He told Rommel that a strong bridgehead in Africa was a 'political necessity' and must be held regardless of cost, '*Koste es was es wollte*.'[31] In different mood he told him – an oft-told tale – that supplies would now be greatly improved. He ordered the Commander-in-Chief of the Luftwaffe, Reichsmarschall Goering, to accompany Rommel to Rome, to see the Duce and the Italian authorities, and to make really effective arrangements for better materiel support. Eventually he dismissed Rommel, curtly ordering him to wait outside, an angry and disillusioned man, for his orders.[32]

Rommel realized that Goering shared the current view that he was the victim of pessimism, had lost his spirit. He loathed Goering, whom he regarded as ambitious, unscrupulous, vain and uncomprehending. They travelled south in Goering's special train, stopping in Munich to pick up Lucy for a blessed relief from the pains of the moment and for a long private conversation. By the time they reached Rome Rommel had despaired of making Goering see what he thought of as sense. Goering appeared impervious to his explanations of the operational realities in Tripolitania. Rommel noted with contempt Goering's sole apparent interest at such a time – purchases for his art collections – and his disinclination to exert himself for Panzerarmee Afrika. He recognized Goering as an enemy.

At a meeting at Commando Supremo in Rome Goering took the chair and Rommel's arguments made no headway. Rommel attempted to affect optimism for the longer term. He said that if the Axis forces could be united in Tunisia and win a victory there it might conceivably be possible to operate eastward again thereafter, into Tripolitania. He recognized that to at least some of the Italian command the idea made, if not sense, better sense than standing and losing.[33]

But, at the end of the meeting, Rommel was still formally committed to holding Tripolitania. Kesselring, who met them in Rome for discussions, opposed Rommel's proposal to withdraw (and soon) the whole Panzerarmee to Gabes; he reckoned that to take the front back to west of Tripoli would increase the air threat to Tunisia, a view he had pressed at the meeting at Arco dei Fileni. To Rommel this was all fantasy – the front would be forced back by superior enemy ground and air power, whatever was ruled in Rome or Rastenburg. He managed, however, to get a more permissive instruction from Mussolini that afternoon, a Mussolini impelled by a desire to save the Italian infantry from the fate of Alamein. He could, Mussolini agreed, construct a rearward position at Buerat, east of Tripoli on the Gulf of Sirte. He could begin to move the Italian non-motorized formations back to it, despite Mussolini's earlier order. But – Goering's *caveat* – Rommel should also begin planning an *attack* eastward again from Mersa El Brega!

Rommel flew back to Africa on 2 December in deep dejection. His sense that his superiors were living in a world of make-believe would have been heightened had he known of a telephone conversation between Hitler and Field Marshal von Manstein, commander of the German 'Army Group Don', on the night of 29 November. It was already clear that at Stalingrad Paulus's Sixth Army was cut off by Russian counterattacks and would be almost impossible to supply or to rescue. To the south another Army Group – Army Group A – was deep in the Caucasus. Manstein protested to Hitler about the dangerous situation threatening this latter Army Group (not under his command) as well as the peril of Sixth Army, whose eclipse would complicate, to say the least, any ultimate withdrawal from the Caucasus.

'Field Marshal,' Hitler said, concluding the conversation, 'I must remind you of something I have already told you repeatedly. We shall march over the Caucasus next spring . . . You will then join up in Palestine with Field Marshal Rommel's army, which will come to meet you from Egypt. Then we shall march with our assembled forces to India where we shall seal our final victory over England.'[34]

* * *

Rommel began withdrawing his Italian infantry from the Mersa El Brega position on 10 December, before the enemy's attack. He was determined that on no account would he accept the risk of fighting a decisive battle. He knew that if he stood anywhere for long he would either suffer direct defeat by overwhelmingly superior forces or be levered out of his position by a flanking march which he had neither the mobile resources nor the fuel to counterattack.* At a conference with Bastico at Buerat on 17 December Rommel rapped out tonnages, port capacities and logistic details at him,[35] matters to which, *pace* some of his detractors, he was invariably sensitive. On 27 December Mussolini, more persuaded by Rommel's arguments in Rome, perhaps, than he had shown at the time, gave Bastico latitude to withdraw to Tunisia, but as slowly as possible.[36] And by 29 December the Panzerarmee was behind the Buerat position – which Goering had argued must be held indefinitely and 'at all costs', but which could, in fact, always be outflanked and in which Rommel had no confidence whatever. He was still oppressed by what he regarded as unreality in those, like Kesselring, who had every reason to know better. Kesselring had told the Italians that Rommel no longer believed in success, that he had lost confidence in his own abilities.[37] The second opinion was false; but the first was correct. And these days when conferring with Kesselring Rommel often found himself enraged.[38]

The Italian High Command, however, was now showing a more vivid appreciation of the dangers threatening the remainder of the Italian Army of Africa. Bastico – a fundamentally decent and courageous man – did his best to act the part of honest broker between an exigent Duce and a Rommel of whose realism he was now persuaded. The British were taking their time, and German war diaries were still referring to the enemy's unwillingness to operate with imagination.[39] Nevertheless, the timing of the procession of the Panzerarmee westward would be dictated by Montgomery. After a major attack on 15 January 1943, in which the Panzerarmee destroyed a considerable number of British tanks, Rommel slipped away to a new position covering Tripoli.

This position, too, was attacked in strength on 19 January, by an Eighth Army which Rommel described as acting with considerably

* In his letter to Brooke of 13 December, already quoted, Montgomery said the enemy 'is very windy and is starting to pull out'. This may have been inelegant, but it was not unjust.

more energy than hitherto. The position was a strong one, and had Rommel possessed anything of the strength – and the fuel – of earlier times he knew he could have made it an expensive business for the enemy. As it was, the British outflanking manoeuvre which levered him out of what he called the Tarhuna–Homs line developed with commendable speed. Montgomery, understandably and consistent with his own intelligence, supposed that Rommel hoped and would strain to hold as far east as possible, as long as possible. Such were, indeed, his orders; but they were not his intentions. Unless he withdrew early he would sacrifice a large (Italian) part of the army.

Once again Rommel was ordered to stand firm for longer than made any operational sense; and once again he had to tell Commando Supremo that the choice lay with them to lose or save their troops. He spoke personally and forcefully to Cavallero who visited him at Tripoli on 20 January, although in fact it was Kesselring who had suggested giving Rommel specific time limits within which to hold successive positions. From Cavallero Rommel received an equivocal reaction – he was to preserve the army but 'gain as much time as possible'.[40] On 22 January Rommel ordered the evacuation of Tripoli (saving 95 per cent of the stores there), and began moving the Italian infantry back to Tunisia to work on a fortified line at Mareth, south of Gabes; a line which itself, he said, could not be held by the Panzerarmee without significant reinforcement.[41]

In these withdrawals Rommel's timing was criticized as being premature. Cavallero wrote on 26 January that he was clearly giving up ground without resistance.[42] At times it may, in the event, have been so – exactly to judge a withdrawal so that an enemy is held to the last moment but a force is successfully extricated with minimum casualty is no easy task, and Rommel, unlike Montgomery, had not the advantage of reading his opponent's code. But more significant than the exact period of delay he imposed on his pursuers at any particular point was the strategic issue which he had so insistently raised. What was the object of imposing delay? Where was it all to lead? Surely nobody could by now suppose that there could be any alternative to a concentration in Tunisia, even if the later, and to Rommel inevitable, consequence of evacuation were unmentionable? And for all the talk about time needed to prepare the Tunisian defences, the most effective way of fighting successfully in Tunisia would be to concentrate the German–Italian forces there with the Panzerarmee. Rommel had not lost his boldness. He had lost his aim. He would have reacted with justifiable cynicism had he been present at a Führer conference at Rastenburg on 20 December, at which Hitler declared that retention of control of North Africa was vital. A withdrawal to *Festung Europa* would indicate that the ability to win the war had disappeared.[43]

On 26 January, in pouring rain, Rommel established Panzerarmee headquarters west of Ben Gardane in Tunisia. On the same day he was told by Commando Supremo that he was to hand over command to an Italian general – the Panzerarmee was now, formally, the 'German–Italian Panzer Army' – after taking up position in the Mareth line. A new Italian army command, First Army, was to be formed and this would subsume Rommel's command. The reason given for immediate relief was the state of Rommel's health. Mussolini had proposed this relief in early January, opining that Rommel was 'no strategist' – a view, often to be echoed, presumably deriving from Rommel's rejection of the strategically desirable (a protracted defence of Tripolitania) because it was operationally impracticable (since it invited certain and earlier defeat by Montgomery).[44]

Rommel had had word of this proposal two weeks earlier, and had sent Berndt once again to East Prussia to discover how the land lay. Berndt had had a long personal talk with Hitler for over three hours – a Hitler who had apparently recovered from his rage with Rommel and who spoke of his trust and his intention of giving the supreme command in North Africa to him, provided he were fit and provided plans matured to create one Army Group of all Axis forces in Tunisia. 'Berndt brought me,' Rommel wrote to Lucy on 19 January, 'warmest greetings from the Führer, whose undiminished trust, as in the past, I have.'[45]

Rommel's health was, indeed, precarious, with sickness, fainting fits, low blood pressure, bad headaches and insomnia; but he knew that his frankness, and his sometimes intemperate language, were responsible for Cavallero's dissatisfaction with the command of what was again, numerically, a largely Italian army. He was, by now, not particularly sorry at the prospect of relief. He was deeply concerned over the fate of the Afrika Korps, but once back at Mareth the German forces in Africa would be almost united, with the Tunis and Mareth fronts only two hundred miles apart. The Panzerarmee would be part of a larger whole and there would be new arrangements. And Rommel believed that a united force concentrated in Tunisia and supplied through Tunis must have better prospects in the short term. He had been asked to transfer 21st Panzer to the

Tunis front on 13 January, and had done so – indeed had himself made the suggestion.

Other changes were now made in the Italian Command. Rommel was sad at the departure of Bastico on 31 January. He had had furious differences with him, and had privately mocked him as 'Bombastico', but he knew he owed Bastico, who had often supported him when it had been most unpopular to do so, a good deal. Rommel heard with no sadness at all, however, of the dismissal of Cavallero.

On 12 February – having decided to give up command only when directly ordered – Rommel noted that it was the second anniversary of the first arrival at Tripoli of the Afrika Korps. The band of 8th Panzer Regiment struck up in greeting outside his caravan at eight o'clock that morning, and he gave a small party for twenty of his longest-serving Africa veterans in the evening, at which Graf Sponeck, 90th Light's commander, made a presentation of a map of Africa, signed by all present; and Rommel replied.[46] Three days later the rearguard of the Afrika Korps took station at Mareth. Tripolitania was lost.

A victory parade had been held by the British in Tripoli on 4 February; and German reconnaissance troops had identified, arriving, the distant figure of Churchill himself. On 3 February the German people had been informed that the resistance of Sixth Army in Stalingrad had finally been overborne. The city and over ninety thousand prisoners, including the Army Commander Paulus, now a Field Marshal, were in Soviet hands. Together Stalingrad and Alamein with its aftermath marked the watershed of German military power.

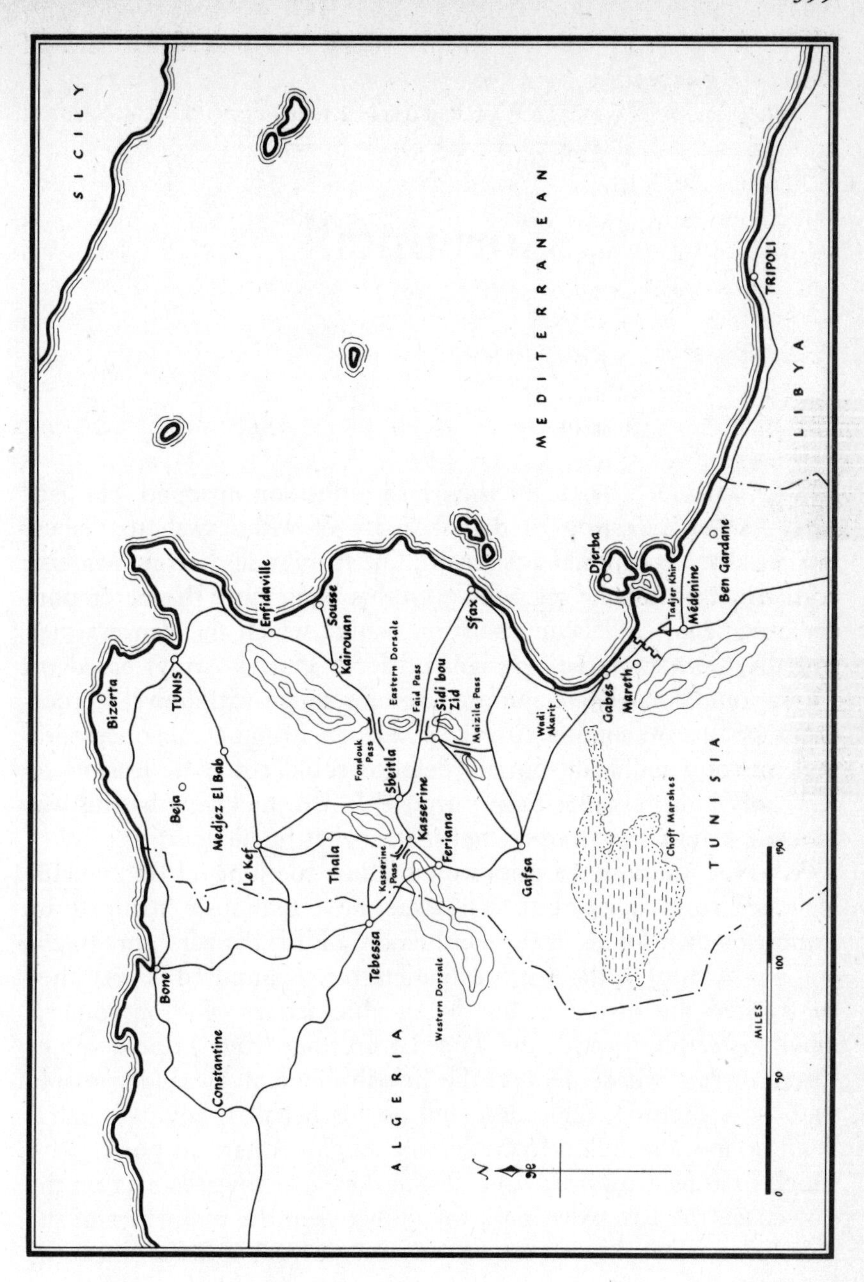

XVII Tunisia. Area of operations, January–March 1943

CHAPTER EIGHTEEN

Curtainfall

ROMMEL THOUGHT the Mareth position unsound. He had, when referring to the necessity of withdrawal 'to Gabes' always proposed a defensive line forty miles beyond Mareth, at Akarit. Privately he regarded even this as no more than a temporary expedient – a delaying position behind which the Panzerarmee and the German forces (designated Fifth Panzer Army) based on Tunis could consolidate and prepare subsequent withdrawal in good order before evacuating Africa. It was, to Rommel, inconceivable that a protracted build-up and defence could either be feasible or serve any long-term strategic purpose. In this he knew that his was to some extent a lone voice, but it was what he believed.

As to feasibility, the armies were holding too long a front, including the front at Mareth. The line now tenuously manned by Rommel's own forces in the south and by Fifth Panzer Army (General von Arnim) in the north extended to four hundred miles. They confronted the disunited but formidable armies of Montgomery, moving steadily towards the Tunisian frontier from Tripoli, and of Eisenhower – whose deputy, the British General Alexander, would soon be exercising field command on his behalf – holding central Tunisia and assembling considerable Anglo–American power with which to attack towards Tunis in the north, or towards Sfax on the east coast, the latter driving a wedge between the two wings of the Axis forces. Despite a theoretic advantage from interior lines, the German–Italian forces had, Rommel reckoned, wholly inadequate strength with which to hold such a front indefinitely. There would have to be contraction; and ultimately evacuation to Europe would be absolutely necessary, unless the armies were to be sacrificed. Time could be won by hard fighting, by well-chosen positions and well-judged withdrawals, but it would only be time, and it would be limited. And how would that time be used for the fundamental strategic objective of the defence of the Reich? Rommel, so often accused of being a tactician impervious to the claims of strategy, found the shoe on the other foot. It was he who reckoned he needed and was not vouchsafed a grand design. It was his superiors who seemed to have little concept beyond the juggling of inadequate forces on a distant map – for the coordination of the Panzerarmee (now to be renamed First Italian Army) and Fifth Panzer Army was initially attempted by Kesselring in, or occasionally flying from, Rome.

In this Rommel probably did the German and Italian authorities less than justice. There was defensible strategic argument for holding the Anglo–American forces to a fight in North Africa. It kept them from other and more important projects, it gave a sort of depth to the defence of southern Europe and it complicated, even if it could not prevent, Anglo–American east–west passage of the Mediterranean through the Sicilian narrows. Mussolini, Rommel knew, regarded the foothold in Tunisia as crucial for European – particularly Italian – defence, and as a delaying ploy against Anglo–American designs on the European mainland. The case was tenable. But to Rommel this could only be argued if Tunisia, with resources available, could actually be held and supplied. It couldn't, except for a short while. Once again the arguably desirable on strategic grounds was operationally near-impossible. It was not, therefore, practical politics. There was no strategic merit in standing at the right place on the map and thereby losing an army.

Rommel's last weeks in Africa were, therefore, weeks of dejection, and their narration one of melancholy. He had little faith in his task. His entire North African enterprise had ended on the threshold of disaster. He was pessimistic, and his letters showed it, about the outcome of the war itself. His confidence in the leadership of his country was greatly diminished. His poor health – and without doubt he was unfit and should not have returned to duty during Alamein – had a profound and lowering effect on his spirits: he had desert sores, felt unwell much of the time, had lost sparkle and resilience. He had given of his best, valiantly, during the retreat. Now he had, for a while, little more to give: or so it seemed. On occasion he showed an extraordinary, a hardly believable, lack of power of decision. It was a black period, wholly untypical, without precedent. Nevertheless, although with part of himself, as before Alamein, he longed for relief, he wrote to Lucy on 7 February: 'Everything in me resists the idea of leaving this theatre of war while I can still stand upright!'[1]

Professionally, Rommel was deeply unhappy. It seemed to him that his opinion on military essentials had recently been disregarded or overruled. He felt that he was no longer trusted. The stigma of defeatism, of instability, of being unable to ride out temporary setback – this wounded him, and he said so. He regarded it as wholly unjust: worse than unjust – he knew that it implied a frame of mind in the one making the accusation which was itself remote from the military realities of the situation, and which thus boded ill for the Panzerarmee and for the German cause.

The local command arrangements, too, were unsatisfactory, indeed absurd, although they were about to change. Now that the two Axis armies were, as it were, back to back, they clearly needed unitary control by one authority. At present that authority was Kesselring, exercising it from Italy: yet a united military operation, by both armies westward against the Anglo–American forces, had been agreed at a conference on 9 February which Rommel attended and which had set far-reaching objectives. It was to aim at the complete destruction of the Americans facing Fifth Panzer Army.[2] This would require one overarching concept, and firm, effective operational command. Instead, there were to be two apparently co-equal army commanders, doubtless with strong and not necessarily harmonious ideas on how matters should go. And coordination – or resolution of argument, or plain direction – would, it appeared, be exercised, if at all, from a great distance. It had now been ordained that 'Army Group Africa' should be created, to command both First Italian and Fifth Panzer Armies, but it would not be effective before 23 February. Meanwhile von Arnim commanded the bulk of the German troops and Rommel, a German Field Marshal, led First Italian Army, under deferred orders to hand it over to the Italian General Messe. It was common knowledge that von Arnim was to have the command of Army Group Africa, although the matter was not yet official. Rommel knew that there were many in Tunisia who would be glad to see the back of Erwin Rommel.

There would, however, be one more sadly short-lived flash of the old Rommel spirit before the curtain fell on the African drama.

Rommel's old 21st Panzer Division (Colonel Hildebrandt: General von Randow had been killed in December during the withdrawal) had been passed to the Tunisian command from the Panzerarmee, had been re-equipped and was now the most experienced formation in Fifth Army. On 1 February the division attacked and seized the Faid pass, an east–west pass through the mountains known as the Eastern Dorsale which ran from north to south parallel to the coast and some seventy miles inland, in central Tunisia. The Anglo–Americans west of the mountains might attack eastward to the coast, cutting off von Arnim from Rommel; to do so they would need to traverse the mountain barrier, and in the central region two passes – at Faid and Foudouk – could enable this. 21st Panzer Division's occupation of the Faid pass pre-empted such an allied move. It also gave von Arnim a sally-port from which to advance west. An even more direct threat to cut the narrow corridor between von Arnim and Rommel would be an Allied thrust eastward or south-eastward from Gafsa, towards Sfax or, to its south, Gabes. This, too, needed pre-emption.

The concentric German–Italian operation agreed on 9 February began five days later with an attack by the 10th (General Freiherr von Broich) and 21st Panzer Divisions from the Faid and Maizila passes towards the small Arab village of Sidi Bou Zid. It went well, driving American troops from their positions, outnumbering and outgunning them locally, inducing a good deal of panic. Forty-four American tanks, twenty-six guns and nearly one hundred other vehicles were abandoned or had been destroyed by the time the pincers of a German attack met west of Sidi Bou Zid. Named Operation *Frühlingswind*, this constituted the right-hand punch – von Arnim's – of the agreed operation. Rommel's was to be the left-hand punch: Operation *Morgenlust*.

Rommel had two principal reasons for welcoming battle. First, he knew that it should ease his anxieties at Mareth. He did not relish the prospect of fighting another defensive battle with Montgomery while, behind his right shoulder and at no great distance, an unbroken Anglo–American force was free to attack him, to interfere with his communications or – the ultimate, unnerving possibility – cut off his retreat: if that enemy could be dealt a sharp blow, could be disrupted and put on the defensive, Rommel would stand a better chance in his forthcoming fight with Eighth Army. He might even be reinforced. He had proposed an operation on these lines on 4 February.

Rommel's second reason was one he nurtured more confidentially, although he had discussed it with Kesselring. Rommel's part in the concentric operation was to attack, with German and Italian troops from First Italian Army, towards Gafsa, from the east. Gafsa, seventy miles inland from Gabes, is on the road running north-west from the east Tunisian coast across the interior to Constantine in Algeria, a distance of three hundred miles. The road passes by Gafsa, Feriana and Tebessa, through that spur of the Atlas mountains known as the Western Dorsale, which lies at right angles to the Eastern Dorsale. Gafsa, a considerable place of ten thousand inhabitants, was held by troops of II United States Corps (General Fredendall). Forty miles north of Gafsa was Feriana: and fifty miles further along the road was Tebessa, known to be a major Anglo–American logistic base, west of the Algerian border.

Rommel was happy with the prospect of attacking Gafsa, as the left-hand punch complementing von Arnim's assault through the passes to Sidi Bou Zid. He had, however, bolder private ideas. He thought it possible – not, perhaps, likely, but possible – that a really determined attack would send the enemy scurrying as he had so often managed before, and that then – also as so often before – he could exploit and harry and drive deep and turn a tactical success into an operational victory. With this in mind he decided to be ready to advance after taking Gafsa, to Feriana and on to Tebessa. This was certainly consistent with the object of the whole exercise, as agreed on 9 February, but it went well beyond the agreed operation. It would be, once again, an unrehearsed and perhaps triumphant exploitation. It could totally disrupt the Anglo–American front in central Tunisia, induce panic and even lead to the enemy's westward withdrawal, into Algeria. It would remove all threat to Rommel's right rear at Mareth. And it would win time – considerably more time – for that orderly constitution of a defensible front round Tunis and Bizerta he believed was the only rational medium-term aim; followed, ultimately, by evacuation to Europe without loss.

Rommel, therefore, believed in what he was doing. On the afternoon of 15 February the Afrika Korps (General Freiherr von Liebenstein, with a group of seventy tanks from 15th Panzer, and the Centauro Armoured Divisions) entered Gafsa. It had been evacuated by the Americans. In the north, after some misgivings about object and objectives, the right-hand punch, von Arnim's – directed by his own deputy, General Ziegler – was advancing on Sbeitla. Next day, Rommel told von Liebenstein to march on Feriana, on the road to Tebessa. He himself drove towards Gafsa, where American destruction of an ammunition dump had caused considerable casualties among the civilian population and where he was greeted by enthusiastic Arab crowds shouting 'Hitler!' and 'Rommel!'

Almost by accident von Arnim's right-hand operation was already further to the north-west than had been envisaged, moving towards Sbeitla. Sbeitla and Feriana in German hands would provide parallel start points for a massive advance through the Western Dorsale, for that operation Rommel had dared suppose possible. That evening, in a telephone conversation, von Arnim agreed to go as far as Sbeitla. The Afrika Korps entered Feriana on the following morning, 17 February. The nearby airfield at Thelepte was overrun, with American aircraft still on the ground.

Rommel was now up with the Afrika Korps, among fighting troops, and fighting troops who were advancing. His spirits began to lift in a remarkable manner, and just after midday on 17 February he wired a proposal to Rome. If all three Panzer divisions – 10th, 15th and 21st – assembled in the region of Feriana under his command, he would advance on Tebessa with a view to overrunning the Anglo-American rear areas, and invading Algeria. The opposition was crumbling and panicky, the opportunity existed. To his unconcealed joy Kesselring's reply was favourable. That night, Rommel ordered a bottle of champagne, a rare indulgence; and told Berndt, who relayed it in a letter to Lucy, that he felt like a warhorse hearing the trumpet.[3]

Next morning, however, he met difficulties, and they stemmed from lack of authority. Von Arnim demurred. He had, unbeknown to Rommel, planned a different and limited westward advance further north, in the Tunis sector, and these operations in the mountains would delay or abort it. Furthermore Rommel's intended axis into Algeria by Tebessa was ambitious and distant. Would it not be better to thrust on a parallel route further to the east, from Sbeitla towards Le Kef, a place south-west of Tunis and only sixty miles from the north coast?

Rommel objected vigorously. He knew nothing of von Arnim's planned separate operation and nobody enlightened him. He had regarded Fifth Army's exploitation on the right after Sidi Bou Zid as sluggardly. He suspected that he, whispered against yesterday as a defeatist, was now again supposed to be gambling, overreaching, clutching at a final chance of recovering past glory. He had never gambled, Rommel wrote indignantly afterwards, but there were now high stakes and very little time. It was not until after midnight on 18 February that the dispute was resolved, by compromise. Rommel was to take the two Fifth Army Panzer divisions, 10th and 21st, under his command as he had proposed, and to advance from Feriana and Sbeitla; but his objective was to be Le Kef, confirmed as such by a signal on 19 February from Commando Supremo.[4]

To Rommel this dilution (another delay) was probably fatal. It meant that instead of a deep, scything, operational movement the enveloping threat to the enemy would be shallow and would lead to no decisive result. Making the best of what he regarded as a bad job, he set his Panzer forces marching north next morning, 19 February, exercising very personal command for the last time. Von Liebenstein had been wounded on the seventeenth and the Afrika Korps was under the temporary command of Ziegler. There were three almost parallel routes: on the left the road to Tebessa, which Rommel had wished to be his *Schwerpunkt* but to which he now allocated a probing force; on the main Sbeitla–Le Kef road, 21st Panzer; and between them, in the centre, under Rommel's experienced engineer commander General Buelowius, a strong armoured group, advancing northward to Thala through the Kasserine pass. Rommel reckoned he would see how each advance went, and reinforce success as and where it came.

The battle of the Kasserine pass was Rommel's last victory in Africa. The left- and right-hand advances of his triple thrust made slow progress, and he switched emphasis to the centre on the second day of the operation. With the Kasserine pass in his hands Rommel could build up enough strength north of the Western Dorsale to develop his advance in any direction and certainly to smash any flank threat to that advance from whatever quarter. He could feint and lunge as he chose. It was the sort of situation he liked.

The pass itself, between precipitous mountains rising four thousand feet from the plain, is narrow – half a mile across. It continues for only a mile until debouching into a basin surrounded by wooded hills where the road divides, one branch running north to Thala and another, the left-hand branch, then a dirt track, ultimately joining the road to Tebessa. The Americans had laid a large number of mines south of the pass and across the roads. Their main defensive positions were in the basin north of the pass, covering the exits from it. These positions were manned by American engineers and infantry, supported by artillery and mortars from three armies, American, British and French; but in no great strength.

In the late afternoon of 19 February Buelowius had begun levering the defenders out of the pass. In time-honoured fashion he had put his infantry to climbing the hills, infiltrating between and getting behind their opponents. When Rommel decided to make his main effort at Kasserine the defence had been loosened and was, he assessed, nervy; but it was intact. He had set 10th Panzer (General Freiherr von Broich) on the march towards Kasserine, and had ordered up the Centauro Division from Feriana. Next morning he drove into Kasserine. By mid-morning his foremost troops were through the pass towards Tebessa, 10th Panzer and the rest of Buelowius's group were beyond the narrows and Rommel was poised for exploitation. His leading elements were reporting no resistance on either the Tebessa or Thala roads; and at seven o'clock in the evening of 21 February Rommel, riding with 10th Panzer,* entered Thala.

Within a day Rommel had called off the attack. It was, perhaps, the most resounding anticlimax of his career.

The decision was entirely his own. Kesselring had visited him on 20 February, enthusiastic for an advance on Tebessa, Rommel's original proposal and apparently now feasible with German forces north of the Kasserine pass. Von Arnim – uncooperative and sceptical – had been induced by Kesselring to pass most of his Panzer divisions to Rommel, although he had managed to retain the enormously powerful Tiger tanks in the northern sector, away from the battle in the mountains.

Rommel had moved among the forward infantry as they worked their way through the pass and beyond, had drawn strength from their obvious happiness that he was amongst them, had himself for a little been happy. Kesselring, buoyant as so often, was urging him to continue, to believe in his star. Kesselring now proposed that Rommel formally assume command of Army Group Africa (set up at that very time but without a formally designated commander) and thus take into his hands the legitimate authority which the army had supposed was about to be assumed by von Arnim, with Rommel departing. The idea stemmed from Hitler himself.

But Rommel declined to continue. He was uncertain about the Army Group command – it had been offered to von Arnim and he was sceptical as to whether this offer really conveyed restored faith in him, Rommel, or whether it was a temporary expedient to tide

* 10th Panzer's newly-arrived Ia was a certain Lieutenant-Colonel von Stauffenberg, shortly to be seriously wounded.

over an anomalous command situation. But whether or not he was Army Group Commander he was determined that the attack should be stopped. He continued to protest to Kesselring on 22 February about the Le Kef decision and by now Kesselring agreed with him: but this was spilt milk. The operation must be halted where it was.[5]

Rommel had his reasons, and they were good. The opposition was hardening. This was no rout, no pursuit. The Americans had shown, at first contact, all the signs of inexperience and shock which might have been expected, quitting the field in many cases in a good deal of disarray: but this had passed and Rommel noted that they had then fought well. He noted, with a practised eye, American resilience, their ability to make a quick recovery and learn from their own mistakes, their relative flexibility; and he also noted, gloomily but without surprise, their profusion of first-class equipment, a good deal of which had been destroyed or abandoned. Now enemy armour, in considerable strength, was reported on the move towards the battlefield.

Because the opposition was hardening, to pursue his more ambitious object, to drive deep into the Allied rear and attempt a major operational victory, a Gazala, would now need more time than Rommel had hoped might be necessary; and might therefore be impracticable. Time was short; there had been loss of it at the start of the operation – divided counsels, lack of authority, had contributed to that. And some of the subordinate commanders, in Rommel's view, were slow in initiative and execution, losing time again. He was riding a horse he imperfectly knew, with less of a turn of speed than he needed. And, as so often before, both fuel and ammunition were low. Then, again, the weather had been foul, and flying inhibited, which was good; but this could not long continue. The British had recently been using Hurricane aircraft firing rockets, with deadly effect against armour.

The shortage of time was also critical because Rommel inevitably fretted about the situation at Mareth, the situation of the army facing Montgomery. Montgomery would soon be in a position where, even by his own demanding standards of material superiority, he would be able to attack. If he attacked while most of what had been Panzerarmee Afrika was engaged in central Tunisia – even in pursuit of a beaten Anglo-American enemy, if things went triumphantly and surprisingly well – he would without question win. Rommel reckoned it was essential to try to knock Eighth Army off balance with a spoiling attack before they could put forward their own prepared and organized attack on the Mareth line. To arrange this he would need time, and unless he marched back now there might not be enough. The offensive stroke at Kasserine – his last – had been tactically successful. He had, for a few brief hours, enjoyed once again the sense of successful battle, of decision, of an enemy vanquished by the energy and skill of his own troops and their commander. But to take it further would have been folly. There was not time. It was the shadow of Montgomery which aborted *Frühlingswind* and *Morgenlust*.

Of course Rommel could have appreciated the time factor before the operations started. He could have pointed out the incompatibility of a foray into central Tunisia with the preparation of a spoiling attack at Mareth, all before Montgomery was ready. He did not do so. It had been conceivable that if the combined operation with Fifth Army had gone really fast and really well he could have managed to win and to follow victory with a forced march to another front, like Harold of England after Stamford Bridge, beating Hardrada's Norsemen and then marching to confront William of Normandy on the south coast. But William conquered.

In a short while Rommel was again at Mareth. At six o'clock in the morning of 6 March he attacked the British at Medenine. Operation *Capri*.

For a little more than a week Rommel had been in nominal command of Army Group Africa – a commander without a properly constituted Army Group headquarters; a commander who, as the whole world knew, had not been first choice for the position – that had been von Arnim; a commander who was expected, very soon, to be leaving Africa for reasons of health; and whose authority seemed already suspect.

For on 26 February von Arnim had launched a separate operation in Northern Tunisia, *Ochsenkopf*, towards Beja and Medjez-El-Bab, an operation he had been planning for some time and of which Rommel first learned only on the twenty-fourth, the day after he formally became Army Group Commander.[6] It was not in any way coordinated with what had been happening in central Tunisia, and it certainly owed nothing to Rommel's ideas or inspiration. Rommel, from what he heard, regarded it as clearly beyond the powers of the forces assigned. If such an operation were to take place it should have been synchronized with the advance to Kasserine and beyond: as it was the enemy was undistracted and *Ochsenkopf*, a three-pronged drive with a considerable number of tanks, was ordered by Rommel to be called off after three days, by which time there were only five tanks left on the main Fifth Army axis towards Beja. Rommel felt the irritation of a man saddled with responsibility for fighting he had in no way planned.

Furthermore both von Arnim and Messe, Rommel's two Army Commanders, were dealing direct with Kesselring and with Rome. Messe presumably had duties as the senior Italian General in the theatre (he had, in fact, made a start on combing the rear areas in order to economize in the number of Italian soldiers – eighty-eight thousand out of 120,000, he claimed – employed there whom he could better use at the front[7]). The case of von Arnim, although obviously an irritant, is easy to understand – he had been, until 23 February, an Army Commander directly responsible to Kesselring; and he had been given to understand that in very few days he would be so responsible again, as an Army Group Commander. In the short, uncomfortable interim there was interposed between them the awkward, exhausted and abrasive figure of Field Marshal Erwin Rommel.

The command relationships were indeed awry – they were described as 'remarkable', '*Merkwurdigen Befehlverhaltnisse*', in a letter from Berndt to Schmundt. Rommel himself exchanged courtesies with Messe who, in a speech on 2 March, referred to the pride with which Italian soldiers had served under Rommel: while Rommel, replying with whatever sincerity of emotion, spoke of having fought for two years shoulder to shoulder, '*Schulter an Schulter mit den Italienischen Kameradenen*'. In all of this there was a certain unreality. Rommel's situation for his final days somehow seems to have more shadow than substance, words and forms rather than flesh and blood.

Rommel's *Capri* attack at Medenine was a total and deserved failure. No day throughout his career is less characteristic of his method of fighting, and that he launched battle in such circumstances still has power to surprise. Montgomery had very recently deployed in a strong position at Medenine four hundred tanks and five hundred anti-tank guns, the latter dug in and supported by a formidable mass of artillery. Montgomery – who knew Rommel's intentions perfectly well from ULTRA – was ready and keen for the fray. He was completely confident that he was in a position to fight and win a defensive battle and that its upshot, as at Alam Halfa, would be a weakened German–Italian army. Montgomery's positions, well-sited as ever, formed an arc of six strong brigades projecting westward north of

Medenine, itself twenty miles south-east of the Mareth line. In reserve was 7th Armoured Division. The right of the position rested on the sea and it was known that, as so often before, Rommel would seek to demonstrate frontally and then break into and round the defenders' left flank.

Rommel told his subordinates that success would depend on speed and surprise; and he wrote later that the only hope would have lain in striking before Montgomery was ready, a sentiment with which Montgomery would have agreed. But Montgomery *was* ready. Rommel's excursion into central Tunisia had provided additional time; ULTRA meant that surprise was out of the question; and Montgomery's material superiority was, once again, so formidable that unless he made some fatal mistake it would be irresistible.

Montgomery did not make fatal mistakes. Methodical, balanced, never overreaching, he had moved inexorably along the North African littoral; seldom within sight of trapping his opponent but only accepting battle (rightly) when his own strong suit, his enormous numerical and supply superiority, could be played. He exerted his strength when – and generally only when – the impetus of his drive could be maintained both spatially and temporally. He knew that at Medenine, if Rommel attacked, such a moment would soon follow.

Rommel's attack was carried out with wholly inadequate forces to deal with such an enemy. It was also carried out by a half-hearted commander. Rommel had long decided that North Africa should be abandoned by the German–Italian forces. More recently – and in this wholly agreeing with von Arnim – he had judged that the best medium-term prospect for defence, covering ultimate evacuation, lay in a withdrawal to a much shortened line, anchored on Enfideville. He had signalled this recommendation to OKW. An Enfideville line might be held for a little longer. A Mareth line could not.

Rommel himself did not argue, after Medenine, that he had been half-hearted; but it is difficult to find in the Rommel of Medenine the Rommel of earlier days. He referred afterwards to the time-consuming debates and arguments which accompanied determination of the plan – peculiar circumstances for a Rommel! And when the plan was finally agreed Rommel (correctly) conceded the executive decision to Messe, now Army Commander of First Italian Army and formally responsible – for an attack made almost entirely by the Deutsches Afrika Korps! It is impossible to avoid sensing relief in Rommel's abandonment of the battle by five o'clock in the afternoon of its first and only day.

It certainly had to be abandoned. As at Gazala, a frontal demonstration on the left, near the coast, had been made on this occasion by the Spezia Division and 90th Light; and the three German Panzer divisions had attempted to break in on a ten-mile front towards and south of a prominent feature, the Tadjera Khir, itself north-west of Medenine. The Afrika Korps numbered about 150 tanks, against Montgomery's emplaced five hundred anti-tank guns, including the new and excellent British seventeen-pounder which could reach out and penetrate Rommel's Panzers at battle-winning range. The enemy front was protected by mines. Montgomery knew exactly how Rommel – or General Cramer, now newly commanding the Afrika Korps* – would move, must move. There were few alternatives, although Rommel later (and unconvincingly) argued that had he divided his armoured forces and effected a pincer movement, a convergent attack from both north and south, the issue might have been different. As it was Rommel, entirely out of character, drove his depleted Panzer divisions against an enemy in well-sited defensive positions, an enemy in greater strength and better equipped than ever before. The outcome could not be in doubt. At the end of the day the Afrika Korps had lost a third of its tank strength and achieved no tactical success of any kind.

Why did Rommel do it? A pre-emptive attack was defensible – he did not know for certain that Montgomery was ready, and he might not have been. The loss of time was regrettable and not all – although in part – Rommel's own responsibility. He would have argued, did argue, that simply to wait at Mareth was to await inevitable defeat by superior forces in Montgomery's own time. Could he have done what he did better and more successfully? Given the balance of forces, given ULTRA, surely not; by that day the Afrika Korps, indeed First Italian Army, could not possibly assemble the strength to overcome or seriously damage an Eighth Army in position. This was no longer the Eighth Army which had sometimes been scattered by forces numerically inferior but tactically dominant. Rommel was, again, confronting a commander who would never fight a badly designed battle, or lose one fraction of his grip upon it. Medenine either had to be fought and lost by Rommel, or never fought at all. Only an unfit Rommel, a man sick in body and mind, could have undertaken so unpromising a venture. The whole of Medenine, including his later comments on it, betrays a Rommel

* He had taken over from Ziegler on the previous day.

damaged and different from the victor of Gazala to a striking degree.

It was a moment of truth. On the day after the battle a reply arrived from Hitler at OKW to Rommel's recommendation for a withdrawal to a line resting on Enfideville. Unsurprisingly, it was an indignant repudiation of any such idea.

It was generally known that, once again, Rommel was flying to see Hitler, and was taking sick leave. One of his most experienced subordinates, von Luck, came to say goodbye and found the occasion emotional. Rommel, sitting in his command vehicle as so often seen since early 1941, was ill and exhausted-looking, but there was still, said von Luck, 'that unique sparkle in his eyes'.[8] There were tears in those eyes, too, as Rommel gave him a memento photograph. Two days later, on 9 March, Rommel handed over command of Army Group Africa to an equally pessimistic von Arnim (who remained formally Rommel's deputy) and flew to Rome. Ostensibly he was to return when his cure was complete. He would never see Africa again.

In Rome Rommel had a discussion with General Ambrosio, who had taken over from Cavallero, and then went with him to see Mussolini. Mussolini, Rommel realized, had come to share the prevailing view of Rommel as defeatist; the Duce was cordial rather than warm. Rommel had always admired Mussolini, a taste not every German shared. He had reckoned that until a late hour the war in Africa had not been taken very seriously in Rome, but Mussolini, flamboyant and often victim of illusion though he was, had always shown appreciation for Rommel's efforts and achievements. Rommel thought that Italy owed much to the Duce, and he had been greatly impressed in North Africa by the level of economic and cultural development brought by the Italians, with Mussolini's enthusiastic support, to that arid shore. But now, Rommel recognized, the Duce had no greater hold on the reality of the North African war situation than had the Führer. Indeed, he had not had since Alamein.

Next day, 10 March, Rommel flew to report to Hitler at his headquarters in the Ukraine. He found Hitler utterly depressed in the aftermath of the calamity at Stalingrad, the loss of the entire Sixth Army. Everywhere the Red Army was becoming stronger, was showing its recovery from the disasters of the first eighteen months of war. Motor transport had been made available from the United States in huge quantities. The Soviets now enjoyed a considerable degree of mobility. Soviet tank output was such that the Wehrmacht was enormously outnumbered – and by tanks of excellent quality. The German forces on the Eastern Front, by contrast, were now ill-equipped and short of men. In January there had been less than five hundred tanks on the whole front, and divisions had been so reduced in units and in overall strengths that the order of battle conveyed little truth. In the east Germany was standing on the defensive, and it would not be easy to sustain it.

Rommel presented himself to Hitler at six o'clock in the evening of his day of arrival and for the next three days saw him frequently, attended his conferences and had a number of long talks with him. Hitler was still living in a world largely shuttered from military reality. He spoke of Rommel leading an expedition against Casablanca when his health was restored – a project so utterly remote from the facts of the strategic situation that Rommel could hardly believe his ears. Rommel thought that he had at least persuaded Hitler that there must be a withdrawal in Tunisia, a shortening of the line to Enfideville, or at least to Akarit, from Mareth. He wrote to von Arnim with guarded optimism, for whatever differences they had had, he and von Arnim saw the Tunisian situation through the same eyes, and saw it as dark. Rommel noted how Hitler was (understandably) easier to deal with when in a state of depression. It was his insane and unrealistic bursts of brutal optimism which his counsellors feared and his armies had reason to regret.

The Enfideville concession lasted only a short time, and events in Tunisia over the next two months took the course which both Rommel and von Arnim had predicted. It was no longer possible to supply Tunisia. There had, from the first week, been a formidable airlift, but Allied attrition of both air and sea transport meant that, of a monthly requirement of ninety thousand tons, airlift and port capacity could only underwrite about half by the time Rommel left Africa.[9] The armies there were massively outnumbered in every item of equipment which mattered, and the Allies ruled the skies. When von Arnim himself surrendered two months later, on 12 May, 238,000 prisoners passed with him into enemy captivity; of these 100,000 were German, a larger loss even than at Stalingrad. Meanwhile on 11 March Rommel was decorated by Hitler with the highest and rarest order of the Iron Cross – the Swords and Diamonds to the Knight's Cross. It was then made plain to him that he was not to return to Army Group Africa. Goebbels, however, wrote on 12 March of Rommel's talk with Hitler going 'wonderfully', so the curious sympathy between two such dissimilar characters was clearly still in evidence despite failure, disillusionment and ill-health.

Exactly two months later Rommel saw the final signal sent by the Afrika Korps before his old 'Afrikaners' marched into the prison cages. The Deutsches Afrika Korps had fought, the signal ran, to the condition where it could fight no more. It would rise again. '*Heia Safari!*'

Rommel's command in Africa ended ingloriously. It had lasted just over two years, for all but the last six months of which, despite ups and downs, it had been the stuff of legend. The campaign in North Africa between the German–Italian forces and the forces of the British Empire will for ever be associated with the name of Erwin Rommel.

Rommel had his detractors during as well as after that campaign, and he has continued to have his critics – critics of his military judgement, of his achievements, of his method of command, of his character. The first and most enduring criticism is that he either failed to understand or paid inadequate attention to logistics. Rommel's unceasing lamentations over his supply situation – primarily in terms of fuel – are taken as the cries of a man stuck in a hole of his own digging. Why, his critics demand, complain about a perfectly foreseeable situation? Rommel knew well that his line of supply was vulnerable to maritime and air interdiction, both between Italy and North Africa and – increasingly – to the enemy's air power on the long haul forward from base. He knew – a vital statistic – how short he generally was of wheeled transport, and how the wear and tear on that transport (and the fuel consumed by it) increased with every mile the Panzerarmee advanced. The latter factors were a matter of simple calculation, and the losses at sea were a question of prudent rather than excessively optimistic assumption. Why blame anyone or anything but himself?

The charge has to be answered, but it does not stick. Rommel understood perfectly well that his operations were constrained by supply, and to what degree – hardly a day passed without his considering the matter. He was periodically assured of particular consignments by particular dates, and he was periodically – indeed frequently – disappointed. Being human he complained, speculated that others (principally the Italians) were not making best endeavours, that his priorities were understated by those who alone could allocate the carrying capacity, the sea escorts and the air effort to supplying him. But he had to make assumptions; he chose to make favourable rather than pessimistic ones and to base his operations thereon. This was bold rather than cautious. It led, on occasion, to avoidable embarrassment.

Yet it also led, often, to victory. Rommel's dramatic successes in the desert were achieved with a tiny margin of logistic insurance, and often not that. He has been accused of making his plans, his decisions, and then telling his quartermaster branch to cope, somehow. In a sense the accusation is fair: Rommel believed – and the belief has frequent historic justification, not only in Panzerarmee Afrika – that to set the pace and scope of operational ambition primarily by the calculations of supply may be to risk little but is often also to achieve little. Again and again his belief was justified. He calculated boldly, sometimes over-boldly. He acted. And it worked.

But not always. That Rommel's tactical plans were sometimes over-ambitious is certain. Was he, on this count, also culpable in respect of those operations which led to Egypt, to Alamein, to ultimate defeat? Until then he could counter every criticism with a tally of achievement, but when he was nearest to the Nile and to a triumphant culmination of the campaign the odds against him were at their longest, his line of supply was at its most extended and the supply situation was bleak. Could he have averted this?

The decision did not lie only with Rommel. The advance into Egypt was absolutely supported by the authorities in both Rome and Berlin. It was, after hesitation, supported by Kesselring. It was not thrust upon a dubious high command by an over-eager Rommel – far from it. He had seen his opportunity – and it *had* been an opportunity – after Gazala, in June. He had thought to destroy a beaten enemy, consummate a victory. Auchinleck had frustrated that – and Rommel paid him tribute for it. Thereafter he was strongly encouraged to hold on to all he had gained, to prepare his strength for one more effort. Plan Orient was still alive. Hitler's own eyes were on the Caucasus – had been since the preceding winter; and if the right wing of Plan Orient retracted, the plan began to lose an essential component. In making a doomed attempt at Alam Halfa, in standing at Alamein, Rommel was acting as a man under authority. But he was profoundly uneasy, an uneasiness which stemmed principally from the logistic situation and the air situation. He was pessimistic about his chances at Alam Halfa and surprised some of his closest associates when he called off the battle. When he returned from sick leave to resume command at Alamein he knew that battle was almost certainly lost. Rommel did not believe in playing safe or in excessive insurance, but he saw logistic as well as other factors perfectly clearly and he made his decisions boldly but rationally in their light.

It has been said that Rommel's strategic sense was limited, that he was essentially a battlefield commander at none too high a level with little comprehension of larger aspects of warfare. It is impossible to contemplate Rommel at war, to read Rommel's own writings or conversation, and to maintain this view, unless 'strategist' is used in the peculiar sense of a master logistician or supreme military intellectual rather than with its dictionary meaning of one versed in the art of directing the broader military movements and operations of a campaign. On the contrary, he harmonized to a remarkable degree the ability to expound and practise tactical skill and to apply the same skill, enlarged, to operational concepts of a wider sort. His mind was broad. It was severely practical rather than theoretical – it always had been. He may have resented not belonging to the General Staff, may have felt at a periodic intellectual disadvantage: certainly, when he first had to handle large bodies of troops he showed certain defects of method which formal training might have obviated. But he learned from his own mistakes, he recovered his balance, with remarkable speed. He thought matters through and, as all men in all matters, he improved with practice. This was the same Rommel who, as an instructor of young officer students, had long ago been used to say: 'Don't tell me what Clausewitz thought. Tell me what *you* think!'

The highly-educated General Staff officers who served on his staff soon recognized a commanding intelligence of rare capacity. Rommel could have commanded at any level, with flair and distinction. His personally drafted report on the North African fighting was clear, accurate, objective and generous. The lessons he drew were unmarked by bitterness or self-justification.[10] Those, furthermore, who remain sceptical of Rommel's broader judgement should recall that in November 1942 he already knew and said that the war would be lost, that Germany's best hope lay in a withdrawal to defend the frontiers of the Reich for as long as it took to make peace: peace under a new leader.

By the same token Rommel has been called an indifferent planner, a commander half in love with that improvisation at which he excelled. There is surely something in this – he knew that he and his Afrikaners generally had the best chance when the battle required instant and fresh decisions in an unforeseen situation; and he revelled in such situations. He chuckled at the elegance of solutions evolved with hindsight: 'The best plan,' he once observed to Manfred, 'is the one one makes when battle is over!'[11]

Yet Rommel did plan; though not in excessive detail, for he thought that unrealistic and poor soldiering – 'No plan survives contact'; not always infallibly (especially as regards timing), for he was human as well as bold; but he planned, he thought ahead with clarity, he had a vision, a concept of battle before he launched it. And he shared it with his subordinates, he put men in his mind. This was far superior to attempting to preordain their every action.

But the myth of Rommel as 'no strategist', stemming from Africa, has been to some extent created not only by his emphasis on (yet flair for) the tactical battle and its necessary improvisations; not only, even, by his deliberate risks taken with logistics. It has also attached to specific major decisions. The record, it has been said, is poor and shows a commander of limited perceptions.

This is difficult to maintain strongly, if Rommel's principal moves in North Africa are considered. To attack almost immediately on arrival and win Cyrenaica thereby was bold but it was right. To attack Tobruk immediately thereafter was unsuccessful – a mistake, essentially a tactical rather than a strategic failure, for Rommel would have profited from delay and reconnaissance and reinforcement. He decided – not for the first or last time – to balance the chances offered by speed, shock, instant action earlier than an enemy might expect, against the advantages of a more deliberate approach. The essay failed, a culpable failure; but that had little to do with Rommel as strategist.

And captains of Rommel's stamp make mistakes. During Crusader, the 'dash to the wire' was certainly a mistaken decision – Rommel believed that the enemy situation was worse than it was and that he had an opportunity which was not there. But the decision to give up Cyrenaica after Crusader was as sensible, given the relative state of the opposing forces, as the decision to take it again, well ahead of any prognosis, was brave, shrewd and correct. The decision to attack at Gazala and the decisions taken during the battle were bold and effective. That Rommel's proposed timetable on that occasion was excessively optimistic was certainly true but not vastly important in assessing his overall performance.

After Gazala Rommel's decision to advance to cross the frontier, to try to destroy a beaten (as he saw it) enemy was surely commendable. Whatever the outcome, it would be a timorous military theorist or historian who would forgive hesitation at that point. Rommel might have withdrawn – it would necessarily have been a long way, since he knew he could be outflanked on the frontier – after he had been checked, indeed beaten, by Auchinleck at Alamein; but by then his own optimism and confidence had infected his superiors and none would hear of withdrawal. Alam Halfa was his last chance and he knew it; and fought it without confidence. Alamein was described by Rommel as without hope before it began, so formidable were the odds; a stalemate might have been forced against a less professional opponent than Montgomery, but Rommel had flown back to Egypt with a heavy heart as well as a failing body.

During the retreat there was room for plenty of dispute on its conduct. Montgomery wrote at the time that Rommel made a mistake not to hold longer at the strong position of Mersa El Brega.[12] That this was tactically possible may well have been so, although the judgement probably neglects the materiel weaknesses of the Panzerarmee at that point; but Rommel, right or wrong, was thinking as the strategist his detractors declared him not to be. For him there was no point in winning days or weeks if the price would be – as it would have been – risk to the Panzerarmee's existence and risk above all to the prospect of its uniting with the German forces in Tunisia for a joint effort.

Kasserine, a tactical success, was an operational failure because Rommel was never likely to have the time to turn it to more lasting account, a factor he knew or could have foreseen before the operation began: but at that stage Rommel needed to try any chance which might win the Panzerarmee time and security for the battle against Eighth Army, and perhaps there was, just, a chance. By the time of Medenine – a half-hearted and wholly uncharacteristic battle showing poverty of imagination as well as resources – Rommel knew that the war, as well as the campaign, was lost. Little of this – including the defeats – indicates a commander deficient in strategic sense.

Rommel was criticized as a gambler, and angrily repudiated it, writing that a bold and calculated acceptance of risk differed totally from a gamble – an act which, if it went wrong, left no rational course remaining. The point is a fair one, although the borderline may sometimes be narrow. But Rommel's 'Afrikaners', while admiring his enterprise, did not think of him as a bold or mindless adventurer. He was, they said, 'not a player, not a chancer. He was very much a calculator, cool, steady, thoughtful,'[13] and although he demanded a lot – '*Kein bequemer General*' – he demanded of himself most of all.[14]

Rommel had his detractors for a certain exaltation of his own record, and the charge is a fair one. Like many commanders, to Rommel the most important theatre of war was the one where he was serving. But those same detractors incline to the view that Rommel's reputation has been inflated and his weaknesses minimized by his enemies; by the British, in particular, in order to make more of their own prowess in ultimately defeating him. They point out that to Germany North Africa was, in essence, a sideshow; that tiny German forces were engaged – at the most five divisions, compared to the hundreds on the Eastern Front; that Rommel was, in the last analysis, an inconsiderable figure; and that his own published writings and those of his admirers conceal his limitations and falsify the record. 'One should not distort history, after the event, to the greater glory of Rommel!' Guderian said with some asperity, and many have echoed him.[15]

As to the argument from scale, it will not do. Military reputation does not simply depend upon the numbers commanded: for the commander it can be as easy to deploy an army corps as a battalion – often easier. There is nothing in Rommel's record to indicate that the smallness of his forces in the least diminishes his reputation – rather the reverse. There were, after all, some twelve divisions under his command at Alamein. And as for the 'sideshow' argument, while it may affect the light in which we perceive Rommel, it does not affect his achievements. Nor is the argument true without qualification. While Plan Orient held its peculiar sway over the minds of Hitler and his staff the Middle East was a great and attainable prize, the Caucasus was an essential objective, and the closing of the Mediterranean was a prerequisite for that global confrontation necessary for Britain ultimately to concede victory.

This was far from foolish. There has often been argument on the Allied side that the Mediterranean was for Britain an undesirable diversion of forces – particularly maritime forces – which were more urgently needed in the Atlantic and, after December 1941, in the Far East; and that their lack in the latter was fatal. But for the British to resign themselves to loss of the Mediterranean – even apart from their obligations, and the mighty factor of oil – would have transformed the history of the war. Italy would have remained in the field, and stronger. The German campaign against Russia could have been supported by an Axis powerfully and without interruption established in the Middle East. The very outcome could have been different. On this Churchill, Mussolini, Hitler, Raeder and Brooke were all at one.

Such German dreams were dead by the end of 1942, and perhaps should have died several months earlier. But while Paulus was advancing on Stalingrad and Rommel on Alexandria they were alive, and Rommel's part in them was great.

As to the general assertion that Rommel's faults have been minimized, it can only be said that his failings, like his achievements, were considerable; but that the latter surely predominated.

Rommel's triumphs in North Africa have been ascribed to the Nazi propaganda machine. His command methods have been sternly criticized as excessively personal, and as inappropriate to a higher commander; and it is true that they were, on occasion, unpredictable and maddening to staff and subordinates alike. It has been said that without the favour of Hitler and Goebbels he might have been unlikely to command even a regiment[16] – a judgement which, if it conveys anything, speaks well of the perspicacity of the Führer and his Propaganda Minister! Germany, it has been said, needed a hero, unsullied by the whispered misery and barbarity of the Eastern Front, a modern man portrayable as in harmony with the National Socialist revolution. Erwin Rommel, it was suggested, fitted the role.[17] There is undoubtedly something in this: in wartime, in every nation, military leaders with a charisma which the public media can exploit are intrinsically desirable; and Rommel had that charisma. But it was based upon reality, upon achievement, and upon a famous and wholly genuine empathy with his troops. The flaws and excesses were offset by the genius and inspiration which flowed from the man. Such did not depend upon propaganda, although propaganda used it.

Rommel was criticized for too mercurial a temperament – now up, now down; excessively optimistic in the aftermath of tactical victory, a pessimistic defeatist after reverse. No accusation stung him more. The appreciations which led to the imputations of pessimism – his view that Cyrenaica should be given up after Crusader, his realization that North Africa (and more) was lost after Alamein and after the Anglo-American landings in Algeria – were, he argued stoutly, the fruits of realism. As for his optimism, he would point to the outcome. Rommel, like all generals, made mistakes; but the overall record was one of brilliance, and above all brilliance in battle.

Rommel powerfully imposed his will and his character upon both his army and his enemies. As to the former, he generated confidence and trust. He constantly preached that the essence of generalship was to win with as few casualties as possible. Germany, he often said, would also need men after the war was over,[18] and no charge is more unjust than that which imputes to him a lust for glory regardless of human loss.[19] Generals of that kind are invariably detected by their troops, with infallible instinct: Rommel, on the contrary, was trusted by Germans and Italians alike. It is true that at 'first Tobruk' he was rebuked – and fairly rebuked – by a subordinate for expensive rashness and for wasting lives, but the error was uncharacteristic of Rommel and was misjudgement, not callousness. He felt that to push fast, even at the cost of lives, would save more lives later. On that occasion he was wrong; but in the main no commander had closer bonds with his men. Furthermore, the discipline of the Panzerarmee was of a piece with the ungrudging respect accorded its chief. Few generals had less recourse to punishment, and it was said that Rommel never signed a court-martial order throughout his time in North Africa.[20]

To his enemies, the Rommel name was unquestionably powerful in its psychological effect. Nobody can doubt, when he slipped away westward after Alamein, that Montgomery's caution in pursuit owed much to the caution of his subordinates – and of Eighth Army; and that that, in turn, owed much to the reputation of Rommel. He had been beaten – inevitably beaten, in the event – by an excellent organizer, employing overwhelming force and determined that victory should be consolidated, should never be put at risk by excessive ambition to exploit it. Rommel could not complain of this. War is not simply the art of battle and manoeuvre, does not only demand energy and flair. It is also a business, in which calculation, organization and investment are required for sustainable dividends.

But what Rommel did well he did superbly, and in a way superior to all who confronted him. For nearly two years he dominated the desert scene. He was opposed by an enemy supported by one of the most remarkable intelligence triumphs – ULTRA – in the history of warfare, a support which again and again enabled the British not only to predict his actions and read his intentions but to destroy the convoys feeding supplies to the Panzerarmee. He fought, in the later and critical stages of the campaign, against overwhelming air superiority. Yet, again and again, he saw, decided, acted, with an energy and speed which shocked and routed his enemy. He imprinted his personality upon the German–Italian Army of North Africa in a way no member of it, no old Afrikaner would ever forget. More stolid than sometimes supposed,[21] but nevertheless intolerant of opposition, restless, urgent, gallant, indefinably gifted in the conduct of war, he led, they followed.

And he fought, except at rare and brief moments, against great numerical odds. His economy of force was striking. Admirers, bitter in defeat, contrasted his genius with what one commentator called the 'soulless calculation'[22] of the enemy commanders. This is comprehensible, but although war may be considered an art, what matters is to win, which certainly requires calculation, soulless or not. Nevertheless, the same commentator may be forgiven the harsh observation that, 'If one considers what the German Marshal could have achieved with the superiority enjoyed by his opponents one can judge the complete mediocrity of the latter.'[23] The point must be taken, although the judgement does not necessarily follow. None can doubt, however, that Erwin Rommel, standing on the frontier of Egypt with air superiority, massive supplies and a numerical edge of two to one, would have turned history in a very different direction in the autumn of 1942.

PART 5

1943–1944

CHAPTER NINETEEN

The Sunray Lamp

ROMMEL'S FIRST task on his return to the Reich from his African tragedy was to recover his health in body and mind, and this took some weeks. During that time, when not in hospital, he occupied himself at Wiener Neustadt in writing his own record and lessons from the campaign. The atmosphere and prospects for Germany were remarkably bleak, and as the first months of 1943 slipped past they grew bleaker.

On the Eastern Front, for OKH the essential front from first to last, the loss of Paulus's Sixth Army had been a shattering blow; and it had been followed by a Soviet advance which took the battle lines to some five hundred miles west of Stalingrad. The Germans then mounted a limited and successful counter-offensive, under way as Rommel returned from North Africa, which recovered the city of Kharkov and left two large east-pointing salients in the front, north and south of Kursk; or, as it appeared to the Red Army, a dangerously vulnerable west-pointing Soviet salient, the Kursk salient. Behind the shoulders of this salient, in German hands, were the cities of Orel in the north and Kharkov in the south.

Hitler, with some reluctance, had decreed that the Wehrmacht should pass to strategic defensive in the east. It would take time to make good the appalling losses of the winter both in men and equipment. In many cases 'divisions' were little more than cadres; the mid-summer tally of army and SS divisions on the Eastern Front showed 182,* but the figure conveys an exaggerated impression of strength. Some divisions were reinforced and adequately manned,

* At this time there were thirty-eight German divisions in the west, fifteen in Scandinavia and twenty in the Mediterranean, including fourteen in the Balkans.[1]

but others were not, and demographic indicators showed that Germany's manpower potential had already peaked and would henceforth decline, despite strenuous and ruthless actions to comb the occupied countries for industrial labour – either 'slave' or under various material inducements.

The equipment situation was also ominous, although extraordinary efforts were being made. In tanks, the strength on the whole Eastern Front was about five hundred at the end of January – less than half the armoured force with which Montgomery had started the battle of Alamein three months earlier on a frontage of thirty miles. At Stalingrad alone, equipment sufficient for forty-five divisions had been lost.[2] Materiel losses far exceeded production, although the latter was being significantly expanded – the number of tanks on the Eastern Front had been brought up to about 2700 by the summer.

In spite of this haemorrhage of strength in the east, however, it had been decided that the best hope for a successful strategic defensive during 1943 would lie in a limited offensive against the Kursk salient, attacking it from the shoulders, from north and south. This offensive, agreed by both Army Group Commanders affected, Field Marshals von Kluge in the north and von Manstein in the south, should, both commanders opined, take place as early as possible. In the limited operations of March in which the Germans had taken Kharkov and Belgorod the enemy had suffered heavily and it was, von Kluge and von Manstein believed, essential to strike before Russian recovery, before the Russians had had time to strengthen the defences of the Kursk salient, and before they got wind of the operation.

The offensive was named 'Citadel', and upon its success would probably depend German fortunes in the east for the rest of 1943, if not longer. Its objects were limited to the destruction of the Red Army in the Kursk salient; and it was planned – planning began in March – to be launched if possible in April. In early May, for various reasons, Hitler postponed it, primarily, he said, because it would be better to await the arrival of the new 'Tiger' and 'Panther' tanks – the first, already in service in Tunisia, carrying the 88-millimetre gun, the second, slightly less massive, with the excellent long 75-millimetre. Citadel was not to take place until the beginning of July, and its protagonists believed then and later that the delay was probably fatal.

The situation on the Eastern Front, therefore, was one of menace, with little hope of victory, but with the possibility of staving off ultimate defeat if a limited success could be achieved in mid-summer. On the other vital front on which Germany was fighting in 1943, however, the enemy offensive was intensifying every week, and there appeared little chance of turning the tide. The Allied air attacks on the industrial and population centres of north Germany began, in March, to achieve greater accuracy (although certainly not amounting to precision) than hitherto, and an ever greater weight of bombs and number of aircraft were directed, most nights, at the Reich. These attacks, although they attained nothing like the terrifying destructive capacity of the following year, were already reducing the hearts of some German industrial cities to uninhabitable rubble.

The Anglo–American air offensive in March 1943 was concentrated on the Ruhr, and then in the summer its full weight was turned on Hamburg; but throughout the period other cities were bombed simultaneously, so that all the time the population, the government and the home-based armed forces of the Reich felt threatened and distracted. This distraction had very direct military effect, besides the physical destruction wreaked. By late 1943 900,000 men – nearly one third as many as the total on the Eastern Front – were employed manning the anti-aircraft defences of the homeland; and these amounted to twenty thousand guns, the effect of whose deployment with the field armies may be imagined. In the east and in the air, whatever the palliatory hopes of Citadel, Germany was already at bay.

There had been one consolatory counter-offensive bringing hope, and in May this hope, too, was to be extinguished. Admiral Karl Dönitz, *Befehlshaber* of the German U-boat service,* had introduced a highly centralized and highly successful command system for his submarine 'wolf-packs' in the Atlantic. This success had, like most military successes, largely turned upon the acquisition and bold use of timely intelligence and its denial to the enemy. New cypher machines had been brought into use in February 1942, and until March 1943 the U-boat war had led to such a favourable exchange rate of Allied merchantmen sunk in the Atlantic against U-boats destroyed that on that front, at least, Germany could claim success; strategically an offensive success, since loss of the war at sea could

* From 30 January 1943 he was Commander of the *Kriegsmarine* in succession to Raeder.

cripple the maritime powers whose coalition was aligned against the Reich.

Success turned to failure in May, the month Rommel's 'cure' at home ended and he returned to duty. The British code-breaking agencies, in the first months of 1943, broke the German submarine control cyphers, and Dönitz's masterly but centralized command involved correspondingly greater vulnerability to signal interception. From March the U-boats were reporting ever-greater and more successful (and, very evidently, better-informed) air operations against them. By mid-May the rate of exchange in a number of mass attacks against Allied Atlantic convoys was recognized by the *Kriegsmarine* as disastrous. It reached the point that, in one particular operation, as many U-boats as merchantmen were destroyed. On 24 May Dönitz – as he hoped, temporarily* – called off the battle of the Atlantic and ordered all U-boats back to base.[3]

There were two more fronts, one as yet inactive but both certainly menacing. One was in western Europe itself. Sooner or later the Allies would invade the coast of France. While the battle of the Atlantic went well for Germany this invasion was, it was appreciated, most unlikely; and the appreciation was sound. Continental east–west railway links were good, and German ability to switch troops was considerable. It was reckoned, therefore, that there would be sufficient time to react to a changed and more threatening situation in the west, provided that coastal defences were well-planned and adequately manned. But, more importantly, invasion would surely never be attempted by the enemy unless he reckoned the trans-Atlantic reinforcement and supply routes were comparatively secure.

The benefit of this calculation was now at risk. Increasing Allied air attack would inevitably have a definite and designed effect on rail communications and the availability of rolling stock; and the situation on the Eastern Front clearly made the creation of central reserves – or the withdrawal of forces from east to west – a matter which might well, in the event, vary between dangerous and impossible. But above all the enemy could now use the Atlantic with relative impunity. Sooner or later during 1943 the Western Front would need significant strengthening.

Then there was the Mediterranean.

* * *

* A new offensive was launched in September; by October it had been smashed.

Rommel, in early May, received a letter from the faithful *SS Oberführer* Alfred-Ingemar Berndt, dated the third of the month. Berndt was back in Goebbels's *Reichsministerium fur Volksaufklaring und Propaganda*. His heart would always be with the Afrika Korps, with the Panzerarmee and its leader, and his ear was close to the ground. He enclosed the pages of Rommel's diary, which he had kept, for the period between December and the end of March. Berndt wrote that he was delighted to learn the *Herr Generalfeldmarschall* was again to be given an appointment. It was apparently as yet unclear exactly what this would be or where; but, Berndt added somewhat mysteriously, he had calculated that the sphere of Rommel's future operational interest would span the whole of Europe and cover some fifteen thousand kilometres.[4] The implication could only be of some sort of roving commission responsible (as it ultimately turned out) for much of the coastline of western Europe; even at that one wonders whether Berndt's typewriter slipped a zero.

Whatever Berndt intended, and whatever Rommel made of it, Berndt's letter continued with concrete suggestions which, however peculiar as directed from a *Ministerialdirektor* in the Propaganda Ministry to a Field Marshal on sick leave, indicated not only a nostalgic loyalty to the Afrikaners but an actual sense that Rommel would soon have some sort of renewed and direct involvement with operational matters. Might there not, Berndt suggested, be constituted a special unit, formed from certain selected former officers and soldiers of the Panzerarmee, capable of incorporating its traditions, carrying forward its expertise and available for special tasks? A sort of ex-Panzerarmee elite commando? A *Sturmbataillon*? Berndt suggested its composition – a unit of six companies, including heavy weapons; he even proposed names of previous members of Rommel's command for particular assignments within the unit, a unit of battalion size: Behrendt and Armbruster were both mentioned. It would be called the '*Afrika-Urlauber Bataillon*',* and could surely be organized very quickly (and perhaps temporarily). At all events the Field Marshal, Berndt understood, would shortly need a small staff for his next task, and Berndt had been charged with making contact with some Afrikaners for it. There were plenty of officers and soldiers who had been evacuated at some point from Africa and had escaped death or captivity.

Rommel's absence from the African scene – and the débâcle there

* *Urlauber*: men on leave.

– had been concealed from the German public. Both Hitler and Goebbels wanted to preserve his name, reckoned a publicity asset, clear of the stigma of defeat. It was, therefore, a task requiring some subtlety to publicize (two months after he actually left Tunisia) Rommel's return to Europe. The communiqué which stated that he had not been in Africa since early March was actually composed by Hitler, and released on 10 May.

For final disaster in Tunisia had come. The British entered Tunis on the day, 7 May, that Rommel probably read Berndt's letter. On 8 May he received a telephone call to report to Berlin and shortly after midday on 9 May arrived at Tempelhof airport. At one o'clock he was in the presence of Hitler.

Hitler wanted to talk about the Mediterranean. 'I should have listened to you before,' he told Rommel.[5] He also wanted to reassure Rommel, to renew the bonds of trust and devotion which he knew had, in the past, held this vigorous and headstrong commander to his service. It is remarkable to what extent he succeeded. During the next two months Rommel was constantly in Hitler's company. He often sat next to him at lunch or dinner. When the Führer's peripatetic court moved from Berlin to Rastenburg, the east Prussian '*Wolfsschanze*' from which he directed affairs on the Eastern Front, or back to Bavaria, to the Berchtesgaden mountain retreat where he felt most at home, thither too moved Rommel. Rommel attended the daily 'situation conferences', heard the news on all fronts as reported to Hitler and witnessed Hitler's own speeches on strategy, on geopolitics, on the characteristics of various nations and their leaders, speeches often rambling and diffuse, but sometimes shot with rays of curious insight despite their turgidity.

During these months Rommel undoubtedly again fell to some extent under Hitler's spell. He had exclaimed bitterly in the aftermath of Alamein that Germany's only hope lay in Hitler's removal and in the cessation of policies, immoral, cruel policies, associated with his name (however inaccurately; like most Germans Rommel clung for a long time to the belief that any really evil things done by the Nazi regime were the work of over-mighty and unprincipled subordinates, the Bormanns, the Himmlers of the Third Reich – especially the Himmlers – but not of Hitler personally). Rommel had, with passionate indiscretion, called out in November for change. He had been impelled not by some sudden, Damascene conversion of conscience but by disgust at the unreality and obstinacy of Hitler's strategic orders, orders unnecessarily costing the lives of soldiers.

'Fliegerführer Afrika, General von Waldau, was already referring to the bewildering changes of plan.' The battle of Gazala, May 1942.

'Rommel treated his Italian higher commanders in a somewhat cavalier fashion.' Rommel with Cavallero and Bastico on one of Cavallero's visits to Africa.

'Rommel knew that within Eighth Army Auchinleck had relieved Ritchie on 25 June 1942.'

'Since 15 August 1942 Eighth Army had been under a new commander, General Montgomery.'

Field Marshal Kesselring – 'a shrewd calculator, albeit by temperament an optimist' – with Rommel on one of his visits to North Africa.

'Rommel joined the headquarters of the Afrika Korps, led by General Nehring.' Nehring with Bayerlein, Chief of Staff.

'General von Salmuth, Commander Fifteenth Army, recorded that "a new phase began when Field Marshal Rommel appeared".'

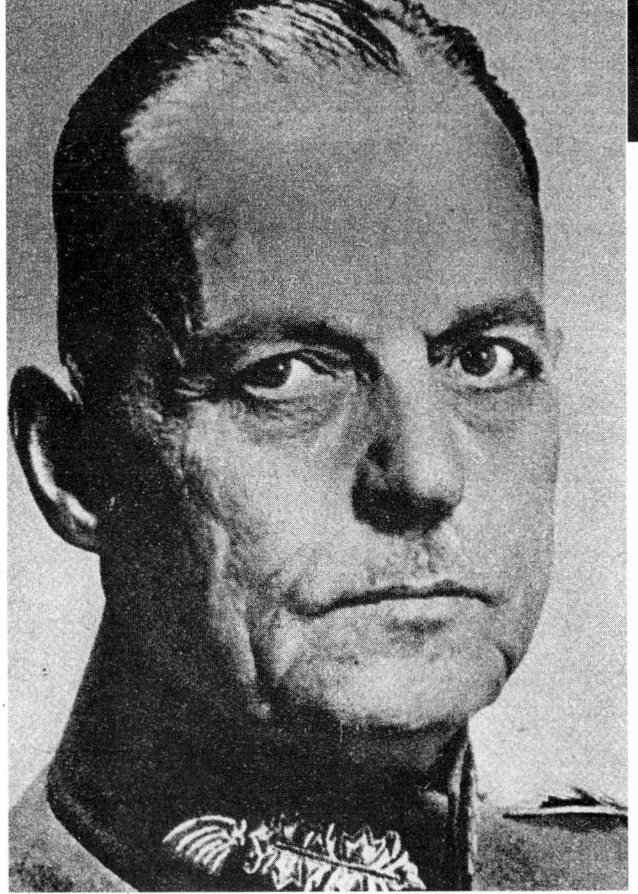

Field Marshal Gerd von Rundstedt, 'an archetypal Prussian cavalryman, cultivated, intelligent, courtly and reserved, the doyen of the German Generalität'.

'Normandy and Brittany were the area of Seventh Army (General Dollmann).'

'General Geyr von Schweppenburg reckoned that a force prepared to deal a powerful, united blow would need to be deployed in reserve, centrally.'

'One officer wholly sympathetic to the conspiracy was General Hans Speidel, Chief of Staff of Army Group B.'

'Kluge told Rommel bluntly that, although a Field Marshal, he must obey orders.'

RIGHT: 'When it was certain that the plan had worked, complementary action was to be taken in the west – in France under the auspices of Befehlshaber Frankreich General von Stülpnagel.'

BELOW: 'Rommel told Burgdorf he was unfit to travel – he had an appointment with the specialist in Tübingen.' Rommel leaves Tübingen after his last appointment.

LEFT: Field Marshal von Rundstedt, Hitler's representative at Rommel's funeral, arriving at the Rathaus of Ulm.

BELOW: Rommel's baton, sword and decorations being carried from the Rathaus.

But now, again in Hitler's presence, again consulted by Hitler as a sort of resident Field Marshal-in-waiting until a definite new appointment took him away, he found something of the old faith returning. He wrote with pleasure in his diary of how obviously Hitler liked having him with him, how repeatedly he showed him his total trust. He wrote with obvious devotion of Hitler's better moods, of his assurance amidst troubles.[6]

And Hitler not only showed confidence: he imparted it. For this extraordinary creature, capable of covert actions and orders of barely believable inhumanity and wickedness, could still inspire, still console, still attract powerful affection and admiration. Rommel had lost one sort of faith in him when ordered to stand fast at Alamein and sacrifice the Panzerarmee. He had lost one sort of affection when Hitler had stormed at him on his winter visit from Africa and told him that his men had failed to fight, and that he, Rommel, lacked resilience and will. But now, watching Hitler working on his problems, sharing with Hitler the agony of those problems, asked by Hitler for advice and sometimes seeing it taken – now Rommel, once again, clearly found loyalty and subordination not irksome but congenial. He could, once again, believe: or at least half-believe. In a revealing phrase Rommel said to von Manstein at the first meeting of the two men at Rastenburg in July 1943 – Manstein, with Kluge, had been summoned from the Eastern Front at the crisis of Citadel – 'I am here for a sunray cure. I am soaking up sun and faith!' Manstein had never met Rommel before; the circumstances were particularly informal since Manstein had been using an unexpectedly free afternoon to bathe in a lake, and, unprepared for the opportunity, was naked. On swimming to the bank he had found Rommel and a number of others laughing cheerfully, all due to attend the Führer's evening conference. Manstein was not a man to be abashed by those or any other circumstances, and had asked Rommel what he was doing at Rastenburg. He understood the answer, and when he asked if they would meet later he also understood Rommel's 'Yes, under the sunray lamp!'[7]

The sunray lamp! There may have been irony in Rommel's phrase, but this was indeed the effect Hitler could have on those around him. He could warm; and they would, if only for a little, feel transformed. Of course, in Rommel, this may be ascribed to his naïvety; and in many matters he was naïve. It may be attributed to Hitler's power and subtlety of flattery; and to Rommel's susceptibility and vanity. The charge would surely have much substance. But for a hard-bitten, successful and principled man to succumb as Rommel periodically succumbed to the enchantment of Adolf Hitler implies strong magic; magic, furthermore, operating at a time when Rommel was writing his own very personal account of the tragedies and lessons of North Africa, an account packed with criticism of German conduct of the war.[8] And the magic was indeed strong – strong enough to bewitch for a considerable time a large part of the German people. It could do so because it was rooted in actual past achievement as well as in promises and prejudice; in a memory of deliverance from guilt and poverty as well as in the imposition of present tyranny and pervasive fear. There were certainly some in Germany and in the German army who never felt the magic, who despised the magician for his lies, for his lack of principle, for his ignorant assumption of strategic wisdom at the cost of good men's lives, for his entourage of self-seeking and merciless sycophants, for – as they in at least some cases perceived it – his unrepentant wickedness. To Rommel this manifested itself only slowly and partially. He owed loyalty to the head of state and to the Supreme Commander. He had eaten his bread, received his trust. He would not find it easy to betray it.

Furthermore, Rommel still believed in the cause, in the essential justice of the war. He was a patriot. It is easy for Germany's opponents of those days to suppose that the aggressive war Hitler began was impossible for a man of principle to support, but this was not in the least so. To Rommel the situation of Germany in 1939 and 1940 had justified the campaigns in Poland and in Norway and in France, as either corrections of the intolerable consequences of Versailles or as pre-emptive moves against an implacable coalition of enemies. The campaign against the Soviet Union was widely seen as forestalling Soviet attack westward – a proposition with little supporting evidence (which troubled Hitler little); while the subsequent visions of German settlement in the east following victory were seen as possible resolutions of an ancient problem – the security of the eastern marches – and of modern problems, of contagious and destructive Communism as well as of economic space for German prosperity.

As the eastern campaign went on, of course – and went from bad to worse – the issue, for every German including Rommel, became one of the defence of Europe and ultimately of the German homeland against what was regarded as a savage and uncivilized foe. Rommel, like most of his compatriots, reckoned Germany's cause was good; the cause of a Europe rationally ordered and defended, very largely, by the energy and self-sacrifice of Germans. In western and southern Europe Rommel thought that reasonable calculation of their own self-interest by the Italians and the French – and the British – would always ultimately lead to friendship with Germany. When Rommel, therefore, renewed his faith in Hitler and stilled for a little his angry doubts about the quality of Hitler's leadership, he was not recovering confidence in a Führer whose fundamental purposes he had come to reject, at least in foreign and strategic policy; while of that Führer's domestic and racial policy his knowledge was still small.

Rommel had, at this time, the opportunity for several general talks about the war situation, and he tried to emphasize to Hitler what had by now become his own settled conviction – that, whatever the rights and wrongs of her cause, Germany was now opposed by too great a military, industrial and manpower strength to hope for outright victory. Rommel had come to this conclusion the previous winter, since the twin calamities of Stalingrad and North Africa; and his belief was now daily reinforced by the air and sea campaigns from which Germany was emerging defeated. Hitler at least listened to him; and did not, it seemed, fundamentally disagree. It was during this period, when Rommel saw much of him, that Hitler, in a melancholy phrase which undoubtedly stayed in Rommel's mind, said 'Nobody will make peace with me.'[9] Later, reflecting, Rommel suspected that Hitler was already perfectly aware that the war was lost. Hitler would be able to deceive himself and others; and the challenges of crises stimulated him for short whiles: but he already lived under the shadow of impending disaster.

Rommel also sensed, at times, a fatalistic, ruthless and self-immolatory instinct in Hitler which shook him deeply and which, he afterwards confided to Lucy, made him feel that Hitler was no longer 'quite normal' (a revealing expression: when had Hitler been quite normal?). The occasion was in July, when Hitler referred to the actual possibility of defeat and said that if the German people lost the war the survivors could rot: a great people, he said, 'must die heroically'.[10] Rommel had been appalled, but from such moments his faith, 'warmed by the sunray lamp', had nevertheless for a little recovered.

He always, however, retained his determination to be his own man, to be loyal but not under excessive personal obligation. When it was delicately put to him after his African victories that Hitler had in mind to make him a present, a farm, Rommel made it clear that he would in no circumstances accept. Others had done so. Guderian had done so. But Rommel said (in 1943) to a brother-officer that Guderian had erred in his acceptance; 'I'd like a farm too, but I've always said "no" and refused.'[11] He must be free to speak unpalatable truths.

Rommel's new assignment was at first ill-defined. It depended upon the Italian situation. After the Tunisian tragedy, it was necessary to consider what Germany's policy in the Mediterranean should be. At their first meeting on 9 May Rommel had emphasized to Hitler the inferiority in quality of the Italian armed forces, a line he took with equal emphasis when talking to Goebbels the following day.[12] Little should be hoped for from that quarter.

And there was the political dimension. Most Germans knew that in the aftermath of the Tunisian débâcle influential circles in Italy would like to make peace with the Allies and desert Germany. Hitler believed that Mussolini, personally, wished to remain faithful; but he had few illusions about the extent to which Mussolini could hold the Italians to the German alliance if the going got rough – or rougher. Rommel made it perfectly clear that in his view the Italians would not fight if the Anglo–Americans invaded Italy or any part of it. There was also, of course, the possibility that the Italians would not only cease to fight, but would actually change sides and cooperate with that enemy against Germany.

Rommel's pessimistic view of the Italians was not universally held. On one occasion he took the young and perceptive Constantin von Neurath of the Foreign Service with him to see Hitler. Ribbentrop, the Foreign Minister, was present. Von Neurath, who was particularly well-informed on the atmosphere and intrigues in Italy, and Ribbentrop gave totally opposing views, and Rommel afterwards told Neurath that in his opinion Ribbentrop was the greatest idiot about foreign affairs he had ever encountered – this was the Foreign Minister of the Reich! Neurath's opinions had been disquieting (and, in the event, accurate), and Keitel, also present, had characteristically confined himself to saying reprovingly later that it was necessary to spare the Führer disagreeable news.[13] Meanwhile, in the aftermath of Tunisia, a limited number of German troops were already committed to Italian counter-invasion defence. There were two Panzer grenadier divisions in Sicily, one of them the 'Hermann Goering' Luftwaffe Division.

It was to deal with the Italian situation that Rommel had been summoned to see Hitler. Now that North Africa was lost, an enemy invasion of some part or parts of southern Europe must be anticipated during the summer. One obvious possibility lay in Greece and the Balkans where there were significant Italian – and smaller German – garrisons. Another, however, was Italy itself. Any invasion of southern Europe, anywhere, was likely to act as a distraction from the Eastern Front, from Citadel's start in early July; and invasion of Italy, particularly if it led to the defection of Italy, was likely to bring the Anglo–Americans – and the Anglo–American bombers – closer to the southern frontiers of the Reich; from which it followed that such an enemy offensive should be held as far away as could be achieved, provided that this made operational sense on the ground.

Two German plans were made by OKW at this time, and Rommel was directly concerned in both. The first, Operation *Alaric*, was intended to infiltrate considerable additional German forces into north Italy, ready to defend Italy against Allied invasion or to hold such invasion as far south as possible for as long as possible – on the reasonable assumption that the Italians could not be relied upon to do it themselves.

For *Alaric* Rommel was instructed to prepare plans for the gradual 'dribbling' into Italy of a force, assembled under his command in Austria and Bavaria, amounting to some twenty divisions; or such was Hitler's thinking. On 22 May Hitler, Rommel's diary noted, signed the order for this new task. The preliminary 'infiltration' would be of four German divisions, if and when the Allies invaded and the word was given, followed by a further sixteen. It was clear that *Alaric* would need sensitive political handling. During the planning and preliminary stages, in June and early July, nothing could be done to give Italy reasonable grounds to object to a planned and uninvited intrusion on to her soil and into her business. The proprieties had to be observed. The German–Italian alliance was by now not of a kind which permitted frank joint planning on both the military and political levels. But then it never had been, although the mistrust by summer 1943 was sharp indeed.

The second OKW plan, *Achse*, was a very different affair, and the need for secrecy was even more imperative. *Achse*, too, would be Rommel's responsibility. The plan envisaged actual Italian defection, and the consequent urgent necessity to disarm the Italian armed forces, take possession of their equipment and, if need be, crush or capture them as enemies of the Reich. On the success of *Achse*, if it became necessary, would obviously depend the continuing German ability to supply and control their own troops already committed south to the defence of Italy – and possibly, by that stage, considerably increased in numbers in response to one or more enemy operations. It was clear that the detailed implementation of *Achse* would depend on circumstances; and that it would probably follow *Alaric*, the troops involved in which would have to be responsible for *Achse*. Following or simultaneously with both operations would come the purely operational question of how to conduct an Italian campaign against an Anglo–American enemy.

Rommel quickly assembled a small staff, some of whom, as Berndt had foretold and helped arrange, came from his old Afrikaners. General Gause, the steady, principled east Prussian, joined, as did Colonel von Bonin, also from the Panzerarmee. Rommel spent a good deal of time in June flying between Berlin, Rastenburg and his new headquarters installed in a barracks near Berchtesgaden. On 7 June he had a brief meeting with Guderian, at Guderian's request, in a Munich hotel. Guderian was now – after a long period in the shadows after disagreement with Hitler – restored to office as Inspector-General of Panzer troops and was busy trying to bring rationality to the generally chaotic system of industrial priorities within the Reich in a way which would help a smooth flow of tank production. Rommel, at this time, was convinced that the emphasis of German production should be on defensive weapons and anti-tank guns. A good many anti-tank guns could be produced with the metal and money, let alone the expensive technology, required for one tank. What would henceforth be required, he was convinced, would be defensive firepower and, above all, *numbers*; especially in the east.[14] It was a striking conclusion by the master of manoeuvre. And periodically Rommel managed to get a day or two at home at Wiener Neustadt.

He was there on 27 June; and one week later, on the afternoon of 4 July, Citadel began, with forty-three German divisions, including eighteen Panzer or Panzer grenadier, launched in the attack.

For the first few days the operation seemed successful, but it was soon evident that the weeks of delay imposed by Hitler had not worked to German benefit. The Russians had fortified their lines in considerable strength. The two attacking armies – from the north, General Model's Ninth Army from Army Group Centre and, from the south, General Hoth's Fourth Panzer Army from Army Group South – quickly found that the Soviet defences were not only in great depth and laid out with exemplary thoroughness, but were covered by mines (including both anti-tank and anti-personnel) laid to a density of over five thousand per mile of front. This was only about a third of the German mine density at Alamein, but it was nevertheless formidable. On 9 July Rommel, visiting Rastenburg, recorded that the attack seemed to be going well; but in truth it had already lost momentum. Furthermore the Red Army had assembled a considerable operational reserve in the east of the Kursk salient ready to intervene against the flanks of the German attacks when the tide began to ebb, as it soon did.

Next day, 10 July, there came the news for which all had been waiting but which all had hoped would not come until Citadel reached a successful conclusion. British and American troops had started to land in southern Europe from air and sea, in Sicily. This was the enemy's Operation 'Husky'.

Husky marked the return of the British and the introduction of the Americans to Continental Europe. The first major German decision was to send more German troops to Sicily, and XIV Panzer Corps headquarters (General Hube) arrived five days later, followed by a Panzer grenadier division and two regiments of parachutists; these would make up a significant German force in Sicily. The second major decision was based on the realization that Sicily could not possibly be long held: operations must be conducted in such a way that the troops could withdraw to a defensible bridgehead in the north-east of the island and cross the Straits of Messina to the Italian mainland, when ordered, intact. The third major decision was that the battle must be conducted under German orders. Nominally, an Italian general was in overall command, but in practice Hube was soon directing operations and would ultimately (one month later) effect a highly successful withdrawal. Relations between the Germans and Italians were watchful and wary, and German mistrust of the Italians and of their readiness to continue the struggle was not confined to the High Command.

Citadel was faltering. Three days after the news of Husky reached Rastenburg Hitler recalled from the Eastern Front the Army Group Commanders concerned, Kluge and Manstein, and told them that because of the threat to Italy – the threat, indeed, of total collapse – the Kursk offensive must cease and troops be withdrawn to their original start lines. With the calling off of Citadel came the final loss of German strategic initiative on the Eastern Front. It had been designed as a spoiling and limited offensive, to establish the conditions for a successful German defensive campaign. The design failed.

That evening Kluge and Manstein talked freely with Rommel before going to bed. It was the evening of Rommel's first meeting with Manstein: the bathing encounter.

'Manstein,' Kluge said, 'the end will be bad. I am prepared to serve under you.'[15]

He left Manstein alone with Rommel. Rommel spoke his own mind. The end, he said, would be disaster. He had been preaching his pessimism to Hitler whenever opportunity allowed during the previous two months. Now he voiced it to Manstein, the most respected strategic brain in the German army. He said that he envisaged total catastrophe, that when the Allies landed in Europe the whole 'house of cards' would collapse.

Manstein said that perhaps the Führer would agree to give up the command which he was exercising so disastrously. Then, perhaps, it would be possible to work towards a stalemate and the negotiation of an acceptable peace. Rommel disagreed. He knew Hitler better than Manstein, and Hitler, he said, would *never* give up the supreme command. Manstein, 'strategist of genius' though he was, was cherishing an illusion. And, like Kluge, Rommel said: 'I also am prepared to serve under you.' Rommel's personal devotion to Hitler may have to some degree returned. His military judgement, however, was unaffected.

On 15 July, the day Hube reached Sicily, Rommel was formally appointed to the command of 'Army Group B' – the name now given to the formation staff he had assembled in Bavaria. It was envisaged that sooner rather than later – for the loss of Sicily was assured, and indeed intended – the main enemy invasion would threaten the mainland. Army Group B would in that case assume responsibility for central Italy. But this could not happen until events had moved on; and it could not overtly happen without Italian concurrence. Or desertion.

The battle of Sicily ground on, far from Rastenburg. On 18 July the Russians began a major counter-offensive along the whole Eastern Front, and on 20 July there were reports of a major breakthrough, reports which turned out to be exaggerated. On the twenty-fourth Rommel met his Army Group B staff together for the first time and addressed them in Schloss Muhlhof in Payerbach in Bavaria. It was still unclear when, where and for what mission they would first be needed; *Alaric* and *Achse* were held ready, but on the previous day,

23 July, Rommel had been given a startling new assignment. He was to fly to Greece to assess the situation there. Army Group B might be used not for Italy but to take over command of the defences of Greece. After Sicily the enemy, instead of heading for Italy, might make a descent on Greece.

Rommel flew to Salonika on the twenty-fifth and started discussing and investigating. He was there for less than twenty-four hours before urgent recall to Rastenburg. In Rome the Fascist Grand Council had met – a body which for years had done little but act as rubber stamp to the Duce's policies. Now, in a well-engineered initiative, it had confronted Mussolini and taken a vote on his future. The vote had gone against him – eighteen out of a total of twenty-eight had opposed the Duce's continuance in power. When Rommel arrived back at the *Wolfsschanze* at midday on 26 July he found considerable confusion. Hitler's most reliable ally had been voted out of office by his own associates and was apparently under arrest! The Italian Government was in the hands of Marshal Badoglio. Within two days Rommel's diary was referring to news from Italy of a hunting-down of Mussolini's supporters, a *Fascistenjagd* under Badoglio, which did not sound promising for the future of Italy's alliance with Nazi Germany.

There was a significant German presence in Italy. Kesselring was still in Rome as *Oberbefehlshaber Sud*. Seventy thousand German soldiers were fighting in Sicily who would need sustaining and ultimately withdrawing. Between them and the Reich was a long, mountainous country with difficult communications and, it seemed, a most unpredictable political orientation. Hitler, outraged, was proclaiming at his conferences that the Italians were wholly untrustworthy; they were declaring that nothing had changed, that Mussolini's displacement was a purely domestic matter, that Italy was loyal to the Axis cause and would fight on. Hitler said that he didn't believe a word of it; and nor did Rommel. Hitler was in favour of instant action. Rommel, however, believed that the situation was complex, and demanded deliberate and well-prepared measures. Hitler's was the impulsive, Rommel's the cautious voice on this occasion, as Goebbels noted in his diary.

Alaric was now ordered. The move to Greece was forgotten, and Rommel began urgent discussions on the security of the Alpine passes into north Italy and the pretexts under which these might be occupied. It was one thing to infiltrate troops; it was another to ensure a secure line of communication to them thereafter. And all this – still – required preservation of the pretence that it was to assist the Italians, despite any protestations which might be made that the latter had not asked for assistance. 'Although the Italians,' Rommel's diary noted plainly, 'will obviously betray us, it's not politically possible to march in.'[16] Information was received daily that contacts between the Italians and the Allies were taking place in Lisbon and elsewhere, and that Italy would request an armistice in less than two weeks.[17] On 30 July troops started moving south, after discussions between Rommel, General Feuerstein commanding the German troops, and the Gauleiter of the Tyrol about route security and the attitude of the Italian people.[18]

Rommel realized that to secure the Alpine passes should pre-empt an impossible situation later, the situation which might arise if the Italians were actually to change sides – or object so vigorously to the German initiative that they took military action; but he was very alive to the political nuances. Everything must be done to preserve as long as possible correct dealings with the Italian authorities, Italian officials and Italian troops. He issued orders to all German troops about the supreme importance of preserving good relations with the local Italian population.[19] Rommel, according to OKW instructions of 1 August, would have freedom of movement and command of all troops throughout Italy on *Achse* being issued, if it were (instructions which circumstances were to modify).[20] Meanwhile he himself had been forbidden to enter Italy, and he was faithful to the order.

Rommel insisted that all soldiers must understand that the Germans were entering Italy as guests, to bring urgent help to Italy.[21] He himself set a most careful example, conscious that to some (in both Italy and Germany) he had the name of an '*Italiener-Hasser*' and angrily rebutting what he called a dangerous slur, '*dieses gefährliche Renommee*'.[22] But the position was psychologically difficult to maintain. Despite Italian protestations of unswerving devotion to the alliance it was known to German intelligence that two-thirds of the King of Italy's Royal Council had voted to end the war at a meeting on the evening of 7 August; and Rommel's trusted associate from the Foreign Ministry, von Neurath, brought news of contacts between the Italian Government and General Eisenhower, the Allied Supreme Commander, which accorded with all the reports reaching Army Group B daily and which Rommel found only too persuasive. On 11 August Rommel was with Hitler at Rastenburg and found him in complete agreement as to the unreliability of the Italians, as well as in general agreement on the successive defensive lines which should be manned in an Italian campaign. On 13 August Gause, now Army Group B's Chief of Staff, gave a presentation on the Italian situation which put very clearly the mistrust he and all of them felt. By then the equivalent of about six German divisions, including 26th Panzer and the *SS Leibstandart Adolf Hitler* had crossed the Austrian–Italian frontier, 26th Panzer entering Bolzano to a tumultuous popular welcome.

The question of higher command in Italy was, perhaps inevitably, hanging fire. Kesselring was somewhat out of favour as having been inordinately sympathetic to the Italians.[23] The matter, was, obviously, vitally affected by whether the Italians were to be assumed still in the war, and if so on which side! It was also crucially affected by assessment of the Allied threat to Italy. The German withdrawal from Sicily was complete by 16 August, twenty-seven thousand men and seven thousand tons of stores being brought safely across in the last five days. Rommel, before the fighting in Sicily ended, had proposed a unified command for the whole of Italy, with one army area of responsibility in the north and one in the south (an arrangement which ultimately took place), and in July Hitler had certainly considered Rommel for that command.[24]

If the enemy were to land in the far south (thought unlikely), there should clearly be planned and successive defensive lines from south to north. Rommel suggested the general lines of Cosanzo (that is, across the ankle of Italy), Salerno, Cassino and the Apennines, but he never believed that too much should be invested in attempting to defend southern Italy. It would be logistically laborious and – to him and to most of the German command a compelling factor – if the Allies invaded Italy at all, they were bound at some time to use their superior maritime power to land forces in the centre or north and perhaps on both the Mediterranean and Adriatic coasts. Too much weight deployed too far south could become a hostage to fortune. Against that were the attractions of holding an enemy – and that enemy's bombers – as far south as possible.

Rommel was now allowed by Hitler personally to enter Italy and to bring *Alaric* from the shadows into the open. The Italians were perfectly aware of what was happening, but if the German moves could be discussed with them under some sort of cloak of normality the moment of crisis, the moment when they had to declare their hand, might be delayed, even averted. On 15 August Rommel flew to Bologna, as *Oberbefehlshabler Heeresgruppe B*, and was greeted at the airport by an SS guard of honour. In a villa outside the city he took part in a conference with the Italian military authorities intended to settle – as far as possible consonant with a pretence of normal allied arrangements – the deployment, the quartering and the communications of Army Group B; as well as the essential matter of the defence of Italy and who was to be responsible for what parts of it.

Inevitably the whole question of Italian governmental policy was introduced. The Italian side was led by the Chief of Staff of the Italian army, General Roatta, anxious to defend the Italians against any suspicion of unenthusiasm or double-dealing, quick to protest at any words which could carry that imputation. On the German side General Jodl, Chief of Operations at OKW, defended the Germans against charges of unfriendly suspicions; he pointed out the crucial nature of German supply lines to their own troops, troops brought to Italy both from Germany and from Sicily to help defend Italy. It was natural, Jodl said on a different and credible tack, that Germany should feel disquiet at the displacement of what all had recognized as a completely pro-German government, that of the Duce, by one whose sympathies were as yet unclear.[25] This led to more Italian protests; their word had been given, Italy was, exactly as heretofore, loyal to the Axis and a devoted partner in the war.

Italian demands – predictable demands – that all German forces in Italy should be under the ultimate authority of Commando Supremo were rejected as unacceptable, and no 'agreements' were made. The conference had been convoked to settle matters of the greatest importance if Italy was to be defended and if the Wehrmacht was to play a vital part in that defence. In fact, and inevitably, it consisted of verbal fencing by men whose mutual mistrust was almost without limit. Indeed, the Italian Government's first formal overtures to Eisenhower were made on that very day.

Next day Rommel flew to Innsbruck with Gause and confirmed the movement details for the rest of Army Group B's movement into Italy to join the divisions already deployed. Plans for *Achse* were reviewed and confirmed on a contingency basis,[26] and it is difficult to believe that his contacts with Roatta had persuaded Rommel that *Achse* was unlikely to have relevance.

Rommel's headquarters were to be established on Lake Garda. He had heard that Kesselring, on learning that Rommel was to hold the united command of all German troops in Italy, had asked permission to resign. Rommel had always regarded Kesselring as unduly receptive of Italian views; he knew that the Italians, or some of them, would oppose any move to give him, Rommel, superior responsibility in Italy – Ambrosio complained about the possibility on 17 August.[27] Rommel was indifferent to such views: his conscience was clear about being an *Italiener-Hasser*, but he had not the smallest doubt that within days or weeks Germany's ally would betray her. All this was play.

Meanwhile the air attacks on Germany were intensifying, and Rommel was becoming anxious about his family and possessions at Wiener Neustadt, and contemplating moves to reduce their vulnerability. The first really big raid on Berlin took place on 23 August, and on the twenty-fifth Rommel read in a Swiss paper that Wiener Neustadt itself (site of an aircraft factory) had been attacked. All was well with his family, but he fretted.

On 30 August SS General Wolff and a party of senior SS officers arrived at Rommel's headquarters 'to study internal security requirements in Italy'.[28]

On 3 September Anglo-American forces started landing in southern Italy, at Reggio, opposite Messina. Next day Rommel flew to Rastenburg for a meeting with Hitler, and lunch. Hitler seemed calm – he made, Rommel wrote in his diary, '*einen ruhigen zuversichtlichen Eindruck*'.* He agreed with Rommel's assessment of the importance of coastal defence in Italy and the probability that the Allies would attempt further to exploit their maritime ascendancy.

On 8 September the announcement of an armistice between Italy and the Allies was made over Rome radio and greeted with considerable popular acclamation. Henceforth the Italians would be neutral in the struggle or, at worst, actually hostile. If the war was to be fought on Italian soil it would be fought by Germany. The long-awaited dénouement had come. At ten to eight on the morning of 8 September *Achse* had been ordered. The ability of the Italians to damage the German forces was to be broken.

In the early hours of 9 September the enemy began to land forces in the bay of Salerno, south of Naples.

In northern Italy Rommel had, by now, eight divisions in Army Group B. His plans for *Achse* had been laid thoroughly and were soon swinging into action. Within two days he received news of Italians actually siding with the enemy against German troops in the south. His own orders under *Achse* were to surround and disarm all

* A quiet, confident impression.

Italian troops and to evacuate them to Germany as prisoners of war, and on 19 September, eleven days after the codeword was given, Army Group B reported that eighty-two Italian generals, thirteen thousand officers and 402,600 soldiers had been thus dealt with, of whom 183,300 had been already transported to Germany.[29] It was now generally and accurately presumed that the Italians had not simply dropped out of the war: they were changing sides, and would soon seek 'co-belligerency' status.

There was also, of course, the question of the Italian civilian population, among whom a significant – and, here and there, very formidable – anti-German partisan movement took shape, ultimately tying down many German troops. Then there were Allied prisoners of war incarcerated in Italian camps. These were to be secured and sent to the Reich and any Italians helping them were to be dealt with mercilessly. There was also always the threat of systematic or sporadic sabotage. Two young Italians were arrested near Garda attempting an act of sabotage and were condemned to death, a verdict Rommel promptly overturned.[30] A hard man when necessary, he took little pleasure in punishment if it could be avoided, and he was popular on Garda-side. Nevertheless Rommel was angered by the Italian change of sides, however expected. On 23 September he issued and signed an order to the whole Army Group saying that the Italian who turned on his previous comrades had forfeited every right to lenient treatment and must be dealt with with the harshness appropriate to a creature who suddenly used weapons against his own friend.[31]

Meanwhile battle had been raging in the south. German troops had lost no time in moving and taking over defence of key ground and installations from the Italians. By the time of the Allied landings at Salerno six German divisions, composing Tenth Army under General von Vietinghoff and including Hube's men from Sicily, were deployed south of Rome. But by 16 September the Allied beachhead had resisted every attempt to throw the invading forces back into the sea, and the Germans had withdrawn.

There was going to be an Italian campaign. The enemy were established on the mainland of Europe. The German army was fighting on two fronts, one in Italy narrow and essentially defensible, the other, in Russia, vast and with little natural defensibility; but two fronts nevertheless. And on the Channel coast a third front was certainly an event, potentially a decisive event, waiting to happen.

* * *

Rommel's remaining ten weeks in Italy were frustrating and undramatic, except for an acute appendicitis followed by a successful operation in mid-September. He was in the north, far from the scene of action. He believed it essential that the whole Italian theatre be under one unified command, and he was encouraged from October onwards to believe that Army Group B would form this command and that he would have supreme responsibility. He believed – he had consistently believed since first addressing the Italian operational problem some months earlier – that it was a mistake to commit too much strength to holding south of Rome and that a more durable defence, given time, could be built in the Apennines south of the Po – in what ultimately became the Gothic Line. Even that, in Rommel's view, should be but an outpost line.

In this his view had always been affected by his estimate of Allied ability to land from the sea in the north; and it insufficiently allowed for Allied reluctance to undertake amphibious operations save within the cover of land-based air forces. Because Rommel was still hugely influenced by his desert experiences of enemy air superiority, he underestimated the caution the enemy would show. He also underestimated their determination, especially that of the British, never to fight except when supported by that same superiority. To Rommel Alamein and Alam Halfa had been traumatic, their effects ineradicable. To the British Dunkirk had been equally traumatic, its effects equally ineradicable. Although they might feint with, attempt to deceive by, even discuss, operations north of the range of fighter cover, Rommel's opponents would not seriously commit forces to them.

Under whatever arguments, and despite Rommel, the decision to hold strongly south of Rome was made. It was consistent with Kesselring's – still *OB Sud*'s – recommendation; the mountains between Naples and Rome offered many immensely defensible positions. The decision was immediately known to the enemy via ULTRA: and it set the character of the ensuing Italian campaign.

Whether it was wise from the German point of view turned on so many variable assumptions that the debate is not yet ended. Certainly in purely operational terms the defence of the Italian peninsula became more difficult for the Germans as it broadened north of Rome. Certainly if the object was to hold the enemy as far as possible from the Reich the strategy was defensible. And certainly Rommel's principal objection to it – the fear of being cut off by amphibious operations further north – proved insubstantial, and perhaps could have always been assessed as insubstantial in view of the limitations aircraft range (and, of course, landing craft shortages, less apparent to the Germans) imposed on Anglo-American operations. There was a further point. The Wehrmacht was heavily committed to the Balkans and would be more heavily committed now that Italian troops there would need replacing. An enemy descent on the Balkans could be the easier if that enemy held southern and central Italy undisputed.

On the other hand – and, again, the debate is not ended – the enemy, too, welcomed the German decision to stand and fight south of Rome. There were and there remain sharply differing views on the Allied side about the benefits and the significance of the contest in Italy, but to Brooke (for instance) Kesselring's policy and Hitler's endorsement of it could only lead to a more extensive German commitment in Italy, while a German withdrawal to north Italy might have indicated an intention to reduce that commitment to the minimum (as Rommel wished).[32] The longer the German line of communication, the more vulnerable and laborious it would be, and the more resources committed to it. The object of the Allied invasion of Italy was to embroil there the maximum number of German troops and to keep them from reinforcing other fronts. This could only be achieved by a sufficiently prolonged campaign, fought with sufficient intensity. German commitment south of Rome promised this. The Allies, of course, were also making a major commitment – much more so than the Germans, because of the expensive nature of establishing and supplying by sea a base far from home, apart from the economy of force for which a defending army can hope on defensible terrain. But the Allies could afford the investment the more; and – the crucial factor, which Rommel certainly never ignored – what mattered was not the totality of the investment but where troops were committed. German troops committed to Italy could not simultaneously repel attacks elsewhere; and when the crunch, the third or north-west European crunch, came and the decisive battle began there as it must, it would be Germany, not her enemies, that would be short of soldiers.

In all this Rommel's instincts accorded with the views of the most perceptive among his enemies. He recognized the operational and tactical arguments for holding well to the south, and he certainly recognized the Balkan factor; but he was sure – had been sure for nearly a year – that the ultimate battles would be for Germany itself, and that to achieve the hope of a tolerable stalemate in those battles maximum force would, sooner or later, need concentrating in east and west, not south. The war could not possibly be won – the balance of resources and of manpower made that by now certain. As he had said to Manstein, catastrophe threatened. If, however, the available forces were concentrated at the points of decision it was, he thought, possible that an initially successful defence in the west might leave Germany capable of imposing some sort of stalemate in the east and negotiating with the Western Allies some sort of peace. He had given up much hope by now of wholly persuading Hitler to this view, but he still clung to the belief that Hitler – despite his illusions, his power of self-deception, his manic obsessions – could be brought to see strategic realities if presented clearly by a man he trusted; or had trusted once. Rommel believed – wrongly and to a sadly late hour – that he had powers of persuasion to which Hitler could yield if they were personally deployed. He also believed that he might possess the power to outface, to reduce, to oppose Hitler. Obscurely, he said to a friend at this time that he knew he, Rommel, alone was capable of 'undertaking something against Hitler'. What this might have meant will never be clearly known.[33]

These matters had not yet found a catalyst. Partisan activity created much work and many 'cleaning operations' – including in Istria, abutting Croatia, for which Army Group B was also responsible and where the partisan bands were gathered now under the control of the *Kroatischer Bandenführer* Tito who, it was reported, had tried to establish authority over all parties with a promise of Slovene independence.[34] On the main front two Panzer corps were manning a front from Mediterranean to Adriatic, between Gaeta and Pescara, and south of Cassino. Rommel, in the north, had nine divisions organized in four corps over which an army commander – General von Mackensen, Fourteenth Army – had now been set. He continued to prepare plans to repel further seaborne invasions, and he grumbled away at the absence of an overall Italian command – he still understood that this was intended, and intended for him, but he realized that his scepticism about German strategy was unlikely to work in his favour, and when he heard that Kesselring, instead of being posted to Norway as planned, was to assume supreme responsibility in Italy, he shrugged his shoulders.

Winter was approaching. Everything Rommel heard from the southern battlefront confirmed his view that German strength was being consumed there to no good purpose.[35] He had tried his best to bring Hitler to his way of thinking but without success; and he supposed that the appointment of Kesselring was yet another indicator that his, Rommel's, views were still relatively unvalued.[36]

Rommel's short time with Army Group B at his headquarters on Lake Garda was, therefore, militarily frustrating and personally depressing. He had a taste of intolerable SS and Party behaviour, of looting by senior officers; of atrocity – the drowning by the SS of some Jews in Lake Garda;[37] and of attempts to interfere with his own responsibilities, as, for instance, by the appointment of a Gauleiter for Istria and Dalmatia who would have responsibilities in conflict with Rommel's operational plans for the head of the Adriatic – the subject of a furious telephone call to Jodl in November.[38] Rommel reacted fiercely to outrages when he heard of them, and he was angered by information about German soldiers trading prosperously on the Italian black market and sending illicit purchases to Germany on a large scale, ordering the severest disciplinary measures.[39] He watched a campaign unfold from which he expected no benefits to come, without playing any positive part in it; and it was unfolding in a war whose outcome he regarded with the deepest pessimism.

Mussolini had been rescued from his Italian captors by a brilliant German airborne commando operation, ordered (with some chivalry) by Hitler, and was living on Lake Garda with an SS guard. Rommel called on Mussolini one day in October, and took the opportunity to tell him some unpalatable truths about the North African campaign. Rommel's feelings about the Italians – including the Italians of Mussolini's regime, under whose nominal direction he had fought in Africa – were far from tender at this time, particularly since (after *Achse*) he had discovered immense quantities of Italian military equipment in store, all dating from the period when he had been assured by Commando Supremo that they had nothing with which to improve the materiel of their men in North Africa, that the cupboard was bare.[40] Rommel had always supposed the Italians unwilling to take the needs of war seriously, and he supposed so until the end. To a large extent this was disillusion born of experience; but the anger of a German officer who found yesterday's friends today's enemies should not be underrated. Rommel knew – and, being human, hated – the fact that the Italian switch of allegiance had been long planned and had lain beneath the surface of many of his own cordial conversations and arrangements.

Then came new orders. Rommel, with the staff of Army Group B, was to move to France and a fresh task. He was to inspect the defences of the west and make recommendations. He left Italy for good on 21 November, without regrets. Lucy had moved from Wiener Neustadt to a house near Ulm, made available to her by the widow of the owner of a brewery killed in an air raid, and from which she would move, early in the following year, to a house in Herrlingen owned by the city of Ulm – the erstwhile property of an 'emigrated Jew'.[41]

Rommel had been disappointed at not being given the supreme Italy command, but he would have had inescapable difficulties with the conduct of a strategy in which he did not believe and for a purpose of which he was sceptical. When he said goodbye to those of his entourage remaining he spoke with indiscreet frankness and feeling. The war, he said, was as good as lost. Hard times must lie ahead. The enemy was increasing in strength daily. Talk of 'wonder weapons' was propaganda and bluff, and above the German soldier were 'people' whose fantasies amounted to delusion.[42] With which side, a particularly trusted staff officer, asked him boldly, do we make peace? Rommel answered him briefly: there could be no question of peace with the enemy in the east. The inference was utterly clear.[43]

CHAPTER TWENTY

Invasion

ROMMEL SPENT a week at home from 22 November 1943, and was able to see certain extensions to the Herrlingen house which had been authorized, take some long walks, and to think, without pressure of instant decisions.

His next task, a special *Führerauftrag*, was to be an examination of the defences – primarily the coastal defences – in the west, reporting directly to Hitler. This would involve clarifying his mind on the operational philosophy appropriate against the obviously imminent invasion of western Europe: but it also involved clarity about the strategic object, the war having reached the point it had in the winter of 1943.

Rommel was entirely convinced – had been convinced for over a year – that the war could in no conceivable sense be won by Germany, as victory had been imagined hitherto. The demands of the Eastern Front were so great in men and materiel, and the losses already suffered there were so crippling, that the only hope – a slender hope – was that enough strength could be built up to hold the Red Army on some defensible line east of the actual borders of the Reich. In Italy Rommel reckoned that it could only be a matter of time before the Wehrmacht would find itself driven inexorably northward. From the air the German population and war-making potential were being ruthlessly pounded, at ever greater human and material cost.

Across the English Channel, presumably poised to invade the north-west European coast when conditions were most favourable, was a huge Anglo–American army of whose overall strength and composition the Germans were reasonably well-informed, as evinced in all Army Group B situation reports (although as the months passed enemy strength became increasingly exaggerated). This huge army would be tactically supported – and its operations would undoubtedly be preceded – by the unremitting activity of equally huge air forces. At sea the battle of the Atlantic had been lost by Germany, and nothing would prevent a free flow of troops, munitions and supplies from North America to the European theatre. Germany's situation was deteriorating all the time.

The surest recipe for the worst military disaster would, in Rommel's view, be a land war on two major fronts, east and west – for he reckoned that the Italian front was containable for a while by well-conducted defensive tactics, so suitable to defence were the successive positions northward to the Po valley: and north of that valley lay the Alpine barrier. But north of the Alps there could be no possible hope of stemming the floods lapping the borders of the Reich if the Anglo–Americans were allowed to establish and then reinforce a fighting front in north-west Europe.

The greatest danger to Germany, almost unthinkable in its horror, lay in successful invasion by the Red Army from the east. If a western front were allowed to develop this horror would, sooner or later, materialize, for the Wehrmacht could not possibly man, for any length of time, two major defensive fronts against enemies whose strength would be maintained by, it was supposed, near inexhaustible reserves of men and materiel. The only hope of success – and by 'success' Rommel, like most intelligent Germans, now meant a sufficiently stable strategic situation from which to negotiate a tolerable peace – lay in the defeat of one enemy so that all resources could be concentrated against the other and a stalemate possibly achieved. That must be the strategic aim. In the circumstances of winter 1943, and with a cross-Channel invasion near-inevitable in 1944 this aim meant, to Rommel, that Germany should concentrate every resource on defeating the western invasion decisively and quickly. If it were allowed to become a protracted 'Western Front' the cause was lost. If, however, it could be defeated early – on the beaches – it would surely be some time before the Anglo–Americans would recover from the set-back and try again. Germany, thereafter, would be able to switch major resources eastward and establish a really effective front – even carry out a limited ground-winning counter-offensive – in Russia. And with that strategic situation achieved there might be hope of a peace, even (Rommel often said) a peace within which a united Europe could seek security and a common defence against the Soviet barbarian in the east.

Rommel had been thinking on these lines throughout the summer, especially since the invasion of Sicily and the failure of Citadel.[1] He therefore accepted his new assignment with energy and alacrity. Pessimistic though he had long been about Germany's future he saw the only immediate hope in defeat of the coming Anglo–American invasion. It could, of course, have been (secretly) argued that Germany's best hope might lie not in defeating that invasion but in giving it as easy a success as possible, while retaining maximum strength on the Eastern Front. But Rommel knew perfectly well that immediately the invasion turned into an actual fighting front there would be inevitable attempts to man both fronts, to switch resources between them, with consequential weakness everywhere, and, ultimately, defeat and the overrunning of the homeland from both directions. He still believed that he could convince, perhaps had convinced, Hitler of the same – that once a western front were set up the war, already lost in the longer term, would be lost in the most disastrous way and the shortest time. He spoke frankly on these lines, particularly to old friends he now found again in his command.[2] The only chance lay in rapid and early operational success in the west, in subsequent strengthening in the east, and thereafter in the hope of peace with the west against the background of a strategic stalemate. It might be a thin hope, but Rommel saw no other. And there were (perhaps illusory) rays of light. Hitler was still able to talk reassuringly about great impending increases in armament production.[3] Success in the west might not be wholly impossible.

The operational deductions from all this were, to Rommel, clear and he did not find reason to modify them during the next six months. As first priority, the coastal defences should be as formidable as men, materiel and time could possibly make them; and Rommel, immediately after his leave, set out on a series of tours of the entire coastal area of threatened western and north-western Europe as ordered by Hitler. He found a great deal wholly defective, and his ruthless energy was now bent on transforming the situation in whatever weeks or months the enemy would allow.

Rommel started in Denmark, where he arrived on 30 November for a ten-day visit. The staff of Army Group B, which was supporting him in what was at first an inspectorate assignment, had been ordered to establish itself in a château at Fontainebleau, preparatory to taking over actual operational responsibility for a part of the Western Front. The whole front, from the tip of Holland to the Pyrenees and along the French Mediterranean coast, was under the authority of *Ober-*

befehlshaber West, Field Marshal Gerd von Rundstedt. Rommel was to report direct to Hitler on the entire coastal defence situation, keeping Rundstedt informed: and was then, on 15 January, additionally to assume direct command responsibility for the northern of two Army Group sectors, a sector including Holland, Belgium and northern France. This was to be the area of Army Group B.

Gause, still Chief of Staff, accompanied by three staff officers, met Rommel on a command train and toured with him. Also of the party were the air and sea commanders concerned; the sea commander, Admiral Ruge, was Rommel's naval deputy and also became a firm and trusted friend. He had joined Rommel in Italy, and was a man in whose company the Field Marshal always took pleasure, to whom he could talk frankly and freely.

During December and January Rommel continued his inspections, covering mile after mile of coast in tireless reconnaissance; turning out coastal battalion after coastal battalion for inspection, march-past and peremptory interrogation; grilling commanders, from generals to corporals; laying down his own priorities. Wherever he went he preached the same doctrine. The enemy must be defeated in the coastal sector. He must be defeated soon after, or even before, struggling ashore. The main battle line must be the beach.[4] This meant that the coastal defences must be enormously strengthened. Coastal artillery was never adequate or adequately protected, but there must be a huge effort made in mining and fortification, and everywhere he went Rommel laid down detail, galvanized, explained, inspired. There should be, in most places where landing was possible to envisage, several parallel minefields, each several miles wide, forming a zone up to five miles in depth and requiring many millions of mines which Rommel and his indefatigable engineer commander, General Meise, set about negotiating from every source, French and German, which could be found to manufacture them. The minefields would be covered from fortified strongpoints, sometimes including stationary tanks and requiring a major constructional effort. Dummy positions, to deceive the invader, were also to be prepared, and fictitious staffs, movement tables and so forth to be coordinated in an Army Group deception programme.

In the sea itself there were to be four belts of underwater obstacles, one in six feet of water, one at low tide, one at half tide and one at mean high tide. Against airborne attack – it was anticipated that there would be extensive enemy airborne operations, both parachute and glider-borne, aimed at disrupting communications behind the

German front – Rommel initiated the erection of stakes in fields to inhibit gliders landing, stakes named *Rommelspargel* by the troops. Airborne operations would, of course, also demand counterattack, but Rommel believed that if the coastal area could be held airborne assault could be dealt with at comparative leisure. Everywhere he went he produced new ideas and experimentation. He was, said Meise admiringly, the greatest field engineer of the Second World War.

Rommel's inspections of the whole threatened area, both before and after he assumed direct operational responsibility for part of it, were unremitting. After his first foray to Denmark, he spent a few more days at home and then joined his staff at Fontainebleau on 18 December, where they had arrived the previous evening to open the headquarters. His first meetings with Rundstedt were cordial. 'Lunched with R. today,' he wrote to Lucy on 19 December. 'He is very charming and I think everything will go well.'[5] From 20 December until the end of the year he was in the area of Fifteenth Army (General von Salmuth): Calais, Boulogne, the Pas de Calais, the mouth of the Somme. This was the northern of two army areas in what was to become Army Group B's area of responsibility. From 2 to 5 January he was on the Dutch and Belgian coast, the Hook of Holland, Rotterdam, Walcheren, the Scheldt estuary; this was *Wehrbereich Nederlande* (General Christiansen), also in Army Group B's area. Between 16 and 20 January he was at Trouville, Honfleur, Fécamp, Le Havre, the mouth of the Seine: scenes of triumph in 1940. From 22 January he was touring Normandy and Brittany, the area of Seventh Army (General Dollmann), returning there and surveying Cherbourg at the end of the month. At the beginning of February he was back in the Pas de Calais. Wherever he went he preached his doctrine to commanders: we must defeat the enemy at once, here, on the beaches; or, if he gets ashore, drive him instantly into the sea.

Rommel was irritated by what he thought was a remarkable lack of urgency in most of those he met. He thought that the comfort of life in Denmark, and to some extent in France, compared to the stringencies of bomb-battered Germany, had not been conducive to military efficiency. He found that very little in the way of fortification or minelaying had hitherto been done, while his own concept demanded an enormous outlay of materiel and commitment of man-hours. The materiel would present problems until the end, but Rommel's personality and power to galvanize and inspire nevertheless worked wonders. Manpower, too, was in short supply. The

Organisation Todt, responsible (at least in theory) for construction and fortification, had only limited capacity. French civil labour could be used, although Rommel insisted that it must be the labour of well-rewarded volunteers only; he was always very sensitive to relations with the French population. But a great deal of the labour, inevitably, was provided by soldiers.

Rommel made high demands and, being Rommel, some found them exorbitant. Von Salmuth (characterized by Rommel's staff as somewhat idle[6]) expostulated vigorously with Rommel on one of the latter's return visits, saying that the Army Group Commander's requirements made such physical demands on the troops (civil labour being limited) that they would be good for nothing when the invasion itself came. Rommel reacted at his most violent and voices were raised. Commanders, Salmuth said, were being driven mad by the intensiveness of Rommel's fortification programme.[7] Eventually peace was made; but this was the old Rommel, the tyrant of the desert, the restless driver of men and events. All commanders acknowledged the astonishing impact their commander-in-chief made, and Salmuth (by no means an admirer of Rommel before they met) recorded that a new phase began 'when Field Marshal Rommel appeared'.[8] As ever Rommel moved like a whirlwind, and had something of the same effect; and as he drove his large Horch car through the French towns and villages there were cries frequently heard – '*C'est Rommel!*'[9]

Rommel's own area of responsibility – Army Group B's domain after 15 January – covered the Atlantic and Channel coasts north of the mouth of the Loire. Within this enormous region there were, as subordinates from north to south, Christiansen (*Wehrbereich Nederlande*), von Salmuth (Fifteenth Army) and Dollmann (Seventh Army). The boundary between Fifteenth and Seventh Army ran southward, west of the Seine, to Le Mans, and gave the Pas de Calais and eastern Normandy to Fifteenth, the rest of Normandy and Brittany to Seventh Army. Within Army Group B area, excluding the Netherlands, were deployed for defence thirty-two infantry divisions,* including eight Luftwaffe or parachute divisions nominally part of the Luftwaffe and formed from Luftwaffe personnel. These

* The total increased as the months passed. At the beginning of June there were a total of fifty-eight divisions, including eleven Panzer or Panzer grenadier under *OB West*, and these were reinforced during the campaign. Fifty-one divisions, including thirteen Panzer or Panzer grenadier, were ultimately to take part in Army Group B's area of operations.

thirty-two were divided, with seventeen in Fifteenth and thirteen in Seventh Armies.

The quality of these divisions varied greatly. During the previous two and a half years strategic interest had been concentrated elsewhere, and the priority of the Western Front had been deemed low. The west had been used as a rest area and a posting destination for the over-age, the medically unfit, and those requiring a rest from the demands of the Eastern Front. Occupation duties had been light and generally agreeable, most divisions were at reduced strength and their effectiveness was often minimal. The infantry divisions had little except horse transport, and not much of that; and were generally below establishment in equipment of all kinds. SS divisions, on the other hand, were often first to be equipped with new weapons and were generally of considerable strength.

Rommel's second priority in the west, therefore, after his transformation of the attitude to coastal defence and his electric effort on all concerned with it, was to obtain an improvement in the numbers and quality of the troops. He profited from the fact that his arrival coincided with Hitler's decision to make the Wehrmacht's foremost task the defeat of the forthcoming invasion. Hitler's Directive No. 51 of November 1943 was emphatic on this. Its terms furthermore – taken with an earlier directive on the subject in March 1942 – were perfectly consistent with Rommel's own views about how that invasion should be dealt with.[10]

Rommel harvested the benefit and was tireless in pressing his advantage. Drafts now included more young men, increasing their proportion of the whole. Divisions in the west were numerically strengthened. Another Panzer division (2nd) was assigned and there were considerable increases in the number and calibre of anti-tank guns – by mid-summer a total of nearly two thousand tanks, assault guns and self-propelled anti-tank guns were deployed. A spirit of confidence began to permeate a command in which, hitherto, there had been a good deal of apathy. In all this Rommel's energy and presumed influence with the authorities was credited with achieving much. Army commanders recognized a transformed situation. Salmuth – who confessed to earlier prejudice against Rommel because of what he had regarded as excessive personal advertisement – soon came to regard this advertisement as having been more a propaganda device of the Party than inherent in the man himself; he was captivated by Rommel's clear understanding and sympathetic, helpful attitude, and although he rebelled against the hard driving he

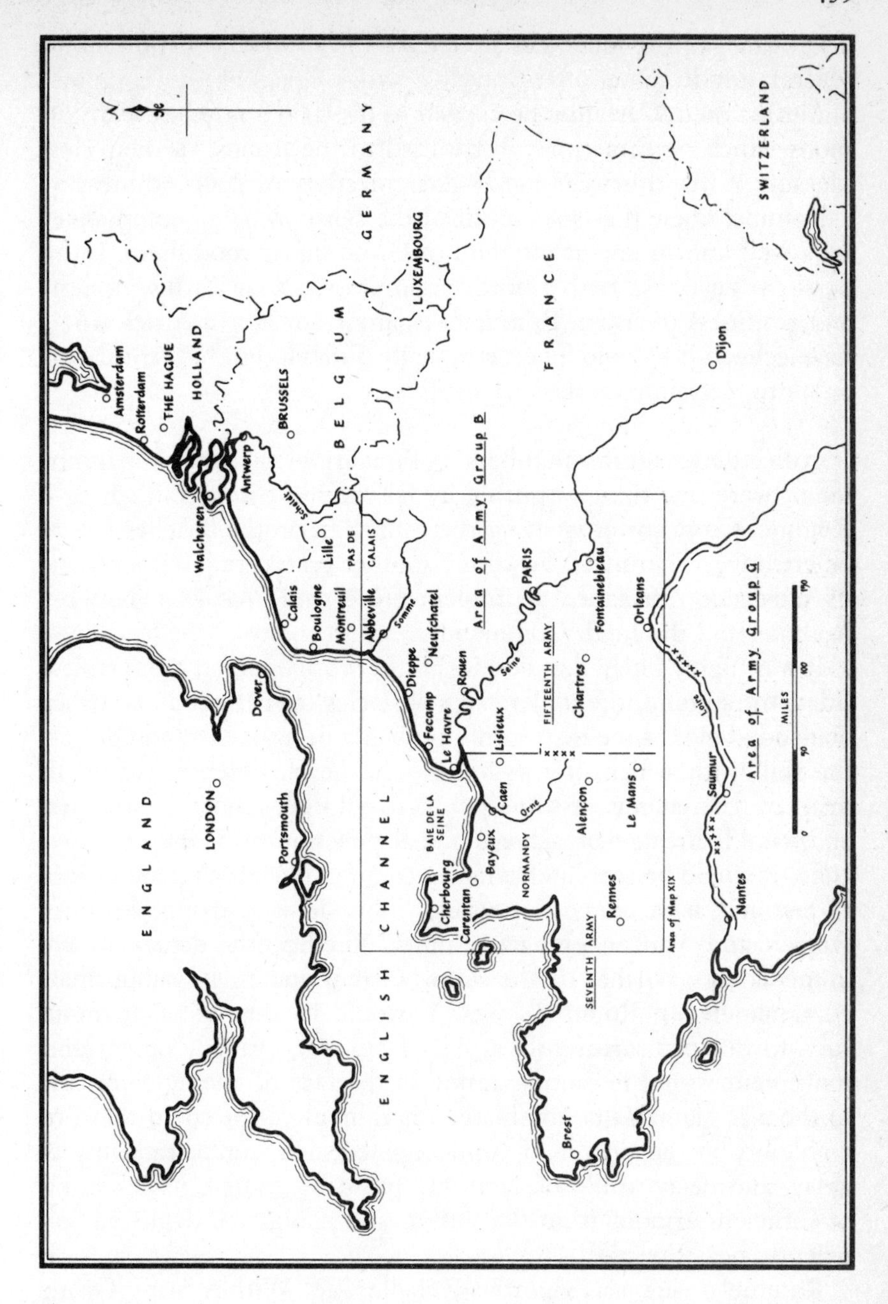

XVIII Western Front, 1944. Army Group B area of responsibility

acknowledged the remarkable effect of the Field Marshal on all.

Certainly Rommel often travelled with a squad of photographers and press men. Certainly he carried in his train a number of accordions which he sometimes distributed to the troops, to their clear pleasure.[11] But this was quickly recognized as a calculated measure – Rommel knew that soldiers liked the sense of being commanded by a well-known personality and one who understood them. In his personal dealings, furthermore, he impressed all with his modesty and readiness to listen. Rommel – many from North Africa would have echoed it – could be exceptionally difficult, but his arrival, said Salmuth, certainly marked a *plus punkt*.[12]

Fortifications, and the numbers and quality of the troops manning them, were one thing, arguably by far the most important thing if Rommel's concept of winning the campaign on the beaches was to be effective. There was, however, another crucial requirement, and its satisfaction provided a problem, never fully resolved, between Rommel and the High Command.

It was highly likely that despite all the work done on the fortifications, the mining, the underwater obstacles (to perfect all of which demanded more time than would probably be available), and despite the skill with which, it was hoped, the beach defences would be manned, the enemy, at some points, would get ashore in strength, and would threaten or achieve actual penetration of the defensive zone. It would be absolutely vital to counterattack such penetrations as fast and as strongly as possible. This demanded mobile units, Panzer and Panzer grenadier units, intelligently deployed and immediately available to the Army Group and to its subordinate commanders. In Rommel's view it would be impossible to move such formations sufficiently rapidly from afar – timely operational movement would be impracticable in the face of enemy command of the air. Alam Halfa dominated his thinking. Nor could there be ambiguity in the matter of control, since such would lead only to delay, and delay would be fatal. He must, he argued, have control of sufficient armour from the outset, and it must be deployed sufficiently near the coast.

Rommel's view was vigorously challenged. Within Army Group B's area, for the most part, were deployed the Panzer and Panzer grenadier divisions allocated to *OB West*, but within Rundstedt's command there was also – and directly responsible to him – 'Panzer Group West', a formation headquarters having the status of a Panzer army, under the command of General Freiherr Leo Geyr von Schweppenburg. Geyr von Schweppenburg, another Wurttemberger, was a clever, reflective '*grand seigneur*' like his superior, Rundstedt; a man with considerable knowledge of other countries,* a well-known horseman and race-rider in his younger days, who had been one of the earliest Panzer regimental commanders when the German cavalry was converting to armour.[13] He was a member of the General Staff; and he considered the Rommel doctrine defective.

Geyr accepted – everyone accepted – that it would be necessary to counterattack strongly and in time; but he believed that such counterattack must be in sufficient strength. He believed that available resources made it inconceivable that coastal defences, however vigorously improved, could prevent a major enemy deployment ultimately taking place on the mainland of France; and he believed that the best and only way to deal with that would be by manoeuvre with the maximum concentrated armoured strength, strength reserved for that manoeuvre and not frittered away on operations of only local significance. He pointed out that the poor quality of infantry divisions manning the 'Atlantic wall' made it only a flimsy defence, and further argued that the coastal zone was but an outpost line, and main battles should not be fought on outpost lines. He had visited OKW at Berchtesgaden, and managed to get support for the principle of an OKW Panzer reserve. This should be Panzer Group West.

Furthermore, nobody could confidently predict where the main invasion would come, and the length of threatened coastline was great. German intelligence was blinded by Allied control of the air (as well as confused, in the event, by a developing Allied deception plan); 'I know,' Rommel said bitterly, '*nothing* for certain about the enemy.'[14] There was little to go on, despite the verbiage of *Fremde Heere West*, except guesswork based on topography, climate and tidal data; and an overall estimate (exaggerated) of the total enemy forces in England. The latter was not of huge significance; it was obvious that the enemy would have great numerical superiority, by whatever margin, if, and only if, he could deploy and establish his forces ashore. ULTRA told the British that Rundstedt produced an appreciation on 8 May which forecast an enemy assault with twenty divisions 'in the first wave'; this, exactly expressed, proved a considerable overestimate. But by 'first wave' Rundstedt probably meant the initial days rather than the first hours across the beaches; and, in the event,

* He had been Military Attaché in Belgium, Holland and England.

twenty enemy divisions were to be in contact extremely soon.

It was therefore crucial to the argument, in Geyr's view, that Panzer divisions deployed for immediate counterattack behind one sector of coast would lose essential time if necessarily switched to quite another. It would be far better, he contended, to accept that the enemy would inevitably be able, somewhere, to line up forces in France; and then, at the optimum moment, to move against him with concentrated strength, trusting to well-tried German superiority in manoeuvre. Geyr also took the view that Rommel's anxieties about the effect of enemy airpower were exaggerated. By using dusk and darkness, despite the real hazards, mobile forces would, somehow, be able to move operationally.[15] Rommel, said Geyr, proposed to use the Panzer divisions purely tactically. This was a breach of accepted military theory – there should be concentration and a major operational effort at the right time. Rommel, he said freely, did not understand the principles of armoured warfare.

There would ultimately be a total of eleven Panzer and Panzer grenadier divisions in the west,* and the argument about their deployment and about control of them went on for some time. The three most probable areas of enemy landing were, it was appreciated, in the Pas de Calais, astride or north of the Somme: between Somme and Seine; or in Normandy, and especially in the Cotentin Peninsula. It was assumed that any landing would be aimed at securing a major port quickly, for follow-up forces and the introduction of heavy armour in quantity. The first probability assumed an attempt on Boulogne or Calais; the second Dieppe or, more probably, Le Havre; and the third Cherbourg. The Germans did not know of the Allied artificial harbour, 'Mulberry', and underestimated the capacity of landing craft to bring heavy equipment over beaches.

Geyr von Schweppenburg – supported in his general thesis by Guderian, Inspector General of Panzer troops, and, more remotely, by Rundstedt – reckoned that a force prepared to deal a powerful, united blow against an enemy invading under any of these three main options would need to be deployed in reserve, centrally, somewhere not far from Paris. He recommended two concentration areas, in woods north and south of the city.† From Paris to Montreuil at

* Reinforced after the campaign began.

† Although Geyr always stuck to the principle of his argument the detail – perhaps less importantly – varied. In one post-war article he argued that half the Panzer reserves might have been deployed 'north of Alençon', which is certainly more convenient for Normandy.

the heart of the Pas de Calais is 125 miles; to Neufchâtel, between Somme and Seine, eighty-five; and to Caen, the key to Normandy, about 140. Each of these named places is less than twenty miles from the coast, and to reach it, and thus the area of possible coastal counterattack, would take the march time of a Panzer formation moving an average of 115 miles – a considerable distance if, as all acknowledged, there were to be major interference by enemy air forces. On the other hand movement, in Geyr's view, *could* take place although with delays; and because of the central placing of the reserves a major and overwhelming blow could be struck, whereas divided counterattack forces would throw smaller and ineffective punches. And the time to reach each of the battlefields *at worst* was less than the time which forces deployed in, say, the Pas de Calais would need to reach Normandy (or vice versa), a mean distance of 170 miles, crossing en route the major rivers of Seine and Somme whose bridges would all too probably have been destroyed by enemy air action or by sabotage, despite best efforts. It would, furthermore, be actually advantageous to allow the enemy to advance with significant strength and expose himself to a decisive counter-stroke, which the German Panzer forces could inflict provided they had not been frittered away in penny packets, probably in some terrain (like the Normandy *bocage*) wholly unsuitable for major operations by armour.

There was a further point. Geyr – there was little contradiction – believed that the invasion would be accompanied by large-scale airborne operations, designed to cut Wehrmacht communications and isolate the battlefield. He argued that concentrated Panzer forces, deployed in depth clear of the immediate invasion area, would be well-placed to deal with these quickly and effectively.

The argument continued throughout the first months of 1944. Rommel contended that although the principle of concentration was impeccable – had he not always preached and exemplified it? – the practice would be negated by the air situation. Reserves held at distance would never arrive or would arrive too late to be relevant. The crucial hours would be the first hours, when the enemy was still struggling ashore and feeling his way from the beaches; and those hours were short. Counterattack forces to deal with the opening situation were essential, and in those first hours time was more important than strength. That meant Panzer troops – even if of theoretically insufficient quantity – on the spot, for immediate action, close to the coast. Admittedly the possibilities were so stretched that this meant an initial blow in no great strength – at least three major contingencies must be covered and the reserves divided accordingly; but there would, everywhere, be something; and thereafter every nerve would need to be strained to reinforce.

As to where these reserves should have their main weight, clearly a view had to be taken on probabilities: Rommel's own instinct was to place that weight astride the mouth of the Somme, ready to take strongest action south of Boulogne or east of Dieppe. Increasingly, however, he was impressed by the possibilities of Normandy, where the beaches would be more sheltered. The essential, however, was to strike early, with something, wherever the first blows fell.[16] As to the airborne threat, Rommel accepted it; but he argued that if the coast were held and the battle of the beaches won the battle could not be decided by a subsidiary enemy effort, as airborne landings would be. The latter could be mopped up at leisure and would never develop enough offensive momentum *divorced from the ground action of sea-landed troops* to have a dramatic effect on the campaign. Rommel conceded that there should, additionally, be some central 'operational' reserve, and agreed that geography indicated the Paris area for it: but it should be comparatively small.

The debate swung to and fro. The Army Group B record shows visits and discussions by Geyr on frequent occasions, starting on 8 January. At a conference with Jodl on 20 January it was agreed that the Panzer divisions must be so placed as to be able to intervene in battle as quickly as possible – a pretty unstartling conclusion. Five days later Rommel learned that OKW had refused to sanction armoured deployment in the coastal sectors, on grounds of the 'general situation'. On the following day Geyr lunched with Rommel, and next morning there was renewed discussion.[17]

Rommel continued his inspections for the next three weeks, including a visit to his southern neighbour and old friend General Blaskowitz, commanding Army Group G, based in Bordeaux. Rommel's attention to coastal defences so far from the vital area facing England has been criticized, but he had been given a continuing overall inspectorate responsibility, and both Mediterranean and Atlantic coasts were threatened – indeed, an enemy landing would take place in the south of France in August, an expedition which had once been mooted for the Bordeaux area as an alternative. On 17 February Rommel was back in Paris for a Panzer Group West war game, at which Guderian was present.

To the supporters of Geyr Rommel was now preaching dispersion and heresy, as well as ignoring those virtues of mobility which he had once so vividly exemplified.[18] Guderian said that Rommel had reverted to type and become an infantryman again.[19] After much reflection several of those involved, having initially opposed Rommel, supported him.[20] After a conference of those principally affected with Hitler on 21 March Rommel believed that he had won his point.

He had not fully done so. The crucial issue was not only deployment, but operational control, the ability to give timely orders, and Rommel had by no means achieved all he wanted. As is not unusual in such a tug-of-war between strong personalities the High Command – in this case OKW, for Hitler himself was, of course, determined to be involved while by no means showing comparable power of firm decision – attempted compromise; to give Rommel a little of what he wanted while retaining some authority over some of the Panzer divisions until or unless specifically released. It was hoped that differences could be resolved by this. The hope was vain. Rommel noted after another conference with Geyr on 9 April that there had been no resolution of their *starke Meinungsverschiedenheite*, their strong differences of opinion;[21] and he got little satisfaction from an interview with Rundstedt the following day. In a letter to Jodl of 23 April he wrote that the decisions of 21 March, as he had understood them, were not being implemented. The Panzer divisions were, in some cases, far from the coast and they were not under his command. Geyr did not believe in the possibility of a successful defence in the coastal zone.

Rommel's pleas were not accepted, and relations became bitter. Rommel wrote to Lucy on 27 April of 'Geyr von Schweppenburg, to whom I have recently been very rough because he wouldn't concede over my plans'.[22] On 28 April both Geyr and Guderian visited Rommel for an evening conference, described in the Army Group diary as including 'exhaustive discussion of the employment of assigned armoured forces'.[23] It did not lead to a satisfied Rommel.

It may seem curious that the commander-in-chief of all these forces, *OB West* himself, did not make a binding decision, or at least a firm recommendation, in this matter so crucial to the forthcoming operations and to his own authority. Field Marshal Gerd von Rundstedt had been Rommel's remote superior as commander of Army Group A in the heady days of 1940, of the *Gespensterdivision*. In Barbarossa he had commanded Army Group South, which had advanced into the Ukraine and ultimately to Rostov and the Donets basin. Despite these achievements he was described by Geyr as largely deficient in understanding of the principle of Panzer operations – a large assertion.[24] Geyr thought Rundstedt weak with Rommel. Rundstedt was now sixty-eight years old, an archetypal Prussian cavalryman, cultivated, intelligent, courtly and reserved; the doyen of the German *Generalitat*.

Rundstedt held the traditional view of high command. He believed that his function was not to immerse himself in detail but to operate from a large-scale map, to contribute an overview of a theatre of campaign, to possess a sound sense of priorities, to anticipate, to delegate and to trust the enterprise of subordinates, whom he visited infrequently.[25] He believed that his own superiors might on occasion be advised, but not contradicted. His judgement was rooted in great military knowledge and historical understanding, as well as in considerable experience of high command and in a stern, imperturbable character.

Rundstedt had been in the west for two years, two years in which strategic interest had largely been focused elsewhere. His age militated against energetic new thinking of the sort Rommel had injected; and, together with his realization that in Hitler's Germany the Führer and supreme commander would inevitably have the last word, it probably militated against the strong resolution of disputes between subordinates. Rundstedt realized the deficiencies of his command. He had, in an appreciation of October 1943, summed them up in uncompromising terms and Hitler had thereafter given priority to the west and given Rommel his *Führerauftrag*. But Rundstedt had no illusions, and little zest.

Rundstedt's headquarters were at St Germain. He himself lived in comfort – too much comfort – in the Hôtel Georges V in Paris, and in Rommel's view the staff of *OB West* were lethargic, while the pleasant and relaxed character of occupation duties in the French countryside seemed, all too often, to have softened the German soldier to a dangerous degree. Rundstedt had high qualities, but his own demeanour had not been such as to infuse energy. He envisaged the forthcoming struggle in orthodox terms. There would be an encounter battle on or near the beaches. If the enemy could be contained and beaten there, as his new and impossibly energetic Army Group B commander was determined to believe, well and good. If not, it would be essential to try to win a battle of manoeuvre in western France, a battle which would, as heretofore, demand the skilful use of concentrated mobile forces. His subordinate, Rommel, seemed to believe that if that stage were reached the war would have been lost. Perhaps, but perhaps not.

Rundstedt, therefore, inclined to side with Geyr in the debate, as did Guderian, who visited frequently. Guderian's view was clear and uncompromising, and it was true to the principles which had always guided him. Concentration of effort dictated that with limited resources – and the German Panzer forces were certainly limited – control should be exercised *in the first instance* by the overall responsible commander. If Rundstedt was to be able to affect the battle he needed initial control of the reserves – operationally, of course, he could delegate as the situation demanded. Guderian, supporting Geyr, had proposed that all Panzer and Panzer grenadier divisions should be massed in two groups, one north and one south of Paris.

The ultimate deployment of armour in the west represented compromise. There were eleven Panzer and Panzer grenadier divisions available – the latter without tanks but with a strong complement of assault guns and fully mobile. These divisions were of varying strength, with the *Waffen SS* divisions the strongest.* Of these eleven, three were in the area of Army Group G (General Blaskowitz), Rommel's neighbour south of the Loire, responsible for the Atlantic and Mediterranean coasts of France; one, 19th Panzer, was in Holland; one, 17th Panzer Grenadier, was deployed near the inter-Army Group boundary in the region of Saumur and Niort, immediately south of the Loire; and six were in Army Group B's area. Of these six, three – 1st SS Panzer in Belgium, 12th SS Panzer at Lisieux and Panzer Lehr (commanded by Rommel's old subordinate, Bayerlein) at Chartres – were to be in OKW reserve and formed part of Geyr's Panzer Group West, as did 17th Panzer Grenadier; while Panzer Group West itself was initially – and somewhat mysteriously – subordinated to Dollmann's Seventh Army. The remaining three divisions – 2nd Panzer in the Pas de Calais, 116th Panzer near Rouen, and 21st Panzer (reconstituted in May 1943, and with a proud name) in the area of Falaise and Caen – were under Rommel's operational control for his own immediate purposes and reactions. Even so, divisions could only be repositioned with the approval of

* Overall the Panzer formations were described as probably representing no more than 30 per cent of the effectiveness of a comparable division of 1940,[26] but the latter had, anyway, been established at far greater strength, and the description did less than justice to the divisions concerned in 1944.

OKW so that Rommel, although he had control when battle started, was constrained in his preparations. Attempts to detach further divisions from Panzer Group West (Rommel wished to deploy 12th SS Panzer west from Lisieux) failed.

The compromise did not please. It failed to satisfy Guderian, who regarded the whole deployment as a fatal dispersion, an *Aufsplitterung*. It did not satisfy Geyr, whose reserve was weaker than his own theories demanded and was, furthermore, dispersed rather than concentrated. It did not satisfy Rommel, in whose view too many of the mobile reserves were too far from what would be the critical area, whether it were in Normandy, the Pas de Calais or somewhere in between. Immediate counterattack could never now be undertaken by more than one Panzer division. It might not be enough. Jodl assured him on 7 May that he could count on the OKW reserve being released once the enemy invasion *Schwerpunkt* became clear, but such assurances can be varied and do not impart the confidence a commander derives from unequivocal authority. And the command system resulted, in the event, in Geyr receiving conflicting orders from Rommel, Rundstedt and Hitler on occasion.

Hindsight is sometimes permissible. If, as to threat and defensive deployment, Rommel had had his way on all counts, there might well have been at least one other Panzer division immediately available to counter the Allies on D-Day, and that might have been decisive. There might also have been effective mining, for which Rommel had pressed, in the Baie de la Seine. On the other hand his eyes were fixed also to some extent on the area between Somme and Seine, leading to enemy capture of Abbeville and possibly Le Havre; and remained so fixed for some time after the actual invasion of Normandy. Rommel believed that there would probably be several separate and perhaps successive landings: in the matter of prediction his guesses were not enormously better than anybody else's. He always believed that the Allied landing would be largely influenced by the shortest distance to Paris, by operations after landing; and this drew his eyes north of the Seine, although the comparative shelter in the Baie de la Seine drew them back to Normandy. Nevertheless Fifteenth Army's sector remained stronger than Seventh Army's, where the blow actually fell; and the pattern of Allied bombing (intended, *inter alia*, to deceive) led Army Group B to report on 3 June that attacks on the Dieppe–Dunkirk sector confirmed that the focal point of any large-scale landing would probably be in that area.[27] The V-weapon launch-sites, when they became operative, also provided argument for expecting Allied emphasis to be placed on invading the Pas de Calais as early as possible.

Rommel's emphasis on winning the battle on the beaches was certainly rational, and his estimate of the effect of enemy air power on operational movement was largely – although not completely – justified by events. On the other hand his defensive works were a good deal less effective than hoped. His programme had, as some had predicted, been over-ambitious in view of the limitations on time, labour and materiel, although there was little doubt that he had transformed the situation from what had been there before. The Allied resources for putting forces ashore and for neutralizing obstacles had been underestimated.

What would have happened had Geyr had his way, and German armoured forces been largely kept clear of the initial battles, must be wholly speculative. Presumably the invasion forces, unthreatened by such armoured counter-moves as took place – or threatened by much less of them – might have assembled a force capable of the planned march to and across the Seine a good deal earlier. Whether they would have done so must be uncertain: undoubtedly Montgomery would have established a sufficiently strong logistic base before major advance and that, regardless of German reactions or lack of them, would have taken time. As it was, although the initial Allied progress after D-Day fell well behind schedule, the ultimate advance ran close to prediction. It seems unlikely that a very different pattern of advance would have developed although it would, of course, have been undertaken by forces comparatively unscathed. The German Panzer forces committed to oppose the bridgehead were to inflict a good many casualties, and the Allies who ultimately advanced had already suffered.

As to the results of an imagined clash between comparatively fresh Allied and German forces somewhere in north-western France, none can say. To achieve a successful manoeuvre the defenders would have needed to hold – somewhere, however tenuously, however temporarily. Manoeuvre needs a framework. If that had been the German concept it is hard to believe troops could have been found for it, unless the coastal defences had been so weakened as barely to present an obstacle. And the whole question of manoeuvre begs, of course, the question of air power, whose effect so dominated Rommel's mind. The defenders would have been operating blind as to reconnaissance, and have been attempting tactical successes under threat from the enemy's hawks, ready always to concentrate and to swoop.

Such imaginings have no conclusion. Geyr's thesis appears largely although eloquently theoretical. His arguments from history for the placing of mobile forces on the outer flank, ready to strike an enemy from that flank, are elegant but of questionable applicability to the situation in Normandy in 1944. On the whole, Rommel's instinct – that in open warfare the material odds, both on the ground and in the air, were so greatly against the Wehrmacht that open warfare must not be allowed to develop – compels acceptance. And his main thesis – that if a western front continued for any length of time Germany would be engulfed from both west and east – surely stands.

Geyr stuck firmly to his views to the end. He believed Rommel exaggerated enemy air power and betrayed the principles of mobile operations, and that there could have been, in western France, successful operations with concentrated forces. He blamed Rommel for not trying it – and for failing to constitute a strong Panzer reserve after battle was joined, by relieving Panzer with fresh infantry divisions in the forward line (although where Rommel would have found the latter is unclear). He grumbled, in retrospect, at Rommel's 'pessimism and lack of strategic schooling';[28] yet Geyr himself, not postwar but to Guderian on 15 June, said, in terms, that daylight movement had been rendered impossible, and referred to the influence of enemy air power as being so tremendous that all command decisions must be based accordingly. Hostile air forces, he wrote of operations on 7 June, had so interrupted road communications by bombing such bottlenecks as villages, bridges, culverts, that detours were scarce and hard to find.[29]

Geyr, indeed, was to refer to all movement as necessarily taking place by night; and a counter-offensive by armoured forces whose movements are largely restricted to the hours of darkness has little effectiveness. Geyr wrote that it was no longer possible to employ a Panzer formation above company strength on the invasion front without heavy losses. Precisely so. In such sentiments his views were with Rommel, and it is hard to credit the successes which he later argued might have attended concentrated manoeuvres had Rommel not forbade them.

More defensible were Geyr's views that the post-invasion situation demanded significant operational withdrawals – he himself wished to evacuate the whole *bocage* country, which he thought minimized the effective advantage of German armour; and by that stage all commanders and staffs objected to Hitler's obsessive refusal to grant any 'operational freedom' whatsoever. But this lay in the future, when things would be, in Rommel's view, hopeless. The immediate priority was how to plan the counter-invasion battle. Criticism of Rommel as deficient in strategic understanding in this business fails. He was simply over-endowed with the practical sense of what will or won't work on the ground in the conditions of war. He always had been. He saw clear.

Initially Rommel's views were close to Hitler's. Hitler was determined – like Rommel – to defend on the beaches and in the coastal zone, but he was criticized for setting his face from the start against giving the commanders in the west any latitude. When invasion came this attitude led to absurd and unsuccessful attempts to control a tactical battle in detail from a bunker far away. The attitude – and the criticism – echoed the situation on the Eastern Front, where Hitler's savage determination not to allow space to be traded had cost the Wehrmacht dear and pinned it to more than one bloodily indefensible position.

But this, of course, was not Russia. The coast of Europe was defensible in a way provided by no Russian river. And in Russia there was depth – great depth – behind the German advance, until the tide had turned and ebbed a long way. There was comparatively little depth in France between the northern coasts and the frontiers of the Reich. Wherever the reserves were deployed, once the enemy landed and were established, Hitler would have demanded early counter-offensive manoeuvres, probably linked to a command to hold some forward and impossible line. It seems improbable that different operational concepts would have led the Wehrmacht to more success than was achieved in the battle of Normandy. The truth surely was that if the Allies established themselves strongly ashore the game was up, however the last tricks were played.

Rommel had moved from Fontainebleau on 9 March and established his personal section of Army Group B headquarters in the noble and ancient château of La Roche Guyon, on a loop of the Seine near Bonnières. Great tunnels had been blasted out of the cliffs overhanging the north bank of the river, forming shelters for the officers and men of the headquarters. The château was the home of the Duc de la Rochefoucauld, with whom and with whose family Rommel and his small team of staff officers were soon on excellent terms. Rommel loved France and the French countryside, and was insistent that relations between his soldiers and the French civilian population should be based on mutual respect and be as cordial as possible.

Neither he nor any other field commander, however, had control over the *Sicherheitsdienst*, the internal security agencies who received orders not from the Army Group, nor from *OB West*, nor from the army static command in France (General von Stülpnagel, in Paris), but from the Reichsführer SS, Heinrich Himmler, directly. That there was a need for internal security none could deny; the activities of the French Resistance – very naturally labelled terrorism by the occupying forces – were such that constant surveillance and defensive precautions were necessary. Bomb alerts and warnings of assassination attempts on prominent personalities (including Rommel) were frequent. Security measures also had to be agreed with the *Krimanalpolizei*. There had been information of an impending attack on Rommel's train as early as 9 December, during the Denmark tour, and there had to be perpetual awareness. Counter-measures were the responsibility of the SS, and Rommel could, on the whole, relax with the sense of being safe among a well-disposed and not especially discontented people.

His own rooms in the château gave on to a rose garden, he worked at a Renaissance desk (on which Louvois had signed the revocation of the Edict of Nantes in 1685), and he took frequent and happy walks in the surrounding countryside. Occasionally he went for a ride. He acquired two dachshunds, which soon became part of his small personal household, their tricks and habits often recounted in letters; they were soon joined by a large sporting dog, Ajax: 'We really must pay dog tax for Ajax,' he wrote to Lucy on 30 March, adding that the younger dachshund was 'really too funny, but *not* yet quite housetrained'.[30]* Sometimes there were hare shoots which he much enjoyed. He was not, however, lulled into supposing that Hitler's Germany was generally popular. 'How peaceful the world seems,' he wrote in his diary on 23 April, 'yet what hatred there is against us.'[31]

Rommel was away on tour more often than not, visiting and revisiting the defensive sectors of his huge command, and still sometimes – as between 23 April and 3 May – travelling to inspect the Atlantic, Pyreneean and Mediterranean defences in the area of his neighbour, Blaskowitz, with the general oversight of which Hitler's order charged him. Everywhere he went he injected energy, con-

* Ajax, to Rommel's grief, died in May.

stantly reminding commanders of his theme – that the war would be decided in the west, and soon. Sometimes he was pleased; progress had been made, his confidence mounted. Sometimes he recorded depression; efforts had been inadequate, comprehension lacking. Success, as always, depended on the quality of commanders, mainly divisional commanders; and this was variable.

At La Roche Guyon Rommel's military 'family' were few. These were the replacements for the old friends of his African *Kampfstaffel*. The chief operations officer, the Ia, was Colonel von Tempelhoff, a trusted companion from Italian days, who often accompanied Rommel on his tours of inspection. Tempelhoff, who had an English wife, had known Rommel since 1938, since the days of Rommel's attachment to the *Hitlerjugend*, when Rommel's reputation in the army was of possessing more enthusiasm for National Socialism than was shared by many of Tempelhoff's background. But in Italy Tempelhoff had often talked with Rommel about peace, had listened to him when he declared it essential and urgent; and Tempelhoff witnessed his superior's alternations of ebullience and depressed return to reality after his meetings with Hitler – as after Rommel's encounter of 21 March, of which he said exultantly to Tempelhoff as they drove away that he had gained every point with Hitler, only to exclaim within half an hour 'What has he *really* given me?'[32]

With Tempelhoff – indeed with all of them – Rommel was in sympathy. It was a happy group, and when Rommel was taken from them they missed him sorely. They missed his stimulating but never domineering lead in conversation – Rommel could listen as well as he talked. They missed his humour, often directed at himself: when Ruge once boldly commented on a red spot on the Field Marshal's face Rommel said it always appeared when his skin was exposed to hot water – and chuckled at the thought that now they could always tell if he had washed his face.[33] They missed the immediate and active interest he showed in new ideas: if somebody suggested a technological device conceivably worth exploring he would telephone next morning with his own proposals for some improvement on it. He was open-minded, interested and companionable; and it was generally agreed amongst them that only those in Rommel's own circle could really know his worth. His health was better and his physical fitness seemed again formidable, although sometimes he suffered from lumbago.

Tempelhoff was assisted by Colonel Freyberg, concerned with

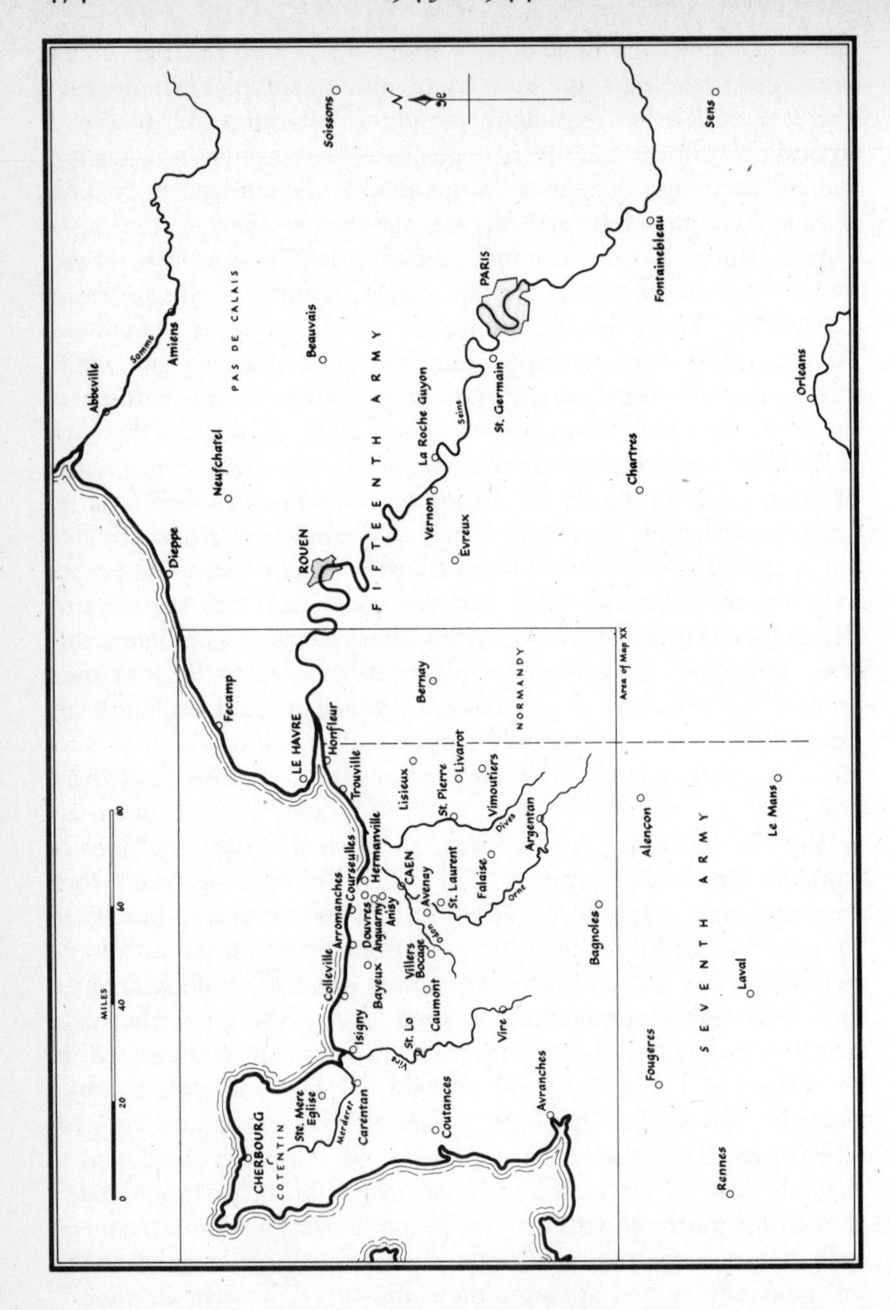

XIX Army Group B area, 1944. The northern sector

personnel questions. The Ic – intelligence – was Colonel Staubwasser, who had been serving in the 'English' section of *Fremde Heere West* in OKH before joining Army Group B. Staubwasser had been Rommel's student at Dresden. He, too, was congenial to the Field Marshal, accompanying him on his walks with one or more of his dogs in the woods around La Roche Guyon. 'Look at the Seine valley,' Rommel said to him on one occasion, pointing to its tranquil beauty in that summer before the invasion, and then spoke of the contrasting devastation of Germany which was going on nightly, and of the need for peace.[34] Rommel also observed that the Allies would never negotiate with Hitler – as Hitler himself had bitterly observed to Rommel.

Also at La Roche Guyon were the chief engineer, General Meise, whose regard for Rommel's ingenuity in Meise's own field was, as has been noted, particularly high; and Colonel Lattmann, chief of the weapons branch and senior artilleryman. Lattmann was an old family friend of the Rommels and to him – as to most of them in degrees – Rommel talked with considerable frankness. The admiration of Lattmann and his wife for Rommel was profound; and later, when members of their family were in serious trouble with the Nazi authorities, Rommel did his best to help them.

The commander of all reconnaissance troops, General Gehrke, also lived at La Roche Guyon, as did a number of aides-de-camp and personal staff officers; one of these, Captain Lang, kept Rommel's personal diary, to the Field Marshal's dictation. Two diaries show the day-to-day activities, the comings and goings of Rommel and his visitors at this time. These were the official record of the Army Group Commander's activities, the *Tagesberichte des Oberbefehlshabers*,[35] which often summarized the more personal Rommel diary,[36] sometimes using the same words; but in the personal diary (albeit dictated rather than in his own hand) Rommel recorded from time to time more human reflections, emotions and anxieties.

A liaison officer from the Luftwaffe lived in the château; he must have often had a difficult time because Rommel sometimes used savage words about the Luftwaffe command (this was reflected in his loathing for Goering). There was also the congenial figure of Admiral Ruge, the representative of the *Kriegsmarine*. Rommel liked Ruge: and Ruge liked and admired Rommel, admired his essential modesty, his simplicity and lack of pretension. The atmosphere was harmonious. There was no accredited National Socialist liaison officer.* Often Rommel talked about the ultimate rebuilding – and necessary peaceful unifying – of Europe.[37]

The staff worked under the Chief of Staff – *der Chef*. When they first moved to La Roche Guyon this was the reliable, knowledgeable Gause, who had seen Rommel under every conceivable circumstance, in triumph and disaster; who had seen him elated and all-conquering, and seen him in the depths of depression. On 15 April Gause was relieved by General Hans Speidel, another Wurttemberger, requested by Rommel, a man of considerable intellect and General Staff experience. Rommel made an excellent speech at the farewell party given for Gause on 20 April. Speidel had reported to Jodl at OKW before joining Army Group B and had been warned against Rommel's *Afrikanische Krankheit*, his 'African sickness', or pessimism.

Rommel's energy supplied something which had been largely absent in the west under previous regimes – coordination. The various organizations concerned with defence works were under different authorities. The army worked under its commanders. The navy and Luftwaffe did the same. The vast *Todt* organization worked under the Munitions Minister, Albert Speer. The *Militarbefehlshaber Frankreich*, von Stülpnagel, had certain authority which interacted with the others, especially insofar as the civilian population and measures affecting them were concerned. The security troops were, as recounted, under the authority of Heinrich Himmler; and for action requiring, however nominally, the cooperation of the French administration of Marshal Pétain there was a German Ambassador in Paris, Otto Abetz, with certain responsibilities. The amount Rommel – or Rundstedt – could simply ordain was strictly limited, and coordination within a complex structure was essential. There had previously been too little, and Rommel's driving enthusiasm together with Speidel's dexterity and intelligence greatly improved matters. Preparations to receive the enemy invasion gathered pace and on the whole Rommel found more to praise than to blame on his visits. Between 9 and 11 May he toured both the Pas de Calais and Normandy, showing particular pleasure at the impression made by 21st Panzer (General Feuchtinger). Often he drove through country across which his *Gespensterdivision* had raced in 1940: and on the evening of 17 May, after inspecting 91st 'Luftlande' Division,

* National Socialist 'leadership officers' had been assigned to all units since December 1943.

Rommel dined by candlelight with the officers, reminiscing of those heady times, of how a French general had clapped him on the shoulder at St Valéry and said, 'You're much too fast!'[38] Later he talked of the lost opportunities in Africa.[39]

He inspected a Tatar battalion next day – there were a number of Russian soldiers serving in the Wehrmacht on the Western Front – and noted their huge enthusiasm.[40] Rommel always responded to enthusiasm, to the palpable effort of troops and to their good spirits; he inspired them and he drew inspiration from them. During his tours he deprecated any special hospitality, as often as not taking coffee or eating in a *Soldatensheim*, chatting and joking with the soldiers. At such moments he wanted to be nowhere else. France, as he told everybody, was where the war would be decided, and he was glad that his post was there, preparing for the ultimate battle. His quick verbal presentations during visits were models of force and clarity. He noted in his diary on 13 May that he had once been written off by some people as unfit, just as the French coast had been written off as indefensible. Now he was in the process of falsifying both views. He had been entrusted with the task by his Supreme Commander, despite the sceptics: '*Der Führer vertraut mir, und das genügt mir auch.*'*[41] Of course it is possible that, since the diary was dictated, impeccable sentiments of this kind were made 'for the record', and were disingenuous, aimed to establish loyal credentials. It seems unlikely. Rommel was not that sort of man, and such subtleties were not in his style, although a certain amount of self-preservative humbug was not uncommon in Hitler's Germany.

Every piece of consolatory rumour of affairs in England was seized on by every German, and Rommel was no more immune than most to the enchantments of wishful thinking. 'In England,' he wrote to Lucy on 26 April, 'the mood is bad, one strike follows another, the cry of "Down with Churchill and the Jews" and in favour of peace is getting louder. Poor omens for so chancy an offensive operation!'[42] In many conversations and letters at this time he clearly struggled to give an impression of confidence in his task and, once again, of confidence in the Führer's judgement, judgement which would disclose to Adolf Hitler the best way to exploit a military success for political ends. He never wavered in his view that the political end must be an early peace.

Confidence on both counts was unjustified. There were serious

* 'The Führer trusts me, and that's enough for me.'

deficiencies, and although Rommel showed a brave face and wrote home of his increasing optimism he was apprehensive, in particular about the air situation; and the matter of the Panzer reserves was still far from satisfactory. His *Fingerspitzengefuhl* contradicted his self-induced optimism. To an old acquaintance he said brightly that conditions were better than in front of Tobruk, were they not?

'"Yes *Herr Marschall*, but the problems are still with us!"

'"You're right!" Rommel said. And he muttered of his presentiment – of nothing good.'[43]

On 19 May Rommel received especial compliments from Hitler on the work which, the Führer learned, had been accomplished; and although the compliments were undoubtedly warming, the deficiencies remained. He nevertheless believed that if, once again, he could see Hitler personally he could persuade him to lend his authority in certain directions. He noted in his diary on 28 May his impression that '*die Männer um den Führer*', the people round Hitler, often kept the truth from him. Rommel still believed that he might be the person to put that right.

Meanwhile it was clear that time was short. Rommel's sympathy for French sufferings under the ever-increasing Allied air attacks of the early summer was sincere. He recognized that these were the inevitable precursors of the battle to come, but he felt for a population exposed to an onslaught meant rather for their unbidden German guests. 'The Anglo–Americans,' he wrote to Lucy on Whit Sunday, 29 May, 'in no way let up from their incessant bombardment. The French suffer from it terribly – in forty-eight hours there have been three thousand dead among the people. Our own losses are moderate only.'[44] It was not the misery of the French, however, but the deficiencies of his own command which caused him the deepest anxiety.

Rommel had always maintained good relations with General Schmundt, Hitler's senior Adjutant: and he hoped that, through Schmundt, he could secure an interview. Lucy's fiftieth birthday was approaching, and the intelligence forecast was that the early June tides made the period unpropitious for an enemy invasion attempt. Rommel decided to take one or two days off, to take Lucy a pair of Parisian shoes, and to be available at home at Herrlingen so that he could be 'fitted in' by Schmundt to the Führer's programme. Hitler was at Berchtesgaden, and it would not take long to drive to Hitler's

Bergdorf from Wurttemberg. Schmundt agreed that if Rommel were at home he, Schmundt, would do his best, and Rommel arranged to set off from La Roche Guyon by car at the beginning of June if no crisis intervened. He reported to Rundstedt on 3 June and received permission to depart next day; his intention was to press for two more Panzer divisions, for a 'Flakkorps' and for a rocket brigade in Normandy[45] under his own control. He planned to take with him Tempelhoff, who could visit his wife in Bavaria, and Lang.

Rommel had always taken opportunities to talk to prisoners, face to face. He believed he could detect something of the intangibles – enemy spirit, atmosphere, opinions – from such encounters. Even now he sometimes reminisced about Brigadier Clifton, the great New Zealand escaper, and others.[46] When he learned from Fifteenth Army staff on 20 May that an enemy officer had been captured off the coast, clearly involved in some act either of reconnaissance or sabotage, he immediately ordered that he be sent to La Roche Guyon. A charge of sabotage could have meant execution. A few hours later the officer in question, Lieutenant George Lane, was brought into the presence of Rommel.

Lane, a native of Hungary, had joined the British army and been commissioned. He was now serving in the Commandos, the special units trained for raiding enemy coasts; and in a very special unit indeed, X Troop of 10 Commando, formed from gallant men who had escaped from German-occupied Europe and were prepared to serve Britain – generally under assumed names, for the hazards if captured were self-evident, and a number of them were in any case Jewish. Had Lane been taken under his real, Hungarian, name (Lanyi) and identity he would almost certainly have been collected and shot by the security troops. As it was he had British nationality and prisoner-of-war status – provided that he were treated as a soldier and not a 'saboteur'.

X Troop, 10 Commando, had been undertaking a number of operations of considerable difficulty and danger on the French coast, aimed at exploring the underwater and shore obstacles. A new type of German mine had been reported and was to be investigated by volunteers – an assignment from which it was reasonably thought that the odds on the return of the volunteer were small. On this occasion Lane (whose fourth operation to France it was) and an Engineer officer had landed in an inflatable dinghy near the mouth of the Somme. They had run into two German patrols, evaded capture and paddled out to sea again – to find their parent dory had gone. Soon they were captured by a German patrol boat, and initially told that they would be shot as saboteurs. A blindfolded Lane, however, was put in a vehicle and driven a considerable distance, at the end of which he found himself shaking hands with a German officer in immaculate uniform whose English was perfect: Tempelhoff. After a while Tempelhoff suggested Lane 'clean himself up a little' as he would soon be taken to meet a 'very important person'. In reply to Lane's question he said: 'Field Marshal Rommel.'

Lane's recollection of Rommel afterwards was of a man who, in different circumstances, would have been instantly congenial; a man friendly, wholly without pomposity, entirely devoid of any suggestion of intimidation or even discourtesy. Rommel was sitting at his desk at the end of a long room, and when Lane entered he at once rose and moved down the room to shake hands rather than using the superiority of the master to make his visitor approach him, to overawe. Rommel talked naturally, even amusedly (Lane spoke German but, consistent with his assumed identity, concealed the fact, so that the conversation was interpreted). Rommel spoke – but without menace – of 'saboteurs', and Lane replied that if the Field Marshal supposed him a saboteur he presumably would not have invited him (they were sitting at a table and having tea). Rommel smiled a question.

'Invited?'

Lane nodded, and said that he was greatly honoured. Rommel then spoke of the need for England and Germany to fight side by side. When Lane rejoined that this might be difficult, since many German policies repelled the British, Rommel looked interrogative. Lane mentioned the Jews.

Rommel remarked, without making much of it, that every country had its Jews: 'Ours are different to yours.' Overall, Lane's impression was of a man of charm and decency, with a pronounced sense of humour. Rommel spoke agreeably of France and the French people, saying that Lane would, if he used his eyes, see a population at ease with their temporary occupiers (Lane reminded him that occasion to use his eyes had been limited by his blindfold!). Rommel – who sometimes noted in his diary his impression that the French preferred the Germans to the English – pursued the point. The French had seldom been so contented, he said.

The conversation was friendly and relaxed.[47] Thus Erwin Rommel, with one wholly within his power, on the eve of those events which would decide the fate of his country and himself.

Within the German army conspiracies against Hitler and against the Nazi regime had been envisaged from almost the first day of the dictatorship, but these had, on the whole, had brief and undistinguished lives. None of them (such as the proposal to arrest Hitler when the order was given to march on Czechoslovakia in 1938) had got far, being undercut by Hitler's remarkable successes and by a very comprehensible lack of support for such revolutionary endeavours. German officers are not natural rebels. The oath exacted by and to Hitler was taken with understandable seriousness. The tradition of *putsch*, of *pronunciamiento*, of governmental takeover by military cabal, is not a German tradition. And after Hitler had given Germany victories in Poland and in France, in campaigns where his own bold genius was held to have triumphed over the nervous prudence of his professional advisers, or some of them, Hitler was near-unassailable. Nor should it be forgotten that to most senior officers of the Wehrmacht, including many who loathed Hitler and despised the Nazi Party, the general thrust of foreign policy had been just and justified. Germany, they believed, had needed security and the rectification of its frontiers; and it was widely supposed that the attack on Soviet Russia had anticipated a Soviet aggression. Since the tide had turned in the east, of course, Germany had felt threatened by what was perceived as a merciless and barbaric foe, now on the offensive. It was a time for unity.

To conspire against the head of state and Supreme Commander of the armed forces was, therefore, a step of extraordinary moral and physical courage or, as opponents would say, of near-unbelievable treachery. Motives for such a step were mixed, and varied with the conspirator. To some – the Moltkes, the Bonhoeffers, the idealistic youths who went to the scaffold after distributing anti-Nazi leaflets as early as 1942 in Munich – the issue was essentially moral. The words, actions and policies of the Nazi Party were, quite simply, criminal and evil; and this was before the huge extent of that evil and that criminality was known. Such people knew enough; they made their protest; they died; or, before death, they conspired. Numbers of army officers, too, had seen enough, heard or experienced enough at first hand on the Eastern Front, to be so disgusted that they could not breathe the same air as their Supreme Commander without a sense of contamination. To such men it was Hitler who was the betrayer of Germany, and they resolved to risk anything in order to set a term to that betrayal. They came to recognize the terrible truth that in defending Germany they were simultaneously defending mass murder. It was of these men – and women – that Churchill, after the war, declared that they 'fought without help from within or from abroad, driven only by the restlessness of their own conscience'.[48]

There was certainly no help from abroad. The Allies' handling of such secret approaches as were made was distinctly cool. The sentiment was well-rooted that the Germans were incorrigible, that Hitler was simply the latest (albeit the nastiest) of a line of Teutonic aggressors, that any German 'Peace Party' was simply out to buy terms in the face of obvious disaster, that the leopard was unlikely to have changed his spots, that any anti-Hitler conspirators were anyway unlikely to be able to deliver success. Such views – from which Churchill was a convert rather than one opposed at the time – were encouraged by those who wished, perhaps covertly, to further the ambitions of the Soviet Union and thus to see Germany smashed and the vacuum of Central Europe filled by Soviet power. These views, albeit unjust, sometimes ill-motivated and painful with hindsight to contemplate, were understandable.

Many of the conspirators came to their conclusions by a different reasoning – political, in that the Nazi system was inherently corrupt; theoretical, in that the Hitler state offended their ideas and ideals of what a state should be; and, above all, pragmatic – founded in their clear-sighted vision of imminent disaster. Those whose motives were essentially moral tended – as have later generations – to regard these latter and late-come adherents to the anti-Hitler cause as opportunist rather than principled; and the comment has certainly attached to Rommel, albeit often without much understanding of where he stood.[49]

It is not, however, to deny a moral dimension to many of the conspirators, particularly in the Wehrmacht, to say that whatever the origins of their views, their actions were also and largely impelled by the destruction they believed Hitler was bringing to Germany. They perceived – the perception was strategic – a war inevitably lost, amidst circumstances of mounting horror. They blamed Hitler for many of the strategic decisions which had produced so fearful a catastrophe. They recognized that Hitler was a pariah in the international community, and had infected his entire country in all men's eyes. They deduced that the crying need was for peace, a peace which could only be obtained after the removal of Hitler since none would make peace with the man – as he himself acknowledged. Hitler, therefore, must go.

A number of the conspirators had agreed for some time that in the repressive, violent circumstances of Nazi Germany, this meant that Hitler must die. He would never abdicate. He was surrounded by self-serving or dedicated acolytes dependent on the regime's survival. No 'Fascist Grand Council' could be convoked, of the kind which (somewhat astonishingly) had deposed Mussolini. The formal offices of government had been largely overtaken by the machinery of the Party. Security was in the hands of a ruthlessly directed SS, a state within the state. If Hitler were to be eliminated, only assassination would do. It was a view reached with appalling difficulty. Germans tend to dislike revolutionary chaos, to be repelled by disorder, to appreciate the benefits of authority even when it is periodically misused. Tyrannicide has its defenders on moral grounds, but is not natural to Germans. It was ceaselessly argued within the secret circles of the 'opposition'. The case had to be terrible indeed.

The case was terrible indeed. It has been a matter of criticism of some of the gallant men and women of the German resistance that they only acted tardily and when it was obvious that Germany would lose the war; that, by implication, they were content to profit from Hitler's successes despite his crimes, and only turned against him when he began to fail.[50] The charge is comprehensible but unperceptive. Not only were many of the conspirators – and most Germans – sincerely in sympathy with much of Hitler's foreign policy, for understandable if misguided reasons; they were also necessarily constrained by the general climate of German opinion, and while Germany was in the ascendant Hitler was unassailable – no attempt could have succeeded (although some, heroically, were devised), and no successor regime could have governed Germany. It was his strategic failures which both condemned Hitler on pragmatic grounds and provided opportunity for his opponents. This may be less than the heroic self-immolation for principle which critics appear to demand, but in no society would it have been otherwise. And, it must be repeated, the sheer scale of evil being done in Poland and the east was fully known to comparatively few.

There was a further complication to the moral dimension. Few German officers objected to a harsh – but not a criminal – treatment of the vanquished and occupied, as in Poland. As resistance to German authority developed the borderline between harshness and criminality became more difficult to draw clearly. 'Terrorist' outrages demanded effective response if the war effort itself were not to be impeded. The taking of hostages, and the shooting of some of those hostages in retribution for the murder of German soldiers, was periodically enforced – and enforced by the military authorities not only in the east but in Italy and France; ironically, among those enforcing such enactments (criminal under most interpretations of international law) were some prominent conspirators against Hitler. Rommel was fortunate in having no service in the east, where the army often became almost inextricably mixed up in atrocity, even if only as having guilty knowledge of the crimes of others; and he had little direct responsibility for civil order in the west. When such matters were referred to him in Italy his record was generally and predictably humane.

Whatever their private feelings, most of Germany's leading soldiers, and those who served and admired them, left conspiracy to others and immersed themselves in the difficult business of war. Nevertheless there were some – heroes to a man – who were both so repelled by Hitler and his philosophy on moral grounds, and so depressed by what they foresaw as ultimate national disaster, that they sought each other in circumstances of inevitable danger and began to plan action – even when the tide was running for Germany, even before Stalingrad, before Alamein. Indeed an early attempt on Hitler's life (it never happened) was planned for August 1941 when he was visiting Army Group Centre on the Eastern Front.[51] Another attempt (also after a visit to Army Group Centre) was prepared for March 1943, involving a bomb in Hitler's aircraft on his return flight to Rastenburg. It failed; the detonator cap of the bomb had not responded to the striker and Hitler landed safely.[52] Other attempts were made to kill Hitler at close quarters – once when visiting an exhibition, once when inspecting new uniforms. Every plan miscarried; and most of them involved the likely or certain death of the assailant himself, prepared to sacrifice his life for Germany and for decency. Many of these noble idealists had friends or relations among the most senior ranks of the Wehrmacht; and accompanying each plan to eliminate Hitler there were, necessarily, plans to secure the support of sufficient commanders and troops both at the front and at home. The SS would probably be loyal to the regime, and in the chaos following its violent overturn there would inevitably be internal disorder, as well as a need to hold the fronts, pending developments. All this demanded expert, intelligent and thorough intrigue; and the Gestapo were everywhere.

It is in the nature of conspiracy that internal dissension about objects, methods and personalities often flourishes; and such dissension cannot, by the circumstances, be resolved in open debate. So it was in Germany. There was not one monolithic '*Widerstand*' but a collection of different parties, some of them mutually antipathetic and mistrustful.[53].

In the spring and summer of 1944 yet another conspiracy was running. The course of the war was daily more serious, and the destruction of Germany by aerial bombardment was daily more intolerable. Some of the leading members of the German opposition to the Nazis, whether overt or covert, had been arrested by the Gestapo – Bonhoeffer in April 1943, Moltke in January 1944, and several others. Within the Wehrmacht General Oster of the Abwehr had been moved and was under surveillance, and Admiral Canaris, its head, had been retired; both were opponents of Hitler, Oster vigorously so since the earliest days. Time was running against the conspirators and against Germany. One officer wholly sympathetic to their cause was General Hans Speidel, now Chief of Staff to Army Group B.

On Sunday 4 June Rommel, accompanied by Lang and von Tempelhoff, set off for Germany. Next day he telephoned from home to Schmundt at the Berghof, and learned that it would be possible for him to have an interview with Hitler on Thursday, the eighth. Meanwhile two blessed days at home promised. Lucy's birthday was on the Tuesday, 6 June, and early that morning, at half past six, Rommel was downstairs in his dressing gown arranging her presents in the drawing room before anyone else was about. The telephone rang.

It was Speidel. There had been extensive and so far successful enemy airborne operations in Normandy; it was not yet clear whether this was the long-expected invasion itself or not.

Rommel telephoned him back at ten o'clock. By then the matter was certain. This seemed to be the real thing. Assault from both air and sea. Invasion.

CHAPTER TWENTY-ONE

The Last Battle

THERE NOW began Rommel's last campaign, a campaign which claimed him as a casualty and which ended in the total defeat of the German armed forces. Like Napoleon's, his last campaign, his last battle, ended in catastrophe.

Rommel conducted it with a good deal of internal torment. Throughout the weeks until he was carried wounded from the field he was a man within whose mind and heart there were deep divisions; and he found it decreasingly possible to heal them. He was three different persons simultaneously. With part of himself he was simply the good, brave, disciplined soldier he had always been, fighting – and losing through no fault of his own – a battle against the odds, doing his intelligent best during that battle, reacting with skill and fortitude to depressing military circumstances, maintaining men's spirits as far as he could, never giving in. The operational dilemmas were comparatively straightforward, but they were appalling, and Rommel faced them with the mounting sense that his superiors were refusing to recognize the truth of the situation as it evolved.

With another part of himself he was the patriot with a perfectly clear strategic sense, who realized – as he had realized when first formulating his philosophy for the defence of France; as, indeed, in Africa and then in Italy – that Germany's only hope lay in peace, that the material balance of power was so hostile to Germany that the nation could be saved ultimate and total destruction only by reaching an arrangement with the Western powers. He believed that this arrangement could best be – perhaps could only be – secured when the Wehrmacht had won some sort of temporary success in the west, had eliminated for a little the immediate threat of land war on two fronts; and the belief strengthened his efforts, unavailing efforts as it turned out, to win the defensive battle on the ground. But when it was clear that that battle would shortly be lost his belief in the necessity for peace became urgent, and he faced the grim fact that it must be snatched not from military stalemate but from imminent military disaster.

With the third part of himself, in those weeks, Rommel drew inevitable conclusions from the second, from the patriot who saw the strategic situation clear. These conclusions were, it may be contended, tardy. They were certainly unpalatable. The final obstacle to peace lay in the person and character of the Führer himself. Hitler himself, clear-sighted in this at least, had said that nobody would make peace with him. Hitler had, as Rommel had by now and unwillingly learned in part, placed himself beyond the international pale, was adjudged in much of the world not simply as an enemy statesman but as a criminal. Hitler, whom Rommel had angrily condemned in the aftermath of Alamein; Hitler, in whose genius Rommel had recovered confidence when again exposed to that extraordinary personality; Hitler, who had shown to Rommel such friendship and trust – recently reiterated; Hitler, who for ten years had been Rommel's Supreme Commander and in whose apparent service to Germany Rommel had found so much gratefully to admire – Hitler was Germany's bane. With Hitler ruling, Germany could expect nothing but a continuance of war, of pounding to pieces by the Western air forces; could ultimately expect only the overrunning by the Red Army from the east. As June became July Rommel struggled with all three parts of himself.

Rommel's perceptions, like the battle of Normandy itself from his standpoint, can be considered in three periods of time. The first of these ran from the hour of invasion until 17 June – twelve days – when Hitler called the commanders on the Western Front to conference near Soissons.

After taking Speidel's second telephone call on 6 June Rommel drove back from Herrlingen to France. He stopped at Nancy and telephoned La Roche Guyon to check that 21st Panzer Division, the only Panzer division within reach of what appeared the enemy's main landing area, had been committed to counterattack. It had; at 7.30 a.m., five minutes after the most easterly (and nearest) enemy landings from the sea had touched down.

21st Panzer, however, was considerably dispersed. Its four Panzer grenadier battalions had been deployed on either side of the Orne river, around and north of Caen, to stiffen the static division (716th) responsible for that part of the coast. Panzer grenadiers possessed mobility and heavy weapons; they constituted a counterattack force, and their presence in that sector undoubtedly helped it, but the Divisional Commander, General Feuchtinger, reasonably complained that the dispersion – ordered by Rommel – meant that his strength was everywhere dissipated.[1] In the conditions of Normandy, and according to Rommel's convictions, dissipated strength mattered less than the ability to respond instantly wherever the enemy's initial blows fell. By this light the strengthening of 716th Division, by giving to its sector at least some power of instant and aggressive local response, was defensible. It meant, however, that 21st Panzer *as a division* was dispersed and less effective than it might have been. It possessed 127 Mark IV Panzers and forty assault guns, but, as Rommel had forecast, movement was desperately difficult and traffic chaos considerable throughout that and the next day and throughout the area of operations.

When Rommel reached La Roche Guyon at half-past nine in the evening of 6 June he was briefed on the situation as far as it was known. There had been, on 5 June and earlier, indications through British BBC transmissions which German intelligence (having cracked certain Resistance cells) was expecting, that the French Resistance were to stand ready for invasion. These indications, however, gave no clue as to where the blow might fall, and were not regarded by *OB West* as sufficiently reliable to alert the troops. Army Group B had only a small field intelligence staff and was in no way tasked to collect and collate from sources of that kind.[2] In the absence of positive warning from Rundstedt's headquarters, therefore, Army Group had ordered and Seventh Army had undertaken no special precautions. Then parachutists had started drifting down from the sky.

With dawn had come the immense armadas from the sea. In the eastern part of the enemy's assault area, north of Caen, 21st Panzer had attacked from west to east (but not until early evening) on 6 June against a British force of infantry and Sherman tanks advancing south from the coast and established near Hermanville. The division had lost thirteen tanks and been halted, although a small party had reached the coast further to the west. North of Caen the bridges over the Orne were in enemy hands, fallen to airborne assault. Further to the west the enemy had established themselves on a front of some ten miles in the sector of coast between Arromanches and

Courseulles, and had pushed inland to a depth of about six miles, in the western part of 716th Division's sector, north-east of Bayeux. Further west again there had been two other large-scale landings; one north of Colleville, where the enemy had suffered significant losses and 716th Division had hemmed the attackers in beneath the cliffs which rose from the beach; but one still further west, in the improbable area of coast west of the Vire estuary – improbable because a great area of damp lands, unpromising for armoured exploitation, lies around Carentan at the base of the Cherbourg peninsula – where a large and threatening enemy beachhead in the sector of 709th Division was now expanding. Meanwhile the two nearest Panzer divisions in OKW reserve – 12th SS Panzer at Lisieux and Panzer Lehr at Chartres – had been released, during the afternoon, to Rundstedt's and thus to Army Group B's command.

The enemy was ashore; ashore in shallow beachheads, separated from each other, but ashore. He had not been dislodged or driven into the sea by immediate counterattack, but nor had he been able to establish himself at certain key points such as Caen.

In dealing with these events commentators on both sides of the line have periodically criticized both Rommel's dispositions and his absence. His deployments have been blamed for reducing the impact of 21st Panzer's counter-action, by dispersing the division; and for denying to the other sectors targeted by the Allied invasion any possibility of armoured counter-stroke at all, since no Panzers were within reach. But Rommel had known perfectly well that not every mile of threatened coastline could be within easy reach of Panzer reserves; 'immediate' or 'timely' counterattack must mean action when the enemy was still within the coastal defensive zone, and time here could and must be procured only by that strengthening of the coastal defences on which he had placed priority. Had he been given operational control of more of the OKW reserve in Panzer Group West he would have been able to deploy some tanks further west, able to counterattack early towards Bayeux or Carentan: but his own resources had not run to that – that which alone might have tipped the scales.

As to Rommel's absence in Germany and the time lost before he reappeared, it is true that some on the German side have believed that Rommel, on 6 June, might have somehow inspired earlier counter-action; that he might perhaps somehow have persuaded OKW – probably the Führer himself – to earlier commitment of the Panzer reserves;[3] that precious hours were lost. Speidel apparently failed to

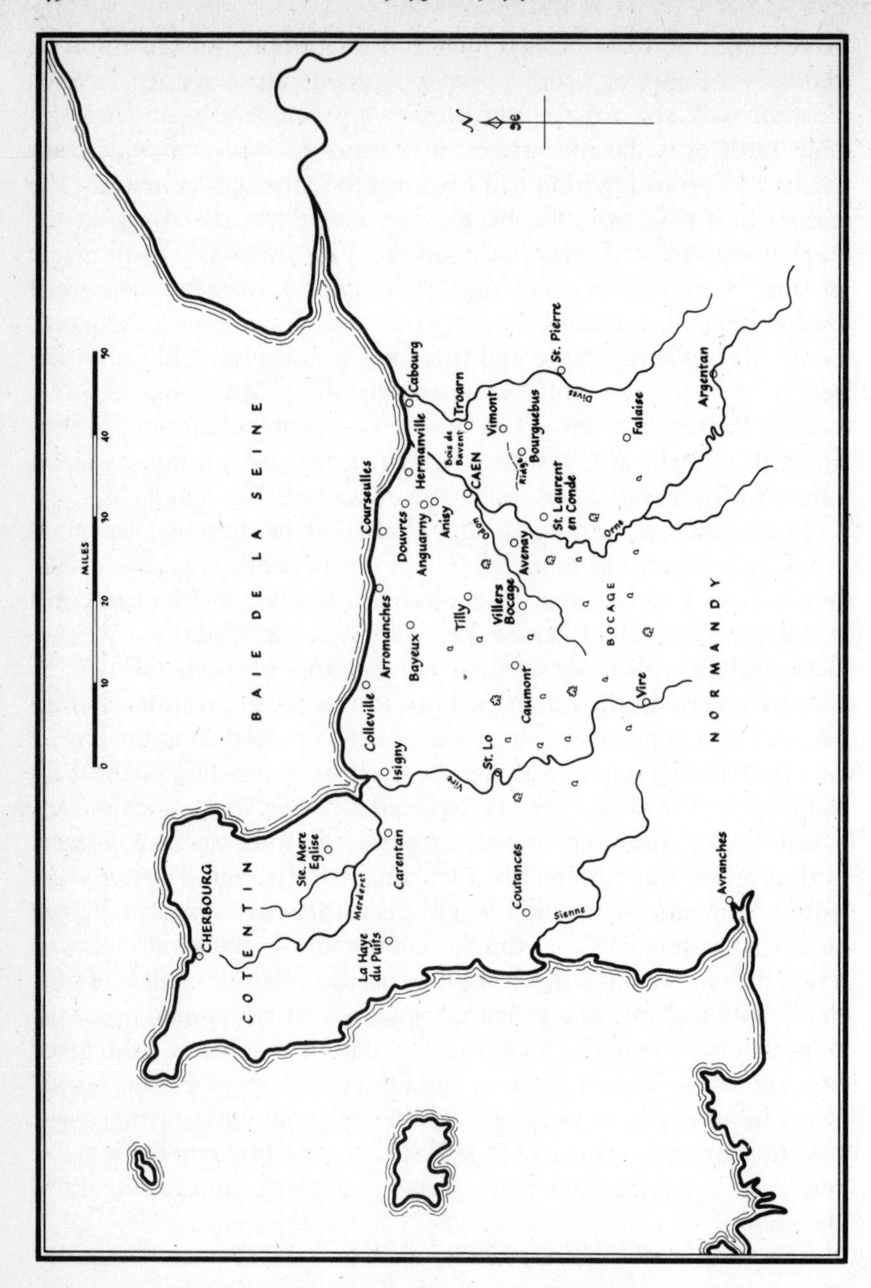

XX The Normandy battlefield, 1944

move Jodl, despite appealing in Rommel's name and with Rundstedt's endorsement. Rommel in person might also have accelerated the actions of 21st Panzer, whose intervention was delayed not only by dispersion but by confused counsels and changing orders. Given the fact of Allied air power, however, given the delay imposed on every daylight German operational move, given the deployment areas of the Panzer reserves (12th SS Panzer at Lisieux, thirty miles from Caen, and Panzer Lehr at Chartres, too distant to come into the debate), and given the fact that Rommel could not have sought a decision until he was *sure* that this was a major assault, it is difficult to believe that his presence, however persuasive, could have decisively affected events in respect of Panzer counterattack on 6 June. The morning was cloudy and visibility imperfect so that, conceivably, a successful call to OKW could have procured the arrival of 12th SS Panzer by early afternoon; but it must remain speculative.

It is surely also true that, given the weight of Allied assault, the weight of aerial bombardment and ship-to-shore bombardment (often stunning the Germans with its accuracy and lethality), the strength simultaneously at several points, together with the extremely successful airborne assault – even a heavier and earlier counterattack by 21st Panzer Division would have only affected one local situation. The Allied invasion was named 'Overlord', and it was destined to overcome. Material superiority, ingenuity, practice and planning – together with human courage and human resourcefulness – triumphed. Only wholehearted acceptance of Rommel's 'heretical' proposal, to deploy more Panzer divisions under his own operational control and very near the coast, might have changed history.

One disappointment for Rommel was the relatively minor delay which his obstacles both underwater and ashore had imposed on the enemy assault; had he had more time in the west it might have been a different story. In some areas, notably the beaches in the Courseulles sector, it was clear that many landing craft were being entangled and scuppered by Rommel's mines and underwater steel obstacles, while some of his concreted strongpoints proved difficult nuts for the invader to crack, but by the end of what the enemy called 'D-Day' France had been successfully invaded. What Rommel had hoped to forestall had happened. The battle, however, was still confined to the coastal zone or near it. The penetration towards Caen was only about four miles. The invaders might yet be driven into the sea; and, had he known it, Rommel's defending forces had checked their advance many miles short of the objectives Montgomery, commanding all the Anglo–American invasion forces, had set.

There was also the question of whether this Normandy operation represented the enemy's main effort. The Allied deception plan, codenamed 'Fortitude', was intended to lead the Germans to deduce a fresh army, not yet committed, in south-eastern England and probably threatening the Pas de Calais;* this had a great deal of success. Throughout June and July German intelligence appreciations were referring to '*Heeres Gruppe Patton*', supposed to be deployed in eastern and south- eastern England between Brighton and the Humber and consisting of anything up to thirty formations, '*starke Verbande*' – perhaps divisions. Whatever the bait deliberately set and swallowed, however, there were known to be considerable numbers of fresh American divisions, whether in the United States or Britain, which could, sooner or later, be deployed; Allied resources in landing ships and craft of all kinds had been shown as enormous; the terrain in the Pas de Calais was perfectly adapted to invasion, and the subsequent attack axes thence towards Paris – and perhaps intended as an operation into the German flank as they defended against an enemy advancing from Normandy – all surely seemed tempting. Rommel kept his eyes on the possibility of another enemy landing until his last day in Normandy, in mid-July. This has been criticized as an obstinate aberration, one manifestly proved wrong by captured enemy documents,[4] but it was one he shared with OKW; it was specifically discussed by Hitler on 17 June and later; and to Rommel it was self-evidently a contingency which remained dangerous. He made two visits of inspection to the sector near Le Havre in the first week of July, and it was wholly reasonable to do so. German intelligence, served as it was by virtually no air reconnaissance, without benefit of ULTRA or anything like it, and often bamboozled by a highly professional enemy deception service which had had a long time to perfect plans, was virtually blind. Estimates of enemy intentions could be based on little but knowledge of terrain, forecasts of weather, realization that the enemy commanded sufficient resources to attempt almost anything he chose, and guesswork. Miraculously, they were sometimes pretty good.

On 7 June Rommel remained in his headquarters, conferred with Geyr von Schweppenburg and took stock of the situation. Geyr,

* General Patton was put through a well-orchestrated programme of high-profile visits to lend credence to his command of this (non-existent) formation.

after the event, criticized the OKW decision to release the reserve Panzer divisions to Rommel; he complained, with some justification, that orders to move 12th SS Panzer and Panzer Lehr Divisions had been given without reference to him; and he argued that, if the reserve divisions had been kept in hand, they could have carried out a concentrated operation northward on about 24 June.[5] By that time they had been sucked into the 'holding' battle; but there were no more infantry divisions available and without the support of armour the position would have cracked and a rout begun. Meanwhile, unless wholly extraordinary successes could be achieved in the next few days, Germany on 7 June was facing what Rommel had always regarded as certain to lead to ultimate disaster in the field; a protracted land war on two main fronts.

During the next ten days Rommel's movements and preoccupations can largely be charted from the diary kept of his personal activities at Army Group B headquarters[6] and the reports written daily by his accompanying staff officer about his visits.[7] Both, of course, were to a large extent sanitized, and naturally excluded the informal exchanges, the cut and thrust of vigorous and often agitated talk between commanders under enormous physical and mental pressure. The general course of Rommel's activities, however, is on file, as well as the instructions he actually gave and situations he recorded. Imperfect though these extracts perhaps are, they provide a more precise guide at least to his formally expressed views and decisions than recollections after the event, no matter by whom. The same is true of the actual record of his encounters with Hitler – again, edited though that may be. Minutes of meetings, in every country, may tell the truth, but not the whole truth.

The two principal areas of concern to Rommel were the sector round Caen, in the east, and the Cherbourg peninsula in the west. While the enemy could be denied any significant progress in these two directions there was a chance of smashing him by counterattacks, necessarily local though these might be. Denied Caen, the invader's left flank had no firm hinge and was vulnerable; while without Cherbourg he had no major port. During 7 and 8 June there was heavy fighting both west and east of the Orne river north of Caen. 12th SS Panzer Division had now joined the battle and on the night of 8 June attacked strongly, using its powerful Panther tanks, towards the sea. Rommel had visited the Caen front that day. He called on the redoubtable *Obergruppenführer* Sepp Dietrich, commander of I SS

Panzer Corps, a roughnecked Nazi who, despite Rommel's aversion to the SS, gave his Army Group Commander total loyalty. Rommel remarked bitterly that had the Panzer divisions – and in this instance he meant both 21st and 12th SS – been permitted actually to deploy their tanks nearer the coast as he had desired, there could have been a stronger attack two evenings previously. This was spilt milk. The essential now was to contest the ground round Caen and to prevent the enemy from cutting off the Cotentin peninsula. Next day Rommel drove to see Dollmann of Seventh Army at Le Mans.

That day, 9 June, Panzer Lehr Division arrived in the Caen sector from Chartres. Two more infantry divisions, 346th and 711th, had sidestepped west from the Seine basin to strengthen the westward-facing German defences east of the Orne; and there were now three Panzer divisions ringing Caen. Geyr planned a limited thrust northward from Caen astride the light-gauge railway, a short push to reach Anisy and Anguarny south of Douvres, to drive a disruptive salient into the British front. In the west the enemy had cleared Isigny but his progress was slow in the difficult *bocage* country south of the place as well as in the reclaimed marshland between the coast, Carentan and La Mère Église. The German front was holding.

But the enemy front was now near-continuous, beachheads clearly soon to be linked and being reinforced all the time. By 10 June the Americans had reached a line only just short of the Merderet river, west of Carentan, while next day a British division – identified as 51st Highland, familiar to Rommel from Alamein days – attacked southward in the sector east of the Orne and north of the Bois de Bavent, an attack which was smashed within hours by local counter-attacks, short jabs which Rommel was sure the situation everywhere demanded and which were all his resources – and the air situation – permitted. East of the Orne, nevertheless, Panzer casualties were heavy.

Rommel visited Geyr on 10 June at Panzer Group West headquarters in an orchard at Le Caine, twenty miles south of Caen. They discussed the fuel and ammunition situation. Both were bad – replenishment convoys were having to make a round trip of 125 miles, so serious had been the enemy's attrition of dumps and communications. The Luftwaffe appeared totally absent. All movement – including, more often than not, Rommel's own movement – was hazardous and sometimes impossible. Rommel left Geyr, having planned to visit 12th SS Panzer Division, and found the journey impracticable unless he were minded for suicide; and that afternoon

Geyr's own headquarters were bombed – whether betrayed by civilians or detected by radio fix, Rommel wrote, was unclear (it was, of course, from ULTRA). Most of Geyr's staff were killed and any offensive planned by Panzer Group West was halted. Next day Rommel conferred with Rundstedt.

The battle was turning, once more, into a *Materialschlacht*; Rommel's forces were being pounded to pieces by the enemy's weight of fire from ground, sea, and air, and by the enemy's seemingly inexhaustible supplies of ammunition. Rommel remained as sure as he had been from before the beginning that this was not a situation which could be remedied by concentration, by an operational offensive – the air situation and much of the terrain made such concepts wholly impracticable. The only resource was to patch up a stretching front, to strike locally and hard whenever the enemy gave the opportunity, to keep him as far as possible hemmed in. Often the intensity of enemy fire made even local and limited counter-moves impossible, and casualties in men and equipment were mounting alarmingly; but there were occasional rays of light. The enemy often seemed sluggish, slow to exploit tactical success, and to have been checked – indeed driven back with remarkable celerity – on those occasions when the Germans had been able to show fierce local strength. The Tiger tank with its 88-millimetre gun – here, as in Africa, master of tank-versus-tank combat – had achieved some astonishing successes, even when committed singly or in tiny numbers; indeed only tiny numbers were available (the Tigers were organized in special *Abteilungen*), but they had undoubtedly spread alarm. During the first fortnight in July they began to arrive in larger numbers, and new heavy tank battalions joined Army Group B, with considerable effect.

On 11 June the British 7th Armoured Division, another old acquaintance of Rommel's, began moving round the west of Caen, and on 13 June entered Villers-Bocage, fifteen miles south-west of Caen itself; but here one Tiger tank moved from the south into the town, shooting up everything in its path, and the British advance was checked and turned. Ringing Caen were now four Panzer divisions. Panzer Lehr and 12th SS had joined 21st Panzer on the ground, all under I SS Panzer Corps, and on 13 June they were reinforced by 2nd Panzer from the Pas de Calais, after a most laborious journey. Rommel's armour was near concentrated, and there was more to come. He could still react north of the Seine with 116th Panzer if a new threat developed. He visited General Graf Schwerin, the

Divisional Commander, on the afternoon of 12 June, but on the thirteenth he was back with I SS Panzer Corps in the Orne sector.

On the same day, 13 June, the two western enemy beachheads linked up, and on the previous day Carentan had fallen. In that sector the defenders were being pushed westward, and it was clearly only a matter of days before the Cotentin peninsula was cut off. On 14 June Rommel made a difficult journey from visiting XLVII Panzer Corps (General von Funck) in the west of the Cotentin and drove to meet 2nd Panzer at Bremoy near Villers-Bocage. He found that Panzer Lehr had defeated a British tank attack and destroyed twenty-five of them in that sector.

Rommel's visits enabled him to gauge the spirit of formations and their commanders. On the whole he found morale surprisingly high among the forward troops, but he was getting reports of misconduct in the rear area, of the looting of bombed houses, and of men skulking in the villages of the hinterland, claiming that they were unable to rejoin their units but in reality showing no intention of doing so. He noted that *Feldgendarmerie* detachments were necessary on the roads and in the villages behind the front.[8] The German army never hesitated to use the strongest possible measures against backsliders, promptly and energetically, and the *Feldgendarmerie* were with reason feared. In general, however, Rommel found the heart of the troops commendably sound, although there were, as ever, differences between one division and another. The *Waffen SS* were undoubtedly good – highly motivated, generally at a higher strength in men and equipment than army divisions, and ruthlessly led. 12th SS Panzer – the '*Hitler Jugend*' Division – were admirable, although Rommel noted that relations between members of the *Waffen SS* and the army were sometimes imperfect, with SS men seeking to disregard the orders of army officers and observing a very different code.

But the troops were being exposed to a terrible battering, and here Rommel questioned his own orders. He had been wholly in sympathy with Hitler's initial determination not to give ground; that the battle must be won in the coastal defensive zone had been Rommel's own message to all from his first day. Now, however, the enemy had – tenuously but indubitably – won the first battle and established himself ashore. Every effort having been made and continuing to be made to eject that enemy, it was now essential therefore to redeploy troops where necessary to fight another day and to minimize the effect, in particular, of naval gunfire. This implied some operational freedom; it implied certain local withdrawals; but there had been (as so often before in Rommel's experience) an order from Hitler himself. No withdrawals. This meant that virtually every movement of troops needed OKW approval. It was a ludicrous interference from on high in tactical detail. A redeployment of 77th Division, for instance – proposed by Rommel to plug a gap near the base of the Cotentin – had to be referred to Rundstedt, who, in spite of Rommel's urging, declined to authorize it on his own and suggested that the situation was less critical than Army Group B supposed.[9] In particular, Rommel wanted the freedom to withdraw troops when necessary into Cherbourg from the south. It was the port which mattered, not the ground. A *Führerbefehl* was received in the afternoon of 16 June. No withdrawals into Cherbourg.[10]

Rommel, in these days, spent much time visiting the various corps headquarters, and Panzer Group West. Sometimes, but not always, accompanied by the appropriate corps commander, he visited divisional headquarters, at every level briefed by commander or chief of staff. Although he paid visits to both Seventh and Fifteenth Army headquarters he did not restrict his interventions to dealing only through them (nor, for that matter, did Montgomery in comparable situations). The Normandy battlefield was comparatively small. Huge masses of men and steel were opposing each other in a restricted space. Decisions, when needed, were needed instantly; and it was Rommel's way, in any case, to inspect, decide and express his wishes without delay. But in Rommel's visits at this time there is no particular sense of the commander who on occasion usurped his subordinates' roles. Each side's team, in Normandy, was necessarily fighting in very close collusion with neighbours, and if the German chain of command had originally left a good deal to be desired proximity, as far as Normandy went, now tended to correct the error; proximity left little scope for confusion or inappropriate interference; senior commanders were inevitably involved in the tactical battle on so confined a field.

On 17 June Rommel drove to a temporary headquarters Hitler had established at Margival near Soissons, in Champagne, in quarters originally prepared for the Führer during the 1940 invasion of England which never happened. Rommel welcomed the summons.

Tactically, Rommel had been able to fight the battle so far more or less as he wished. It had not gone successfully because the coastal defensive zone on the beaches and behind them had not imposed the delay or done the damage to which he had aspired. The concept was surely right, but those critics among his subordinates who had told him that his programme was too ambitious for the available time, manpower and resources had been partially justified. Nevertheless his defensive systems and dispositions had held the enemy for crucial days, had certainly made the invasion a difficult business and its exploitation slower than planned (and expected by the more pessimistic German assessments), and had here and there inflicted serious losses. The cost to the defenders had, however, also been inevitably and alarmingly high.

The counterattacks had at least checked any ambitious enemy advance. They had been limited in scale, as Rommel had ordered and of the necessity of which he was convinced. Movement had been gravely impeded by enemy air power – much as Rommel had prophesied, and he remained unrepentant at the deductions he had drawn from this factor; but movement – slow, laborious, here and there expensive in vehicles lost to direct air attack – had, in the event, been possible and, beside the four Panzer divisions in Normandy, a fifth, 1st SS Panzer, was due on 18 June from Belgium and a sixth, 2nd SS Panzer, was on the march from the south of France. Furthermore Rommel had been assured that Hitler was authorizing a move to the west from Russia of two more Panzer divisions, 9th and 10th SS; and within a week they would arrive. By three weeks after the invasion there would be, by any reckoning, a formidable concentration of German armour in Normandy, a concentration which showed little signs of being inhibited by (very real) anxieties over further enemy landings elsewhere, whether induced by Allied deception or not. Panzer division losses had been considerable but they had undoubtedly scored highly against the enemy's armour.

Matters could only have been better, tactically, had Rommel been permitted to hold more Panzers in Normandy, nearer the coast; had he or *Fremde Heere West* guessed more accurately where the invasion was coming, and that it was coming unsupported by a second landing; or had he had more time. As things were the situation was grim, but this did not result from tactical mistakes or shortcomings; and it might conceivably improve.

The grimness of the situation, which he hoped to impress on Hitler, derived, in Rommel's view, from two self-evident factors, and although he had probably underestimated one, he had given fair weight to the other. It resulted from the enemy's total air superiority, which he had forecast; and from the enemy's huge materiel superiority, which had been deployed ashore with great energy and ingenuity. Normandy was rapidly turning into an Alamein. It was as it had been in Africa, but considerably more so; and there was an added tactical factor – naval ship-to-shore gunnery – which was tipping the odds even further against troops remaining within range of it. All this led Rommel to renewed pessimism in strategic terms. Convinced as he was that the only hope for Germany would have lain in some sort of strategic stalemate in the west, providing a rational basis for negotiation and permitting reinforcement of the Eastern Front at least in the short term, he had pinned this strategic hope on the tactical defeat of the enemy invasion. That hope had almost died. An established western front – the phenomenon he had feared – was near reality; and it could not hold long.

At the conference on 17 June Hitler made a surprisingly robust impression, and was able for a tiny while to communicate something like optimism to his hearers – Rommel was accompanying Rundstedt, both with their chiefs of staff, and Jodl and Schmundt were among those present. Rommel opened proceedings with a pretty accurate indication, on the map, of enemy and German positions and strengths. He drew the analogy with Africa – the enemy was using his huge materiel superiority to steamroll the way to success. The greatest German lack was air power and air defence – it was this deficiency which reduced movement and thus replenishment to pitiable proportions. The troops were fighting well against great odds but the battle was being lost in the air. Rundstedt made a few observations and produced the one detailed request he wanted to carry with Hitler – leave to withdraw troops in the northern part of the Cotentin peninsula towards Cherbourg.

Then Hitler spoke. The Cotentin would now, inevitably, be cut in two, he accepted that; but Cherbourg itself must be strengthened and must for a while be held at all costs. An exceptionally able commander must be appointed – Cherbourg must certainly hold out until mid-July. Without Cherbourg the enemy's supply situation would remain difficult and could become critical. This, Hitler said, was the key to the whole situation; and he gave promises of a new mining programme, undertaken both by Luftwaffe and *Kriegsmarine* between Le Havre and the east coast of the Cotentin, which would isolate Normandy from sea supply. As to the rest, Hitler contented himself with saying that the situation east of the Orne appeared particularly important to clear up.

To Rundstedt's brief survey of possible enemy landings on other parts of the coast Hitler observed that the British had now deployed in Normandy all their most experienced divisions, among others.

This showed that the enemy were committed to exploitation of Normandy, to prime emphasis upon it, although a landing in the Pas de Calais was still possible. Rundstedt suggested certain anodyne principles for agreement; that the enemy bridgehead must be contained; that 'tactical adjustments should depend on the local situation' (surely an optimistically crafty formulation); that reserves should be held ready for commitment against attempts to break out of the bridgehead. Rommel, finally, warned against leaving the Panzer divisions within range of naval gunfire; and emphasized that to use them in large-scale attacks would wear them down. As far as could be managed infantry divisions should be deployed holding ground and the Panzers be withdrawn to reserve, on the flanks of possible enemy break-out axes and out of range of enemy ships' guns.

Hitler, in concluding, again emphasized that the northern part of the Cotentin, and thus Cherbourg, must be held as long as possible. A purely defensive battle in the rest of the bridgehead – from Caen to Carentan – was an unacceptable long-term concept because the enemy could build up too great a materiel superiority. He must, therefore, be attacked; and the way to attack him was by interdicting his supplies through the air and sea measures which he, Hitler, had already spoken about. And with this his hearers had to be content.[11] Rommel had hoped that Hitler (and the senior men at OKW) would visit the front and hear from some front-line commanders for himself, after the Margival conference. This had been promised, but was cancelled; and it was one of the many bitter reproaches levelled by Rommel and others at the High Command that throughout the Normandy fighting they never came near it. Too many plans were made and orders given 'from the green table'.[12]

After the war Jodl reported Rommel, at this meeting, as angering Hitler by asking him what he really thought about the future development of the war. It may be so; it may be that Jodl, in good faith, confused Rommel's report on this occasion with that twelve days later, when the atmosphere was tense indeed.[13] Jodl also reported Rommel as publicly accusing some of the SS units of alienating the French with atrocities, an accusation which did nothing to mollify Hitler.

By a bitter coincidence, the Margival conference was immediately followed by a ferocious four-day storm which disrupted for a while Allied supply (just as Hitler had promised would be accomplished by the *Kriegsmarine*, rather than Divine intervention); but the German forces were in no condition to take the offensive and exploit this circumstance.

The second period of Rommel's trial lasted from 18 until 29 June, when he had another session with Hitler. During this period the situation in Normandy worsened remorselessly. On 18 June, as it happened, the enemy command had itself judged the bridgehead was firm, and that the first phase of Overlord had been accomplished. In the west American progress in the Cotentin was relentless. In the *bocage*, the close hedgerow country in the centre of the Normandy front, there was an unending sequence of enemy pushes, both small-scale and large-scale, each accompanied by intensive artillery and air attack; this was fighting of a particularly expensive kind, high in casualties, small advances purchased with a good deal of blood both of attackers and defenders. Tactically, the men of Army Group B still felt that they often had the best of it, but their strength was being sapped by attrition. The front was not long and two million men were contesting it. Rommel on his visits still got a good impression of the fighting troops, but he knew what they were suffering.

Rommel still had one eye cocked north of the Seine as well as on Normandy, and on a visit to 116th Panzer Division on 19 June he again mooted the possibility of an Allied landing between Seine and Somme[14] – the *Fremde Heere West* estimate of the number of troops in England was still wildly exaggerated. He visited the division again two days later and shared with Schwerin some tactical impressions of the fighting so far. Wherever he went in these days, Rommel 'expressed wishes', 'made suggestions', 'conveyed impressions', rather than rapping out orders to a subordinate's subordinates. The impression is of a higher commander reasonably confident with his subordinate commanders, riding on a light rein, imparting encouragement as far as he could, sharing impressions, listening more than shouting. This, of course, was consistent with his position. It also, perhaps, indicated something of his frame of mind.

For as the last ten days of June passed, Rommel felt ever more acutely the imminence of catastrophe, and sometimes spoke to intimates about it, about his feeling of *déjà-vu* as he watched the advance of a calamity he could do nothing to avert but for which he felt responsible.[15] When he had left Hitler on the afternoon of 17 June he had been initially almost buoyant. Something of the old 'sunray lamp' had no doubt warmed him. Visiting Geyr on the following day he had told the sceptical commander of Panzer Group West what had passed. But as the days went by, and the enemy was known to be all the time reinforced, and the enemy aircraft flew sortie after sortie, and the bombs and shells fell, and men died and died, Rommel found that it was impossible any more to believe in tactical successes, except occasional, purely defensive and very local; and if that were all that could be hoped at the tactical level, what could be hoped strategically? And that meant 'What could be hoped politically? What could be hoped for Germany?'

There were occasional shafts of light. When he visited Sepp Dietrich on 21 June, Rommel heard that Dietrich was tolerably confident of holding any British push with the Panzer divisions of I SS Corps. The long-expected V-weapons had been launched against England on 12 June, and a ripple of optimism had run along the German front. Since manpower was failing it was agreeable for Rommel to be told by Feuchtinger of 21st Panzer that he could enlist two thousand French volunteers, keen to fight against the British, if it were permitted.[16] Such things gave hope. But no observer with eyes in his head could fail to see that time was running out for the defenders, that the trend of events was all one way and almost certainly irreversible.

Rommel discussed the situation with Rundstedt on 26 June and telephoned him again next evening. Would Rundstedt agree to him, Rommel, making a special journey to Berchtesgaden to seek another interview with the Führer and to lay before him the full seriousness of the situation? On 17 June they had been given promises of miraculous methods by which the enemy's supplies would be interdicted. It was perfectly plain that even if these measures came about they would take time to be effective – and there was, anyway, no sign of them. Meanwhile the situation on the ground was getting worse daily. Rommel did not believe Hitler was told the truth by OKW, or some of its members. He had seen General Thomale, Guderian's Chief of Staff of the Panzer arm, visiting LXXXIV Corps sector (General Marcks) on the previous day and besought him at all costs to form his own objective impressions and to tell Hitler what they were. They could not be other than frightful. On 27 June Cherbourg had surrendered, albeit after vigorous defence and with plans under way which completely destroyed its function as a port for a further four weeks. Rommel's mind had moved a good way since 17 June. So rapidly deteriorating a tactical situation must lead to grave strategic conclusions; and although Rommel as yet only said this to a few (he would soon say it to several), to grave political conclusions as well.

Actual and complete defeat in Normandy would soon threaten, and defeat in Normandy must mean the end for Germany.

Rundstedt agreed to Rommel's proposal and said that he would accompany him. Both men set out on 28 June and met by arrangement near Paris on the main road to Germany, Rommel being accompanied on this occasion by Tempelhoff's assistant, Major Wolfram, as well as Lang.

Rommel and Rundstedt spoke together for some time, and Wolfram could hear the conversation, or some of it. 'Herr Rundstedt,' Rommel said, 'I agree with you. The war must be ended immediately. I shall tell the Führer so, clearly and unequivocally.'[17] Rommel knew very well what that implied – had Hitler not said to him, 'Nobody will make peace with me'? But the recent fighting and simple logic had persuaded him that this was not a matter which could be deferred. The stalemate from which – unrealistically – he had hoped negotiation might be developed was not going to happen. The fighting in the west had to be stopped. Every day would make matters worse for Germany. And as he drove on towards Germany Rommel quietly said to Wolfram what was in his heart: 'I feel myself responsible to the German people.' He said that his burden could not simply be regarded as that of a military commander. The whole world was in arms against Germany. Victory was out of the question. The enemy had now won his foothold in the west. The war must be ended.[18] During the day he heard that Dollmann, Commander of Seventh Army, had had a fatal heart attack.

Rommel spent that night at home and drove next morning to Berchtesgaden. There he found both Goebbels and Himmler, and decided to try a word with them before reporting to Hitler. After a talk with the ever-friendly Goebbels he supposed – and said to Wolfram – that he had gained an ally for his project; the project of telling Hitler the unvarnished truth and asking that conclusions be drawn. Wolfram was unconvinced; nor was he convinced that Rommel had turned Himmler into an ally by dilating on the fighting performance of the *Waffen SS*. Himmler, according to Wolfram, remained 'opaque'.[19]

Before the conference with Hitler Rommel also managed to talk for an hour and a half with Guderian – a conversation recorded in the diary report (surely incompletely) as being concerned with such matters as re-equipment of the Panzer divisions, and concepts of future operations; Guderian was still Inspector General of Panzer troops. At six o'clock the session with Hitler began. Besides Rundstedt, Keitel and Jodl from OKW were present; and after two and a half hours they were joined by Goering and Admiral Dönitz, commanders of the Luftwaffe and the *Kriegsmarine*, as well as by the large and stertorously breathing Field Marshal Sperrle (commander of Third Air Force in the west) and many others.

Ultimately Hitler summed up with a number of points, a harangue dignified by the term 'directive', of startling banality, falsity and irrelevance to the real needs of the situation. The most important task, the Führer said, was to halt the enemy offensive, preparatory to clearing up the whole Allied bridgehead. This would largely be accomplished by the Luftwaffe, but in addition mining at sea would decisively interrupt Anglo–American supplies. A variety of special weapons – and one thousand aircraft from new production – were about to come into service. A large number of torpedo boats and submarines would soon be operating in the Channel – in, at the most, four weeks. And fleets of new transport vehicles would soon be moving west from the Reich.[20]

So far the official record: but at the start of the conference matters had gone as no staff officer or stenographer recorded. Hitler had asked Rommel to speak first; and Rommel had begun, as he had vowed, by saying he thought that day the last moment at which he would have the chance to lay the whole situation in the west before the Führer. '*Die ganze Welt*,' he continued, '*steht gegen Deutschland, und dieses Kraftverhaltnis* –'*

Hitler interrupted him sharply. Would the *Herr Feldmarschall* please concern himself with the military, not the political situation. Rommel rejoined that history demanded he dealt with the whole situation. Hitler again rebuked him, telling him to deal with the military situation only.

Rommel did so, and the conference took its unrealistic course. Before it ended, however, he made a final attempt. Having castigated the Luftwaffe for its inadequacy, he said that he could not leave without having spoken to Hitler 'about Germany'. '*Herr Feldmarschall*,' Hitler responded icily, '*ich glaube Sie verlassen besser das Zimmer!*'† Rommel left the room.[21] He never saw Hitler again.

* * *

* 'The whole world stands arrayed against Germany, and this disproportion of strength –'

† 'Field Marshal, I think you had better leave the room!'

Henceforth Rommel's mind was clear, and during the last two and a half weeks of his active service, the third period of his trial, he spoke it to a good many who, shattered by the course of events, asked him what was to happen to the army and to Germany. He did his duty assiduously, visiting, cheering, correcting errors, proposing tactical improvements with as much energy as ever. To do less would have been to betray his men. He was their commander and Germany was still at war. As far as it lay with Rommel, the enemy's attacks would be defeated and the enemy's progress halted, and in these final weeks Rommel was everywhere. But Germany must have peace, and he could not believe that the Supreme Commander was minded to help bring that peace about – in which case it must fall to somebody else, with power to act. To Ruge, on one of their walks, he had said soon after the campaign began that it was obviously vital to seek peace while the Germans still held some territory. Time had nearly run out.[22]

Rommel visited Geyr at Panzer Group West the day after returning from Germany and told him – or told him some – of what had transpired. They discussed the Caen sector, on both sides of the Orne. The enemy was very likely to try a major break-out there; and during the next fortnight Rommel visited that area frequently. Geyr wished to hold the three divisions of I SS Panzer Corps over twenty miles south of Caen, in woods near St-Laurent-de-Condé, poised to strike any massed enemy advance on advantageous ground. Rommel disagreed. True to the philosophy he had held throughout he believed that at least part of the Panzer force should remain immediately east and south-east of Caen, should stiffen the front. To withdraw most of the armour as far as Geyr proposed would be to leave the infantry, unsupported, to be rolled up by the enemy when they attacked – and he agreed that it was in this sector they would certainly attack, if only to race by the shortest route eastward towards the V-1 sites.

It was his last conversation with Geyr. Hitler had heard that the commander of Panzer Group West was as pessimistic as the commander of Army Group B – and even more anxious to withdraw to positions in depth – and had ordered his replacement. (Geyr had produced an unvarnished report and it had been sent to OKW. The report recommended withdrawal from the 'Caen salient'; taking up a defensive line from Aveney in the east through Villers-Bocage to Caumont, and along the Orne; and the adoption of 'elastic' tactics[23]).

Succeeding Geyr in command of the Panzer Group was General

Eberbach. Visiting him on his first day, 5 July, Rommel again discussed the Caen sector.[24] The enemy, when they attacked southward east of the Orne, as they undoubtedly would, must be broken up by anti-tank artillery fire, *Nebelwerfer* (rocket launchers) and Panzers. The defence must be organized in great depth.

On the night of 7 July a great enemy air raid destroyed much of Caen and killed a large number of civilians, but did little to the defending forces. On 10 July Rommel was again in the same area, the area of LXXXIV Corps and of I SS Panzer Corps. The ground was held by men of 16th Luftwaffe Division, and they did not all inspire confidence in Rommel; casualties, however, had been especially heavy in the senior ranks, three of the division's battalion commanders having recently been killed.

All divisions, including the Panzer divisions, were now very short of men as well as equipment, and divisions from other areas – like Fifteenth Army's – were arriving in Normandy without a full complement.[25] Rommel returned to this sector, east of Caen, on two further occasions, 12 and 15 July, suggesting adjustments, aiming to give maximum depth to the front, replacing Panzer units committed to ground-holding with infantry as far as humanly possible. He heard frequent requests for withdrawal and refused them. In his orders he was entirely true to the *Führerbefehl*. In every sector, he told II Parachute Corps (General Meindl) at Vire, the front must be held: '*Front muss unter allen Umständen gehalten werden*.'[26]

But Rommel was under no illusions about what the troops were suffering. 'Our soldiers,' wrote one of his commanders, 'enter the battle in low spirits at the thought of the enemy's enormous material superiority. They are always asking "where is the Luftwaffe?" The feeling of helplessness against enemy aircraft operating without any hindrance has a paralysing effect and during the barrage this effect on inexperienced troops is literally soul-shaking.'[27] Many a British veteran of Dunkirk would have agreed with him; and losses were appalling from artillery as well as air strikes. Even the smallest enemy attacks appeared to be preceded by saturation bombardment. On 14 July Rommel found one parachute regiment wherein, of a total of one thousand reinforcements received since battle began, over eight hundred had already fallen. 'The morale of the troops is good,' the Chief of Staff of Panzer Group West told Rommel, 'but one can't beat the materiel of the enemy with courage alone.'[28] This was the faithful Gause, Rommel's former *Chef*, now with Eberbach. Everywhere the mood of commanders was despairing. In the east a great

Russian offensive had started on 23 June in the central sector, north of the Pripet marshes, and by the beginning of July the Red Army had torn a huge gap in the German front, taken Minsk and were marching across Poland. They would soon reach the Vistula. Sometimes Rommel talked of the possibility of suicide, but only to reject it. In these circumstances suicide was merely desertion.[29]

Old acquaintances, officers who had long served Rommel and knew him well, put the matter to him plainly, and he to them. 'Lattmann,' Rommel asked his artillery representative on 10 July as they drove together to visit LXXXIV Corps, 'what do you think about the end of the war?'

'*Herr Feldmarschall*, that we can't win is evident. I hope we can keep enough strength –' He meant to continue: 'to reach a decent peace'.

'I will try to use my reputation with the Allies,' Rommel said frankly, 'to make a truce, *against Hitler's wishes*.' He still envisaged, quite unrealistically, a future in which the Anglo-Americans would agree to help hold the line against the Soviet Union, and he spoke of this to Lattmann.[30] A few days later he was talking to Colonel Warning, Ia of 17th Luftwaffe Division. Warning had been on Rommel's staff in North Africa – had, indeed, been at Alamein when Rommel had received Hitler's 'No withdrawal' order and had been told by Westphal on that occasion to go and keep Rommel company, as the Field Marshal needed another human being in his solitary agony.[31]

Warning was thus an old friend. On 15 July, a hot day, he was wearing a thin brown desert tunic. He and Rommel were old 'Afrikaners'. When they were out of earshot of others (his divisional commander was away visiting the front), Warning put the question frankly: 'Field Marshal, what's really going to happen here? Twelve German divisions are trying to contain the whole front.'

'I'll tell you something,' Rommel replied. 'Field Marshal von Kluge and I have sent the Führer an ultimatum. Militarily, the war can't be won and he must make a political decision.'

Warning looked at Rommel with astonished disbelief.

'And what if the Führer refuses?'

'Then,' Rommel said, 'I open the west front. There would only be one important matter left – that the Anglo-Americans reach Berlin before the Russians.'[32]

To his old subordinate, Westphal, also serving in Normandy, Rommel spoke in the same grim sense.[33] Later he told his son

Manfred that the time to 'open the west front' would have been when, as must inevitably happen, the Anglo–Americans ultimately broke out. Then it should be possible unilaterally to abandon resistance and let the impetus of purely military events set the pace of history, since political initiatives had been neglected.[34] But by the time of this conversation all had moved on.

Rommel's 'ultimatum' with Field Marshal von Kluge was in fact signed on the following day, 16 July. Geyr had not been the only casualty of the meeting with Hitler on 29 June. After it the Führer had decided that the elderly Field Marshal von Rundstedt had also better be replaced, and replaced by an officer who could drive the defeatism – and the insubordination – out of the commander of Army Group B. Rundstedt had sent a pessimistic report to OKW urging, like Geyr, withdrawal at least from the Caen bridgehead, and quoting Rommel – fairly – on the necessity for this degree at least of operational freedom; the report had not pleased Hitler. Asked by Keitel what he really proposed, Rundstedt had said shortly, 'Make peace!'

Kluge was Hitler's choice to replace Rundstedt as *OB West*. A Prussian, he had been Rommel's army commander in 1940, driving Fourth Army triumphantly through France. In the terrible winter of 1941 he had relieved Field Marshal von Bock in command of Army Group Centre on the Eastern Front; and it was after visiting Kluge's headquarters in 1943 that Hitler had narrowly escaped death at the hands of the conspirators who placed a bomb in his aircraft. Before taking over from Rundstedt Kluge had been briefed at OKW; and he knew, of course, from his experiences in 1940, the headstrong character of Erwin Rommel, now characterized as a pessimist.

Their first meeting was disagreeable. It took place on 3 July, immediately Kluge assumed command, in the presence of Speidel and Tempelhoff at La Roche Guyon. Kluge told Rommel bluntly that although he was a Field Marshal he must get used to obeying orders. The rebuke, unequivocally, was for disobedience; and it could only derive from Kluge's visit to the Berghof before travelling to France.

Rommel was angry. He had, he knew, faithfully obeyed orders. His crime was speaking bluntly about the true facts of the situation. On 5 July he sent Kluge a situation report[35] stating that Cherbourg and the Cotentin peninsula could not possibly have been held for long (which he assumed was a particular reproach). He also set down the measures he had earlier – repeatedly – requested, and the results of not meeting them. All this – and the consequences for resupply of the air situation – Rommel reiterated to Kluge; and he repeated his angry complaints about the early denial to Army Group B of control over the divisions of Panzer Group West. It was this which had produced the present appalling situation, not Rommel's disobedience. He sent the report under cover of a brief note in which he simply said that Kluge's remarks, in the presence of Rommel's own staff officers, had hurt him deeply. He asked *OB West* what possible grounds there were for the accusation. And he sent the report (but not the covering letter) to Hitler by the medium of Schmundt, who acknowledged it courteously.

Rommel's first encounter with Kluge had been unpromising, but Kluge was a shrewd, experienced commander, and it did not take him long to form his own impressions. They entirely confirmed the view of the situation Rommel had presented to him. It was unimportant how it had happened and whether it could have happened less disastrously; the situation now was one of imminent catastrophe. The front could not possibly hold for more than a few weeks. The troops were being ground to pieces in a *Materialschlacht* of monstrous proportions. The enemy now had some forty divisions in the field, but a simple count of formations meant nothing – what mattered was that the Anglo-Americans could call on seemingly limitless reinforcements of men, equipment and supplies while the Wehrmacht was wasting away. Kluge saw Rommel repeatedly during the first fortnight in July, discussing the situation and dining at La Roche Guyon on the twelfth. And on the sixteenth Rommel sent him what he had described to Warning as his 'ultimatum'.[36] He assumed that it would immediately be sent to Hitler.* The Wehrmacht had lost 117,000 men, including 2700 officers, since 6 June, and had received only ten thousand replacements. It was time for truth.

In this brief and stark document Rommel said that the ultimate crisis in Normandy was now approaching. The troops were fighting heroically but the strength of the enemy, above all in tanks and artillery, the paucity and slow arrival of replacements for the hideous losses, the inexperience now of many of the German formations and the intensity of the air and ground bombardment to which they were incessantly exposed, meant that the enemy must break through into

* When it was forwarded, Kluge's covering letter said he had concluded that 'the Field Marshal [Rommel] was unfortunately right'.[37]

the French hinterland in the near future (‘*absehbarer Zeit*’). Rommel did not put an exact term on this, but the message was clear, and he knew that Kluge, from his own observations, endorsed it. There were wholly inadequate Panzer reserves to deal with the situation. The Luftwaffe had been conspicuous by its absence. Rommel had added to the two-and-a-half-page draft a sentence in his own hand: ‘It is necessary to draw the political conclusions from this situation.’

Speidel and Tempelhoff, no doubt mindful of Hitler’s explosion when Rommel had tried to advert to the situation inevitably facing Germany as a consequence of the strategic position on the Western front, persuaded Rommel to delete the word ‘political’. He did so. And signed.

On the previous day, 15 July, Rommel had yet again visited the ground east of Caen. The next major enemy attack, he told the commander of 346th Division, in position north of the Bois de Bavent north-east of Caen, would come in the Caen sector. The enemy’s assembly of armour must make that likely.

It would probably come two days later, on the seventeenth or eighteenth. On 16 July he repeated that assessment on visits to Panzer Group West and to LXXXIV Corps, deployed in the same area east of Caen and commanding the static defences there. The enemy, Rommel said, would attack in this sector, simultaneously with an attempt to get Caen itself. Depth must be the key to the defence – the enemy must meet line after line. 21st Panzer must hold its Panzer grenadier regiment close behind the front for immediate intervention. The Panzer divisions must take responsibility for infantry units in their sectors as operations developed.[38] And on the day following his missive’s despatch to Kluge he set out to visit II Panzer Corps (General Hausser) and two divisions near Villers-Bocage which had experienced heavy casualties.

Thence he decided to drive to I SS Panzer Corps, to Sepp Dietrich; once again to the threatened sector east of the Orne. He left Dietrich at four o’clock in the afternoon and took the road through St Pierre, en route to La Roche Guyon. Enemy aircraft were active everywhere, and to reduce risk Rommel’s driver, *Oberfeldwebel* Daniel, was told to take a minor road before Livarot. The intention was to rejoin the main road a few miles north of Vimoutiers, and thereafter to work eastwards towards the Seine and home.

On 15 July, the day before Rommel signed his ‘ultimatum’ and sent it to Kluge, the chief of staff of the German reserve army, the ‘Home

Army' (with offices actually in the Bendlerstrasse OKW building), Colonel Graf Schenk von Stauffenberg, flew from Berlin to Hitler's headquarters at Rastenburg. He arrived at eleven o'clock in the morning and telephoned General Olbricht in Berlin. Olbricht, according to plan, then began issuing orders for the movement of troops into Berlin. This was in fulfilment of an official 'Alert plan', codenamed 'Valkyrie', by which bodies of troops from the various training schools near Berlin would be mobilized and brought into the capital. The ostensible aim was to secure government offices and various key points if there were information of an imminent insurrection by the millions of foreign forced labourers working in Germany, a danger sometimes mooted. There was no information of an insurrection by foreign forced labourers. Stauffenberg, in his briefcase, had a bomb.

Stauffenberg, a man of magnificent presence, dynamic force and with a brilliant soldierly record, had acted as something of a catalyst in the conspiracy. Some thought him volatile and unpredictable but all recognized his courage. He was, essentially, of a heroic type. He believed in action. He had examined his conscience on the morals of tyrannicide and had reached his conclusions. Now he was the mainspring of all.[39] He was due to attend a conference with Hitler and he intended to fuse the bomb and time it to detonate during the Führer's conference, killing (at least) the Führer. This was Stauffenberg's second visit to Rastenburg with a bomb. The first had been on 11 July, when he had returned with it to Berlin because Himmler (thought by some of the conspirators also to be an essential target) was not there.

When it was certain that the plan had worked, complementary action was to be taken in Berlin (hence the call to Olbricht, and 'Valkyrie') and in the west – in France, under the auspices of *Befehlshaber Frankreich* General von Stülpnagel. In Germany a new government would be proclaimed, headed by General Beck as head of state. Field Marshal von Witzleben was to be commander-in-chief of the Wehrmacht while the Reich Chancellor would be Dr Goerdeler, a former Lord Mayor of Leipzig (who, fearing arrest, was about to go into hiding). Government would include a number of survivors from earlier and more decent times, all written down in an agreed list of state appointments. All had in common a loathing of what Nazism had become, a horror at the prospects facing Germany and a readiness to share responsibility for Hitler's death.

The coup and the change of government would be followed as

rapidly as possible by overtures for peace extended by the field commanders in the west to their military adversaries in the Allied camp. Rommel had made clear in numerous conversations his conviction that this latter course must, at the right time, be adopted; and Speidel, together with other selected officers in both *OB West* and Army Group B headquarters, knew, at least in principle, the way it was hoped events would run after any 'replacement' of the Führer.

At one o'clock in the afternoon the conference at Rastenburg began. Stauffenberg again telephoned Olbricht and returned to the conference room to find that Hitler, after the opening few minutes, had decided to leave it. Back in Berlin Olbricht, on receiving yet another telephone call from Stauffenberg, signalled to all units to stand down and return to barracks. The 'Alert' had been an exercise, it was explained, and had served its purpose in testing their ability to turn out promptly. The next Rastenburg conference at which it was again ordered that Stauffenberg be present to give details of replacements for the fighting fronts being formed under orders of the Home Army in Germany was scheduled for 20 July. And once again it was planned that Operation Valkyrie would, if matters went as they should, be put into effect.

On 16 July the Commander 'Special Air Service' in England applied for approval to commit a small and special group, to be dropped by parachute in German-occupied territory with the mission 'to kill or kidnap and remove to England Field Marshal Rommel or any senior members of his staff'. SAS headquarters had received information about Rommel's presence at La Roche Guyon from a strong party already dropped into Burgundy and tasked to interrupt German westward reinforcements to Normandy; the commander of this latter party had been befriended by a member of the French Resistance who owned land near La Roche Guyon and who was well-informed of circumstances there – including Rommel's habits, his walks, the routes he took. After refusing a request by the SAS Commander, Major William Fraser, to carry out the anti-Rommel mission with his own party, SAS headquarters obtained permission to send another and very small party – two officers and four soldiers. The operation instruction was signed on 20 July and the signal announcing the drop (scheduled for the night of 25–26 July) was issued on the twenty-third. Opinion at staff level in the British 21st Army Group in Normandy had moved towards 'killing rather than capturing the gent in question'.[40]

*　*　*

On 17 July Rommel's car reached the main road, leading south towards Vimoutiers. An air sentry was riding in the back of the car, *Obergefreiter* Holke. The accompanying staff officers were Major Neuhaus and Captain Lang.

Suddenly Holke yelled that there were enemy aircraft heading in on the road they were taking; they were coming from behind, low and fast. There was a shout to the driver, Daniel, to speed up, to race three hundred yards to where it looked possible to pull off the road and take cover. Before they got there the leading enemy aircraft opened fire. The car went out of control and ended up in a ditch on the left of the road. Rommel had already been hard hit even before the second attacking aircraft came in, strafing the wrecked car and the prone bodies of its occupants once again.

The nearest military hospital was a *Luftwaffenortlazarett* at Bernay, on the road to Rouen, twenty-five miles away. Rommel was first attended by a French doctor in a French hospital – the property of a monastic order – in Livarot, near the scene of the wounding; but even getting him there had involved a wait of three quarters of an hour before Lang could get a vehicle. Thence an unconscious Rommel, together with Daniel, was borne to Bernay. Rommel's skull was severely fractured, and there were wounds to the temple and face. Daniel died of his injuries.

Next day, 18 July, the British Second Army attacked over the ground which Rommel had repeatedly visited and where he had consistently anticipated the next major operation – east of Caen, east of the Orne. The attack was preceded by one of the greatest air bombardments in support of ground forces which the war had seen. The bombardment went on for three hours, between 5.30 and 8.30 of a beautiful morning. The British VIII Corps (General O'Connor), with three armoured divisions and an infantry division on the left flank, then moved southward. At the same time II Canadian Corps (General Simonds) attacked in the city of Caen itself.

By the end of the day, although their forward positions had been overrun with considerable casualties, the Germans were still in control of the Bourguebus ridge, high ground south of Caen, and had destroyed a significant number of enemy tanks. This was no enemy break-out, no defeat. The German assemblage of anti-tank guns, supported by the Panzer regiments lying back just behind the forward positions, had done great execution. The German line was holding. Perhaps, unlike Napoleon, Rommel did not lose his last battle after all.

CHAPTER TWENTY-TWO

'For the Honour of Germany'

AT EIGHTEEN minutes to one in the afternoon of Thursday, 20 July 1944 a huge explosion wrecked the hut in which Hitler was conducting a conference at his headquarters at Rastenburg in East Prussia.[1] Several of those attending, including Rommel's friend General Schmundt, Hitler's Military Secretary, were killed or died of their injuries. There was chaos in the hut. Hitler survived.

Stauffenberg had placed the bomb. For the third time he had brought it from Berlin by air in his briefcase and, just before the conference, had broken a capsule to enable the timed mechanism to start working. Stauffenberg's hand had been replaced by an artificial limb after severe wounding in Tunisia, but he had practised and become adept at the manipulation required. He had placed the briefcase, fuse already in progress, under the conference table (his position was only two places from Hitler), had left the room 'to take a telephone call' a few minutes after the conference started, and had waited until the explosion before driving off. At the perimeter of the headquarters compound he spoke sufficiently convincingly to be allowed by the guard to pass. Within twenty-five minutes he had reached the airfield and was soon in the air, en route for Berlin.

Within the hut Stauffenberg's neighbour at the table had, without design, shifted the briefcase with his foot to the far side of a stout oak table support, which probably reduced some of the direct explosive effect. To Stauffenberg, seeing and hearing the explosion from a short distance away, it was inconceivable that a man as near to it as he knew Hitler to have been could survive. But survive he did.

The first report to reach Berlin was at one o'clock. It stated that Hitler was dead. Within minutes Keitel, present at the conference and also a survivor, ordered a message to be sent to Berlin to correct this: the Führer was alive. Immediately thereafter lines went dead.

In Berlin the conspirators believed, with desperate optimism, that the corrective message must be bluff. The first call had been from General Fellgiebel, chief of communications and privy to the plot. It had been agreed that all signals should start with the simple message conveying Hitler's death, '*Führer tot*,' and this had happened. But the second call purported to come from the office of the Field Marshal, Chief of Staff of OKW.[2] There was confusion. And now communications with Rastenburg appeared broken, as Fellgiebel had undertaken to arrange for a while after the deed was done. Now it was necessary to set matters in hand in Berlin and to send out the prepared messages to every command and military district: '*Führer ist tot!*' Some communications were sent, some calls made. To those expecting the message, hoping for the message, it seemed conclusive. Kluge's *Oberquartermeister I*, Colonel Finckh, telephoned Speidel at La Roche Guyon in the early afternoon: '*Attentat!** *Führer ist tot!*'[3] Later in the afternoon General Blumentritt from *OB West* also rang Speidel, with one word: '*Tot*.'[4] But in Berlin there was agonized doubt. Operation Valkyrie was deferred. Was Hitler really dead?

At quarter past three that afternoon Keitel, communications from Rastenburg now restored, managed to speak briefly to General Fromm, Commander of the Reserve Army and Stauffenberg's superior. Keitel made plain to him that the Führer was alive and completely in control of the situation – Mussolini was visiting Rastenburg and Hitler, having changed his damaged clothes, was playing the host with confidence and composure. Meanwhile, he, Keitel, had naturally agreed with Himmler that the Security Service and Gestapo must have an entirely free hand in interrogating whomsoever they wished, whatever his rank or service. Nobody could be protected. The intended crime was horrific.

In the preliminary messages already sent out by the conspirators the first agreed appointments within a new regime had been announced. (The first of these was that of Field Marshal von Witzleben as Commander-in-Chief of all the Armed Forces of the Reich.) These messages – picked up in Rastenburg – were soon contradicted by signals from the *Führerhauptquartier*. Only orders emanating from Rastenburg were to be obeyed.

In Berlin there was still no Operation Valkyrie until four o'clock when, after agitated debate, General von Hase (the Military Com-

* Assassination attempt.

mandant of Berlin and one of the conspirators) summoned Major Remer, commander of the *Wachbataillon Gross Deutschland*, to his office. There had been an unfortunate accident to the Führer, he said, and emergency measures were to be taken, similar to those ordered a few days previously as an exercise. Remer was to take three companies and secure the *Regierungsviertel*, the official quarter of the city. A cordon must be established and nobody, whatever his rank, no minister, no general, was to be allowed through it. Remer returned to his barracks and gave out orders. The time was near half past four.[5]

Somewhat unfortunately for the conspirators an official of Goebbels's Propaganda Ministry, Dr Hagen, was visiting Berlin and giving a lecture to some of Remer's officers that day. On returning from von Hase Remer saw him, and Hagen mentioned that he believed he had seen the former Commander-in-Chief, Field Marshal von Brauchitsch, in uniform in Berlin earlier that day (he had not). Something in the juxtaposition of events, the rumour of a disaster to the Führer, the sudden orders to the *Wachbataillon*, the appearance of a notable military figure from the past, all prompted Hagen's suspicions. He told them to Remer, who agreed that Hagen should instantly go to Goebbels himself, who was in Berlin.[6]

Remer's troops were now climbing into vehicles. Orders had been given and the position of the cordon laid down. A fourth company had been ordered to muster and move, as a reserve. Remer reported again to von Hase's headquarters. While there he overheard a conversation between two officers which included reference to the necessary arrest of *Reichsminister* Goebbels. This made Remer extremely thoughtful, but Hagen was by now on his way.

At twenty to five that afternoon Stauffenberg, euphoric, reached the Reserve Army Headquarters in the OKW building on Bendlerstrasse – which, because of the functions of the key members of the conspiracy, had become its nerve centre. He arrived believing that the *Attentat* had succeeded and presuming that the measures laid down for Operation Valkyrie must be already well advanced. Instead they had barely begun. Fromm – keeping his distance from the conspirators and kept at arms' length by them – had received his call from Keitel over an hour before.

But the preliminary messages (all soon to reach Gestapo desks) had already gone out or were in process of doing so. These messages had established the authority of von Witzleben; and at six o'clock further messages were sent to all military districts ordering the immediate arrest of leading Nazi officials, concentration camp commandants, the hierarchy of the SS. They were sent in the name of Fromm (who had now been placed under temporary arrest by the conspirators) and were authenticated by Stauffenberg. By now the conspirators realized that there was appalling doubt about the whole business but, recognizing that matters had gone too far to be disavowed, they managed to believe that if Berlin itself were secured, and the military commands – including very particularly the high command in the west – were persuaded that the regime had been overturned, the coup might still, somehow, become reality. Messages continued to go out from the Bendlerstrasse, and continued until twenty-five past nine in the evening, now in the name of General Hoepner, a leading conspirator who had been formally nominated by Witzleben to the Reserve Army command in place of the now-discredited Fromm. The last of the huge number of messages to all military districts and commands still started with the words '*Der Führer Adolf Hitler ist tot!*'[7]

At half past six Remer's men had, as ordered, sealed off the government quarter, and he himself again went to von Hase's office. He had expected to find Hagen, with some clarification of a situation which still seemed perplexing, but instead he received a message from Hagen who, it seemed, feared to go to the *Kommandantur*. He feared arrest. And he had an urgent message for Remer. Remer, personally, was to report to Dr Goebbels.

Remer told von Hase's staff that he had been summoned by *Reichsminister* Goebbels; and was immediately given a counter-order by von Hase: 'Remer, stay here!'[8] To his own adjutant Remer said that he had been ordered to see Goebbels and forbidden by General von Hase to comply. He managed, nevertheless, to leave the *Kommandantur* somehow and shortly afterwards he was brought into the *Reichsminister*'s presence, alone. In answer to a question Remer said that he was wholly loyal to the Führer.

Goebbels gazed at him. They shook hands. Within a minute Goebbels handed him the telephone. Hitler himself was on the line; and Hitler gave Remer his orders. In a few decisive sentences in a well-recognized voice the Führer gave him full authority to take all measures, however drastic, which he judged necessary to save the government of the Reich.[9]

Remer returned to his own headquarters and gave orders for the *Wachbataillon* to assemble in the garden of Goebbels's house in Hermann Goering Strasse. There, at half past eight that evening, he paraded his men and Goebbels addressed them. This battalion had an historic mission, Goebbels said; and Remer explained to them that he had received a personal charge from the Führer himself.

Not only the *Wachbataillon* had been involved in Valkyrie. Remer now heard that a Panzer unit had assembled at the Fehrbelliner Platz, under the orders of General Guderian. He made contact with it – he had Hitler's authority to take command of all forces, of whatever sort – to be told that Panzer troops would respond to the orders of Guderian alone. This could be difficult – clearly any armoured units would have the edge were shooting to start; but at that moment a previous commander of *Wachbataillon Gross Deutschland*, Lieutenant-Colonel Gehrke, appeared and persuaded the Panzer men that all this was in order – they were, he presumed, loyal to the Führer. Remer had already sent for reinforcements and heavy weapons from the reserve '*Wach*' brigade at Cottbus but they were not, in the event, necessary.

Remer had now established in his own mind that the centre of dissidence, the traitors' headquarters, seemed to be in the OKW building in the Bendlerstrasse. Thither he sent a company under *Oberleutnant* Schlee to secure the building and investigate; and when Schlee – cordon established and entrances held – entered it he found himself requested to report to General Olbricht, unbeknown to him the leading conspirator.

Schlee told his men that if he failed to reappear in twenty minutes they were to storm the building and search it. When he reached Olbricht's ante room a colonel (it was Mertz von Quirnheim, a key figure in the *Attentat*) forbade him to leave the room. Schlee, however, slipped away. It was all perfectly clear to him – and to Remer – now. This, as Goebbels had told them, was a military putsch, timed to coincide with the murder of the Führer. And its nerve centre was here in the Bendlerstrasse. It was quarter past nine.

In Paris they had been waiting. At some time during the afternoon a telephone message from Berlin had been taken by Lieutenant-Colonel Casar von Hofacker to the effect that Hitler, Himmler and Goering were all dead. This was the signal – disappointingly late, for rumours and unconfirmed messages had been humming since soon after one o'clock – which was to spark the western end of the conspiracy, the *Westlösung*.[10] Hofacker, on Stülpnagel's staff, was a cousin of Stauffenberg. A reserve officer of the Luftwaffe, a dedicated opponent of the regime, he had been taken off virtually all other duties by Stülpnagel in order to act as a liaison officer and coordinator between Berlin and Paris; between the *Berlinerlösung* and the *Westlösung*.

For there were two very separate arms to the conspiracy, two complementary plans for revolutionary action. Action at the centre of the Reich was aimed at taking over the instruments of government and instantly putting into effect completely new policies, policies which had been discussed and formulated and to some extent committed to paper in the months or years of plotting. This, the first action, the *Berlinerlösung*, would include getting rid of Hitler, by whatever means. The *Westlösung*, essentially dependent on the first, was aimed at the army in the field in the west. It was intended to secure its loyalty in a post-Hitler Germany and, as soon as possible, to negotiate an armistice, as between soldiers.

The centre of this second, western, arm of the conspiracy was Stülpnagel's headquarters, the headquarters of the *Befehlshaber Frankreich* in Paris. Not all those who knew some or all of the plans made for the *Westlösung* had knowledge – certainly not knowledge in detail – of what was intended within the *Berlinerlösung*. It was supposed that some sort of coup was envisaged in Germany, as a result of which Hitler would be somehow removed (there was only partial realization and in some cases absolute disapproval of the idea that this implied murder). The removal of the Führer, by whatever means, would remove the obligations of the personal oath of loyalty which all in the armed forces had sworn. The problem thereafter – as seen in France – would in the first instance lie with the SS (1200 in Paris alone) and its subordinate instruments the *Sicherheitsdienst* (the SD) and the Gestapo. If these could be rapidly neutralized there would at least be no organized counter-revolutionary appeal to the troops.

Clearly a revolution at the centre would be frustrated and any attempt at peace negotiations be aborted if the enormous numbers of Germans under arms at the fighting fronts refused to recognize that revolution, and continued to proclaim loyalty to Adolf Hitler or some other designated or self-designated Nazi. On the Eastern Front the situation could look after itself – the armies were locked in combat with the advancing Russians and no German government, revolutionary or otherwise, could break off that struggle. But in the west there was still a front tenuously holding, there was hope, there must be negotiation. That meant that the troops must be ready to follow their commanders, and those commanders must support the new order. Essentially this meant *Oberbefehlshaber West* and the commanders of the Army Groups. They would play the lead parts, while the scene was to be set by Stülpnagel, the essential link-man with Berlin.

Immediately word came from Berlin, therefore, the leading personalities of the SD and Gestapo in Paris were to be arrested and brought before a court convened under Stülpnagel's authority; all of which had been prepared, and all of which, in the late afternoon of 20 July, began to be set in hand. Arrests were made. The German regime in Paris was shortly to be in the undisputed hands of the *Verschworerclique*, the conspirators. For a remarkably short time.

At La Roche Guyon there was terrifying doubt that evening. It was known to Speidel and one or two others that dramatic happenings, intended to lead to negotiation and peace, were pending; and when, that afternoon, Speidel had taken Finckh's call and heard the words '*Führer ist tot*,' it seemed that all was clear. It was presumed that some of the first contradictory signals from Germany had been put out for effect – that revolutionaries and what appeared to be powerful voices of counter-revolutionaries had both attempted to persuade the military commands of the Reich that matters had gone their way and that messages or orders in a different sense were false and could be disregarded. By this reading, even a broadcast from Germany by Goebbels at seven o'clock that evening might be merely a desperate propaganda effort. In the broadcast Goebbels had said that a wicked attempt had been made on the Führer's life and had failed, that Hitler was alive. But it might be a lie.

To La Roche Guyon from Paris came Stülpnagel, accompanied by Dr Horst, Speidel's brother-in-law and serving in the military administration, and von Hofacker. It was a sombre evening. Stülpnagel had already spoken on the telephone to Beck in Berlin and gained little comfort. Beck was, at that moment, of the view that whether or not Hitler was dead – and there was a thin hope that Keitel had lied on the subject – they must behave as if he were. The plans of the conspirators were going ahead in Paris and several hundred leaders of the SS were being arrested without a shot fired, but what had happened at Rastenburg, what was happening in Berlin, was still obscure. A meal was eaten in silence.[11]

Hofacker spoke quietly to Tempelhoff. He said that Tempelhoff knew Stauffenberg personally, and knew Mertz von Quirnheim. So did Hofacker, of course, and Stauffenberg was his cousin, but this was Tempelhoff's headquarters. Would Tempelhoff ring Berlin and ask what exactly was happening? Tempelhoff did so from his own office and contacted Stauffenberg – at what precise time is uncertain, but Hofacker, who had returned to Paris by midnight,[12] was still at La Roche Guyon and Stauffenberg was clearly still at liberty in the Bendlerstrasse. The conversation was suddenly ended at the Berlin end, and Captain Dummler, who kept the Army Group War Diary, came into the room and took the telephone from his superior's hand, saying as he did so: '*Herr Oberst, was machen Sie dann? Sie kommen in die Teufelskuche!*'*

It was a merciful act of insubordination. When Tempelhoff returned to the dinner table Kluge looked a question at him and Tempelhoff simply said '*Nein!*'[13] Kluge – who had taken over direct command of Army Group B as well as being *OB West* after Rommel's wounding – said no more. He had often been approached by the conspirators and they had hoped for his cooperation – indeed, like Rommel's, it was essential – in carrying the plan forward to a negotiation with the enemy. But meanwhile, if Hitler were still alive . . .

Kluge then openly dissociated himself from the conspiracy with which some others at the table – notably Hofacker – had boldly and openly identified; indeed, Kluge told Stülpnagel that he, Stülpnagel, should certainly be arrested, and probably shot.[14] Some time before midnight Stülpnagel with his companions drove back to Paris. Next morning he received an order from Keitel: Report immediately to Berlin.

In Berlin, in the Bendlerstrasse, General Fromm, who had been placed under arrest by his subordinate, Stauffenberg, had known since mid-afternoon that Hitler was alive. He knew too that he risked being accused, if not of complicity in the *Attentat*, at least of failing to notify others of what he had had reason to believe was intended, and of which he probably approved. Remer's men, now deployed throughout the *Regierungsviertel*, were clearly on the side not of the *Verschwörerclique* but of authority. After Schlee had escaped from Olbricht's room he had returned with soldiers; and soon they were followed by Colonel Gehrke, he who had pacified the doubts of the Panzer men in the Fehrbelliner Platz and told them where their duty lay. Soon the conspirators within the Bendlerstrasse were disarmed.

* 'What are you doing, Colonel? You are getting into hellish hot water!'

Fromm was again at liberty and, as he immediately made clear, restored to his proper authority as Commander of the Home Army.

Fromm acted quickly, Gehrke witnessing all with approval. He constituted himself a one-man court martial and condemned to death Stauffenberg, Olbricht, Lieutenant von Haeften, who had accompanied Stauffenberg to and from Rastenburg, and Mertz von Quirnheim.* Beck, who had walked into the building in civilian clothes that afternoon to be present when the new regime took shape, attempted suicide unsuccessfully and was finished off with his own pistol. He had been designated as head of state of the new Germany which was intended to start on 20 July 1944 but which never came to birth.

Stauffenberg and the others were shot by a squad of *Unteroffizieren* drawn from Schlee's company in the Bendlerstrasse car park, beneath the headlights of trucks lined up for the purpose. It was half past midnight. At midnight Adolf Hitler had broadcast to the nation and spoken of an attempt to kill him by 'a very small clique of unprincipled, ambitious and criminally stupid officers'. The attempt had failed. He, Adolf Hitler, was alive, was still Supreme Commander of the German Wehrmacht and Führer of the German people. And now justice would be done.

Within hours the investigation started, and almost every day from 21 July *Obergruppenführer* Kaltenbrunner, head of the *Sicherheitsdienst* wrote to *Reichsleiter* Martin Bormann at Hitler's headquarters a very full account of the result of the latest inquiries and Gestapo interrogations.[16] Kaltenbrunner's near-daily communications lasted until 15 September, when he explained that in future a report every three days would probably suffice.

The amplitude of Kaltenbrunner's reports and the evidence collated is striking. The course of the conspiracy itself – in Germany, in France, within the Wehrmacht – was meticulously pieced together, with every small incident, every possibly incriminating conversation, every chance remark or contact recorded, sifted, put in its place. The conspirators had produced a good many documents and these – lists of a future government, proclamations to the German people and to the German Wehrmacht, papers on foreign policy,

* Fromm was himself condemned to death by a people's court on 7 March 1945. The court found that he had acted on 20 July only to save his skin rather than from conviction, and had earlier shared the conspirators' defeatism. He was reported as having said (on 3 July) that it would be best if the Führer took his own life.[15] That was quite enough.

on a future Europe, on educational policy, religious policy, racial policy, constitutional forms, enactments on justice and liberty of the subject and of conscience – were all carefully dissected, made subject of comment and annexed to Kaltenbrunner's offerings to Bormann. For the Führer's interest.

Much of the stuff of these conspirators' documents was brave, intelligent and farsighted; and although the attempt failed wholly and painfully – and was, perhaps, conceived with inadequate discretion and realism – the minds of those who made it deserve all honour, and their uncompromising language almost burned the paper. There was talk in these documents of Hitler's 'bloodstained hands', of his progress which had left behind a trail of tears, grief and misery. They clearly had an effect on even the bitterest of enemies. Kaltenbrunner's critiques of the matter he submitted were painstaking and thorough. It was as if he could not bear to record an argued point of view, however treasonable and hostile to National Socialism, without at the same time recording, almost as if persuading himself of its validity, the proper, surely decent, surely compelling, National Socialist counter-view.

Hitler, from the very first, took a practical and political view of the whole business, as had been apparent from the wording of his broadcast. This was the work of a wholly unusual, wholly unrepresentative and tiny minority. In no way must any impression be given that the *Verschwörerclique* was typical of any section of society, whether military or civil. Hitler may have fumed against 'blue-blooded swine' in private, but just because some of the criminals bore aristocratic names, on no account must there be any *corporate* accusation of the aristocracy; nor, because some were generals, of the *Generalitat*; nor of the officer corps; nor of the Wehrmacht. Hitler had no intention of making unnecessary enemies. On 24 July Bormann wrote to all Gauleiters and Reichsleiters conveying Hitler's express wish in that sense.[17] The perpetrators must be isolated in the eyes of men.

But it was very clear from all the language of inquiry and report that the crime of Stauffenberg and others was not simply attempted murder. It was defeatism – *Defaitismus und Pessimismus*. The words recur again and again in Kaltenbrunner's reports, with reasons given (wholly accurately) and the justifications of the accused propounded. And this, whatever the view of the *Attentat*, was probably an accusation to which Rommel would have entered no defence whatsoever, because it was most certainly true.

* * *

Three days after the *Attentat* Rommel was moved to a hospital at Le Vesinet in St Germain near Paris. He was a difficult patient. In the first days, at Bernay, he had been visited by Speidel, Ruge (who then came to Le Vesinet almost every day until Rommel's discharge), Tempelhoff, Lang and others. He had lain in great pain in a darkened room. Now, at Le Vesinet, he was recovering some vitality by day, although the nights were hard. Already, however, Rommel was thinking about getting back to the troops and to his duty. In Normandy the strained front was still holding, but he knew it couldn't be long before the enemy would break out and come swarming across France, as he had warned must happen. When he affected to make light of his injuries, the surgeon brought a human skull from the pathology department to his bedside and cracked it with a hammer to show Rommel what had happened to his head.

One of the first visitors to Bernay, although not allowed to see Rommel, had been his artillery chief, Colonel Lattmann. Lattmann had been asked by Speidel to bring back the Field Marshal's cap and field service baton, and did so. As he was climbing into his car he was approached by an elderly Frenchman who turned out to be the doctor who had given Rommel the first injection after the attack. The doctor was much moved, and Lattmann realized, not for the first time, how warm was the regard for Rommel among the French.[18] Another early visitor was Lang, who was allowed to see Rommel on 21 July and told him about the *Attentat*. Rommel, it was clear to Lang, was deeply shocked.[19] His first letter to Lucy after the accident was dated 24 July and referred to being shattered by the *Attentat*, and only grateful to God for its failure.[20] Such an expression must have been obligatory, a matter of survival – the censorship would have certainly found suspicious any avoidance of so dramatic an occurrence in Rommel's correspondence. In the same letter he wrote that he had shortly before given to higher authority, '*nach oben*', his own views on the situation – a reference to his 'ultimatum', which he clearly regarded as a matter for open and unabashed acknowledgement.

At Le Vesinet Rommel's visitors were permitted to stay longer, and he became talkative and fretful. Members of his staff and subordinate commanders came – Tempelhoff, Staubwasser from Army Group B, von Salmuth from Fifteenth Army. Ruge read to him daily. An old friend, Kurt Hesse – a colleague lecturer at Dresden, a war correspondent in France in 1940, and now Commandant of the St Germain military district – visited him in early August and stayed for an hour. Rommel spoke freely. Hitler, he said, had learned nothing from Stalingrad. The war was clearly lost and must be ended. To Lattmann, another faithful visitor, he said the same. When fit, Rommel said, he must once again see Hitler and persuade him; he said the same to Ruge.[21] They must, somehow, finish the war. The German people had suffered too much. This was all familiar to Lattmann, to whom, in Normandy, Rommel had spoken frankly about negotiating for peace, Hitler or no Hitler.[22] But the *Attentat* was now overshadowing such talk, and to Lattmann and others these exchanges must have felt more than a little dangerous. Kurt Hesse remembered afterwards that Rommel looked long at him when they parted. 'Hesse,' he said, 'I think it's best that I got it in the head!'[23]

Kluge came, but of anything that passed between them there is no record. Stülpnagel had left Paris on 21 July, had climbed out of his car near Verdun and attempted to shoot himself; he was dragged from a canal, resuscitated and carried back to Germany in poor condition, blinded. Wagner, Quartermaster General at OKH and a frequent visitor at La Roche Guyon, had committed suicide. And now the Gestapo inquiries into the conspiracy were reaching into every German staff, every German military department. Nobody had any illusions about the brutal vigour with which those inquiries would be pressed, and nobody had any illusions about the penalty for those found guilty.

Nor among the mass of officers engaged in fighting the war was there much disposition to object. Views of the leadership of the Reich differed sharply, and many were sceptical; but Rommel's former staff officer, von Mellenthin, probably spoke for many – he was at the time engaged in the desperate battles in southern Poland against the Red Army offensive north of the Carpathians, towards Cracow and the southern Vistula. 'The reaction at the front was unambiguous,' Mellenthin wrote after the war. 'We were dumbfounded to learn that a German officer had been capable of making this attempt, particularly at a time when the men on the Eastern Front were waging a life-and-death struggle . . . the front-line soldiers were disgusted to hear of the attempt on Hitler's life and indignantly refused to approve of it.'[24] By the time he wrote this Mellenthin had learned much and had changed his views of the *Attentat*, but his recollection of the times rings true. In pursuing the conspirators the Reich authorities had much of Germany, both in and out of uniform, behind them. Many in the highest ranks of the army scrambled to dissociate themselves from treason and to align their sympathies anew with the regime. Within the army the Hitler salute was made obligatory. Guderian wrote to all members of the General Staff, of which he was now Chief: 'Every General Staff officer must be a National Socialist leadership officer.'[25]

There was inevitable shock that names distinguished in German military affairs or military history* had – or so it seemed – tried to stab the Fatherland in the back when it was fighting for its life. A special court of inquiry, an *Ehrenhof*, was established, initially under the presidency of Field Marshal von Rundstedt himself, to hear evidence about the involvement of officers – evidence largely deriving from the Gestapo interrogations – and to pronounce on whether or not some should be dismissed from the army and handed for trial to a people's court, as accused common criminals. Kaltenbrunner – unctuously but not necessarily falsely – reported among ordinary people a profound relief at Hitler's miraculous escape, a hatred of presumed 'reactionaries' who formed the enemy within, and determination that *Totaler Krieg* could now be waged.

To Germany's enemies, of course, all this seemed largely a matter of villains falling out among themselves, and there was, until much time had passed, little sense of the lonely and valorous self-sacrifice which had distinguished the conspiracy. The Gestapo was hunting down men and women whose only support lay in each other and in their heroic consciences.

On 8 August Rommel was driven home to Herrlingen, determined as ever to put a brave show on personal adversity. 'Well, Loistl,' he said to *Obergefreiter* Loistl, his soldier-servant, 'so long as one isn't carrying one's head under one's arm things aren't too bad!'[26] But there was grim irony in the joke that August. The Gestapo was doing its work at high pressure, statements were made to the Court of Honour, the *Ehregericht*, demonstrating the involvement of many of the senior officers in France, and on 29 August the blinded Stülpnagel was condemned to death with Hofacker and others, and executed on the same day.

Rommel was deeply distressed to hear of Stülpnagel's fate. He had, in these last two months, suffered more personal bereavement than even in the desert. He felt shattered by the casualties in Army

* Among many others, of every degree, a von Moltke, a von Kleist, a von Alvensleben – the list of condemned, a roll of honour indeed, in some places ran like a record of earlier and glorious Prussian campaigns.

Group B – 140,000 men had been lost in the Normandy fighting – every man, as it seemed, a trusting friend; and among the seniors there were many dead on the battlefield. But Schmundt, his old friend, had died from Stauffenberg's bomb, and now Stülpnagel at the hands of the executioner in Berlin. And Kluge, with whom Rommel's relations had been uneven but who had come to share his views, had taken poison on being summoned to Germany.

Kluge, ambivalent to the last, had addressed a fulsome letter of support to Hitler before taking his own life, assuring the Führer that he was more in sympathy with him than he apparently credited;[27] but Kluge knew that by some of his words and actions he was compromised as a defeatist in Hitler's eyes, probably as an intending negotiator with the enemy, if not as an assassin. Immediately on hearing from Tempelhoff on the evening of 20 July that Hitler was alive Kluge had telephoned a relative at the hospital at Le Vesinet and asked, as prearranged, that a capsule of poison be sent to him.

Three more weeks had dragged by for Kluge. The final defeat in Normandy was marked on 15 August by the fall of Falaise and the beginning of a general and disorganized retreat. Then Kluge, having handed over command to Field Marshal Model, was recalled to Germany. The Gestapo had extracted from Dr Goerdeler, the intended Chancellor of the new Germany who was arrested on 12 August after a month in hiding, that he had talked to Kluge in the previous year and said that it would be for the military leaders to find the best way to get a decent peace from a lost war. Kluge, said Goerdeler, had recognized that and promised to talk further about it to his brother officers.[28] And other suspects had reported Stauffenberg, pivotal figure, as saying that the Army Group commanders would join in once the thing began to move.[29] Kluge, knowing his likely fate, killed himself.

Soon after Rommel's return home his son Manfred, now aged fifteen, was given leave from the anti-aircraft battery where he was serving as a Luftwaffe auxiliary, leave which continued for many weeks so that he could be of assistance to his father. In those weeks he was able to get to know better than ever before the father who had for most of his adolescence been separated from him by war. Rommel always liked teaching; whether in conversation or command he was a natural instructor, perhaps an inherited quality. He tried to teach Manfred the calculus as late as the beginning of October.[30] He was always of a practical, analytical turn of mind. His wound had left him with severe headaches and had injured his sight so that reading was difficult and painful. Ruge had read to him at Le Vesinet, and now Manfred read to him at Herrlingen. And Rommel talked.

He talked with great freedom of what was on his mind, and others visiting the house at that time, or simply overhearing the conversation, noted the lack of caution in Rommel's openly expressed opinions.[31] Hitler, Rommel said, was now out of his mind. He would carry on until he died in the *Wolfsschanze*. To Manfred, Rommel spoke of Hitler's lunatic attempts to interfere in operations and – above all – of the Führer's refusal to understand that Germany had lost the war and by avoiding surrender in the west must inevitably be overrun from the east as well, a fate beyond contemplation awful. Manfred was naturally disturbed at this uncompromising pessimism, this scathing rejection of the regime by a father who had long been a hero-figure to all Germany as well as to his family; but Rommel argued the matter out with him, explained his reasons to his son, and left no doubt in his mind as to where he, Rommel, stood. If he had not been wounded when he was he would – and perhaps could – have attempted to open the western gates into Germany, stopped the war in the west by direct negotiation, despite his orders, despite his constitutional obligation to obey, despite the Führer. From one who had always insisted on the absolute duty of obedience in the soldier this was alarming. 'You will often receive an order,' Rommel had written to Manfred on the latter reaching age for service, 'whose sense you don't understand. Obey unconditionally.'[32] Now the preceptor was preaching and justifying rebellion.

About the attempt on Hitler's life, however, Rommel was equally uncompromising. His views were no longer influenced by personal affection for Hitler – that had been overtaken by his rejection of what Hitler had done to his own troops and to Germany; but Rommel consistently regarded the *Attentat* as an ill-conceived idea and repudiated it. He expressed admiration for the courage of the *Berliner Kreis*, of Stauffenberg and the others; and that personal friends had been involved and paid the penalty grieved him. But he thought that, by any reckoning, a dead Hitler would have condemned Germany to civil war, would have created a martyr, would have bequeathed to future generations another *Dolchstoss* legend. Far better to proceed against Hitler legally, if there were legal charges to be laid. Rommel could not bring himself to believe that Germany could have been steered into peaceful acceptance of a successor regime after the assassination of Adolf Hitler; or that such a regime could have been acceptable as a negotiating party to the Allies. He had come to believe that only physical surrender in the field could end the war, and he had no doubts that even that would be difficult to procure. Rommel knew the German soldier – none better. He knew that the idea of obtaining general support within the Wehrmacht for a coup against Hitler would be doomed to failure.

Rommel was, too, somewhat contemptuous of the methods whereby the *Attentat* had been arranged. It was, he said acidly, characteristic of the whole business that for so delicate a task as priming and setting a bomb-fuse the conspirators had chosen Stauffenberg, an officer with serious physical disabilities. In this sort of criticism Rommel – whose knowledge of the detail was probably imperfect – was uncharitable. There was nothing in Stauffenberg's actions which betrayed incompetence – rather the reverse. But Rommel thought that all he heard of the *Berliner Kreis* end of the conspiracy smacked of what he called amateur plotting in the atmosphere of Berlin drawing rooms; where, he asked, were the reliable troops, intelligently briefed, which must be a necessary element, even given accomplishment of the actual murder?[33]

This, surely, was to underestimate how dangerous and laborious was the planning of a revolution in a society dominated by Heinrich Himmler's SS. Certainly there was hesitation and confusion during both the planning and the execution of the *Attentat*. Certainly there were divided counsels and to some extent divergent aims – perhaps a dangerous mix of the practical preparation of a deadly attempt with idealistic visions of world reform. Valkyrie certainly failed; but everything was against success, and the leading spirits behind the endeavour believed that the very fact that it was made might one day be thought vital for the honour of Germany. Nor were they wrong.

With September came news of Speidel's sudden dismissal from his position as Chief of Staff of Army Group B. Army Group B itself had largely dissolved as an organized force, but the headquarters were kept in being and Speidel had been suddenly relieved.

The Western Front had collapsed. The moment Rommel had anticipated had arrived, and arrived pretty well in the time he had predicted. In the west of the Normandy bridgehead the Americans had at last broken out, had advanced south, wheeled west to occupy Brittany, and with their main body had driven eastward towards

Paris. On 25 August the British, on the left of the Americans, had crossed the Seine very near Rommel's old headquarters of La Roche Guyon and had reached Amiens by the end of the month. On 3 September they entered Brussels. The Wehrmacht was now in full flight towards the frontiers of the Reich and whatever defensive positions could be organized on the Meuse and the Rhine. This was 1940 in reverse, and everywhere there were accounts of enormous numbers of German soldiers passing into Allied prisoner-of-war camps with scarcely the prospect of a battle.

In spite of this huge disaster, however, there were signs by mid-September that some sort of tenuous stability had been restored to the front. Vast numbers of men had been lost, but vast numbers had also made their way east and were being reorganized with urgency and skill by the experienced formation staffs of the Wehrmacht. Certain key strategic areas – notably the banks of the Scheldt, denying enemy access to Antwerp – were prepared for a sustained fight. The Allies were, in places, hesitant in their triumphant advance and it was clear that the very immensity of their success created logistic problems for its exploitation. From mid-September there was almost a breathing space, and with extraordinary resilience the German army, shattered as it seemed by the Normandy campaign, appeared to have turned again to fight.

Rommel followed all this daily, and had plenty of informants. He even heard very early word of a projected German counter-offensive (which ultimately took place in the Ardennes in December), and snorted with contempt at the idea when, as he said, every measure against the Anglo–Americans could now only harm Germany.[34] He had formally been posted as giving up command of Army Group B on 3 September, and three days later Speidel appeared at Herrlingen. Speidel had been given no reason for his own dismissal, and his head was full of plans to get Rommel's support for yet another approach to Hitler, to plead for a renewed peace initiative – an approach perhaps to be made through Guderian, a realistic soldier.

But next day Speidel was arrested at his home. There was a sense of unseen enemies closing in; and Rommel was sure (a certainty shared by his wife and staff) that his house was under surveillance, which could only mean by the SD or some other arm of the SS. Rommel now believed that there were those in the state who would prefer him dead. He was certain they could not proceed by law but they might by assassination. On the walks he took daily, often accompanied by Manfred, he always went armed, and he also armed his son. He arranged with the local army command for a military sentry on the house.

It was a curious situation for a popular Field Marshal who was not, as far as he somewhat naïvely supposed, an object of legitimate suspicion; but Rommel reckoned that the Third Reich had embarked on a course of illegality, and if his removal were to be thought desirable he was in no doubt it would be managed. Because he believed – with justification – that his name stood high in Germany, he thought that the Nazi authorities might hesitate to move against him openly, but might do so in secrecy and darkness. His crime in their eyes (he supposed) was that he must be clearly identified with what they called 'defeatism', with a certainty that the war in the west would be lost. He had said as much openly to Hitler. He had said it to Keitel. For so charismatic a commander to think thus might be thought too dangerous.

None of this in the least modified Rommel's expressed views on the reality of the war situation. Germany, he said frankly, was finished. There had to be peace; and since the struggle in the east was very literally one of life and death there had to be peace in the west, which meant surrender. Such arguments might not in themselves legally be capital offences, but it was probably thought intolerable that a proponent of them should live. Hitler would certainly be unforgiving towards one who openly regarded the continuance of the war as lunacy. Rommel received a warning from the National Socialist *Kreisleiter* of Ulm at this time: the SD were saying that the Field Marshal no longer believed in victory.[35] Since Rommel had, in effect, made exactly that clear to OKW and to the Führer himself he must have shrugged a shoulder; but the warning was ominous.

Dr Strölin, *Oberburgermeister* of Stuttgart, drove to Herrlingen on the day Speidel was arrested to ask if the Field Marshal could do anything to help his former Chief of Staff, their mutual friend. Strölin's own house had been searched on 10 August. Darkness was drawing in. The Third Reich was spinning to catastrophe, round and down in a vortex of implacable savagery and suspicion. Rommel knew that Strölin was an uncompromising enemy of Nazism. They had talked together in the previous winter and Strölin had told Rommel that he, Rommel, was perhaps the only man with the prestige to lead Germany on a new course. Rommel had agreed that the Fatherland was headed for disaster. Without commitment to any leadership role he had also accepted that he had a responsibility for (as he then saw it) making the situation plain to Hitler and urging him to draw the inevitable conclusions.[36] In that he had failed.

Rommel drafted a letter to Hitler at the beginning of October, testifying to Speidel's efficiency and loyalty and speaking of his distress at hearing of Speidel's arrest.*[37] On 4 October the case of Speidel was heard by the *Ehregericht* in Berlin, with Keitel presiding; the case was not found proven against him.

On 7 October Rommel received a message from Keitel asking him to report to Berlin – a special train was to be put at his disposal. When Rommel telephoned to enquire the purpose of the summons he was unable to contact Keitel, but spoke instead to the head of the *Personnel Amt*, General Burgdorf, another colleague of Rommel from Dresden days, who had replaced Schmundt. The summons, Burgdorf said, was in order that the Field Marshal's future employment could be discussed with him. Rommel told Burgdorf that he was unfit to travel – he had an appointment with the specialist who was dealing with him, Dr Albrecht of Tübingen, and Albrecht wished him to keep the appointment and stated that he should not make the journey to Berlin.

On 11 October Rommel was visited by a very old friend, Major Streicher, a brother officer from the First World War and after, whom he had not seen since before 1939. They had a long talk and afterwards Streicher spoke of the Field Marshal being 'very, very serious, much more so than ever before'. Rommel told Streicher that he had refused to go to Berlin. He didn't trust 'them'. His enemies, whoever they were or why, seemed to him to be getting near.[38] On the same day Admiral Ruge, an ever-faithful friend, called for a good dinner and a talk until midnight. Next day Rommel – who had complained of headaches but was showing more fitness than was perhaps discreet – drove with Ruge to Augsburg. To Ruge he also said that he had refused to go to Berlin – that he would never get there alive.[39] And on 13 October he visited an old comrade from *Gebirgsbataillon* days who had been helpful to the family recently, Oskar Farny, and told him bluntly that Hitler wanted to get rid of him, he was sure of it.

To some there may have appeared in all this the smack of an obsessive persecution complex in a still convalescent man, but in truth Rommel's *Fingerspitzengefuhl*, his scent for situations and for danger, had not failed him. On the afternoon of his visit to Farny a telephone message, taken by Loistl, reached Herrlingen. It was the

* The letter was never sent.

response to Rommel's declining of Keitel's summons to Berlin. The message said that two generals would call on Rommel at his home on the following morning, 14 October. These would be Burgdorf and his assistant, the *Amstgruppenchef*, General Maisel.

CHAPTER TWENTY-THREE

'What did Rommel Know?'

To what extent, if any, was Rommel involved in the actual conspiracy against the life of Hitler and his regime?

Contemporary evidence is scanty and, here and there, conflicting. Plotting, in the Third Reich, was a risky business; plans and participation in the *Attentat* were not often committed to paper, and few definite deductions can be made from documentary sources. Euphemism, veiled speech and deliberate ambiguity were used for some sort of security and can equally mislead later generations. Subsequent reminiscence has sometimes, inevitably, been coloured by the changed sympathies of new times. The part played by or the knowledge of Rommel is disclosed chiefly by recollection and impression; often guesswork and simple deduction from character. Much has to be speculation – reconstruction from what is certain about the conspiracy itself and from the nature and mind of Rommel at particular times, as well as from recorded remarks of his own, themselves often made in peculiar and potentially dangerous circumstances.

The question may be considered in three periods of time: before the Allied invasion of Normandy; after it and until the end of June 1944, when it can reasonably be said that Rommel had given up any hope of defeating it or of holding the enemy in the west for more than a few more weeks; and between the end of June and the actual attempt on Hitler's life, the *Attentat*.

In the period before the invasion, Rommel had suffered acute disillusionment with Hitler's leadership ever since Alamein. He had exclaimed to trusted subordinates that the war could not be won, that Hitler should make way for other leaders, that the internal policies of the National Socialist state – or some of them – should be radically changed.[1] This depressed frame of mind undoubtedly stemmed from the military consequences of Hitler's supreme command – primarily from Stalingrad and Alamein, which together marked the watershed of German military fortunes and which had both, as battles, been marked by brutal obstinacy on Hitler's part. And this depression – realistic depression, reflecting only credit upon the objectivity of his mind – remained, waxing and waning, with Rommel throughout his short and unconstructive period in Italy and until his arrival in France. He brought to France a conviction that the war must be ended, but ended in a military situation which preserved Germany from a sustained war (which she must lose) on several land fronts. If this last were allowed to develop it could only, ultimately, lead to the overrunning of Germany.

To what extent, and for how long, Rommel continued to regard this impending (but still avoidable) disaster as inseparable from Hitler's leadership is more questionable. Certainly he realized – few could avoid the realization – that the enemy were unlikely to treat with Hitler, whatever the military circumstances; Hitler himself had expressed as much to him, in stark terms. Rommel believed, from bitter experience, that Hitler was still living with periodic illusions of ultimate victory, was deceiving himself about the seriousness of the military crisis confronting Germany. He thought of Hitler as a man surrounded by sycophants and mediocrities who dared not tell him the truth although he also thought that Hitler sometimes realized the truth, while failing to face it. But during this time Rommel had not discarded his faith that he could, somehow, appeal to Hitler's military understanding (for which he had once had considerable regard), that he could find words to convey the harsh realities.[2]

For – however briefly – Rommel had by the end of 1943 recovered a certain belief in Hitler's political judgement, his timing, his extraordinary ability to snatch advantage from unpromising conditions, his opportunism. Because of his view of the strategic dangers of the overall situation Rommel undoubtedly regarded the coming battle in the west, the enemy invasion of north-west Europe, as the decisive battle of the war. His flicker of encouragement at Hitler's insights helped give him temporary optimism about it. And, like the accomplished soldier and dedicated patriot that he was, he rejoiced at being given a key practical part to play. All his writings and exchanges at that time – before the invasion – evince a man giving his whole heart to the military task and believing in it. It might be only a necessary step to a temporizing peace, but the battle against invasion had to be won.

Rommel, neither at this stage nor later, was blind to moral issues, issues which transcended strategy and politics. At Christmas 1943 he had told his family distressing facts he had learned about certain actions of – or taken under the auspices of – the National Socialist regime. He had learned of these from Dr Strölin. Strölin, a man of high and generous principles, had confided to Rommel appalling things about the fate of Jews 'resettled' from Stuttgart in the east. Rommel had seen a paper, written by Strölin in 1943, arguing against persecution of Jews, a brave paper which had led to threats but to no direct action against him.[3] There were at this time still some Jews working in Germany, although not many; and the dreadful shadow of the holocaust which had been secretly taking place for nearly two years was now etched in suspected outline on Rommel's mind. He had heard earlier from friends, particularly General Blaskowitz, about evil things done in the east, in Poland and in Russia; and when Manfred, with youthful enthusiasm, had suggested joining the *Waffen SS** his father curtly said that he wouldn't allow it. He had, he said, learned of mass shootings, illegal killings, attributed to the SS.[4]

Yet Rommel still probably found it impossible to connect Hitler, personally, with evil of this kind. When Dr Strölin paid another visit to Herrlingen in February 1944 he actually mentioned the necessity of 'eliminating' Hitler, and Rommel told him that he would be obliged if Strölin abstained from such remarks 'in front of my young son'.[5] The rebuke may have been prompted by discretion as well as propriety, but it is consistent with a Rommel who still blamed the crimes of the Nazis on Hitler's subordinates, not on Hitler himself. War, Rommel knew, was a terrible business, and terrible things were done by both sides. These, however gruesome, were details, and were perhaps exaggerated. The Führer's concerns were on a different and higher level. Rommel was clearly disturbed by Strölin's revelations, but he did not yet deduce from them a necessity to renounce Hitler.

* The *Waffen SS*, the military formations fighting under army command, were a different organization from the other divisions of the SS, and notably from the *Einsatzgruppen*, the special squads established, very largely for murder, in the eastern occupied territories. The *Waffen SS* were, on occasion, tarred with the same brush, sometimes but not always justly; but on the whole they conducted themselves as exceptionally brave and skilful soldiers. There was, nevertheless, a periodic question-mark over their loyalty to the army command. Nearly one million men were serving at this time in the thirty-eight divisions of the *Waffen SS*.

Nevertheless, Rommel's reported conversation with Strölin on that occasion certainly bore witness to his pessimism about the military situation and his acceptance that he, personally, might have some extraordinary and disturbing role forced upon him by inexorable events, that he might have to 'commit himself for the saving of Germany'.[6] But what concerned Rommel at this time, before the D-Day landings, was the military challenge, the military task. It is difficult to give heart and soul to such a task while simultaneously promoting plans to remove one's supreme commander.

The two parts of the conspiracy in 1944 – the *Berliner Kreis* action at the centre, and the *Westlösung* in France – occupied different people in different places, and although both were part of one general concept intended to replace the German government and to negotiate peace, there were only tenuous and risky means of communication between the two. Speidel, Chief of Staff to Army Group B, was a key member of the western end of the conspiracy. It was supposed, after the event, that he had been especially charged by the conspirators with bringing Rommel into the business,[7] and certainly Rommel greatly esteemed him; Speidel also admired Rommel, but did not stand in awe of him – a fellow-Wurttemberger, he had known Rommel from early days, and they were close. The prestige of Rommel's name and his popularity in Germany would be a considerable asset to the *Umsturz*, the revolution, and Speidel himself described after the war how during his weeks with Army Group B, between arrival in April and the invasion itself, he colluded with Stülpnagel about the practicalities of a coup in the west, keeping Rommel, he said, informed.[8] At Marly, near Paris, on 15 May, Speidel recorded that he, Rommel, Stülpnagel and Stülpnagel's Chief of Staff discussed what might be done and how. In exchange for an undertaking by the Anglo-Americans to cease their bombing of Germany, Hitler could be arrested (Speidel recorded Rommel as being explicitly opposed to any attempt on his life) and peace negotiated. The enemy would be spared the enormous human costs of invasion and the political impossibility of negotiating with Hitler.

All this was in the first, the pre-invasion, phase. Rommel is said (by Speidel) to have authorized further discussions on the same theme between Speidel, Baron von Neurath (a previous Foreign Minister and 'elder statesman', largely untainted by Nazism) and Strölin, discussions which took place on 27 May at Speidel's home at Freudenstadt. After this Rommel allegedly expressed himself as 'ready'; Germany must be spared further sacrifice. It was furthermore agreed that these revolutionary steps – amounting, as far as the army in the west was concerned, to a military coup against the established order – must, if possible, be taken before the enemy's invasion was launched.

The practicality of all this must have been hard to credit, and the political consequences even more so. The Allied policy of demanding 'unconditional surrender' was well-known – indeed it had been used by German propaganda to considerable effect, as showing the implacability of Germany's enemies and the necessity (because inevitability) of fighting to the last drop of blood. Nevertheless those who believed the war certainly lost had to hope and scheme for something, anything, which could prevent the ultimate catastrophe of a collapse on the Eastern Front and a Russian invasion. And the run of events here described in the first phase firmly implicates Rommel, at least in the 'western' conspiracy, the attempt unilaterally to make peace, while equally firmly excluding him from any complicity in a plot to kill Hitler.

But there are difficulties. It is at first hard – not impossible, but hard – to attribute to the straightforward and dedicated Rommel the duplicity necessary to play with conviction two such disparate parts. On the one hand we see him, by every account, throwing himself with typical energy and enthusiasm into preparing the army in the west to defeat invasion; and believing in the chance of success in that battle. He may, he probably did, have regarded this as a necessary preliminary to peace negotiated from a relatively favourable military position; but he certainly played the part with commitment, and it is impossible to believe his playing insincere. Yet at the same time he was apparently ready to act against his Supreme Commander, preferably before the armies were brought to battle, despite the fact that he was sceptical of getting widespread Wehrmacht support for an anti-Hitler coup.

The two parts do not absolutely exclude each other. Any professional soldier, especially so gifted and dedicated a professional soldier as Rommel, would be bound to tackle the military task to the best of his ability, and might still simultaneously plan how to conclude the whole disastrous business of the war by negotiation and political action. It must have led to extraordinary inner tension and contradiction, but these were extraordinary times. Rommel never believed that the Wehrmacht could easily be induced simply to change sides and go against Hitler, but a peace move, led by respected commanders, might have secured their acquiescence if its conditions and presentation were convincing.

The persuasion and views of others must have played a considerable part. The prime mover in the conspiracy in the west, Stülpnagel, was close to Rommel – they had been together at the infantry school in Dresden in former days and were old friends. Rommel revered Stülpnagel, an outstandingly brilliant General Staff officer who had succeeded his cousin, another Stülpnagel, in the French command and who had seen the disaster endemic in Hitler's character from an early stage. And Rommel certainly valued Speidel, valued his intelligence and his judgement. Rommel was of a simpler, a more straightforward character than either of these; but, while continuing to absorb himself in how to win the forthcoming battle, he may have consented to be aware of their dangerous intentions, to lend his name. Holding the views he did about the necessity of peace, he may have assented thus early to preparations, to pre-planning for negotiation with the enemy. Such dissimulation must be reconciled with his absorption in his daily task at that time, with his mounting confidence (misplaced but surely convincing) in the outcome of the battle and the efficacy of his counter-invasion measures, and with such diary entries as that on 13 May, only two days before the meeting at Marly: '*Der Führer vertraut mir und das genügt mir auch*.'[9] An entry dictated for effect, for the record? Possibly. These were dangerous days, although such prudent humbug was remarkably out of character. It is not easy, but it is not impossible to square Rommel's entire conduct at the time with the proposition that he had been already converted to the possibility and desirability of some sort of early revolution leading to peace. And any sort of revolution, any sort of radical change of regime, would necessarily be accompanied by action in the west, action at the front.

It seems more likely, however, that discussion between Rommel, Stülpnagel, Speidel *et al.* was largely at that stage conducted on a contingency basis. There would, almost certainly, need to be a change in Berlin if negotiations for peace were to stand a chance – with that proposition Rommel, despite his loyalty, would have had no difficulty. Such a change, however procured, could lead to a disputed succession, to attempts at power by other leading Nazis, above all to a question-mark over the attitude of the army, particularly the army on the western front. In such circumstances, conditional on an initiative at the centre, it would be right to have plans prepared, plans to neutralize the SS and the Gestapo (who would probably seek to overturn any non-Nazi succession), plans to move for an armistice, to reassure the army, to conduct a negotiation from a reasonably firm military base. Thus far Rommel's participation is credible – and compatible both with his energetic discharge of his actual duty and his absolute rejection at all times (recorded by all who spoke with him, including Speidel) of any idea of killing Hitler. Rommel's assent to the proposition that all this should be considered and ideally should happen *before* the invasion is also credible, given the wholly unrealistic nature of the conspirators' assumptions in the first place, based as they were on a belief that Britain and the United States might, in 1944, agree to stand fast and give Germany a free hand against the Soviet Union if Hitler were removed.

Then, however, came 6 June and a new phase in the saga. Soon the Allies were ashore and established. There was now no question of a coup pre-empting invasion. Invasion had come.

During the next period, the last three weeks of June, Rommel was almost entirely preoccupied in fighting the battle, a battle which by the end of the month he recognized as lost. Had that battle been won (as Rommel had hoped, and hoped as a preliminary to negotiation) the anomaly, of course, arises that Hitler's position would have been strengthened by the victory. This remains academic. There was no victory.

It was now a matter of establishing the most favourable military situation within which peace negotiations might be envisaged. The best which could be hoped for, after the initial counterattacks had failed, was a temporary stalemate – the holding of a firm line; and this, Rommel believed, should be accompanied by a last effort to persuade Hitler. The Führer, Rommel managed to suppose, might concede to the obvious imminence of military defeat what he would certainly refuse to concede to argument. Rommel's energies, therefore, were directed towards persuasion based on the facts of the fighting. There had been no revolution in Berlin, and it is unlikely Rommel expected one at this stage; nor would he necessarily have welcomed it. There was a large difference between a political upheaval at home when the defending troops in the west were waiting entrenched behind the Channel barrier and the same upheaval when they were locked in battle with the enemy. Rommel may have been prepared to play his part in what he called an 'operation'[10] – in effect a coup – before the invasion, but now matters were different.

Different, but certainly no better, and when on two occasions – on 17 and 29 June – Rommel had tried and completely failed to persuade Hitler towards what he reckoned was realism, he knew that little time was left. The military situation no longer offered the comparative stability of the pre-invasion era. It would not, for more than a few more weeks, offer the smallest chance of continuing stalemate. Peace – in effect surrender by whatever means – had become essential if the whole dreadful sequence was to be stopped and the very worst avoided. Accusations of pessimism or (by Kluge) of insubordination left Rommel untouched. He knew the truth of the battlefield and he knew that the battle situation could not have been improved by him. He had done his best. Afterwards there were suggestions that he had deliberately held back 2nd Panzer Division for action in aid of a coup, but the suggestion has been conclusively rejected by those well-placed to know the truth and it was wholly inconsistent with Rommel's character and reactions.[11] He had fought as hard as he knew how. Rommel has often been criticized as primarily – or 'purely' – a tactician rather than a strategist, but it was his sure and accurate assessment of the progress of the tactical battle which led him to absolute certainty about the strategic and thus the political prospect. Germany had lost.

Thus Rommel came to the third and last phase, the first three weeks of July. It was in this period that Rommel spoke freely – no doubt too freely – about using his own initiative, opening the gates to the invaders, leaving it to the forces of the Western Allies to impose the German revolution.[12] With defensible logic, he judged that the right moment for this would probably be when the enemy inevitably broke out of the bridgehead. Rommel could not envisage ordering men, as yet unbeaten, to lay down their arms and march into captivity, and there is no particular reason to suppose they would have obeyed such an order; but when the Anglo–Americans were in full cry, with their huge mechanized forces and their abundant supplies of fuel, it would be a different matter. He spoke to a number of his officers, including the tough old Nazi warrior Sepp Dietrich, in this sense, and was sure that if he gave the word even such as Dietrich would comply.[13]

There could be methods of making contact with the enemy command – an exchange at Cherbourg involving the repatriation of some captured nurses had shown the way. By this third and last phase in his inward Odyssey Rommel had recognized that he might be reviled by his own countrymen one day, yet knew that he would have chosen the better way.[14] Meanwhile he had to do his best to save lives and honour, to keep on fighting, ready to open an individual negotiation when the right hour struck. Because he now assented, wholly, to the idea of unconstitutional and unilateral action Rommel may be said, by this point, to have become a committed 'conspirator'.

But during this phase, between the end of June and 20 July, plans for the *Attentat* were reaching maturity, the *Berliner Kreis* was bracing itself for the supreme moment of the conspiracy at the centre, men were readying themselves for the murder of Adolf Hitler. What did Rommel know?

Rommel knew, of course, that the conspiracy intended to lead to peace negotiations, a consummation he wholly endorsed, included a number of members who believed it essential to kill Hitler as a precondition of success. He had heard such talk from Speidel, from Strölin, probably from Stülpnagel, and he had always objected. Because he recognized that nobody would make peace with Hitler, and because he also recognized, at least from mid-June, the imminence of disaster, it is unlikely that he would have thought it his duty to report such conversations, and to that extent he probably possessed at least generalized 'guilty knowledge'.

Definite and precise knowledge of what was intended was a different matter. There were plenty of rumours rife in Germany and in the Wehrmacht at that time, and, unless he was specifically told, Rommel had no reason to know that the option of assassination had been chosen. Both before the event and afterwards he had spoken unequivocally and forcefully against the *Attentat*. Not only did he think it politically inept and self-defeating (a view which may have been wholly mistaken but was surely sincere), but it was also contrary to his every instinct. Assassination of the constitutional head of the German Reich was to Rommel an appalling as well as a foolish idea. If Hitler had acted criminally he must be removed and proceeded against by law. Rommel said to several that the man must not be made into a martyr for the German people – he must be acted against openly and correctly. Viewed with hindsight, in the light of the millions for whose deaths – with no sort of legality – Hitler was responsible, such scruples may seem inappropriate, even absurd, but Rommel did not view with hindsight nor in that light.

To him, ruthless soldier though he might be, murder was wrong and to strike down from behind a trusting superior was dishonourable. After his death, and several months after the end of the war in Europe, Lucy made a public statement repudiating any suggestion that Rommel had participated in the 20 July plot – whether in preparation or execution.[15] She did this at a time when such participation was already beginning to be recognized as supremely virtuous. She made her statement, therefore, on the unpopular rather than the fashionable side, and it carries conviction. It is unlikely that Lucy was ignorant of the truth and she boldly proclaimed it.

Evidence in a contrary sense, however, of Rommel's 'guilty knowledge' of the *Attentat* comes from several sources. Tempelhoff, at La Roche Guyon on 20 July, was told of the *Attentat* rumours by Staubwasser. Radio reports, Staubwasser said, were of the Führer having survived an assassination attempt. 'Wrong,' Tempelhoff responded, 'The Führer's dead! The Field Marshal, the Chief [Speidel] and I were in the picture, its all been prepared a long time!'[16] But this, even assuming the accuracy of Staubwasser's reportage and the truth of Tempelhoff's remark, still begs the question of whether 'the Field Marshal' meant Kluge or Rommel. Kluge, having taken over Army Group B as well as *OB West*, was installed at La Roche Guyon.

The second source was Stülpnagel. Stülpnagel was alleged to have muttered the name of Rommel when first resuscitated and brought to Germany for interrogation after his suicide attempt. Certainly Stülpnagel hoped to have coordinated measures with Rommel for the *Westlösung*, for the neutralization of the SD and Gestapo, and for the approach to the Allies if there were an overturning of the regime in Berlin. To that extent Stülpnagel may have led to Rommel, and as far as it went the trail ran true; but it did not necessarily run through complicity in murder. It was inevitable that Rommel would be at the least touched by the suspicion which soon engulfed Stülpnagel. The two were old friends. Their headquarters were not far apart – La Roche Guyon is a mere forty miles from Paris. Stülpnagel had the means of domestic control if it came, as it was intended to come, to dealing with the SS in France; but Rommel was the commander of virtually all the front-line troops on the Western Front, and if there were to be a request for an armistice and a negotiation it was Rommel who mattered. To his interrogators Stülpnagel's role surely implied the cooperation of Rommel. It could not be played else.

On 17 May Stülpnagel had met Beck in the company of Baron von Teichmann, another conspirator; and at that meeting Beck asked both Teichmann and Stülpnagel to talk to Rommel as early as possible and to tell him that in the view of Dr Goerdeler (Chancellor designate) Hitler must be assassinated. Arrest was out of the question – Hitler must die. The record (by Teichmann) of this meeting is consistent with Beck's knowledge that Rommel had already expressed himself strongly against assassination and in favour of dealing with Hitler only by law; and thus that Rommel needed persuasion.[17] This meeting of Stülpnagel with Beck, furthermore, took place only two days after Stülpnagel's meeting with Rommel at Marly, so that the latter's recalcitrance about the *Attentat* would have been uppermost in his mind. There is, however, no evidence that the persuasion which Beck urgently sought was actually exercised; and certainly no evidence, if it was, that it succeeded.

The third source was Speidel himself. Whatever knowledge Speidel shared with Rommel, there was no doubt that he had been a leading spirit in the 'Western conspiracy', and that alone, if proved, was enough to hang him. He always, however, strongly denied specific knowledge of the *Attentat*.[18] He was incriminated by the evidence of another which, he said, was fabricated or mistaken; but he himself claimed after the war to have known only that plans somehow to get rid of Hitler existed, but not to have been aware of details of an actual assassination attempt. In this case he could not have shared guilty knowledge with Rommel.

The key source, also leading to and through Speidel, was Colonel Casar von Hofacker. Hofacker visited Army Group B headquarters in July, accompanied by Dr Horst, Speidel's brother-in-law and an official, a *Regierungsrat*, in Stülpnagel's military administration – he who was to spend the evening of 20 July at La Roche Guyon. Together on 9 July they saw Rommel, and Hofacker spoke on behalf of Stülpnagel about the gravity of the general situation.[19] Hofacker was a pivotal figure in both conspiracies, a coordinator between actions in France and actions of the *Berliner Kreis*, and it was later alleged that he had on this occasion informed both Rommel and Speidel of the coming *Attentat*; on returning from La Roche Guyon Hofacker, perhaps relaying what he wished to believe, told a colleague that Rommel had been told of the *Attentat* (including the intention to kill Goering and Himmler as well, if possible) and had said that he was content to play his own part to bring the plan to success.[20]

The allegation (that Hofacker had informed Rommel of the coming *Attentat*) was ultimately to be shown to Rommel, as conclusive proof of his guilt in the matter. There was also a different (and inconsistent) allegation – that Hofacker had informed not Rommel but Speidel (whom he had seen on the same day); and that Speidel had admitted to the conversation but had claimed to have done his duty by afterwards informing Rommel. Either allegation would have been enough to hang Rommel – the second if Speidel's claim to have passed on the information were accepted, the first as standing alone.

The facts of this particular aspect are obscure. At the proceedings of the *Ehrenhof* on Speidel at the beginning of October (Keitel presiding) Kaltenbrunner produced what has been called above the second allegation – that Hofacker on 9 July told Speidel of the coming *Attentat*, and that Speidel had dutifully informed Rommel. Those at the *Ehrenhof* who were inclined to clear Speidel of complicity in the *Attentat* (very much including Guderian) accepted both the second allegation presented by Kaltenbrunner and – against the views of Kaltenbrunner – the validity of Speidel's alleged self-exculpation.* By telling his superior, Rommel (they said), Speidel had done his duty. Very obviously Rommel – whose conduct was never examined by the *Ehrenhof* – had not.

Speidel later denied the entire business, which rested, he said, on fabrication. There is no extant record of the detailed statements made by Hofacker under interrogation, which would presumably clear up the question. There is no extant record of the statements made under interrogation by Speidel. It must remain obscure exactly what the Gestapo extracted from Hofacker – and, of course, whether or not it was true. Five points, none of them conclusive, must be added.

First, Speidel saw Hofacker when both were being held for interrogation, and wrote afterwards that he bore signs of physical ill-treatment.

Second, the interrogator, Kiessel, subsequently alleged that Hofacker admitted speaking to Rommel on 9 July about the 'general situation', and stated that Rommel had talked of the necessity to force Hitler's hand if he would not act.† Kiessel, however, reported Hofacker as denying any talk between them of the *Attentat*,[22] a denial which contradicted other alleged Hofacker statements.

Third, Horst, Hofacker's companion on the visit to Rommel on 9 July, reported that Hofacker was not particularly elated on their return drive to Paris together; whereas he might have been had he supposed he had successfully persuaded the Field Marshal to approve

* Report by General Kirchheim, a member of the *Ehrenhof*. The *Ehrenhof* appeared content to receive evidence and form its judgements on reports (largely 'confessions') presented to them by Kaltenbrunner rather than themselves directly examining principals.[21] Kirchheim's (post-war) report was contersigned as accurate by Guderian.

† '*Wenn der Führer nicht wolle, musse man ihn zwingen.*'

of the idea of the *Attentat* (although there is independent evidence that he did so suppose[23]). Horst, too, denied that it was ever mentioned.[24] His testimony was given in 1975, and there appears no particular reason to doubt it.

Fourth, and perhaps most significantly, there is no record in Kaltenbrunner's reports to Bormann of any of this. Hofacker was often mentioned – there was no doubt whatever that he was a leading figure in the conspiracy, and he probably never denied it. He was referred to by some who knew him as an anti-Nazi fanatic. He was by every account a brave, intelligent man whose career had been mostly in business, and his father had been Rommel's commanding officer in the First World War, so that his easy access to the Field Marshal is all the more imaginable. Had Hofacker made statements under interrogation which incriminated Speidel – and, by inevitable extension, Rommel – in the *Attentat*, it would have surely made the main stuff of one of Kaltenbrunner's daily reports to Bormann. There is no hint of that. Kaltenbrunner's many references to Hofacker – incriminating enough as they stood – were concerned with his reports that the High Command in the west believed the war lost, believed there must be a negotiation, believed that the enemy would be on the road to Paris in as little as six weeks (accurate, as it turned out). Hofacker was interrogated frequently, first in France, and his moves in the critical days between 9 and 20 July were accurately reconstructed by the Gestapo. After his visit to Rommel at La Roche Guyon he had travelled to Berlin, where he had, wrote Kaltenbrunner, reported the imminence of enemy breakout.[25] This, of course, was *Defaitismus und Pessimismus*, obviously derived from Rommel – and certainly it corresponded with the latter's views. It prompted the conspirators of the *Berliner Kreis* to action, since it showed that time was probably short. To Kaltenbrunner and Bormann it was reprehensible, treasonable stuff. But by itself it was not evidence of Rommel's knowledge of the *Attentat*.

Then, fifth, there was Colonel Eberhard Finckh. Finckh had recently taken over duty as head of the Quartermaster branch at *OB West*, Kluge's headquarters. He visited Berlin on 23 June and spent many hours with Stauffenberg, an old friend whom he had not seen for some time. They had been joined by Olbricht and Mertz von Quirnheim, key members of the conspiracy, and Finckh (by his own testimony) had been asked by them what he thought of the general situation. Finckh later talked to his interrogators of Stauffenberg's powerful personality, of his fanaticism; and Stauffenberg, Finckh said, had scoffed at the *Marschalle* who failed to exert their influence in stressing the seriousness of the situation to Hitler. Finckh allegedly saw Rommel on a visit to Army Group B two days later and told him of conversation about the *Attentat* and of Stauffenberg's involvement.[26] At that stage, of course, details of the forthcoming attempt were not yet established, so Finckh's remarks would have been tentative, if they were made at all.

Finckh, who was certainly involved, was arrested, interrogated and sentenced to death. On at least four occasions Kaltenbrunner's descriptions of him and his evidence reconstruct his conversations with Stauffenberg, his devotion to Stauffenberg, his political innocence. Hofacker told his interrogators that Finckh belonged to the sort of non-political circle of officers who divorced in their minds military discipline from political system.[27] In post-war interrogation by the Allies Keitel thought that he remembered another officer of Kluge's staff, besides Hofacker, whose evidence had shown Rommel to have possessed 'guilty knowledge' of the *Attentat*, and if this were so it may have been Finckh.[28] But, as with Hofacker, there is no indication of it in Kaltenbrunner's reports. Of course Rommel's name would, by order, have been expunged from these reports (although the originals exist) when it was later decided to save his reputation. Nevertheless the flow of Kaltenbrunner's communications to Bormann gives not the smallest hint of suppression of any prominent name.

But there is a sixth point. Hofacker's evidence – however extracted and however false or true – was shown to Keitel when Hitler was discussing with him the awful matter of Field Marshal Rommel's treachery. This evidence allegedly contained the statement that Rommel had asked Hofacker (at La Roche Guyon on 9 July) to 'tell the gentlemen in Berlin' that they could count on him, Rommel.[29] This was implicitly denied by the testimony of Horst, who was present, but even if true, it did not necessarily reveal an exchange about the *Attentat* – Rommel, after all, was by then making alarmingly little secret of his readiness to play a part in negotiating for peace.

Hofacker's visit of 9 July was recalled by Rommel later, when Speidel visited him in hospital after the *Attentat*. He said then that he now saw the conversation with Hofacker in a different light,[30] so perhaps hints had been given which only subsequent events clarified. Whether Hofacker, by wishful thinking, gave an exaggerated idea to his fellow-conspirators – and not only to them – of Rommel's possible support for the *Attentat* may never be known.* What mattered to them was his assessment of how soon military collapse would come, and an indication of his readiness at the right moment to cooperate in negotiation with the enemy.

It is also possible that Hofacker had paid an earlier visit, in June, to Rommel, a visit during which Rommel (according to one report of Hofacker's subsequent account[31]) saw Hofacker alone ('*unter vier Augen*') and spoke of action being desirable 'now rather than later'. The context of Hofacker's alleged account implies that revolutionary action was in mind; but by no means implies murder. Hofacker's statements, however, were clearly regarded as constituting the prime case of guilty knowledge against Rommel, possibly reinforced by acceptance of the allegations made in the case of Speidel. And they were shown to Hitler.[32]

Yet what did this evidence amount to? It turned on the allegation – denied by Kiessel, Hofacker's interrogator; denied by Horst; denied by Speidel; and, of course, most certainly denied by Rommel even when he was within minutes of his own death, and knew it – that Hofacker had discussed the *Attentat* with Rommel, or with Speidel who had reported the conversation. This depends upon some version or other of the reports of Kaltenbrunner; and upon a subsequent account of his conversation given (but not confessed in testimony) by Hofacker himself.

Others have spoken since of Rommel's involvement in more definite terms. Dr Hans Berndt Gisevius, an *Abwehr* officer and member of the German resistance, testified at Nuremberg of the 'very painful impression' made when Rommel 'proposed to have Hitler assassinated'.[33] This 'proposal' by Rommel runs counter to every other piece of evidence. The allegation of 'guilty knowledge', of direct involvement, and – most of all – of initiating the idea of the *Attentat* must be regarded as, at most, inconclusive. Gisevius also referred to Beck's use of Rommel's name – Rommel, Beck allegedly said, who had always been a convinced adherent of Hitler, was now, in July 1944, speaking of the necessity to remove not only Hitler but Himmler and Goering simultaneously from the scene. This was all hearsay, in an atmosphere in which views were being attributed without much evidence; and the conspirators wanted to believe

* Casar von Hofacker, with Stülpnagel and others, was condemned to death by a people's court at the end of August but was not executed until December.

Rommel was with them, even if belatedly. Rommel himself never spoke with Gisevius; or, for that matter, with Beck.

A different *general* conclusion may not be provable but seems far more likely. Rommel had always objected to the idea of assassinating Hitler and knew nothing of the detailed plot to do so. Of course he knew that the idea had been bandied about by a number of people, including several he regarded as friends; and he wished to take advantage of a change of regime and to end the war as eagerly as any. Certainly he regarded – now strongly, now with misgivings – Hitler's continuance in office as the greatest impediment to peace; on the day of his wounding he had said to General Eberbach, on leaving Panzer Group West headquarters for the last time, that the Führer had to 'disappear'.[34] But personal participation in the *Attentat*, in murder, was not for Rommel, and if he had 'guilty' knowledge it was imprecise, dubious, and (in his mind) to be deplored.

Nevertheless the interrogation reports, even if false or inconclusive in their explicit involvement of Rommel with the *Attentat*, must have made abundantly clear, as many an old friend could have confirmed without benefit of Gestapo deduction, that the Commander of Army Group B no longer believed in victory and had been ready to play a part in negotiation with the enemy. Of this there was no doubt; and it was treason. Suspicion, anyway, was running through the whole Nazi system. Distrust was rife. Stauffenberg was reported by a number of those interrogated as having said that the Army Group Commanders would join forces with the conspiracy once the *Attentat* succeeded, and it is probable that the SD formed exactly the same opinion.[35]

On 14 October Manfred Rommel was given short leave to visit home – he had been back at his anti-aircraft battery site for one week. He took a train and arrived at Herrlingen before seven o'clock in the morning. An old friend and staff officer of Rommel's, Captain Aldinger, had been given leave to stay with the Field Marshal, to act as his personal amanuensis and aide: Aldinger, a reserve officer who had served in the same battalion as Rommel in the war of 1914–18, had also been with him in the present war – in the *Gespensterdivision* in 1940, in North Africa and then in Normandy. A near-contemporary, he was a family friend and was living with the Rommels at Herrlingen.

Some hours after his arrival Manfred took a walk with his father. Rommel told him that he was expecting a visit from two generals, allegedly to discuss his future employment. He said he was not sure whether this was the true purpose of their visit or not. They then returned to the house.

At midday Generals Burgdorf and Maisel arrived. There was a gate from the public road to the garden, through which ran the short drive to the front door, and Rommel had told his soldier-servant Loistl to leave it open since he expected visitors. Loistl was surprised to see that in spite of this the visitors' car and driver remained in the road. Loistl, having taken the Generals' coats and announced them, asked the driver – from the *Waffen SS*, he noted – to come into the house, an invitation he declined, saying that he had his orders and he knew what he was doing.[36] Loistl noticed another small, grey Mercedes car was parked nearby, and that a civilian from it spoke to the Generals' driver.

Inside the house Rommel told Manfred to leave the room. For the next forty-five minutes he was alone with Burgdorf and Maisel.

It was Keitel who had given Burgdorf his task, and he had made it clear that the orders came directly from Hitler.[37] Field Marshal Rommel, Keitel said, had been involved in the conspiracy. His guilt was shown by the testimony of Lieutenant-Colonel von Hofacker. The Führer felt deeply wounded by this betrayal: he had always highly esteemed Rommel.[38]

Hitler had decided that the German people should remain without knowledge of the high treason of the especially popular Field Marshal. Rommel's name should, if possible, be kept clean from imputation of involvement in this appalling crime.[39] Rommel, therefore, was to be faced with the evidence (Keitel gave Burgdorf a copy of Hofacker's statement or statements for this purpose) and given an alternative. He could accept arrest, by Burgdorf, and trial for high treason; or he could take 'the officer's way'. In the latter case he would be given a state funeral. His family would not be penalized. His death would be proclaimed as natural.[40] Burgdorf was to take quick-acting poison with him.

Burgdorf and Maisel came out of the *Herrenzimmer*, the main living room of the house, together with Rommel, their host and victim. They went into the garden and walked up and down while Rommel went upstairs to find Lucy in their own room. He paused on the way and spoke to Loistl: 'Send Manfred to me now, and Aldinger after half an hour.' Then he disappeared.[41]

Lucy's subsequent account of their last moments together was as straightforward as it was moving. Rommel told her that he had been given a choice, by Hitler's order; suicide or to appear before a people's court. He had clearly made up his mind instantly, for his first words to Lucy were that he would very shortly be dead. He said that Burgdorf and Maisel had brought the means – a poison which functioned in three seconds.

Rommel told her that the reason was his alleged participation in the events of 20 July; and that the evidence of Generals von Stülpnagel and Speidel, together with that of Lieutenant-Colonel von Hofacker, had accused him. So much, it was clear, Burgdorf had told him, as well as showing him (presumably) the written report of Hofacker's testimony produced by Keitel. There was a further damning circumstance alleged against him, Rommel said. Dr Goerdeler (who had been arrested on 12 August and interrogated*) had mentioned Rommel's name as a future President of the Reich.[42] Rommel had never met nor talked with Goerdeler.

Rommel told Lucy how he had replied to Burgdorf and Maisel. The accusations, he had said, were unbelievable. There was no truth in them and they could only derive from blackmail or some sort of illicit extortion of false evidence. To Lucy he said that he was unafraid of facing a people's court – he could defend his every action honestly and openly. He was convinced, however, that he would never appear in court – he would, somehow, disappear before it came to that. Talk of 'future employment', the phrase used by Burgdorf in their earlier telephone conversation, had been brutally deceptive – every detail of his, Rommel's, extinction had been planned. He was convinced that this was the end. He said goodbye.[43]

Manfred now joined him and to him, too, Rommel briefly described the choice given him and the choice he had made. Again he explained that the alleged evidence of Speidel and Stülpnagel and the suggestion made by Goerdeler had implicated him.[44] Manfred received the firm impression that Rommel believed Speidel's and Stülpnagel's confessions either did not exist or had been extracted by torture.[45] Rommel told his son that the family would not suffer, provided he made the choice he had. Then he said goodbye to Aldinger.

Manfred, like Loistl, had noted several vehicles, which he thought were occupied by armed men in civilian clothes, parked near the house. His father, completely calm, took leave of him. Then, with

* He was executed in February 1945.

Aldinger, Manfred accompanied the Field Marshal to one of the waiting cars. Rommel was wearing his overcoat and cap and carrying his Marshal's field service baton.

At the car Burgdorf and Maisel were waiting. They saluted Rommel with the (obligatory since July) Nazi greeting, '*Heil Hitler!*' Rommel climbed into the back of the car.

Fifteen minutes later the telephone rang in the house in Herrlingen. On the line was a *Reservelazarett*,* in temporary accommodation in the *Wagnerschule* in Ulm. Field Marshal Rommel appeared to have suffered a heart attack. He had been brought to the hospital by two Generals, in a car. He was dead.

* Reserve hospital.

CHAPTER TWENTY-FOUR

A Necessary End

A STATE funeral had been decreed by Hitler, in a message which expressed his personal grief, that of the German Wehrmacht and the whole German people. It was a long message, recounting Rommel's entire career and his heroic services to the Fatherland from 1914 until the end.[1]

Instructions for the funeral were detailed and meticulous. Organization would be the responsibility of the local military district, *Wehrkreiskommando V*, to whom subsequent orders about defrayment of costs were sent two days after the event.[2] A special train for distinguished mourners, leaving Berlin at seven o'clock on the previous evening, would arrive at Ulm at 10.40 on the morning of Wednesday 18 October. The ceremony was to take place at one o'clock that afternoon in the Rathaus of Ulm, where Rommel's body would lie in state, watched over by officers of the Wehrmacht with drawn swords, the Field Marshal's baton, sword and decorations atop the coffin. The Führer's representative, to lay his wreath, utter his condolences and speak the words of the funeral oration would be Field Marshal von Rundstedt. He would leave his hotel in Ulm at seven minutes to one and enter the Rathaus on the hour exactly, saluting the bier and then taking his place in the front row next to Frau Rommel and at the head of a great assembly of military and civil dignitaries. The funeral parade, drawn up to escort the coffin to the Rathaus, would consist of two companies of the German army with military band, and a third, mixed, company of detachments from the Luftwaffe, the *Kriegsmarine* and the *Waffen SS*.

So it was; and all were impeccable. East, west and south the Wehrmacht might be fighting the last, desperate battles of its remarkable career, but its ceremonial drill and appearance, when occasion demanded, were beyond reproach. The arrangements ran to perfection. At exactly one o'clock the *Trauermarsch*, the second movement of Beethoven's Third Symphony, the 'Eroica', itself originally conceived in honour of Napoleon, began.* It was immediately followed by von Rundstedt speaking from the podium.

Rundstedt recited Rommel's exploits: his services to Germany; his tragic wounding in Normandy. He referred to him as a 'convinced National Socialist' at the beginning of the war, a technical falsehood but in spirit not a grave one. He recited Rommel's achievements in France, in North Africa against numerical odds culminating in the taking of Tobruk, his personal courage. This fearless warrior, Rundstedt continued impassively and with decreasing accuracy, was filled with National Socialist spirit: 'His heart belonged to the Führer.' 'Your heroism,' Rundstedt apostrophized the bier, with what sense of irony can only be guessed, 'shows us all, again, the watchword "Fight until victory".'

As Rundstedt laid Hitler's wreath at the bier the congregation sang the ever-moving lament from German soldiers for a departed comrade, '*Ich hatt' einen Kameraden*'. A battery was heard firing a nineteen-gun salute. Rundstedt resumed his place and the ceremony concluded with the *Deutschlandlied*. The programme then prescribed that the Field Marshal would, at that point, turn and formally convey to Rommel's family† Hitler's sympathy and his own. He was remembered as speaking only three brief and not particularly gracious words before quitting his place, giving one last salute to the bier and leaving the Rathaus. There were further ceremonies and music – and an eloquent oration by Baron von Esebeck – at the crematorium whither the bier was transported in convoy through the crowded streets lined with troops. Later Rommel's ashes were interred in the village cemetery in Herrlingen. Lucy, unsurprisingly, had found the proceedings near intolerable.

In March 1945 Lucy received a letter stating the Führer's desire to erect a memorial to Rommel, a project he had placed in the hands of Professor Kreis, architect in charge of design of all German war cemeteries. Drawings were proffered and Lucy's opinion sought –

* Another (superseded) order for the ceremony provided at this point not for the 'Eroica' but for the funeral march from Wagner's *Götterdammerung*. It has been suggested that the latter might have been rejected as a suspect allusion, since Siegfried was killed because of the envy of the powers of darkness.[3]

† Besides Lucy and Manfred, Rommel's two brothers, his sister and other relatives were present.

the suggested choice was a massive lion on a pedestal. To Lucy it was all odious. No memorial was begun. For Rommel, of all men, to leave the world surrounded by humbug was obscene; and the crocodile tears of a Führer who had willed his death were humbug indeed. Her own concluding sentence, in her later statement on her husband's fate, was as straightforward and true as his own character: 'Thus ended the life of a man who had devoted his entire self throughout his time to the service of his country.'[4]

To some, like Keitel and Jodl at OKW, Rommel's choice of suicide proved his guilt beyond question. Knowing as they thought they did Hitler's regard for him, Hitler's reluctance to believe ill of him, they said that had he been innocent Rommel would not have accepted the situation. He knew Hitler's affection – he would, somehow, have personally appealed to the Führer. And he knew his own public popularity – he would have appealed to the German people, whether in session of the people's court or by some other means. That he chose to go in silence showed them (they testified) that it was as von Hofacker had stated.[5] Rommel had had guilty knowledge of the *Attentat*.

It was not so: and Rommel's own forthright statement to Lucy that he did not fear trial, that he was innocent of involvement with the plot to murder Hitler, convinces; as does Lucy's own unequivocal statement after the war.[6] She knew him like no other, there was total trust between them, and men do not lie when about to die. But Rommel had concluded that he was to be eliminated, and that he would probably be killed before being given the chance of defending himself. He would, furthermore, be sacrificing his family, whereas he had been assured that if he chose suicide they would be unmolested. He made his decision.

There was more. The dominant factor, surely, was Rommel's awareness that although he had had neither part in nor specific knowledge of the *Attentat* he had been determined to make his own move for peace, and had been frank about it to many. He knew perfectly well that that was treason, that it would not be forgiven and that under that accusation both he and his family would be condemned. He proclaimed his innocence of the crime of which Burgdorf and Maisel accused him because he had never supported the assassination of Hitler, and he was indeed innocent. But of that other 'crime' he was guilty. There may be curious confirmation of it in the fact that Himmler sent a private message to Lucy (by Berndt) that he,

Himmler, had had no hand in the business of Rommel's contrived suicide.[7] Himmler, with overall responsibility for SD, SS and Gestapo, had no instincts for mercy towards the plotters of the Führer's assassination; but he was already meditating his own personal approach to the Allies. He had seen the writing on the wall. He may have had greater sympathy than his Führer for what Rommel would have tried to do had not fate intervened.

It has been argued that opposition to Hitler's murder was inconsistent and not particularly admirable in one who was determined to flout the Führer's will and to harvest the benefits of Hitler's departure or immobilization.[8] The argument is respectable; but the fact remains that Rommel felt otherwise. He distinguished between *Attentat* and attempted peace negotiation. He was guilty of opposition, of believing the war lost, Hitler's leadership disastrous, and peace essential for Germany, and of this 'guilt' he was unashamed, just as he had always fearlessly given his views, however unpopular. After his 'ultimatum' to Kluge for Hitler's eyes Speidel reported Rommel as saying: 'I've now given him a last chance. If he doesn't draw the right conclusions it's up to us thereafter.'[9] In all this Rommel felt – fairly – that he had always spoken and acted openly and rightly. It may have been naïve, but it was in character. But Rommel's naïvety did not extend to supposing that he could be pardoned for planning a negotiation with the enemy, that he could play the part of Yorck at Tauroggen* and escape National Socialist retribution.

Rommel had felt death close for some time. Death as such he had never feared. He had lived with it hourly through the most dramatic and rewarding periods of his life, and he might credibly have exclaimed, as Shakespeare has Caesar exclaim:

> The valiant never taste of death but once.
> Of all the wonders that I yet have heard,
> It seems to me most strange that men should fear;
> Seeing that death, a necessary end,
> Will come when it will come.

* In December 1812, General Yorck von Wartenburg, commanding a Prussian army in highly unpopular and enforced alliance with Napoleon, negotiated a separate convention at Tauroggen with the invading Russians and enabled them to march against the French unopposed. The convention, although initially repudiated by the King of Prussia, was immensely popular.

Death came to him when his mind was intolerably troubled and his heart torn by misshapen circumstances. He had lived by patriotism, by simple love of country; and now the wells of patriotic feeling had been poisoned by the secret wickednesses being done in patriotism's name. He had reached extraordinary heights as a fighting soldier and he had suffered the rejection of his views and the sacrifice of his men's lives by the manic resolution of a despot feeding on illusion. He had invariably cherished the regular decencies of life and he now knew that they were set aside without scruple by his country's own rulers. He had spent his energies for the honour, the greatness and the security of a Germany which would soon, quite certainly, be at the mercy of a new barbarian invasion, all the more frightful for having been provoked by Germany herself. That 'necessary end' came when life had mocked everything in which Rommel believed. His hopes were ended. It was time to taste of death.

On 20 October a special laudatory order of the day, signed by Adolf Hitler, was published. Rommel was naturally accorded lengthy eulogies in the German press, and his death was felt throughout the country, inured though it was to bereavement and disaster. Through his own achievements, well publicized as they always had been, he was a national hero. Among those who dared to think such thoughts it was whispered that Rommel would have taken over as Supreme Commander if anything had 'happened' to the Führer.[10] The foreign press also carried extensive coverage, *The Times*' obituary in London on 16 October running to more than a thousand words. It accorded somewhat grudging praise to Rommel's tactical ability, showing that churlishness (as Churchill has put it) generally inseparable from discussion of the enemy in wartime, but was inaccurate in ways which would undoubtedly have irritated its subject, connecting him with the Nazi Party from its beginnings (he never was a member) and describing him as a 'storm troop leader, attached to Hitler's bodyguard', possessing a taste 'for the methods of a gangster in civil war'; all, as it happened, nonsense.

After the war Rommel's reputation naturally gained from disclosure of the truth about his death. Interest focused on his attitude to the regime, and any sort of distancing from the Nazi nightmare was a virtuous obligation in an increasingly – and sincerely, and impressively – democratic Germany. To have opposed Hitler to the point of enforced suicide was to be on the side of the angels, and Rommel was given the status of a latter-day hero of the German resistance.[11] That this was something of a simplification the preceding pages have, perhaps, demonstrated; and before long a reaction set in which portrayed Rommel as a convinced adherent of Hitler until the military tide turned – an ambitious opportunist, devoid of any particular principle.[12] This, too, was an inadequate portrayal of the man.

Rommel's part in the anti-Hitler conspiracy was ambivalent. He has been accused of willing the end but standing back from the means – certainly, as a patriot, he ultimately wanted Hitler and his authority gone; but the tug of loyalty was strong. Conspiracy, Corporal Loistl remarked, was wholly alien to the Field Marshal's open character; and a personal servant tends to see clear. Rommel was indeed a patriot and his character was indeed honest and open. And he was a moral man – decent, chivalrous, devoted to the people and traditional institutions of his homeland, scrupulous in his own conduct, generous, merciful and fair-minded. He disliked the destruction which war brings; his temperament was far removed from that of the sort of commander who finds cause for exultation in the fear and suffering inflicted on an enemy. He seldom hated. It may be that a warrior people – and certainly in Rommel's time the Germans could, without hyperbole, be so described – finds it more natural than does a more pacific folk to fight without hatred. The latter need spurring to go to war at all, and thereafter generally need the stimulus of demonizing the enemy, whereas to Rommel's kind war was war, a frequent activity of man, a job to be done without need of hating.

And Rommel loved his country. He was always ready for self-sacrifice for Germany. He regarded the business of fighting for the Fatherland as a supreme challenge and a supreme privilege. And he was born to a generation of Germans who thought, with some reason, that their earlier sacrifices had been abused and nullified by the lack of patriotism, the lack of courage and the lack of energy of others. His was an embittered and disillusioned generation.

Then came Hitler. And Rommel was grateful to Hitler for, as he saw it, rescuing and restoring Germany – rescuing her from internal disorder and disunity and bankruptcy and ignominy; and restoring her to international stature, dignity, greatness. This personal devotion was to be clouded to the point of obliteration by Hitler's perverse strategic decisions, by his refusal to face facts, by what Rommel came to see as his complete mental instability; and by what he learned of Hitler's crimes, although he probably persuaded himself that these were mainly the works of others rather than the deliberate actions of the Führer. The blindness with which the robustly commonsensical Swabian Rommel managed to believe that Hitler was not personally as involved in evil as his subordinates continues to amaze, but it was a blindness he shared with millions of others. Until very near the end, Rommel felt the spark of an earlier devotion. It was, after all, consistent with his concept of loyalty, of disciplined duty. It was consistent with obedience to an oath. And in so feeling Rommel was one with all those decent, patriotic Germans, who were also grateful for, as they saw it, the order and the sense of dawn after darkness which Hitler had brought to the Fatherland. The extent of the evil which these virtuous souls were unwittingly serving was known only to few.

And in his dealings with Hitler Rommel was sincere. Others evaded the point or were cowed, or (rarely but honourably) realized early that no ordinary reasoning or ordinary moral standards could serve in such dealings. Rommel may have been – was – simple and 'taken in' in his dealings with Hitler, but he was frank, honest, straightforward. Naïvely, he gave Hitler credit for being prepared for rational discussion, rational decision; for sharing the ordinary moral principles of civilized man. Ultimately Rommel saw that this was a mental attitude impossible to maintain, but for long he did so.

Part of this derives from Rommel's lack of perception, a lack he shared with a majority of his countrymen although he, unlike them, knew Hitler well. But part of it derives from Hitler's own extraordinary character. The twentieth century would be inexplicable had Adolf Hitler been simply a figure of manifest and recognizable evil, or personally repellent, or vilely cruel in his private dealings, or contemptibly ridiculous. To some – very few – Germans he was one or all of these things; but to most he was different – gifted, of magnetic personality, with extraordinary memory, rapid understanding, wide grasp of affairs, almost supernatural foresight and insight, phenomenal willpower and wholly dedicated love of country. This – not the secret instigator of mass murder, not the deranged creature who dragged Germany down with him like some diabolic Samson in the temple – was the Hitler most of those touched by his influence perceived. This was Rommel's Hitler, until his own disillusion with the course of Hitler's war made it finally impossible to hold the picture any longer in the frame. The disposition of an efficient and disciplined people to obey orders and attend to specific tasks without especial concern for their ultimate purposes – this, perhaps, made

Germans more vulnerable than some to becoming accomplices, however detached, in the crimes of the regime; but it was only the idealized Hitler, for much of his life Rommel's Hitler, who can give sense to the fact that a great nation was propelled to disaster with remarkably few dissentients.

As a soldier, too, Rommel's star waxed and waned. The legend of the tactician who was 'no strategist' took root and grew sturdily, together with the reputation for logistic incomprehension or imprudence – legend and reputation which deserve analysis rather than generalization. Rommel's own writings[13] were reckoned by many to be illuminating, but were regarded by others (with some reason) as incomplete and self-justificatory; and Rommel, both during his lifetime and since his death, has been criticized for incessant carping at the shortcomings of others. The criticism is often fair, but can certainly not be applied only to Rommel among distinguished soldiers. The inflation of Rommel's name both by his enemies and for reasons of domestic propaganda has also been the target of many,[14] but his achievements can be left to tell their own tale, both good and (at times) indifferent.

Beyond dispute, Rommel was a master of manoeuvre on the battlefield and a leader of purest quality. Wherever he appeared he inspired. His speed of perception and decision, his energy of execution and his boldness of concept placed him among the great; and his military exploits have left a footprint in history as clear as that of Prince Rupert, to whom Montgomery once, in a somewhat uncharacteristic flight of imagery, likened him.[15] Certainly he erred badly at times. The first attack on Tobruk was hasty and ill-prepared, the 'dash to the wire' was prompted by a misreading of the situation, Alam Halfa offered only improbable chances of success (and was called off early), Medenine was a disaster. But the victories, generally with the dice loaded against him, display a very recognizable quality of command, a quintessential 'Rommel'. The two conquests of Cyrenaica were marked by it. Gazala was marked by it. Kasserine was marked by it. Even the long, hopeless retreat to Tunisia was marked by it, just as Cosna and Matajur had been. Whether in the saddle in France in 1914, on his feet racing ahead of his men in the mountains of Rumania or the Italian Alps, leading the *Gespensterdivision* headlong towards the Channel in 1940 or sweeping along with him the tanks of the Panzerarmee in the African desert Rommel was, in Speidel's words, '*Unser Rommel – immer derselbe Rommel.*'*[16] He led brilliantly and he led from the front.

But he was more than a tactical commander of bravery and genius. He was reflective. He evolved from his own experience and observation soldierly lessons which he committed to paper and from which all learned and continue to learn. Wherever he went, as has been remarked, he taught: and he teaches still. Rommel was not only a master practitioner; he deduced theory from practice and the military art benefited therefrom.

At the higher level Rommel, so often criticized for having insufficiently wide or profound a vision to be rated a strategist, saw major issues extremely clearly. Certainly in the handling of troops he gave primacy to the tactical battle – he excelled in it, he realized that the most elegant of strategic plans founder unless battles are won by troops at the sharp end of conflict; strategic opportunity, as the saying runs, follows tactical success. But Rommel's sure feeling for the battlefield, his famous *Fingerspitzengefuhl*, his scent for victory, were by no means incompatible with a broader understanding of manoeuvre by larger forces than those he was enabled to handle. Everything he achieved and everything he wrote persuades that in, for instance, Russia he would have been an operational master with as sure a touch as any, and probably superior to most. And his views on grand strategy and the conduct of war, although necessarily speculative and untried, were imaginative and wide-minded. That he believed in the possibilities of 'Plan Orient' hardly stands to his credit; but he shared the belief, at least occasionally, with his Führer, with OKH and with the British Chiefs of Staff.

For Rommel in most things was a realist. He was a thrifty, industrious Swabian, shrewd and practical. He was, latterly, denounced as a pessimist, but he saw the essential elements of a military situation clear and without illusion. He was not always right – no commander is – but his errors seldom arose from self-deception or a refusal to face unpalatable facts. When, as in Normandy, his reason told him that manoeuvre in any major or significant sense was impossible, that Allied strength, especially in the air, denied all possibilities save small-scale, grinding, tactical battles against the odds, he made correct deductions and refused to disguise them. When he saw, in North Africa, that the Panzerarmee was facing such material superiority that

* 'Always our Rommel, always the same.'

it must retreat, must refuse battle, he said so. And when his sense told him that the war was lost he said that, too.

The consequential criticism of pessimism must not be allowed to blur the fact that Rommel's was timely, justified pessimism. Nor should it obscure the other essential fact that, far from lacking a sanguine, adventurous temperament Rommel was so endowed almost to excess. He was a natural, albeit a reasoning, taker of risks in war. He reckoned that war is so uncertain a business, so dependent on a concatenation of unpredictable chances, that boldness, a touch of optimism and above all speed can and generally should do better than attempts at exact calculation. Rommel did not believe in deferring battle until the odds assured victory. Had he done so there would, for better or worse, have been no North African campaign.

Montgomery claimed that his own achievements derived from the fact that he never fought an unsuccessful battle, and for Montgomery this was both an accurate statement and a wise policy. It was, however, a policy only available to one with both time and resources. Rommel, more often than not, had insufficient of either. Nor was he ever in a position to wait until his situation and the odds improved. He fought at a numerical disadvantage again and again, and his exploits can only be measured against that fact. He relied on skill to offset quantitative inferiority. The bitter exclamation, already quoted, comes always to the mind: 'If one considers what the German Marshal could have achieved with the superiority enjoyed by his opponents . . .'[17] War is usually an option of difficulties. Again and again Rommel could choose inactivity or take a calculated risk. He believed that inactivity is seldom forgiven a general by fate.

Of course Rommel, ultimately, was beaten. He lost. But, although what must matter in war is to win, that truism cannot provide the sole criterion for judgement of military talent. War may be considered as a business, open to audit, but its conduct is also an art. Ultimately Napoleon was beaten. So was Montrose. So was Lee. Few could deny their genius. With all his imperfections, as a leader of men in battle Erwin Rommel stands in their company.

APPENDIX

Erwin Rommel: Chronology

15 NOVEMBER 1891: Erwin Johannes Eugen Rommel born at Heidenheim, Wurttemberg

19 JULY 1910: Joins 124th Wurttemberg Infantry Regiment as cadet. (Infanterieregiment König Wilhelm I, 6 Wurttembergische, Nr 124)

MARCH 1911: To cadet school – Königliche Kriegschule, Danzig

JANUARY 1912: Commissioned Lieutenant. To 124th Regiment

1 MARCH 1914–31 JULY 1914: Attached 49th Field Artillery Regiment

1 AUGUST 1914: Regimental duty, 124th Regiment

21 AUGUST 1914–SEPTEMBER 1914: In action at Bleid, in the Meuse valley, on the Verdun front. Platoon Commander

SEPTEMBER 1914: Battalion Adjutant, 2nd Battalion, 124th Regiment, south-west of Verdun

24 SEPTEMBER 1914: Wounded in thigh. Iron Cross Second Class. Hospital

JANUARY 1915: Rejoins 124th Regiment. Regimental duty in the forest of the Argonne

29 JANUARY 1915: Raid on the 'Central Position' in the Argonne. Iron Cross, First Class

JUNE 1915: Major attack in the Argonne

JULY 1915: Wounded in leg

SEPTEMBER 1915: To 1st Battalion, 124th Regiment. Company Commander

SEPTEMBER 1915: Promoted Oberleutnant

OCTOBER 1915: To Königliche Wurttemberg Gebirgsbataillon. Company Commander

29 DECEMBER 1915–OCTOBER 1916: Regimental duty with the Gebirgsbataillon, Hilsen Ridge, Vosges

OCTOBER 1916: Marries Lucy Mollin, in Danzig

NOVEMBER 1916: To the Rumanian front

6 DECEMBER 1916: Bucharest falls to Germans

7 JANUARY 1917: Attack on Gagesti
EARLY 1917–JULY 1917: Gebirgsbataillon to France, Hilsen Ridge
AUGUST 1917–OCTOBER 1917: Gebirgsbataillon returns to Rumania. The Mount Cosna front
AUGUST 1917: Wounded in shoulder
19 AUGUST 1917: Mount Cosna taken
OCTOBER 1917: Gebirgsbataillon to Italy. Isonzo front
24 OCTOBER 1917: German attack, Mount Matajur
26 OCTOBER 1917: Mount Matajur taken
OCTOBER 1917–NOVEMBER 1917: Advance towards the Piave
10 NOVEMBER 1917: Longarone taken
18 DECEMBER 1917: Awarded *Pour le Mérite*
11 JANUARY 1918–20 DECEMBER 1918: To Staff LXIV Corps, Western Front. Promoted Hauptmann
11 NOVEMBER 1918: Armistice, Western Front
21 DECEMBER 1918: To 124th Regiment. Regimental duty
JUNE 1919–JANUARY 1921: Internal security duties, Germany
JANUARY 1921: To Reichswehr Infantry Regiment 13. Company Commander
1924: Commands machine-gun company
DECEMBER 1928: Manfred Rommel born
SEPTEMBER 1929–SEPTEMBER 1933: Instructor, Infantry School, Dresden
APRIL 1932: Promoted Major
30 JANUARY 1933: Adolf Hitler becomes Chancellor of Germany
1 OCTOBER 1933–14 OCTOBER 1935: Commands 3rd (Jäger) Battalion, 17th Infantry Regiment, Goslar. Promoted Oberst-Leutnant
30 JUNE 1934: 'Night of the long knives'
2 AUGUST 1934: Death of Hindenburg. Oath taken by the army to Adolf Hitler as 'Führer of the German Reich and people, Supreme Commander of the Armed Forces'
SEPTEMBER 1934: Rommel meets Hitler for the first time
MARCH 1935: Conscription restored to German Army
15 OCTOBER 1935–9 NOVEMBER 1938: Instructor, Infantry School, Potsdam
MARCH 1936: Demilitarized Rhineland occupied by German troops
SUMMER 1936: Attached to Führer's military escort for Party rally, Nuremberg
FEBRUARY 1937–1938: Nominated War Ministry Liaison Officer to Hitler Youth organization
1937: *Infanterie greift an* published
MARCH 1938: Austrian *Anschluss*
SEPTEMBER 1938: Czechoslovak Sudeten provinces ceded to Germany
OCTOBER 1938: Rommel nominated to command Hitler's field headquarters for occupation of Sudetenland

10 NOVEMBER 1938–22 AUGUST 1939: Commandant Kriegsschule, Wiener Neustadt. Promoted Oberst
10 MARCH 1939: Hitler's ultimatum to Czechoslovakia
15 MARCH 1939: Hitler enters Prague. Rommel commands his escort
23 AUGUST 1939: Rommel commands *Führerhauptquartier* on mobilization. Promoted Generalmajor (effective from June)
1–30 SEPTEMBER 1939: Polish campaign
5 OCTOBER 1939: Victory parade, Warsaw
15 FEBRUARY 1940–14 FEBRUARY 1941: Commands 7th Panzer Division
9 APRIL 1940: German invasion of Norway and Denmark
10 MAY 1940: Operation *Sichelschnitt*, offensive on the Western Front, begins
13 MAY 1940: 7th Panzer Division cross Meuse
15 MAY 1940: Rommel awarded Clasp to Iron Cross (Second Class)
16 MAY 1940: 7th Panzer Division advance through Maginot Line extension
17 MAY 1940: Enter Landrecies
21 MAY 1940: Advance to Arras. Defeat British counterattack
26 MAY 1940: Rommel awarded Knight's Cross to the Iron Cross
27 MAY 1940: Advance across La Bassée canal to Lille
3 JUNE 1940: 7th Panzer Division cross Somme canal
9 JUNE 1940: Advance to Seine Valley
10 JUNE 1940: Occupation of St Valèry-en-Caux
19 JUNE 1940: Cherbourg surrenders
22 JUNE 1940: Armistice concluded between Germany and France
JANUARY 1941: Rommel promoted Generalleutnant
7 FEBRUARY 1941: Tenth Italian Army surrender to British at Beda Fomm in North Africa
FEBRUARY 1941: Rommel appointed Commander German troops in Libya
12 FEBRUARY 1941: Lands at Tripoli
MARCH 1941–APRIL 1941: First Cyrenaica offensive. First attack on Tobruk
15 MAY 1941: British attack on Egyptian frontier
15 JUNE 1941: Second British attack on Egyptian frontier. Operation Battleaxe
16 JUNE 1941: German counterattack in frontier area
22 JUNE 1941: German invasion of Russia. Operation Barbarossa
15 AUGUST 1941: German forces constituted as Panzer Gruppe Afrika
SEPTEMBER 1941: German raid on Egyptian frontier area. Operation *Sommernachtstraum*
18 NOVEMBER 1941: British offensive into Libya. Operation Crusader. Rommel's withdrawal through Cyrenaica
7 DECEMBER 1941: Japanese attack United States fleet and British possessions in South-East Asia. Germany declares war on the United States

21 JANUARY 1942: Rommel awarded Swords to the Oakleaves of the Knight's Cross
21–29 JANUARY 1942: Second Cyrenaica offensive
22 JANUARY 1942: *Panzer Gruppe Afrika* becomes *Panzerarmee Afrika*
30 JANUARY 1942: Rommel promoted Generaloberst
27 MAY 1942–20 JUNE 1942: German Gazala offensive
21 JUNE 1942: Tobruk taken. Rommel promoted Field Marshal
21 JUNE 1942–1 JULY 1942: Advance into Egypt
1–26 JULY 1942: First battles on Alamein line
15 AUGUST 1942: Montgomery assumes command British Eighth Army
30 AUGUST 1942–2 SEPTEMBER 1942: Battle of Alam Halfa
19 SEPTEMBER 1942: Stumme assumes temporary command of *Panzerarmee Afrika*. Rommel to Germany on sick leave
30 SEPTEMBER 1942: Reception for Rommel in Berlin Sportspalast
23 OCTOBER 1942: Battle of Alamein begins
25 OCTOBER 1942: Rommel recalled to Africa. Stumme found dead
3 NOVEMBER 1942: Rommel receives Hitler's order to stand fast on existing positions at Alamein
4 NOVEMBER 1942: Rommel orders withdrawal of *Panzerarmee*
8 NOVEMBER 1942: British and American forces land in French North Africa. Operation Torch
9 NOVEMBER 1942: Soviet counter-offensive opens on Stalingrad front
10 NOVEMBER 1942: German forces begin reinforcement of Tunisia
22 JANUARY 1943: Evacuation of Tripoli by *Panzerarmee*
26 JANUARY 1943: *Panzerarmee* headquarters established in Tunisia
2 FEBRUARY 1943: Final surrender of German Sixth Army at Stalingrad
4 FEBRUARY 1943: British victory parade in Tripoli
14 FEBRUARY 1943: German–Italian operations in Southern Tunisia. Operations *Frühlingswind* and *Morgenlust*
19–21 FEBRUARY 1943: Battle of the Kasserine Pass
23 FEBRUARY 1943: Rommel appointed Commander-in-Chief, Army Group Africa
6 MARCH 1943: Battle of Medenine. Operation Capri
9 MARCH 1943: Rommel leaves Africa
24 MARCH 1943: German *Kriegsmarine* call off battle of Atlantic
4 JULY 1943: German offensive at Kursk begins. Operation Citadel
10 JULY 1943: Anglo–American invasion of Sicily. Operation Husky
15 JULY 1943: Rommel appointed Commander-in-Chief Army Group B
25–26 JULY 1943: Rommel on reconnaissance to Salonika
25 JULY 1943: Mussolini deposed by Fascist Grand Council
30 JULY 1943: German troops begin move over Alpine passes into north Italy. Operation Alaric
15 AUGUST 1943: Headquarters Army Group B established in north Italy
16 AUGUST 1943: German–Italian withdrawal from Sicily complete

3 SEPTEMBER 1943: Anglo–American invasion of Italy
8 SEPTEMBER 1943: Armistice announced between Italy and the Anglo–American powers
9 SEPTEMBER 1943: Anglo–American forces land at Salerno. Army Group B undertakes disarming of Italian forces: Operation *Achse*
21 NOVEMBER 1943: Rommel leaves Italy. Army Group B headquarters moved to France
30 NOVEMBER 1943: Rommel starts extended inspection of western coastal defences
15 JANUARY 1944: Army Group B assumes responsibility for Atlantic and Channel coasts north of the Loire
9 MARCH 1944: Army Group B headquarters established at La Roche Guyon
6 JUNE 1944: Anglo–American invasion of France
6 JUNE 1944–15 AUGUST 1944: Battle of Normandy
23 JUNE 1944: Major Red Army offensive in central sector begins
29 JUNE 1944: Rommel's last conference with Hitler at Berchtesgaden
16 JULY 1944: Rommel signs 'ultimatum' report on seriousness of situation in the west
17 JULY 1944: Rommel wounded in air attack near Vimoutiers. In hospital or at home on sick leave until October
18 JULY 1944: British Second Army attacks near Caen
20 JULY 1944: Attempt on Hitler's life at Rastenburg in East Prussia
AUGUST 1944: Normandy front collapses. German retreat
7 OCTOBER 1944: Rommel ordered to report to Berlin
14 OCTOBER 1944: Generals Burgdorf and Maisel visit Rommel at home at Herrlingen. Rommel's death
18 OCTOBER 1944: State funeral at Ulm

Bibliography

Balfour, Michael and Frisby, Julian, *Helmuth von Moltke* (Macmillan, 1972)

Barnett, Correlli, *The Desert Generals* (William Kimber, 1960)

Barnett, Correlli (ed.) *Hitler's Generals* (Weidenfeld & Nicolson, 1989)

Barnett, Correlli, *Engage the Enemy More Closely* (Hodder & Stoughton, 1991)

Bayerlein, Fritz, 'El Alamein', in *The Fatal Decisions* (Michael Joseph, 1956)

Behrendt, Hans Otto, *Rommels Kenntnis vom Feind in Afrika Feldzug* (Verlag Rombach, Freiburg, 1980)

Bennett, Ralph, *ULTRA in the West* (Hutchinson, 1979)

Blumenson, Martin, *Rommel's Last Victory* (Allen & Unwin, 1968)

Bond, Brian, *France and Belgium 1939–1940* (Davis-Poynter, 1975)

Brooks, Stephen (ed.), *Montgomery and the Eighth Army* (Army Records Society and Bodley Head, 1991)

Bullock, Alan, *Hitler and Stalin* (HarperCollins, 1991)

Carsten, F.L., *The Reichswehr and Politics* (Clarendon Press, 1966)

Carver, Michael, *El Alamein* (Batsford, 1962)

Carver, Michael, *Tobruk* (Batsford, 1962)

Carver, Michael, *Dilemmas of the Desert War* (Batsford, 1986)

Chalfont, Alun, *Montgomery of Alamein* (Weidenfeld & Nicolson, 1976)

Cooper, Matthew, *The German Army, 1933–45* (Macdonald & Janes, 1978)

Cox, Richard, *Operation Sealion* (Thornton Cox, 1974)

Craig, Gordon, *The Politics of the Prussian Army* (Oxford, 1955)

Craig, Gordon, *The Prussian-German Army 1933–45* (Oxford, 1964)

Demeter, Karl, *The German Officer Corps in Society and State* (Weidenfeld & Nicolson. 1965)

D'Este, Carlo, *Decision in Normandy* (Collins, 1983)

Douglas-Home, Charles, *Rommel*

(Weidenfeld & Nicolson, 1974)
Eisenhower, David, *Eisenhower at War* (Collins, 1986)
Engel, Major, *Heeresadjutant bei Hitler* (Deutsche Verlags Anstalt, Stuttgart, 1974)
Ensor, R.C.K., *England, 1870–1914* (Clarendon Press, 1936)
von Esebeck, Hans Gert, *Afrikanische Schicksaljahre* (Limes Verlags, Wiesbaden, 1949)
Fest, J.C., *Hitler* (Weidenfeld & Nicolson, 1974)
Galante, Pierre (with Eugene Silanoff), *Operation Valkyrie* (Harper & Row, 1981)
Geyr von Schweppenburg, *The Critical Years* (Allen Wingate, 1952)
Gilbert, Martin, *The Holocaust* (Collins, 1986)
Gisevius, Hans Berndt, *Bis zum bittern Ende* (Fretz v. Wasmuth Verlag, 1946)
Goebbels, Josef (trans. Taylor), *Diaries 1939–41* (Hamish Hamilton, 1982)
Goebbels, Josef (trans. Lochner), *Diaries 1942–3* (Hamish Hamilton, 1948)
Gorlitz, Walter, *The German General Staff* (Hollis & Carter, 1953)
Guderian, Heinz, *Erinnerungen eines Soldaten* (Vowinckel, Heidelberg, 1951)
Halder, Franz, *Kriegstagebuch* (Kohlhammer, Stuttgart, 1963–4)
Hamilton, Nigel, *Monty* (Hamish Hamilton, 3 vols, 1981–6)
Hastings, Max, *Overlord* (Michael Joseph, 1984)
Heckmann, Wolf, *Rommels Krieg in Afrika* (Bergisch Gladbach, 1976)
Hildebrandt, K. (trans. Falla), *The Third Reich* (Allen & Unwin, 1984)
Hinsley, F.H., *British Intelligence in the Second World War* (vol. 2, HMSO, 1981)
Hoffmann, Peter (trans. Barry), *The History of the German Resistance* (Macdonald & Janes, 1977)
Horne, Alistair, *To lose a Battle* (Macmillan, 1969)
Hunt, David, *A Don at War* (William Kimber, 1966)
Irving, David, *Hitler's War* (Hodder & Stoughton, 1977)
Irving, David, *The Trail of the Fox* (Macmillan, 1977)
Jackson, William, *The North African Campaign 1940–43* (Batsford, 1975)
Jackson, William, *Overlord, Normandy 1944* (Davis-Poynter, 1978)
Kahn, David, *Hitler's Spies* (Hodder & Stoughton, 1978)
Keegan, John, *The Mask of Command* (Jonathan Cape, 1987)
Kesselring, Albert, *Soldat bis zum letzten Tag* (Athenaeum Verlag, Bonn, 1953)
Koch, Lutz, *Erwin Rommel: Wandlung eines grossen Soldaten* (Verlag Walter Gebauer, Stuttgart, 1950)
Koch, Lutz, *Erwin Rommel und der Deutsche Widerstand gegen Hitler* (Vierteljahrshefte fur Zeitgeschichte, Munich, 1953)
Lamb, Richard, *Montgomery in Europe* (Buchan & Enright, 1983)

Lamb, Richard, *The Ghosts of Peace* (Michael Russell, 1987)

Lewin, Ronald, *Rommel as Military Commander* (Batsford, 1968)

Lewin, Ronald, *The Life and Death of the Afrika Korps* (Batsford, 1977)

Lewin, Ronald, *ULTRA Goes to War* (Hutchinson, 1978)

Lindsay, Donald, *Forgotten General* (Michael Russell, 1987)

Liddell Hart, B.H. (ed.), *The Rommel Papers* (Collins, 1953)

Liddell Hart, B.H., *The Second World War* (Cassell, 1970)

von Luck, Hans, *Panzer Commander* (Praeger, New York, 1989)

Ludendorff, Erich, *The General Staff and its Problems* (Hutchinson, 1920)

Mackee, Alexander, *Caen* (Souvenir Press, 1964)

Macksey, Kenneth, *Rommel, Battles and Campaigns* (Arms & Armour Press, 1979)

Macksey, Kenneth, *Guderian* (Macdonald & Janes, 1975)

von Manstein, Erich, *Lost Victories* (Methuen, 1958)

von Manteuffel, Hasso, *Die 7 Panzer Division in Zweiten Weltkrieg* (Cologne, 1965)

von Mellenthin, F.W., *Panzer Battles* (Cassell, 1955)

Montgomery of Alamein, *Memoirs* (Collins, 1958)

Mordal, Jacques, *Rommel* (Historama, Paris, 1973)

O'Neill, Robert, *The German Army and the Nazi Party* (Cassell, 1966)

Overy, Richard, *The Road to War* (Macmillan, 1990)

Pitt, Barrie, *The Crucible of War* (Jonathan Cape, 1980)

Prittie, Terence, *Germans Against Hitler* (Hutchinson, 1964)

Reuth, Ralf, *Des Führers General* (Piper, Munich, 1987)

Rommel, Erwin, *Infanterie greift an* (Voggenreiter, Potsdam, 1937. Translated as *Infantry Attacks*, *Infantry Journal*, Washington D.C., 1944, and published in that translation by Greenhill Books, 1990)

Rommel, Erwin, *Krieg ohne Hass* (translated excerpts in Liddell Hart, *The Rommel Papers*, op. cit.)

Ruge, Friedrich, 'The Invasion of Normandy' in *Decisive Battles of World War II*, ed. Jacobsen and Rohwer (Putnam, 1965)

Ruge, Friedrich, *Rommel in Normandy* (Presidio Press, 1979)

Saurel, Louis, *Rommel* (Editions Rouff, Paris, 1967)

Schmidt, H.W., *With Rommel in the Desert* (Harrap, 1951)

Seaton, Albert, *The Russo-German War 1941–45* (Arthur Barker, 1971)

Seaton, Albert, *The German Army 1939–45* (Weidenfeld & Nicolson, 1982)

von Senger und Etterlin, Frido, *Neither Fear Nor Hope* (Macdonald, 1963)

Speer, Albert, *Inside the Third Reich* (Weidenfeld & Nicolson, 1970)

Speidel, Hans, *Invasion 1944* (Rainer Wunderlich Verlag, Tübingen/Stuttgart, 1949)

Stahlberg, Alexander, *Bounden Duty* (Brasseys, 1990)

Strawson, John, *The Battle for North Africa* (Batsford, 1969)

Strawson, John, *Alamein* (Dent, 1981)

Sykes, Christopher, *Troubled Loyalty* (Collins, 1968)
Taylor, A.J.P., *The Origins of the Second World War* (Hamish Hamilton, 1961)
Terraine, John, *Right of the Line* (Hodder & Stoughton, 1985)
Trevor-Roper, H. (ed.), *Hitler's War Directives* (Sidgwick & Jackson, 1964)
Warlimont, Walter, *Inside Hitler's Headquarters* (Weidenfeld & Nicolson, 1964)
Westphal, Siegfried (commentary), *The Fatal Decisions* (Michael Joseph, 1956)
Westphal, Siegfried, *The German Army in the West* (Cassell, 1951)
Wheeler-Bennett, John, *The Nemesis of Power* (Macmillan, 1961)
Wheeler-Bennett, John, *Hindenburg – The Wooden Titan* (Macmillan, 1936)
Wilmot, Chester, *The Struggle for Europe* (Collins, 1952)
Wiskemann, Elizabeth (ed.), *The Anatomy of the SS State* (Collins, 1968)
Young, Desmond, *Rommel* (Collins, 1950)

Notes

The following abbreviations are used, with document or file reference numbers:
BAMA: Bundesarchiv Militararchiv, Freiburg
EPM: EP Microfilm Ltd, Wakefield, Yorkshire, England (followed by reel number). These documents, including records of interviews, were assembled during research for *The Trail of the Fox* by David Irving, and placed on microfilm.
IWM: Imperial War Museum, London
IZM: Institut fur Zeitgeschichte, Munich
NAW: National Archives, Washington DC

CHAPTER I

1 Winston S. Churchill, *Marlborough: His Life and Times* (Harrap, 1936)
2 G.F.R. Henderson, *Stonewall Jackson* (Longmans Green, 1911)
3 Churchill, op. cit.
4 Clarendon, *History of the Rebellion* (Clarendon Press, 1888)
5 Exhaustively discussed in K. Demeter, *The German Officer Corps in Society and State* (Weidenfeld & Nicolson, 1965)
6 In the author's possession
7 Field Marshal Count Waldersee. G.A. Craig, *The Politics of the Prussian Army* (Clarendon Press, 1955)
8 Ludendorff, *The General Staff and its Problems* (Hutchinson, 1920)
9 The Schlieffen Plan, with its emphasis on an immediate offensive in the west, marked a departure from earlier thinking. The elder Moltke had believed that the Rhine barrier and the territorial gains of 1870 enabled Germany to stand on the defensive in the west (that, indeed, had been the object) while undertaking an offensive in the east; Moltke, who understood politics, knew that a

western offensive demanded either Belgian cooperation or that country's violation, and constituted a high political risk strategy. From 1892 Schlieffen had changed all that.

CHAPTER 2

1 Under Schlieffen's concept the right wing, advancing through Belgium, had been *seven times* stronger than the left, and had been timed to reach Abbeville and brush the Channel at the mouth of the Somme by the thirty-first day after mobilisation – 31 August. German plans, like those of everyone else, were predicated on a short war.
2 Quotations throughout taken from Rommel's personal accounts, NAW T84 277–278 and IWM Misc. 14. These formed the basis for his book *Infanterie greift an* (Voggenreiter, 1937), of which English-language editions have been produced (second edition, Greenhill Books, 1990). I have used and translated Rommel's original text (which he edited and modified).

CHAPTER 3

1 Theodor Werner, quoted in David Irving, *The Trail of the Fox* (Macmillan, 1977)
2 Manfred Rommel

CHAPTER 4

1 Basil Liddell Hart (ed.), *The Rommel Papers* (Collins, 1953)

CHAPTER 5

1 For the exchanges and dramas of this revolutionary time and the military attitudes adopted, see F.L. Carsten, *The Reichswehr and Politics* (Clarendon Press, 1966)
2 Manfred Rommel
3 Sir John Wheeler-Bennett, *The Nemesis of Power* (Macmillan, 1961)
4 Carsten, op. cit.
5 Wheeler-Bennett, op. cit.
6 Hans von Luck, *Panzer Commander* (Praeger, New York, 1989)
7 Desmond Young, *Rommel* (Collins, 1950)

CHAPTER 6

1 Manfred Rommel. Lucy supported the DDP, the *Demokratische Partei*, which had special roots in Wurttemberg. The conservative nationalists Rommel distrusted were the DNVP, the *Deutschenationale Volkspartei*.
2 IZM ED 100/186
3 See, for instance, Albert Speer, *Inside the Third Reich* (Weidenfeld & Nicolson, 1970)
4 Quoted from L. and R. Heston, *The Medical Casebook of Adolf Hitler* (William Kimber, 1979),

by Hugh L'Etang, Leeds University, 1991

5 Carsten, op. cit.

6 Walter Gorlitz, *The German General Staff* (Hollis & Carter, 1953)

7 The late Sir John Wheeler-Bennett, shortly before he died, told the author that he had recently come on new evidence (not specifically described) which showed that at least some of the leaders of the Reichswehr were involved in the business of 30 June 1934 'up to their necks'.

8 Manfred Rommel

9 Author's recollection

10 J. Wheeler-Bennett, *Hindenburg, the Wooden Titan* (Macmillan, 1967)

11 Hans Buckheim (ed. Elizabeth Wiskemann), *Anatomy of the SS State* (Collins, 1968)

12 Young, op. cit.

13 Manfred Rommel, interview, EPM 3

14 Ibid.

15 Frau Kirchheim, interview, EPM 3

16 Manfred Rommel

CHAPTER 7

1 Demeter, op. cit.

2 Irving, op. cit.

3 Ibid.

4 Young, op. cit.

5 Wheeler-Bennett, *The Nemesis of Power*, op. cit.; Gorlitz, op. cit.

6 Heinz Guderian, *Erinnerungen eines Soldaten* (Vowinckel, Heidelberg, 1951)

7 Wheeler-Bennett, *The Nemesis of Power*, op. cit.

8 Alexander Stahlberg, *Die Verdammte Pflicht* (Verlag Ullstein GmBH, Berlin/Frankfurt-am-Main, 1987) (translated as *Bounden Duty*, Brasseys, 1990)

9 Robert O'Neill, 'Fritsch, Beck and the Führer', in *Hitler's Generals*, ed. Correlli Barnett (Weidenfeld & Nicolson, 1989)

10 IZM ED 100/186

11 H. Krausnick (ed. Wiskemann), *The Anatomy of the SS State* (Collins, 1968)

12 K. Hildebrandt, *The Third Reich* (Allen & Unwin, 1984)

13 Carsten, op. cit.

14 Wheeler-Bennett, *The Nemesis of Power*, op. cit.

15 Ibid.

16 Robert O'Neill, *The German Army and the Nazi Party 1933–1939* (Cassell, 1966)

17 Manfred Rommel, interview, EPM 3

18 Manfred Rommel

19 Irving, op. cit.

20 Manfred Rommel, interview, EPM 3

21 Wheeler-Bennett, *The Nemesis of Power*, op. cit.

22 Anita Prazmowska, 'Eastern Europe between the Wars' (*History Today*, October 1990)

23 See for instance Field Marshal Erich von Manstein, *Lost Victories* (Methuen, 1958)

24 Brian Bond, 'Brauchitsch', in Barnett, op. cit.

25 von Manstein, op. cit.

26 NAW T77/858

CHAPTER 8

1 See for instance Donald Lindsay, *Forgotten General* (Michael Russell, 1987)
2 But purely author's speculation
3 NAW TT 77/858. KTB Führerbegleitbataillon
4 Ibid.
5 von Manstein, op. cit.
6 Quoted in Krausnick, op. cit.
7 David Irving, *Hitler's War* (Hodder & Stoughton, 1977)
8 Manfred Rommel. See also Gorlitz, op. cit.
9 Engel, *Heeres Adjutant bei Hitler* (Deutsche Verlag, Anstadt, Stuttgart, 1974)
10 Manfred Rommel, interview, EPM 3. On this reel there are copious recorded interviews as well as personal comments and reminiscences.
11 Hildebrandt, op. cit.
12 von Manstein, op. cit.
13 Engel, op. cit.
14 Quoted in Liddell Hart, op. cit. Some of Rommel's writings in these papers (which were combined, editorially, with extracts from his letters) were published in Germany as *Krieg ohne Hass* (Heidenheim, 1955)
15 Liddell Hart, op. cit.
16 See, for instance, Matthew Cooper, *The German Army 1933–1945* (Macdonald & Janes, 1978)
17 Liddell Hart, op. cit.
18 Irving, *The Trail of the Fox*, op. cit.

CHAPTER 9

1 Leutnant Braun, 25 Panzer Regiment, NAW T 84/277
2 Hasso von Manteuffel, *Die 7 Panzer Division im Zweiten Weltkrieg* (Cologne, 1965)
3 von Luck, op. cit.
4 Braun, op. cit.
5 cf. David Fraser, *Alanbrooke* (Collins, 1982)
6 Liddell Hart, op. cit.
7 BAMA N 117/1. Hanke, whose civilian rank was *Staatsekretar*, incurred Rommel's displeasure by boasting that, in terms of protocol, his rank was higher than that of the divisional commander; allegedly Rommel withdrew a particular commendation because of this impertinence (see Irving, *The Trail of the Fox*, op. cit.). Hanke, nevertheless, was an extremely brave and competent subordinate in battle. Later he was appointed Gauleiter of Silesia, fleeing from Breslau when the Red Army approached it in 1945.
8 von Manteuffel, op. cit.
9 BAMA N 117/1
10 von Manteuffel, op. cit.
11 Liddell Hart, op. cit.

CHAPTER 10

1 Liddell Hart, op. cit.
2 Ibid.
3 'Artillerie nach Vorn', NAW T 84/277
4 IZM ED 100/175

5 J.C. Fest, *Hitler* (Weidenfeld & Nicolson, 1974)
6 von Manteuffel, op. cit.
7 Hildebrandt, op. cit.
8 Wheeler-Bennett, *The Nemesis of Power*, op. cit.
9 Quoted in Michael Balfour and Julian Frisby, *Helmuth von Moltke* (Macmillan, 1972)
10 The *Flachenmarsch* concept, according to Rommel's description and the dimensions of 7th Panzer Division, produced a rectangle. Clearly a variable was the formation adopted within and by particular units – certainly Rommel had Panzer battalions on both flanks as well as leading, and he often referred to his orders to them to fire on the move with outward traversed turrets. The inevitable difficulties for wheeled vehicles meant a good deal of bunching and straggling at bottlenecks; and although the length of the whole division on *Flachenmarsch* (twelve miles), with a frontage of two thousand yards appears to give a large measure of dispersion (about 150 yards on average between vehicles), 'average' would have been rarely attained. Rommel tended, anyway, to encourage tanks to be handled in fairly close order.
11 von Luck, op. cit.
12 Liddell Hart, op. cit.
13 BAMA N 117/1
14 IZM ED 100/186
15 BAMA N 117/6

CHAPTER 11

1 Richard Cox (ed.), *Operation Sealion* (Thornton Cox, 1974)
2 Hildebrandt, op. cit.
3 See Enno von Rintelen, NAW MS B493. General von Rintelen had been in Rome from October 1936. He was responsible to OKW.
4 IWM AL 1349/11
5 Hans-Otto Behrendt, *Rommels Kenntnis vom Feind in Afrikafeldzug* (Verlag Rombach, Freiburg, 1980)
6 Ibid.
7 Lutz Koch, *Erwin Rommel* (Verlag Walter Gebauer, Stuttgart, 1950)
8 Behrendt, op. cit.
9 Ibid.
10 Liddell Hart, op. cit.
11 von Mellenthin, *Panzer Battles 1939–1945* (Cassell, 1955)
12 See, for instance, 'Rommel wie er wirklich war', *Deutsche Soldatenzeitung*, September 1952
13 Hans Karl von Esebeck, *Afrikanische Schichsaljahre* (Limes Verlag, Wiesbaden, 1950)
14 Ibid.
15 See for instance Graf von Schwerin, interview, EPM 3

CHAPTER 12

1 Behrendt, op. cit.
2 Ibid.
3 Liddell Hart, op. cit.
4 Behrendt, op. cit.
5 Martin Middlebrook, 'Paulus', in Barnett, op. cit.

6 To the author
7 von Schwerin, interview, EPM 3
8 Behrendt, op. cit.
9 Ibid.
10 IWM AL 831
11 von Steinitz, BAMA N/117/12, February 1942
12 Behrendt, op. cit.
13 Nigel Hamilton, *Monty, Master of the Battlefield* (Hamish Hamilton, 1983)
14 Behrendt, op. cit.
15 F.H. Hinsley, *British Intelligence in the Second World War*, Vol. II (HMSO, 1981)
16 Behrendt, op. cit.; von Schwerin, interview, EPM 3
17 See extended description in von Luck, op. cit.
18 Irving, *The Trail of the Fox*, op. cit.
19 von Schwerin, interview, EPM 3
20 IZM ED 100/175
21 Streich, EPM 3
22 'Rommel wie er wirklich war', op. cit.
23 Richard Bentley, *The Private Journal of F.S. Larpent* (London, 1853)
24 Robin Edmonds, personal communication
25 Behrendt, op. cit.
26 See, for instance, Guy Sajer, *Le Soldat oublié* (Robert Laffont, Paris, 1967)
27 See Martin Gilbert, *The Holocaust* (Collins, 1986)
28 Ibid.
29 Krausnick, op. cit.
30 Balfour and Frisby, op. cit.

CHAPTER 13

1 Behrendt, op. cit.
2 General-Oberst Halder, *Kriegstagebuch* (W. Kohlhammer Verlag, Stuttgart, 1964)
3 von Taysen, interview, EPM 3
4 IWM 14/542
5 John Terraine, *Right of the Line* (Hodder & Stoughton, 1985)
6 Behrendt, op. cit.
7 21 Pz KTB, IWM GMDS 18572/2
8 Rommel diaries, EPM 9. (These diaries were kept for Rommel at his *Kampfstaffel*, or tactical headquarters, and recorded hourly occurrences.)
9 von Mellenthin, op. cit.
10 von Esebeck, op. cit.
11 IWM GMDS 18572/2
12 IWM AL 897
13 von Mellenthin, op. cit.
14 IWM AL 897
15 Lieutenant-General Fritz Bayerlein, in Liddell Hart, op. cit.
16 W.G.F. Jackson, *The North Africa Campaign 1940–1943* (Batsford, 1975)
17 Ernst Franz, 'An Rommels Seite', in *Der Frontsoldat erzaht*, 1954
18 Behrendt, op. cit.
19 See, for instance, Major-General Freyberg, 'The New Zealand Division in Cyrenaica' (official report)

CHAPTER 14

1 Behrendt, op. cit.
2 Ibid.; see also David Kahn,

Hitler's Spies (Hodder & Stoughton, 1978)

3 Armbruster, interview, EPM 1. Armbruster, who was half Italian, was a junior staff officer at Rommel's headquarters and acted as interpreter at most conferences.
4 Behrendt, op. cit.
5 Franz, op. cit.
6 Armbruster, interview, EPM 1
7 Bayerlein, in Liddell Hart, op. cit.
8 BAMA N 117/2
9 H.R. Trevor-Roper (ed.), *Hitler's War Directives* (no.41) (Sidgwick & Jackson, 1964)
10 Liddell Hart, op. cit.
11 Koch, op. cit.
12 von Luck, op. cit.
13 Franz, op. cit.
14 Liddell Hart, op. cit.
15 Behrendt, op. cit.
16 Author's recollection
17 See for instance Ralf Reuth, *Des Führer's General* (Piper, Munich, 1987)
18 IWM AL 2596
19 Ibid.
20 Rommel diaries, EPM 9
21 Ibid.
22 von Mellenthin, op. cit.; Behrendt, op. cit.; Anlage 3
23 KTB 90 Light Division
24 von Mellenthin, op. cit.
25 The whole matter is fully discussed in Michael Carver, *Dilemmas of the Desert War* (Batsford, 1986)
26 Hinsley, op. cit., Vol. II, Appendix 16
27 Behrendt, op. cit.
28 Robin Edmonds, personal communication

CHAPTER 15

1 IWM AL 2596
2 Armbruster, interview, EPM 1
3 von Waldau, diaries, EPM 1
4 Armbruster, interview, EPM 1; Rommel diaries, EPM 9
5 Sir Edward Tomkin, personal communication
6 Behrendt, op. cit.
7 Tomkin; see also Louis Saurel, *Rommel* (Editions Rouff, Paris, 1967)
8 von Waldau, diaries, EPM 1
9 90th Light Division KTB, IWM AL 831
10 Rommel diaries, EPM 9
11 Ibid.
12 Ibid.
13 Ibid.
14 Koch, op. cit.
15 Rommel diaries, EPM 9
16 Ibid.
17 Koch, op. cit.
18 BAMA N 117/3
19 Armbruster, EPM 1

CHAPTER 16

1 Rommel diaries, EPM 9
2 Behrendt, EPM 3
3 Ibid. The agents' progress had, in any case, been charted by ULTRA.
4 DAK KTB, IWM AHB VII/87
5 IWM AL 2596
6 Armbruster, EPM 1
7 IWM AL 2596
8 Ibid.
9 Ibid.
10 M. van Crefeld, 'Rommel's Supply Problem 1941–42', *RUSI Journal*, September 1974

11 Liddell Hart, op.cit.
12 General Sir Alan Brooke, later Field Marshal Viscount Alanbrooke, Chief of the Imperial General Staff and the only military superior Montgomery venerated
13 Rommel diaries, EPM 9
14 The British had planted a misleading 'going' map, which may have influenced the German decision on thrust line
15 Rommel diaries, EPM 9
16 Armbruster, EPM 1
17 IWM AL 898/3
18 Behrendt, op. cit.
19 IWM AL 2596
20 von Esebeck, op. cit.
21 IWM AL 898/3
22 IWM AL 1349/2
23 Conference minutes, EPM 11
24 Cavallero diaries, EPM 2
25 Gilbert, op. cit.
26 IWM AL 898/3

CHAPTER 17

1 Behrendt, op. cit.
2 IWM AL 898/3
3 Ibid.
4 IWM 2596
5 Rommel diaries, EPM 9
6 Cavallero diaries, EPM 2
7 Ibid.
8 Siegfried Westphal, *Erinnerungen* (Mainz, 1975)
9 Cavallero diaries, EPM 2
10 Behrendt, op. cit.
11 Rommel diaries, EPM 9
12 Ibid.
13 Behrendt, op. cit.
14 Warning, interview, EPM 3
15 Albert Kesselring, *Soldat bis zum letzten Tag* (Bonn, 1953)
16 Constantin von Neurath, interview, EPM 3
17 Rommel diaries, EPM 9
18 Cavallero diaries, EPM 2
19 Ibid.
20 Ibid.
21 Liddell Hart, op. cit.
22 e.g. Hamilton, op. cit.
23 Rommel diaries, EPM 9
24 Franz, op. cit.
25 Armbruster, EPM 1; Manfred Rommel, EPM 3
26 von Luck, op. cit.
27 IWM AL 2596
28 Conference minutes, 22 and 24 November 1942, EPM 11
29 Cavallero diaries, EPM 2
30 von Neurath, EPM 3
31 Rommel diaries, EPM 9
32 Koch, op. cit.
33 IWM AL 1026
34 Stahlberg, op. cit.
35 IWM AL 1026
36 Ibid.
37 Cavallero diaries, EPM 2
38 Armbruster, EPM 1
39 See, for instance, 164 Division KTB, 17 January 1943, IWM AL 881
40 Cavallero diaries, EPM 2
41 Rommel diaries, EPM 9
42 Cavallero diaries, EPM 2
43 IWM AL 1026
44 Cavallero diaries, EPM 2
45 IWM AL 2596
46 Rommel diaries, EPM 9

CHAPTER 18

1 IWM AL 2596
2 Rommel diaries, EPM 9

3 IWM EAP 21-X-14/9
4 Rommel diaries, EPM 9
5 Ibid.
6 Ibid.
7 Ibid.
8 von Luck, op. cit.
9 IWM AL 1025
10 BAMA N 117/7
11 Manfred Rommel, interview, EPM 3
12 *Montgomery and the Eighth Army: Selected Papers of Field Marshal Viscount Montgomery*, Document 35 (Bodley Head, for Army Records Society, 1991)
13 von Esebeck, op. cit.
14 Ibid.
15 General Streich, lecture, EPM 3
16 Kenneth Macksey, *Rommel: Battles and Campaigns* (Arms & Armour Press, 1979)
17 Reuth, op. cit.
18 Westphal, op. cit.
19 See, for instance, Heckmann, *Rommels Krieg in Afrika* (Gustav Lubbe Verlag, 1976)
20 Westphal, op. cit.
21 Schmidt, *With Rommel in the Desert* (Harrap, 1951)
22 Colonel Aleme, BAMA N 117/21
23 Ibid.

CHAPTER 19

1 Albert Seaton, *The Russo–German War 1941–45* (Arthur Barker, 1971)
2 Cooper, op. cit.
3 Correlli Barnett, *Engage the Enemy More Closely* (Hodder & Stoughton, 1991)
4 BAMA N 117/4
5 Liddell Hart, op. cit.
6 Rommel diaries, EPM 11
7 Stahlberg, op. cit.
8 Published as *Krieg ohne Hass*. Reproduced in full in Liddell Hart, op. cit.
9 Liddell Hart, op. cit.; Manfred Rommel, 'Betrachtungen über das Jahrhundert 1891–1991', *Stuttgarter Zeitung*, 1991
10 Liddell Hart, op. cit.
11 Koch, op. cit.
12 Rommel diaries, EPM 11
13 Koch, op. cit.
14 Liddell Hart, op. cit.
15 Stahlberg, op. cit.
16 Rommel diaries, EPM 11
17 NAW T 311/276
18 Ibid.
19 Westphal, op. cit.
20 NAW T 77/792
21 Rommel diaries, EPM 11
22 Ibid.
23 Liddell Hart, op. cit.
24 Goebbels, *Diaries*, 27 July 1943 (trans. Louis Lochner, Hamish Hamilton, 1948)
25 Rommel diaries, EPM 11
26 Ibid.
27 Irving, *The Trail of the Fox*, op. cit.
28 Rommel diaries, EPM 11
29 NAW T 311/276
30 Armbruster, interview, EPM 3
31 NAW T 311/276
32 See, for instance, Fraser, op. cit.
33 Koch, op. cit.
34 NAW T 311/276
35 Koch, op. cit.
36 Ibid.

37 Manfred Rommel, interview, EPM 3
38 Franz, op. cit.
39 NAW T 311/276
40 Koch, op. cit.
41 Loistl, interview, EPM 3
42 Franz, op. cit.
43 von Tempelhoff, interview, EPM 3

CHAPTER 20

1 See Rommel's series of conversations with Bayerlein in Liddell Hart, op. cit.
2 von Luck, op. cit.
3 Liddell Hart, op. cit.
4 Friedrich Ruge, *Rommel in Normandy* (Presidio Press, 1979)
5 IWM AL 2596
6 von Tempelhoff, interview, EPM 3
7 Colonel-General Hans von Salmuth, reminiscences, EPM 4
8 Ibid.
9 von Esebeck, IWM AL 1579
10 Trevor-Roper (ed.), *Hitler's War Directives*, op. cit.
11 Ruge, op. cit.
12 von Salmuth, EPM 4
13 Warning, interview, EPM 3
14 Koch, op. cit.
15 Geyr von Schweppenburg's views, both on preliminary deployment and on the course of the campaign, are very clearly expounded in Panzer Group West report, NAW MS B466, and in an interview, NAW MS ETHINT 13. His views were also published in articles in *An Cosantoir* (Vols. IX and X, 1949–50) and other periodicals, and were repeated in Guderian, op. cit.
16 Ruge, op. cit.
17 BAMA N 117/22
18 Geyr; see note 15 above
19 Manfred Rommel
20 For example Warning, Tempelhoff, Staubwasser, Lattmann, interviews, EPM 3. Von Salmuth believed (EPM 4) that the controversy was particularly sharp between Rommel and von Rundstedt.
21 BAMA N 117/22
22 IWM EAP 21-X-14/9
23 BAMA N 117/22
24 Geyr; see note 15 above
25 Ruge, op. cit.
26 Hans Speidel, *Invasion* (Rainer Wunderlich Verlag, 1949); see also Speidel on 'Ideas and views of Field Marshal Rommel on defence and operations in the West in 1944', NAW MS B 720
27 IWM AL 510/1/3.
28 NAW MS B 466.
29 Geyr von Schweppenburg, 'Invasion Without Laurels', *An Cosantoir*; see note 15 above
30 IWM AL 2596
31 Rommel diaries, EPM 11
32 Tempelhoff, EPM 3
33 Ruge, op. cit.
34 Staubwasser, interview, EPM 3
35 BAMA N 117/22
36 EPM 11. Ruge also kept a meticulous account of daily movements and concerns; see Ruge, op. cit.
37 Ruge, op. cit.
38 BAMA N 117/22
39 Rommel diary, EPM 11
40 BAMA N 117/22

41 Rommel diary, EPM 11
42 IWM AL 2596
43 Koch, op. cit.
44 IZM, ED 100/175
45 BAMA N 117/22
46 Rommel diaries, EPM 11; see, for instance, 17 May 1944
47 George Lane, personal communication. Lane was later awarded the Military Cross for gallantry.
48 Richard Lamb, *The Ghosts of Peace* (Michael Russell, 1987)
49 See, for instance, H.B. Gisevius, *Bis zum Bittern Ende* (Fretz von Wasmuth Verlag, Zurich, 1946)
50 See, for instance, Wheeler-Bennett, *The Nemesis of Power*, op. cit.
51 Terence Prittie, *Germans Against Hitler* (Hutchinson, 1964)
52 Ibid.
53 See Gisevius, op. cit., for a thorough (albeit somewhat one-sided) dissection of the subject

CHAPTER 21

1 See Carlo d'Este, *Decision in Normandy* (Collins, 1983)
2 'Papers on the non-alerting of Seventh Army, 6 June 1944', EPM 3
3 General Siegfried Westphal, Liddell Hart Papers, King's College, London, quoted in d'Este, op. cit.; see also Saurel, op. cit.
4 See note 15, Chapter 20, above
5 Ibid.
6 BAMA N 117/22
7 BAMA N 117/23
8 Ibid.
9 Ibid.
10 BAMA N 117/22
11 BAMA N 117/23
12 Koch, op. cit.
13 Jodl testimony, International Military Tribunal, Nuremberg
14 BAMA N 117/23
15 See, for instance, Ruge, op. cit.
16 BAMA N 117/23
17 Wolfram, interview, EPM 3
18 Ibid.
19 Ibid.
20 BAMA N 117/23
21 Wolfram, interview, EPM 3
22 Ruge, op. cit.
23 NAW MS B 466
24 BAMA N 117/23
25 Ibid.
26 Ibid.
27 General von Luttwitz, quoted in Flower and Reeves (eds.), *The War* (Cassell, 1960)
28 BAMA N 117/23
29 Ruge, op. cit.
30 Lattmann, interview, EPM 3
31 See Chapter 17, p. 382
32 Warning, interview, EPM 3
33 Westphal, interview, EPM 3
34 Manfred Rommel
35 Betrachtungen, 3 July 1944. Copy in author's possession.
36 Army Group B, Betrachtungen zur Lage, 15 July 1944. Copy in author's possession.
37 IWM AL 510
38 BAMA N 117/23
39 Freiherr von dem Bussche-Streithorst. Personal communication
40 HQ SAS Operation Instruction 32. Copy, with associated signals, EPM 3.

CHAPTER 22

1 Pierre Galante with Eugene Silianoff, *Operation Valkyrie* (Harper & Row, 1981)
2 The sequence of messages between Rastenburg and Berlin is fully discussed in Peter Hoffmann, *History of the German Resistance* (Macdonald & Janes, 1977). The second message was also apparently passed by General Fellgiebel from Rastenburg; and communications were never wholly interrupted.
3 Tempelhoff, EPM 3
4 Dummler, interview, EPM 3. Dr Dummler, a reserve officer, kept the Army Group B war diary.
5 Remer, report 22 July 1944, NAW T 84/21
6 Hagen, report, NAW T 84/19
7 NAW T 84/21
8 Remer report, NAW T 84/21
9 Hagen report, NAW T 84/19
10 Walter Bargatzky, one of the conspirators in the military administration, Paris, account, EPM 4
11 Bargatzky, EPM 4
12 Ibid.
13 Tempelhoff, EPM 3
14 Bargatzky, EPM 4
15 Kaltenbrunner's reports to Bormann, NAW T 84/19/20/21
16 NAW T 84/19/20/21. Published as *Speigelbild einer Verschworung* (Seewald Verlag, Stuttgart, 1961)
17 NAW T 84/21
18 Lattmann, EPM 3
19 Lang, interview, EPM 3
20 Liddell Hart, op. cit.
21 Ruge, op. cit.
22 Lattmann, EPM 3
23 Kurt Hesse, interview, EPM 3
24 von Mellenthin, op. cit.
25 IWM AL 1579
26 Loistl, interview, EPM 3
27 Jodl testimony, International Military Tribunal, Nuremberg
28 Goerdeler's first interrogation, NAW T 84/21
29 NAW T 84/21
30 Manfred Rommel, letter, EPM 4
31 See, for instance, Major Streicher, Loistl, Manfred Rommel, interviews, EPM 3
32 IZM ED 100/176
33 Manfred Rommel
34 Manfred Rommel, EPM 3
35 Frau Lucy Rommel, EPM 4
36 Young, op. cit.
37 Liddell Hart, op. cit.
38 Major Streicher, EPM 3
39 Young, op. cit.

CHAPTER 23

1 See Chapter 18
2 Young, op. cit.
3 Ibid.
4 Liddell Hart, op. cit.
5 Manfred Rommel, interview, EPM 3
6 See Chapter 22; and Koch, op. cit.
7 Schwerin, interview, EPM 3
8 Speidel, op. cit.
9 Rommel diaries, EPM 11
10 Speidel, op. cit.
11 Ruge, in *An Cosantoir* (Vol. XI, 1951), quoting also Generals Blumentritt and Zimmermann

12 See Chapter 21, Lattmann, Warning, *et al.*
13 Manfred Rommel, interview, EPM 3
14 Manfred Rommel
15 Frau Lucy Rommel, EPM 4
16 Staubwasser, interview, EPM 3
17 'Bericht des Barons von Teichmann', EPM 4
18 Speidel, interview, EPM 3
19 SS Obersturmbannführer Dr Georg Kiessel, interrogator statement, EPM 4
20 Freiherr Gotthard von Falkenhausen, EPM 4
21 'Dokumente zur Ehrenhofsitzung gegen Speidel, 4.10.44', EPM 4
22 Kiessel, EPM 4
23 See note 20, above
24 Dr Horst, interview, EPM 3
25 NAW T 84/19
26 Hoffmann, op. cit.
27 NAW T 84/20
28 Keitel, interrogation report, 28 September 1945, EPM 4
29 Ibid.
30 Speidel, EPM 4
31 Bargatzky, EPM 4
32 Jodl testimony, International Military Tribunal, Nuremberg, 2 October 1945
33 Gisevius testimony, International Military Tribunal, Nuremberg (Vol. XII)
34 CSDIC (UK), Report GRGG 1347, 19 August 1945, EPM 4
35 NAW T 84/20
36 Loistl, interview, EPM 3
37 Keitel, interrogation, EPM 4
38 Jodl testimony, International Military Tribunal, Nuremberg
39 Ibid.
40 Keitel, EPM 4. Keitel told Burgdorf (confirmed in a note) that Rommel would be tried 'under martial law'. Jodl testified that it would have been before a people's court, not a court martial, and this was Rommel's understanding. It would have corresponded to action taken against other conspirators, including Field Marshal von Witzleben, and some of Keitel's later testimony (e.g. 28 September 1945) confirms it.
41 Loistl, interview, EPM 3
42 Rommel's name does not appear in the (frequently amended) lists and organizational charts planned for a post-Hitler Germany by Goerdeler (see Hoffmann, op. cit.; Prittie, op. cit.; numerous interrogation reports, NAW T84/20). Some authors (e.g. Koch, op. cit.) have him 'sounded out' about ultimate political leadership as early as 1943 and rejecting the idea, saying that he knew nothing of politics. This rings true, but no doubt even then there was a good deal of loose and speculative talk among those disgruntled with the way things were going. This was a long way from *Verschwörung* or *Attentat*.
43 'Bericht über den Tod des Generalfeldmarschalls Erwin Rommel von Frau Lucie-Maria Rommel', EPM 4
44 Manfred Rommel, statement, 27 April 1945, EPM 4
45 Manfred Rommel, letter, 13 November 1974, EPM 4

CHAPTER 24

1 OKW signal, 17 October 1944, NAW T 84/277
2 OKH order, 20 October 1944, NAW T 84/277
3 Manfred Rommel
4 'Bericht über den Tod des Generalsfeldmarschalls Erwin Rommel von Frau Lucie-Maria Rommel', EPM 4
5 See Keitel, Jodl, EPM 4
6 See Chapter 23, pp. 542–3
7 Manfred Rommel, interview, EPM 3
8 See, for instance, Hoffmann, op. cit.
9 Speidel, EPM 4
10 Major Ehrnsperger, EPM 3. Ehrnsperger accompanied Generals Burgdorf and Maisel to Herrlingen.
11 e.g. Young, op. cit., etc.
12 e.g. Heckmann, op. cit.; Macksey, op. cit.
13 *Krieg ohne Hass*, op. cit.; Liddell Hart, op. cit.
14 e.g. Reuth, op. cit.
15 Television interview
16 Speidel, EPM 4
17 See Chapter 18, p. 423

Index